Systems Analysis and Synthesis

Bridging Computer Science and Information Technology

Systems Analysis and Synthesis

Bridging Computer Science and Information Technology

Barry Dwyer

AMSTERDAM • BOSTON • HEIDELBERG • LONDON
NEW YORK • OXFORD • PARIS • SAN DIEGO
SAN FRANCISCO • SINGAPORE • SYDNEY • TOKYO
Morgan Kaufmann Publishers is an Imprint of Elsevier

Acquiring Editor: Todd Green
Editorial Project Manager: Amy Invernizzi
Project Manager: Mohanambal Natarajan
Designer: Maria Ines Cruz

Morgan Kaufmann is an imprint of Elsevier
225 Wyman Street, Waltham, MA 02451 USA

Library of Congress Cataloging-in-Publication Data
A catalog record for this book is available from the Library of Congress

British Library Cataloging-in-Publication Data
A catalogue record for this book is available from the British Library.

ISBN: 978-0-12-805304-1

For information on all Morgan Kaufmann publications visit our website at www.mkp.com

Foreword

Developers of software-intensive systems have yet to achieve a practical symbiosis of formal and informal concerns and techniques. Too often practitioners disdain formal techniques as irrelevant or too difficult to learn and apply. Researchers and advocates of formal methods too often fail to recognise the need for careful study and analysis of the real world that furnishes the system's subject matter and forms the arena in which its difficulties will be found, its effects felt, and its quality judged.

Barry Dwyer's book is an excellent antidote. A concise basis of mathematical structures and techniques is established in one masterful chapter and deployed in every development aspect of the central case study—an administrative system for a university. Everywhere the mathematics is applied with a full concern for the real problem in hand and for the practicability of proposed solutions.

The treatment throughout is broad. Design concerns such as inter-process communication, business rules, database schemas, form design, and input and output interfaces, are discussed in chapters filled with general insights and practical demonstrations of sound established techniques. A chapter on project management discusses high-level cash flow analysis for determining economic feasibility and the detail of critical path analysis for managing its execution; it also contains good advice on documentation standards for programs and other products of the development work. An excellent chapter on dynamic characteristics of a system introduces the reader to queuing theory and probability distributions, and goes on to analyse a simple manufacturing firm's business as a problem in control theory.

This book brings together three ingredients: discussion of software engineering practices and principles; beautifully clear descriptions of essential mathematical techniques; and direct application to compelling examples and case studies. It draws on Barry Dwyer's experience of teaching university courses, on his practice of systems development in several application areas, and on his work as a control engineer in the aircraft industry. But above all it draws on his knowledge and mastery of the formal and informal disciplines necessary for developing dependable systems. If you are a practitioner, a teacher or a student of software engineering, you will benefit greatly from reading and using this outstanding book.

Michael Jackson

25 December 2015

Contents

Preface

Why You Should Read This Book

This book is meant as the basis for a final-year course in *Computer Science* or *Information Technology*. Alternatively, it can be read — with benefit — by a graduate who has come face to face with a real systems analysis and design problem, or even a seasoned systems analyst looking for a fresh and more systematic approach.

Many *Computer Science* and *Information Technology* students graduate knowing a great deal about programming and computer technology without having any idea how to apply their knowledge to the real-life problems of a business. Some — like those who specialise in IT support — might never need to, but many others will be expected to write application programs that directly satisfy business requirements. This can be a daunting task for the newcomer; a modest-sized business can have a system of procedures that seem unbelievably complex to an outsider. Unfortunately, there is rarely anyone who can explain the system. It may have evolved over many years into a complex mixture of computer and manual processes. Senior management may know *what* the system is meant to do, but not know *how* it does it. Middle management may understand parts of the system well, while lacking understanding of the details. Functionaries may know *how* to do things, but have little idea *why* they do them.

Discovering what an existing system does is called 'systems analysis'. Creating a new system is usually called 'system design', but I prefer the term 'system synthesis' because there are methods and algorithms that can be used to *derive* a correct implementation from a system's requirements.

An *experienced* systems analyst will often proceed intuitively — but not necessarily correctly — to an outline of the implementation, then use that as a framework on which to hang the necessary details. There are two problems with this approach: it needs the experience that the beginner lacks, and it often leads to an ill-conceived result. Therefore, in this book, I have used a simple specification language to state the business requirements independently of their proposed implementation. More importantly, I have used powerful mathematical methods from graph theory to derive conceptual frameworks that reveal *every* correct implementation. The graphs help the analyst *make* decisions; they don't merely record decisions the analyst has already made. It is these graphical methods that I consider to be the most important part of the book.

Although I introduce aspects of systems analysis early in the book, its main emphasis is on system synthesis. Only after understanding the convoluted way requirements become embedded in a system's infrastructure can an analyst stand any chance of understanding what requirements an existing system embeds. Analysis is a kind of reverse engineering. One needs design skills to reverse the design process.

System design is a many-faceted subject, and involves topics that are not normally regarded as part of a *Computer Science* qualification. In particular, the chapters on the human interface, system dynamics, and project management are likely to confront a computer science student with new and interesting material. In the early chapters, I have included a summary of ideas that I expect most readers will already know, but biased towards to the specific requirements of the overall subject.

Systems analysis has as many facets as there are types of business. If you want to create a computer accounting system, it is wise to read a book or two about accountancy, if you want to create a job scheduling system for a factory, it is wise to read a book or two about job scheduling, and so on. It is impossible to cover all such topics in detail. In fact, the only case study explained in detail concerns an imaginary educational institution. There are two reasons for this choice: the reader has almost certainly encountered a real one, and it is an application with which I am all too familiar. Even so, the case study has been scaled down to be much simpler than any real system — otherwise the book would have been longer without being any more instructive.

I debated for a long time whether the title of this book should use the phrase "information system" or simply the word "system", because I would argue that every interesting computer system is an information system. Even a micro-controller for a production line is a primitive information system. It knows certain things, such as what product is to be made, and the current availability of parts. It responds to human input through an interface — perhaps as simple as a *Start–Stop* button. It acts in a planned manner, and ultimately it causes useful things to happen. These are all defining characteristics of an information system.

Credentials

After graduating in physics, I began my career as an electronics engineer in the aerospace industry, designing flight simulators for guided missiles and learning about flight control systems. At that time, I could truly claim to be a rocket scientist. I then moved on to designing hardware for head-up displays. During this time I learnt a lot about the human interface as it concerns the pilot of an aircraft.

Finding that I could (at that time) earn more by writing software than by designing hardware, I moved into software engineering — usually concerning file and database systems, but with occasional forays into application programming when more interesting work was unavailable. My last job in London was designing specialised database software for the first computer criminal records system at Scotland Yard.

In the early 1970's, I migrated from England to Australia, where I was in charge of software development for the administration of a tertiary education institution, setting up their first interactive computer systems for book-keeping and student registration.

After a while, I drifted into teaching, and for a period of ten years or so I taught a one-semester course called *Systems Analysis and Project* at the University of Adelaide: Final-year students were grouped into teams of five or six, and each team would find five or six potential clients. An assistant lecturer, Heiko Maurer, and I would select the projects that we thought most practical for the teams to design *and implement* within the rather limited time frame. Clearly, my own experiences as a systems analyst and computer programmer were a help. Heiko and I acted as design consultants, and were vicariously exposed to a wide range of systems analysis problems. The projects proved amazingly varied, ranging from machine-shop scheduling to matching the dental records of fire victims. Almost all of them involved designing a database. We found that students always underestimated the number of data structures they would need, wrongly believing that one or two complex structures would lead to a simpler implementation than a larger number of simple ones. The book therefore devotes a lot of space to database design.

Students were required to hold weekly project meetings. The minutes of these meetings were posted to an on-line database, and the jobs they had assigned one another were displayed by a utility program as a Gantt chart. Alongside the practical work, students received twice-weekly lectures, in which I aimed to deliver the information they needed in time for them to use it.

During the course of a single semester, students would progress from a state of complete bewilderment to one of pride in a job well done. Several groups went on to exploit their ideas commercially.

When I worked as an application programmer, I was convinced that all my problems were caused by systems analysts. When I worked as a systems analyst, I learnt that the problems were instead caused by the *infrastructure* upon which applications were built — be it batch processing, client/server, Internet, or whatever. My own experiments have satisfied me that under 10% of application code may have anything to do with the business rules; the rest is needed to interface with the user and the infrastructure.

During my time, I have seen digital computers increase tenfold in speed and storage capacity every decade — meanwhile steadily decreasing in size and cost. I have worked with hardware, software, and humans in a wide range of applications, and at the end of it, I am convinced that *all* systems are united by a few common principles, and that these principles can be taught.

Plan of the Book

The chapters in this book roughly follow the order of the *Systems Analysis and Project* lectures I gave at the University of Adelaide. For the reader's benefit, the first three chapters include background material that my students were assumed to know, slanting it in a way that is relevant to later chapters. These early chapters condense the contents of a good sized textbook, so the treatment is necessarily sketchy. The reader who is unfamiliar with this material is advised to consult the texts mentioned in the *Further Reading* sections at the end of the chapters concerned.

The later chapters contain material that I could only have included in a full-year course, and every chapter goes into rather more depth than was possible in lectures. Despite this, the chapters remain summaries of material drawn from several sources, and most chapters — and indeed some sections — could fill a textbook. The interested reader is again invited to study the *Further Reading* sections.

- Chapter 1 describes the scope of the book, explains the difference between analysis and *synthesis*, and points out that intractable methods of system design are potentially useless.
- Chapter 2 presents the mathematical background to the ideas from *graph theory* that dominate the book. It also outlines the many ways these mathematical notions can be represented within a computer system.
- Chapter 3 concerns *atoms* and *events*, the fundamental building blocks of systems. Atoms are modelled by finite-state automata that change state in response to events. The chapter also discusses the relationship between states and behaviours.
- Chapter 4 shows how to relate the different kinds of atoms with *functional dependencies*. It is a *graphical approach to database design* that immediately results in a well-designed schema.
- Chapter 5 introduces the reader to a *specification language* designed to describe the business rules that define events — in a way that is independent of their implementation.
- Chapter 6 discusses the way *database systems* work. This is useful not only to a reader who uses a commercial database management system, but also to one who must hand-craft data structures for some other infrastructure. It also introduces the problem of *deadlock* in concurrent processing systems.
- Chapter 7 examines how events control the way data flow between processes, and how *use-definition analysis* can affect system synthesis.
- Chapter 8 concerns the *interfaces* between a system and its environment, especially the *psychological factors* concerning its interfaces with humans.
- Chapter 9 describes how *business rules* can be expressed in ways that are more easily understood and modified than expressing them directly as program code. It includes a discussion of *expert systems*.
- Chapter 10 discusses the dynamic behaviour of systems, including *queueing theory*, modelling and *simulation*.
- Chapter 11 discusses *project management* and control. It culminates with a section that describes how the various tools discussed in earlier chapters can be put together as a successful *methodology* for system development.
- Appendix A explores the relationship between state-based and grammar-based descriptions of system behaviour.
- Appendix B discusses the traditional normalisation approach to database design — if only because its terminology is required knowledge for anyone who uses databases.
- Appendix C gives the answers to the exercises at the end of each chapter.
- The book ends with a comprehensive index.

The chapters are arranged in an order that develops ideas and tools in a logical progression. This may not always be the order in which the tools need to be used in a typical project. In particular, in teaching a project-based one-semester course it will prove necessary to take things in whatever order is needed to complete projects on time. In this context, a preliminary feasibility study would be a luxury; students will first need to know how to structure data, how to design interfaces, and how to schedule their own activities. Once they have made a start on their projects, lectures can fill in the details — in much the same order as the chapters in this book. The content of such lectures will depend on what students can be assumed to already know. For example, the students I taught were already familiar with *SQL* and finite-state automata, among other things.

Boldface type is used in this book whenever a new concept is introduced and defined, either formally or informally. Such concepts appear in the index with the page number of the primary reference set in boldface type. Boldface is also used for programming language keywords. Italics are used for names of all kinds, and sometimes, just for *emphasis*.

Acknowledgments

My thanks go to:

- Sid Deards, my old Circuit Theory lecturer, who first showed me the advantages of synthesis over analysis,
- Paddy Doyle, who first showed me that synthesis works for software too,
- Michael A. Jackson, my mentor in the subjects of software development and project control,
- Heiko Maurer for his help with all my student projects,
- my ex-students, for teaching me how to teach,
- Carole Dunn, Peter Perry and David Knight, who have read drafts of various chapters and told me where I could improve them,
- the various implementors of LaTeX and of *Prolog*, without whom I would never have got this book into shape,
- my wife, Linda, for her patience, loyalty and tolerance.

List of Figures

List of Tables

List of Algorithms

CHAPTER

Systems, analysis, and synthesis

1

CHAPTER CONTENTS

1.1 SYSTEMS

A **system** is a collection of parts, or components, that interact in such a way that the whole is more than the sum of its parts. For example, a bicycle is a system, but if you buy a kit to make a bicycle, you don't yet have something you can ride. You don't have a system. First, you must assemble the parts. Correctly. The connections and interactions between the parts are just as important as the parts themselves.

This book is not about bicycles, but about systems, mainly information systems. Even so, there are useful analogies between information systems and physical systems such as bicycles, simply because they are both examples of systems. Further, the function of an information system is usually to model some kind of physical system, so that the two become analogues of one another. For example, a stock control system models the movement of physical stock in and out of a store, and an accounting system models the movement of money between accounts. (Indeed, the modelling of money has become so well established that actual money is involved less and less as time goes on. The miser no longer has to count his gold. The shopper carries a credit card instead of a purse. Even coins and notes are mere symbols for the gold and silver bullion they replaced.) So, by information system, we mean something that models aspects of the real world. In principle, the medium it uses to do this is unimportant, but we are primarily concerned with the medium of the digital computer.

Some systems are obviously hybrids of a physical system and an information system. For example, a robot, or a guided missile, has eyes and limbs that are physical sensors and actuators, but they are interconnected by an information system. Less obviously, what we may think of as a 'pure' information system usually has eyes and limbs that belong to humans. Data are entered into a stock control system because a store hand perceives that goods have been delivered. When the stock control system issues a purchase order, it is up to a human, somewhere, to supply some physical goods to replace those that have been used.

What makes an information system relevant to the real world is that both must obey the same *rules* or *laws*. Among the most fundamental physical laws is the conservation of matter.[1] Every physical object that enters a stock control system should either eventually issue from it or remain in stock. Money can move between accounts, but a good accounting system cannot allow it to evaporate. If it turns out that the model and the physical system disagree, we suspect wrongdoing.

Using computers to model systems is not a one-way street. If you want to explain a computer sorting algorithm to someone, you may decide to demonstrate it with a deck of playing cards. This suggests that the study of physical systems can enlighten our understanding of information systems, and *vice versa*.

Many books on system design suggest that every software component should have a single well-defined function. Despite this, we mustn't assume that this is a property of *efficient* systems. A bike tyre is part of its propulsion system, part of its steering system, and part of its braking system. It also contributes to the comfort of the rider. It is *possible* to make systems in which each part has a single well-defined task, but it isn't always a good idea. A bike rider *could* lower a skid in order to stop, which would mean that the tyres could be better designed for their other functions, but we don't make bicycles that way.

Design decisions are not universal. If we consider aircraft technology, tyres don't contribute to propulsion in any way, and they contribute to braking and steering only during taxiing. At higher speeds, reverse thrusters or parachutes provide the brakes, and control surfaces do the steering.

Why are systems made of components at all, rather than in one piece? In the case of the bicycle, one reason is forced on us: some parts, such as the wheels, have to move relative to other parts. Why isn't the tyre part of the rim of the wheel? Because it is made using a different technology. Why isn't the rear axle part of the frame? First, because it would be impossible (or at least, very difficult) to install the wheels. Second, because the axle is best considered as part of the wheel. Why? Because its

---

[1] Strictly speaking, 'matter and energy'.

interface with the wheel is more complex than its interface with the frame. Keeping the interfaces between parts simple is an important aspect of system design. Given simple interfaces, it then becomes possible to have different people design and make the parts. For example, a bicycle's wheels may come from a specialist supplier.

These same considerations apply to information systems. Consider a system that allows a user to browse a database across the Internet. Because the user is in one place and the database is in another, the system is forced to have at least two parts: a client and a server. Typically, the client process will use a web browser driven by *HTML*. On the other hand, since the server has to access a database but must also use Internet protocols, it will probably use a combination of *SQL* and *HTML* technologies, and although *SQL* and *HTML* will need to interface with one another (perhaps using *PHP*), they will tend to form separate components because they use different technologies. The *SQL* and *HTML* aspects of the system may well be implemented by different specialists. Making a system from components makes it easier to put the system together. A good designer makes a system of parts that have simple interfaces. This is where a system designer has the most freedom — but has to make difficult decisions. Once these decisions are made, the rest is routine.

The design of efficient systems depends on the current technology, and technology is always changing. A danger facing all system designers is the failure to adapt to technology change, and to continue to use methods that have become outdated. When we see old movies of the original Wright brothers' *Flyer*, we see a machine whose construction more closely resembles that of a bicycle or a kite than of a modern aircraft. We cannot criticise the Wright brothers for this, but we would certainly sneer at a modern airliner that used such out-dated technology.

Despite constant technological change, anyone who has designed systems for any length of time realises that some aspects of past experience *are* transferable to new technologies, and that *some* aspects of system design must therefore remain constant. In this book, we attempt to isolate and teach these fixed principles, as applied to information systems.

An important feature of man-made systems is that they are designed to be useful to *people*, even if it is merely to entertain them.[2] Therefore, anyone who wants to make good systems has to understand people. This area of human-machine interaction is known as 'human factors', or 'ergonomics'. Again, there are principles that govern ergonomics, which this book aims to teach.

1.2 AIM OF THE BOOK

Systems analysis usually concerns information systems, mainly in business. These systems are often complex because of their size or because of their intricacy (i.e., having many processing rules), rather than on account of tricky or time-consuming algorithms.

[2] Indeed, the notion that systems have a purpose is so central to our thinking that we tend to assume that *all* systems have one. We look at the universe and ask, 'What is it all for?'

Systems analysis includes everything from identifying a problem to implementing its solution, except routine programming tasks. The output of the analysis must present no technical challenge to the programmers who will implement the system. Few activities in systems analysis can be formalised as algorithms; most rely on experience or intuition. Once something can be formalised, it tends to be incorporated in the next high-level language or CASE (Computer-Aided Software Engineering) tool. It ceases to be considered as part of systems analysis, which therefore consists mostly of informal methods.

In this book, we present a new approach to the design of systems, with an emphasis on information systems. Despite this emphasis, the techniques taught here are general enough to apply to many other problem areas.

The approach is new in two ways:

First, there are many ways a set of operational requirements can be implemented by a physical system. Here, we express aspects of the operational requirements in a way that is independent of the implementation. The result is a formal specification that allows us to derive systems that are provably correct. They will do what the user wants. Other aspects are less easily formalised, and concern how the system looks and feels, and the technology that underlies it. We discuss these other aspects, but aren't (yet) able to formalise them.

Second, the approach focusses on what we call atoms, a concept similar to objects (as in object-oriented programming). Atoms differ from objects in that they are usually more primitive and don't necessarily map directly onto program objects or other program structures. In other words, there may be an intermediate stage of design between atoms and their expression as objects. The approach is not incompatible with object-oriented design, but it sees it as one of many alternatives.

1.2.1 KERNEL, INTERFACE, AND ENVIRONMENT

Every system operates in an environment. It connects to the environment through interfaces. For example, a bicycle has interfaces with its rider, the road, and the air, which together form its environment. (In a larger sense, its environment includes the road system and its rules and regulations.) Its interface with the rider includes the handlebars, pedals, and saddle. Its interface with the road is its tyres. Its interface with the air is its aerodynamic shape, which is one reason why racing bikes have lowered handlebars.

As time goes on, interfaces tend to become more sophisticated. For example, racing cyclists wear special shoes to fit the pedals, and skin-tight costumes to reduce aerodynamic drag.[3] Even so, the core purpose of the bicycle remains constant: to provide a speedy means of man-powered surface transport. This purpose is the kernel of the system.

---

[3] The international rules of bicycle racing limit the aerodynamic sophistication that is allowed, otherwise racing bikes would reach even more dangerous speeds.

Consider a typical business system. In the distant past, customer orders would have been received by letter, transactions recorded in a journal, information stored in ledgers, and bills sent by mail. In the early days of computing, orders might have been punched onto cards, stored in computer files, and bills printed by a line printer. Somewhat later, the orders might have been typed into a text-based 24-line monitor, stored in a relational database, and bills printed by a matrix printer. Then would have come a mouse-driven graphical interface, with orders received and bills sent by e-mail, then a touch-driven or voice-driven interface, and, in the future, perhaps a thought-controlled three-dimensional virtual-reality experience.

Historically, each change of the interfaces meant that most of the software had to be rewritten. The interfaces are simply[4] an application of current technology. In practice, over 90% of the programming task is concerned with interfaces, either with the environment or with a database. The point is, throughout all these changes, the kernel remained the same: to bill customers for the goods they ordered. The kernel is the set of **business rules**. A specification of a system needs to describe its rules in a formal notation, such as mathematics or a programming language. The interfaces are often best described using pictures or through design standards.

This doesn't mean that interfaces are unimportant. Badly designed interfaces are frustrating to use. Well designed interfaces are fun and rewarding.[5] Fortunately there are some psychological principles that we can exploit to get them right.

1.2.2 SYSTEM DEVELOPMENT MODELS

There are at least three traditional models of system development:

- Classical **life-cycle models** divide system development into a series of well-defined phases.
- **Evolutionary models** (or **prototype models**) develop a system by successive approximation.
- **Transformational models** develop a complex system by transforming a simpler one. This book will focus on a particular transformational model.

All methods involve the same kinds of activity:

Feasibility Study: Deciding which problems are worth solving.

Requirements Analysis: Deciding exactly what the problem is.

Interface Design: Deciding how the system will look to users.

Simulation and Modelling: Exploring the behaviour of a proposed system *before* building it.

[4] Simple, but very time-consuming.

[5] I'm sure you can think of both good and bad interfaces that you have used.

Data Design: Deciding how data will be represented and stored.

Process Design: Deciding what processes will be needed and how they should work.

The first four steps are usually performed in consultation with a client. In large organisations these activities are usually carried out by a team, often with a rigid job hierarchy, passing system documentation down a chain from senior analyst to programmer. In smaller or more democratic organisations, one person may perform the whole job, from feasibility study to programming.

Systems analysts are often ex-programmers, ideally with expert knowledge of the problem area. If that is lacking, **design methodologies** can help the analyst come to grips with unfamiliar problems.

1.2.3 CLASSICAL LIFE-CYCLE MODELS

A typical classical methodology is to divide activity into five phases:

- Requirements Analysis,
- Logical Design,
- Physical Design,
- Implementation,
- Maintenance.

These phases are separated by contracts with the client, based on written specifications. Since the specifications cascade from phase to phase, this approach is also called a **waterfall model**.

At the end of the **requirements analysis** phase, the specification may say what activities the system is to perform, what the benefits of the new system will be, what existing procedures are to be replaced, and how much the system should cost. The client will then check this document and decide whether to proceed with the next phase.

At the end of the **logical design** phase, the client may receive a draft operations manual and revised estimates of how much the system will cost. Again, the client will decide whether to approve further progress.

Physical design has little impact on the client, except in refining the documentation from the logical design. It is mainly directed at solving implementation problems.

At the end of the **implementation** phase, the client will receive a working system, and instructions for using it. Programs will also be documented in preparation for maintenance.

Once the system is installed, it will need to be *maintained* for two major reasons: to correct deficiencies in its design (e.g., bugs), and to adapt to changing business requirements (e.g., changes in tax laws). **Maintenance** accounts for about 75% of all the work done by information systems departments. This may not be a bad thing and may simply reflect the robustness of the original physical design.

Various life-cycle models are broadly similar, but differ in the number of their phases, and the precise documents that define the boundaries between them.

All life-cycle models have drawbacks:

- Divisions between phases are rarely clean. For example, problems may be found during the implementation phase that affect the feasibility of the system or cause the physical design to be changed.
- Clients don't understand what they are agreeing to; even a computer expert finds it hard to understand a complex system from its documentation. Few people have enough imagination to foresee how a complex system will work in practice, especially if the technology is new or unfamiliar.
- It is almost impossible to keep all the documentation up to date, especially once software maintenance begins.

1.2.4 EVOLUTIONARY MODELS

Evolutionary models are developed by building a crude version of the system, which users can then criticise. This feedback is used to improve the design step by step, until it is satisfactory. This approach has been made easier through the use of '4th generation' systems products, often based on a database management system, which allow much more **rapid prototyping** of application programs than is possible in procedural languages. One advantage is that users may start using an incomplete system before its 'luxury' features are implemented.

Evolutionary models have some drawbacks too:

- The first hastily constructed prototype should probably be thrown away, but programmers are reluctant to discard anything in which they have invested time and energy.
- A smooth evolution from the user's point of view may not correspond to a smooth evolution of the system design. There is a tendency to adopt quick fixes, rather than long-term solutions, until the system becomes incomprehensible, and further evolution becomes impossible.

1.2.5 TRANSFORMATIONAL MODELS

Transformational models are based on the view that most of the complexity in systems arises from the need to make them efficient. If efficiency didn't matter, much simpler designs would often be possible. The transformational approach is to design the simple system (a logical design) and *transform* it into the desired system (a physical design). This transformation can be achieved either by hand, through programming tools, or (ideally) completely automatically.

- In the case that the transformations can be both made and *chosen* automatically, the methodology often becomes formalised as a feature of a programming system.
- In a 'strong' methodology, an algorithm makes the transformations, although the designer must choose the correct set of transformations intuitively.

- In a 'weak' transform methodology, the transformations themselves must be made intuitively, but their correctness can be checked formally.
- A methodology in which the correctness of a step cannot be checked is called 'informal'.

Transformational models often evolve from informal techniques, by way of formal techniques, to programming tools. They are areas of active research.

1.3 ANALYSIS

The word 'analyse' means to separate something into its parts in order to understand it better. This especially includes, of course, the way the parts are interconnected. Systems analysis can therefore be similar to reverse engineering — but without the suggestion of commercial espionage. We are given some existing system, and we want to know how it works, usually with the objective of making a new system to do the same things more cheaply or efficiently.

There is danger here: we can sometimes see *what* something does without knowing *why* it does it. We can analyse a bicycle and deduce that the pedals turn a gear wheel that powers a chain that turns a sprocket that turns a wheel. But it is only after we learn that it is not an exercise machine, but a vehicle whose wheels go on the ground, that we have any chance of improving its design, and perhaps inventing the motorbike.

A systems analyst is therefore someone who studies a system in order to understand it, usually with the purpose of improving it. The analyst has an advantage if he or she has expertise in the field in which the system is used. For example, to understand a business accounting system it is well to have had a few lessons in accounting. Although it is sometimes possible to learn about the relevant field by observing the system itself and asking intelligent questions about it, this has three disadvantages:

- The existing system may already be dysfunctional and a poor example to follow.
- It is often easier to find out what the system should do from a textbook.
- The system may do some things that are no longer, or never were, useful.

I am reminded here of an acquaintance who was asked to design a computer system for an insurance company. He wasn't the first to be given that task. Several others had tried and failed. The inner workings of the existing office systems were so complex that no one had managed to understand them. Realising this, he refused to even think about the existing system and spoke only to the company executives and the sales representatives — the people at the top and the people at the bottom — the company's interfaces with the outside world. He went on to succeed where the others had failed.

The reason is plain. He didn't ask 'How?', but 'What?' and 'Why?' The 'How' was related to the technology of filing cabinets and forms-in-triplicate and had evolved into a system of baroque complexity. (For a likely reason, see 7.5.3.) Given computer technology, the existing 'How' became irrelevant. By knowing 'What' and 'Why', inventing the new 'How' became straightforward.

Given that the likely reason for analysing a system is to improve it, a good systems analyst should be familiar with the latest technologies for producing a new system. Too often analysts specialise in one technology, say a particular database system, and use their favourite technology irrespective of the problem. As the saying goes, 'To a hammer, everything looks like a nail'. New technologies are always emerging, and the wise analyst embraces the new, without necessarily rejecting the old.

'Analysis' has a second meaning. We can also analyse by making experiments. These experiments can take place on the actual system being studied or on a model of it.

In an input-output experiment, we can analyse the behaviour of the system without examining how it works. This is how we might learn how to use an unfamiliar piece of software for which we have no documentation. Unfortunately, we can never be certain that we have discovered all its features. We may even invent our own theory to explain to ourselves how it works, but we shouldn't be too surprised when our theory proves faulty.[6]

A second kind of analysis is to build a model. We use models for a variety of reasons. One is that we want to study a system before going to the expense of building it. For example, we may want to assess the quality of the user interface, perhaps simply by making mock-ups of reports or screen-based forms and discussing these with users. We may want to investigate the system's ability to handle heavy workloads by building a simulation of its queues and servers; or, if the system includes some measure of decision-making, or artificial intelligence, we may want to check that it chooses wisely.

An interesting aspect of models is that the same model can describe systems that are widely different physically. For example, in a later chapter we shall find surprising similarities between the mathematical models of a stock control system and of a person using a shower. The advantage of such similarities is that, arguing by analogy, if we can learn to control the shower, we can learn to control stock.

1.4 SYNTHESIS

Many books in this field have a title such as 'Systems Analysis and Design'. Why does the title of this book use the word 'synthesis' rather than 'design'?

An electronics engineer designing an active linear filter knows that such a filter can only be made if its behaviour can be described by the ratio of two polynomials of a certain form. If it can, the engineer can derive two three-terminal networks and connect them in a certain way to an operational amplifier, and the design problem is solved. The details and the jargon aren't important to us; the point is that if the problem can be expressed in a certain way, then it is routine to design a filter that does the job. The resulting filter is correct by design. It is almost a no-brainer. A

[6] Scientific theories are like this.

computer can do it. On the other hand, if the requirement can't be expressed as a ratio of polynomials, no linear filter can possibly do the job — period. The engineer must either approximate the requirement as such a ratio, or use a different technology.

Let us look at an example of synthesis relevant to information systems:

Imagine that we have the following business rules:

Rule 1: If $balance \geq debit$ and $status$ = 'ok' then
set $approve$ to 'yes' and $error$ to 'no'

Rule 2: If $balance < debit$ and $status$ = 'ok' then
set $approve$ to 'no' and $error$ to 'no'

Rule 3: If $status \neq$ 'ok' then
set $error$ to 'yes'

By some intuitive process, we might *design* the following snippet of program code,

```
if (status == 'ok') then {
  if (balance < debit) then
    approve = 'no'
  else
    approve = 'yes';
  error = 'no' }
else
  error='yes';
```

How can we know that this procedure is correct? In this case, it is not hard to prove: we can check that each rule is implemented properly. In general however, especially if a program contains loops, it can be hard to prove that it is correct.

The alternative approach is to *synthesise* the procedure. Algorithm 1.1 is suitable for sets of rules of this type:

Algorithm 1.1 Converting rules into procedures.

Step 1: While the last action in each rule is the same, place it after all the rules, and eliminate it from further consideration.

Step 2: Pick some condition. Divide the rules into two groups: those for which the condition is true and those for which it is false. If a rule does not specify the condition, put it in both groups.

Step 3: Generate a conditional **if**-statement whose *true* branch contains the first group of rules and whose *false* branch contains the second group of rules.

Step 4: Apply the same procedure to each group of rules recursively, but exit the recursion when a group of rules becomes empty.

Let's see what happens to the example:

Step 1 The last actions differ; nothing happens.

Step 2 We pick the condition (*balance* < *debit*).
Group 1 contains,

Rule 2: if *status* = 'ok' then set *approve* to 'no' and set *error* to 'no'.

Rule 3: if *status* ≠ 'ok' then set *error* to 'yes'.

Group 2 contains,

Rule 1: if *status* = 'ok' then set *approve* to 'yes' and set *error* to 'no'.

Rule 3: if *status* ≠ 'ok' then set *error* to 'yes'.

Step 3: We generate,

```
if (balance < debit) then
    {Group 1}
else
    {Group 2};
```

Step 4: Applying the same process recursively to Group 1 we have,

Step 1 Nothing happens.

Step 2 The only condition is (*status* = 'ok').
Group 1.1 contains

Rule 2: set *approve* to 'no' and set *error* to 'no'.

Group 1.2 contains

Rule 3: set *error* to 'yes'.

Step 3 We generate,

```
if (status == 'ok') then
    {Group 1.1}
else
    {Group 1.2};
```

Step 4 Applied recursively to Group 1.1 we have

Step 1 The last actions agree; we generate `error = 'no'` (to be last) and delete "set *error* to 'no'," then generate `approve = 'no'` and delete "set *approve* to 'no'."

Step 2 Nothing to do.

Step 3 Nothing to do.

Step 4 Exit the recursion.

and so on ...

At the end of this process we have,

```
if (balance < debit) then
    {if (status == 'ok') then
        {approve = 'no'; error = 'no'}
    else
        {error = 'yes'}}
else
    {if (status == 'ok') then
        {approve = 'yes'; error = 'no'}
    else
        {error = 'yes'}};
```

How do we know that the resulting procedure is correct? Because somebody has already proved once and for all that if Algorithm 1.1 is followed, the solution is *always* correct.[7]

You will notice that the synthetic solution is longer than the one that was derived informally. On the other hand, if we had picked the two conditions in the reverse order, we would have reached the same solution as the informal approach. Unfortunately, although two conditions can be placed in only two possible orders, as we increase the number of conditions, the number of their possible orderings grows very rapidly.

This example shows an important property of synthetic methods. It can be easy to synthesise solutions that are guaranteed to be correct, but very many solutions may be possible, and it may be hard to choose the best. Therefore, for synthesis methods to be practical, we need also to have some ways of finding optimum or near-optimum solutions.

1.4.1 COMPOSITION AND DECOMPOSITION

There is a subtler reason for using the word 'synthesis' in the title of this book, which concerns the way we approach the question of optimisation. How do we decide from what parts a system should be made? Traditional approaches usually consider how best to decompose a system into parts. This is the direction of analysis: breaking things down. The alternative is to first split the system into as many fragments as possible (atoms), then to decide how best to compose them together. This is the direction of synthesis: building things up.

The reason why composition can be better than decomposition is that when we take a system as a whole and separate off any small fragment, the system almost certainly gets worse. We need to decompose a system into larger parts before we can expect it to improve. Guessing how to do this is often something that only comes with experience, and experience isn't an easy thing to teach, either to a human or to a computer. On the other hand, when we compose two small fragments into a single

[7] A program to convert rules to program code in this way would need to diagnose errors in the set of rules, e.g., when combinations of conditions exist where no rule can be applied.

Table 1.1 The time taken to generate the permutations of a list of numbers overtakes the time taken to sort it as soon as it contains four elements

Length	1	2	3	4	5	6	7	8	9
Sorting	1	4	9	16	25	36	49	64	81
Permutations	1	2	6	24	120	720	5,040	40,320	362,880

component, we often see an immediate improvement. We simply let the components grow. This is an approach we can teach. Even a computer can do it.

1.5 TRACTABILITY

In this section, we discuss the question of whether algorithms have any practical value. An algorithm is quite useless if it takes too long. For example, an instruction such as, 'Assemble the basic parts into components in the way that maximises efficiency', is quite useless, even if the measure of efficiency is well defined. The reason is that there are typically so many ways of assembling things that we could never consider them all. If a system has only 10 basic parts, they might be combined in over 1,000 ways. If it has 100 basic parts, they might be combined in more than 10^{30} ways.[8] It will be our aim to avoid such useless advice.

Formally, we speak of some problems being tractable and others being intractable. These terms refer to how rapidly the time or storage space needed to solve a problem grows with the size of its input. For example, even using a simple method such as bubble-sort, a list of names can be sorted in time proportional to the square of the length of the list, and more advanced algorithms such as Quicksort can do even better. On the other hand, if we are asked to generate all the permutations of a list of names, the time needed grows much more rapidly, in proportion to $n!$, the factorial of its length, n — simply because that is how many permutations it has.[9]

As Table 1.1 shows, the time taken to sort is overtaken by the time taken to generate permutations as soon as we reach four items. Of course, the algorithm for sorting might be more complicated than that for generating permutations; it might take 100 times longer to sort a list of one item than it does to generate its only permutation, so the comparison would no longer be fair. Table 1.2 shows what happens if we multiply all the sorting times by 100. Despite this change, the time taken to generate permutations quickly overtakes the time taken to sort the list, as soon as we reach seven items.

[8] In fact, in many more ways than that. These estimates simply assume that each part can be assigned to one of two subsystems. There can be three subsystems, four subsystems, and so on.

[9] $n! = (1 \times 2 \times 3 \times \cdots \times n)$.

Table 1.2 Even if the sorting algorithm is inherently slower by a factor of 100, the time taken to generate permutations overtakes sorting as soon as the list contains seven elements

Length	1	2	3	4	5	6	7	8	9
Sorting	100	400	900	1,600	2,500	3,600	4,900	6,400	8,100
Permutations	1	2	6	24	120	720	5,040	40,320	362,880

Another way to see the contrast is to imagine the effect of using a faster computer. Suppose that each unit in Table 1.2 represents one microsecond of CPU time for our current computer, so it can generate all the permutations of nine items in 362.88 milliseconds. In about the same time, it could sort a list of not just nine, but about 60 items. Now suppose a new computer is 100 times faster. Again, in the same time, it would be able to sort over ten times as many items, but it would be able to generate the permutations for a list with only two extra items. From this, we learn that there are problems that future technology will do little to help. We call these problems *intractable*, and the remaining problems *tractable*.

Letting n denote the size of the problem — usually the length of its data, Figure 1.1 shows the growth rates for several functions of n on a logarithmic scale. Multiplication by a constant (e.g., 100) merely displaces a line vertically. The plots for n, n^2, n^3, and so on are straight lines of increasing slope. Because of this, the highest power of n will always dominate. For example, the function $f(n) = 100n + n^2$ is dominated by n^2: once $n > 100$, $f(n) < 2n^2$.

In **complexity theory**, we make a clean division between tractable and intractable problems. If the time or storage space needed to solve a problem is never greater than some *finite* polynomial expression involving n, the problem is called **polynomial**, and is considered **tractable.**[10] If no such polynomial exists, it is considered **intractable**, and is often loosely referred to as **exponential.**[11] The distinction between tractable and intractable problems doesn't depend on constant factors (such as the factor of 100 in the sorting example) because, above a certain value of n, an exponential algorithm will always take longer than a polynomial one. An intractable problem is therefore one whose growth on a log-log scale cannot be bounded by a straight line.

We denote the **order of complexity** of an algorithm using **big-O notation**: an algorithm with execution time $f(n)$ has complexity $O(g(n))$ if $f(n) < kg(n)$ for some constant k, and sufficiently large n. Thus $100n + n^2$ has order $O(n^2)$ because $100n + n^2 < kn^2$ for $k = 2$ and $n > 100$.

[10] Although $\log_y n$ cannot be accurately expressed as a *finite* polynomial, since $\log n < n$ for all $y > 1$, a problem with complexity $O(\log n)$ is tractable.

[11] The log function is such that $x = \log_y n$ if $n = y^x$. Logarithmic growth is therefore the antithesis of exponential growth: in exponential growth, time grows exponentially with n; in logarithmic growth, n grows exponentially with time. In complexity theory we often omit to mention y. Its value only displaces the log-log curves for $n = y^x$ and $\log_y n$ by a constant factor.

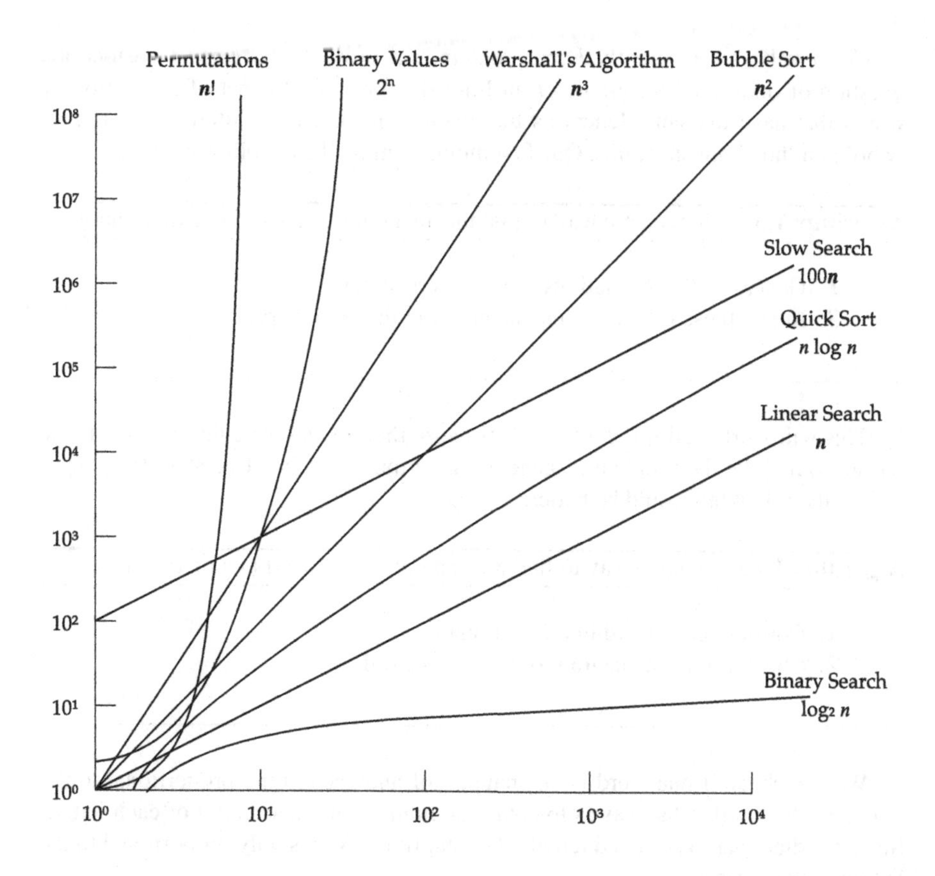

FIGURE 1.1

The growth rates of various functions on a log-log scale: each vertical and horizontal division represents growth by a factor of ten. (Warshall's Algorithm is explained in 2.5.5.)

We can always make an easy problem hard. Consider Algorithm 1.2. Since its first step has complexity $O(n!)$, the whole algorithm is intractable. Fortunately, we know that tractable $O(n \log n)$ algorithms exist for the same problem. We must therefore distinguish the complexity of a problem from the complexity of an algorithm with which we solve it.

Algorithm 1.2 An intractable way to sort a list.

1. Generate all permutations of the list.
2. Choose a permutation in which its elements are in ascending order.

Conversely, an apparently *hard* problem can prove to be *easy*. Consider the question of finding all anagrams of an English word, i.e., the set of all dictionary words that have the same letters as the given word, but in a different order; e.g., 'who' and 'how' are anagrams. Our first thoughts might be as follows:

Algorithm 1.3 An intractable way to find the anagrams of a word in a dictionary.

1. Generate all permutations of the letters in the given word.
2. Reject those permutations that aren't in the dictionary.

This will work well for short words because their letters have few permutations, but we know that the problem of generating permutations is intractable. For longer words, the following would be better:

Algorithm 1.4 A tractable way to find the anagrams of a word in a dictionary.

1. Consider each word in the dictionary.
2. Check if it is an anagram of the given word.

We can check if one word is an anagram of another in time proportional to the length of the word. (One way is to count the number of occurrences of each letter.) Since the dictionary is of fixed length, the total time taken is only proportional to the length of the given word.

How can we know the *inherent complexity of a problem*, that is to say, the complexity of the best possible algorithm that can solve it? In some cases, the answer is easy. In the case of generating permutations, the length of the output sets a lower limit to the complexity of the problem. All we have to assume is that it takes n times longer to output n permutations than it does to output one permutation. On the other hand, if we already *know* a polynomial-time algorithm that solves the problem, then the problem itself must be tractable.

Some problems exist in a kind of limbo. Nobody can prove that the problems are intractable, but nobody has been able to find a polynomial-time algorithm that solves any of them. Some of these problems may actually be intractable; some may not.

An important sub-class of these problems are called **NP-complete**. They have the following properties:

- No polynomial-time algorithm has been discovered for any of them.
- Given the *solution* of an NP-complete problem, we can prove it is correct in polynomial time.
- If anyone ever finds a polynomial-time algorithm to solve any of them, then *all* the NP-complete problems have polynomial-time algorithms.

NP-complete problems have the nature of puzzles: the answers are hard to find, but once found, they are obvious. NP-completeness is the basis of cryptography. One popular method of encrypting (the RSA algorithm) is based on numbers that are products of large primes. Anyone can take two large primes and multiply them, but it is by no means easy for someone else to factorise the resulting number.

It is important to know that this class of problem exists:

I am reminded of one of the first programming assignments I was ever given. It was to find the best sequence in which to insert disks into a drive, given that several programs wanted to share use of the disks, but not necessarily in the same order. For example, one program might need to use disks A, B, and C, in that order, another might need to use B, D, then C, and a third might need to use C, D, E, then B. (In this case, one optimal way to mount the disks is the sequence $[A, B, C, D, C, E, B]$.) I devised several algorithms to deal with this problem, but each had its weakness: I could always devise a case for which it didn't find the shortest sequence. Eventually I came to the conclusion that the only way I could be sure of finding the shortest solution was to try all possible sequences of length 1, then length 2, and so on, until one was found that solved the problem. Sadly, such an algorithm would be intractable, so I grudgingly settled for a method that got good answers most of the time.

That was a long time ago, when I had never heard of NP-complete problems.[12] I know now that the problem was NP-complete, so I don't feel grudging any more. Indeed, if I *had* solved it, a lot of mathematicians and computer scientists would be out of work.

Remember this story the next time you spend weeks solving a problem, only to find that the answer was obvious with 20/20 hindsight![13] Perhaps it was NP-complete.

Almost all optimisation problems are intractable. For example, scheduling sets of jobs in a machine shop to achieve delivery deadlines or to maximise the use of machinery are both intractable problems, as is creating the shortest examination timetable for a university.

How do I know that *my* problem was NP-complete? Because it is virtually the same as a known NP-complete problem, called *Shortest Super-sequence*.

In practice, that is how any problem is shown to be NP-complete. We try to find a way of mapping a known NP-complete problem, K, onto our own problem.[14] If we can do that, then if we can find a way to solve our own problem in polynomial time, then there would also be a way to solve K (by mapping it onto our own problem), and therefore *all* NP-complete problems, in polynomial time — and we shouldn't be prepared to believe that!

Such a step proves that our problem is NP-hard, at least as hard as any NP-complete problem, although it might be harder. To show that it is NP-complete, we should show that our problem can also be mapped onto a known NP-complete problem.

---

[12] So long ago, in fact, that neither had anyone else.

[13] Or 6/6 in metric units.

[14] There are books that list known NP-complete problems. See 1.8.

Why are such problems called NP? 'NP' is an abbreviation for 'non-deterministic polynomial'. If we had a computer that, when faced with a decision, could guess (or non-deterministically decide) the right choice to make, then it would need only polynomial time to find and verify a solution. Unfortunately, given the deterministic nature of actual computers, they must explore alternatives systematically, and may take exponential time to hit on the right answer.[15]

On average, we improve things by using *heuristics*. A heuristic is a rule for making correct guesses most of the time. This can lead to two kinds of algorithms: methods that always take polynomial time, but don't guarantee to get an optimal result; and methods that always get an optimal result, but don't guarantee to take polynomial time. It is characteristic of heuristic methods that, given the heuristic, we can always devise problems that defeat the heuristic. The heuristic is useful only if such cases are rare in practice.

How does tractability affect the problems of systems analysis and synthesis?

First, in the worst case, not only don't we know how the system was designed, but neither can we easily prove it is correct. In such a case, it will almost certainly prove incorrect.

Second, although we don't know how the system was designed, it used a means of construction that makes it relatively easy to prove it is correct.[16]

Third, we may have an algorithm for system construction that *guarantees* the system is correct, without the need for proof — as in the example where business rules were converted into a correct procedure. This is the ideal, *provided the construction algorithm is tractable*.

Unluckily for us, many system design problems seem to be intractable. For example, consider Algorithm 1.1, which turns business rules into a computer procedure. We saw that the order in which the conditions are considered affects the quality of the result. A possible means for finding an optimal solution could be that of Algorithm 1.5.

Algorithm 1.5 An intractable way of converting rules into a procedure.

- Consider every possible ordering of the conditions,
- Generate the corresponding procedure, taking the conditions in the order given,
- Choose the best solution.

Since the first of these steps is intractable, the whole method is intractable and cannot be recommended except when the number of conditions is small. It would be intellectually dishonest to tell you to use such an intractable method. So we shall

[15] This is one reason people are excited about the idea of quantum computers; in principle, they are non-deterministic.

[16] This was the basis of an argument used to promote the use of structured programming, contrasted with the unbridled use of 'go to'.

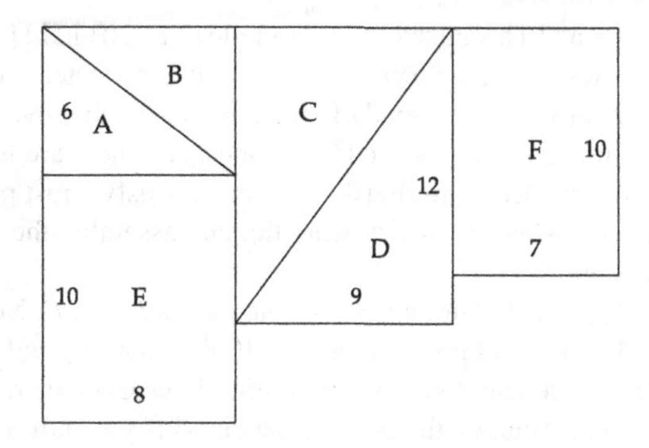

FIGURE 1.2

A puzzle using paper shapes. Cut the pieces from a sheet of paper (not this one!), and reassemble them as explained in the text. Dimensions are in centimeters.

sometimes be forced to recommend heuristic methods. For example, in this problem, it would *usually* be best to deal first with those conditions that appear in the greatest number of rules. Such a heuristic guarantees a quick solution, and usually a good solution, but it cannot promise to always find the best solution.

1.6 MENTAL BLOCKS

Try a little experiment, which I once saw demonstrated by Edward de Bono, of *Lateral Thinking* fame. All you need are a sheet of paper, a ruler, a pair of scissors and a pen or pencil.

Cut out a rectangle 6 cm by 8 cm from the paper, and, in landscape mode, cut it diagonally from bottom right to top left to make two equal triangles. Label these pieces *A* and *B*. Now cut out a second rectangle 9 cm by 12 cm, and again in landscape mode, cut it diagonally from bottom right to top left. Label these pieces *C* and *D*. Now cut a rectangle 10 cm by 8 cm, and label it *E*. Finally, cut a rectangle 7 cm by 10 cm, and label it *F*. (See Figure 1.2.)

The pieces make a puzzle. Assemble the pieces to make the neatest shape you can, and if you need a definition of 'neatest', you can assume it means 'having the shortest perimeter'. The pieces are to be laid flat, on a flat surface, lettered side up, without overlapping. There are no tricks in the way the problem is worded; the pieces fit together like a jigsaw puzzle or a tangram.

Start by assembling pieces *A* and *B*. Now add piece *E*. Next, add pieces *C* and *D*. At this point, most people will have assembled a 16 cm by 17 cm or 16 cm by 20 cm rectangle with a corner missing. Unfortunately, piece *F* doesn't supply the missing corner. See if you can improve your solution, but don't waste too much time, because you can easily find the best solution in a different way.

What has gone wrong? This puzzle is an example of an NP-hard problem, and one can only find the neatest shape by trying every possible arrangement, or by trusting in a heuristic. The six pieces have a total of 20 edges, and each move involves pairing two edges. The first move involves over 150 choices, and there are five moves in all: too many choices to consider. Consciously or unconsciously, most people, given the goal of finding the neatest shape, use a heuristic, and assemble the neatest shape at each stage of the assembly.

Now start over from the beginning. Assemble pieces *E* and *F*. Now add piece *C*. Next, add piece *A*. Finally, add pieces *B* and *D*. If all has gone well, using the same heuristic, you will have assembled a rectangle that is nearly square. If you already got this solution without help, go the head of the class! If you didn't get it *with* help, check the instructions carefully, get it right, and then try the experiment on a friend.[17]

What is the message behind this experiment? The point is that, given the pieces in the first order, even after they are all present, the best solution remains elusive. In the second order, the solution should be obvious. In short, if you try to solve a problem before you have all the pieces, it is a matter of chance whether you will find the solution, and you may block yourself from ever finding it. This is a real phenomenon, and it results directly from how our brains work. The moral is, 'Don't make a theory to fit the facts until you have all the facts, otherwise you will find yourself fitting the facts to the theory'.[18]

The trouble with such advice is that most of us can only think about systems in the ways we are used to and can only absorb facts if we fit them into our evolving design. One problem this book will try to solve is to let you form a working solution without it blocking you from finding the ideal solution. As a means to this end, we need to present ways to *specify* a system without even thinking about its design, and as a means to *this* end, the next chapter covers some mathematics. With luck, most of it will be familiar to you.

This use of specifications leads to another unconventional characteristic of this book. Many books deal with the various stages of system design in the order they are used in practice — perhaps beginning with the process of interviewing the client — but this book is goal-directed. We begin by looking at the *output* of the interviewing process: a relatively formal description of the system, an evolving model. The act of building up this formal model then directs the interviewing process, which we describe later.

[17] I used this puzzle in university classes for several years. The pieces of paper were placed on an overhead projector and their shadows were projected for the whole class to see. The class elected a 'volunteer' to assemble the pieces. He (it always seemed to be a male) never found the correct rectangle, and the rest of the class were then eager to improve his solution and to shuffle the pieces. No one ever found the solution this way. Then, out of a kind heart, I would give the pieces to the original volunteer, but in the second order, and he would find the solution right away.

[18] This is the essence of detective fiction. The author presents you with facts in an order that will make you form one wrong theory after another, while the fictional detective pontificates with some pronouncement like, 'Me, I never make theories, I regard only the facts'.

1.7 SUMMARY

In his 1969 book, 'The Mechanism of the Mind' (ISBN 0-14-013787-4), Edward De Bono gives this analogy of the mind[19]:

> Think of a landscape with streams, rivers, and lakes. When rain falls, it runs along the streams and rivers, making them deeper, flooding the lakes. Think of the rain as stimuli or problems, the streams as thoughts, and the lakes as solutions. New information follows well-worn pathways, and new pathways rarely form.
>
> Think of the brain as the land. Although the land directs the rain, it is the rain that has carved the land. The land has been passive. The brain is merely the *passive* result of its experiences.

This rather depressing analogy suggests that we are totally devoid of creativity. But there is hope: the hope is education, which can dig new channels to direct our thoughts. Sadly, much education is implementation-directed. The building engineer learns how to use an I-section girder; the computer programmer learns to use a particular software development kit or learns what sorts of things a binary search tree is useful for. Thus, we learn to think in terms of potential solutions, the lakes where the rain finally comes to rest. Problems flow by the steepest path directly to the nearest solution. In this book, I hope to dig some channels that lead to reservoirs high in the hills, from which we can direct our thoughts towards the most appropriate solutions.

In his 1967 book, 'The Use of Lateral Thinking' (ISBN 0-14-013788-2), De Bono does some civil engineering, showing us how to build dams and divert flows: we should always question our first thoughts, and maybe our second thoughts too. His book suggests several other tricks that we can use. One, for example, is to pick a word from the dictionary with a pin, and ask how that helps our problem. The brain is passive. We must seek inspiration from outside. I hope to offer you inspiration in this book, by getting you to think about familiar problems in unfamiliar ways.

There is nothing new in this book. It is a kit of parts I have assembled into a reasonably coherent system. This has meant drawing topics from many sources, although none of them very deeply. This seems to be the first time all these topics have all been brought together in one place. I can take no credit for this. It is merely the passive result of my experiences.

One thread that runs through the book is the question of tractability. I have avoided, as far as possible, suggesting that you solve intractable problems. For example, many books on database design will include an instruction such as, 'Find a minimal cover G of the set of FDs F', as if it were no trouble at all.[20] Although an

---

[19] This is my paraphrase, not a quotation.

[20] A cover of F is a set of FDs with the same closure as F. Finding the closure of a set of FDs is an exponentially hard problem. Fortunately, there is an efficient shortcut for finding a minimal cover, which we describe in 4.4.3.

experienced designer can often find such a cover intuitively, experience may be the one thing you don't have.

1.8 FURTHER READING

The complexity of algorithms is discussed much more comprehensively in the first three chapters of 'Foundations of Computer Science: C Edition' by Alfred V. Aho and Jeffrey D. Ullman (1994, ISBN: 0-7167-8284-7). Although this excellent book is out of print, PDFs of the book are available on the Internet.

'Computers and Intractability: A Guide to the Theory of NP-Completeness' (1979, ISBN: 0-7167-1045-5), is a classic textbook by Michael Garey and David S. Johnson, which introduces the reader to complexity theory and lists over 300 distinct NP-hard problems. It is a useful guide to any computer scientist who wants to know if their own particular problem is intractable.

The RSA encryption algorithm was invented by Ron Rivest, Adi Shamir, and Leonard Adleman at MIT and is named after the initial letters of their surnames. It was published in 1978 in *Communications of the ACM* 21 (2): 120–126 as 'A Method for Obtaining Digital Signatures and Public-Key Cryptosystems'. It was the first practical public-key encryption system. Later, it came to light that essentially the same idea had been invented a few months earlier by Clifford Cocks at the Government Communications Headquarters in the UK, but his work was classified and had to remain secret for many years.

Quicksort is one of several sorting algorithms with average complexity $O(n \log n)$ — although its worst-case performance is $O(n^2)$. It was published in 1961 by Tony Hoare as 'Algorithm 64: Quicksort', *Communications of the ACM* 4 (7): 321.

Edsger W. Dijkstra was an early advocate of constructing programs that are provably correct. His 1976 book, 'A Discipline of Programming' (ISBN: 0-13-215871-X), explains how to derive procedures from their pre-conditions and post-conditions. He also showed how to use 'loop invariants' to prove the correctness of procedures containing loops — although the choice of a suitable invariant is sometimes tricky. This book was also influential in promoting the use of structured programming.

1.9 EXERCISES

1. Apply Algorithm 1.1 to the set of rules on page 10 again, this time picking the other condition first. Verify that it produces the same result as the 'intuitive' solution.
2. Suppose you have to sort a list of length l containing integers in the range $1 - n$. Is it possible to sort them in time with complexity less than $O(l \log l)$?
3. A computer has a word length of n bits. Is the problem of listing each possible word tractable or intractable?

CHAPTER 2

Mathematical background

CHAPTER CONTENTS

INTRODUCTION

There are five closely related mathematical concepts that unify the following chapters: propositional calculus, predicate calculus, set theory, the algebra of relations, and graph theory. We won't need to study any of them in depth, but by referring to these concepts, we shall save a lot of long-winded explanation later on. Many readers will already be familiar with these topics, and they will merely need to check how the notation in this book compares with what they are already familiar with. Indeed, if you know what is meant by the strongly connected components of a graph, you already understand a key concept in what follows.

For those readers unfamiliar with the material, I shall explain it in three ways: informally, algebraically, and pictorially. Later on, here and there, I shall venture into some other branches of mathematics, but they will be explained as we go along. I present only notation and a few key theorems here, no formal proofs. This chapter is a *summary*, not a tutorial. Since the material in this chapter could fill a textbook, the reader who needs a more tutorial style is advised to study one or more of the textbooks listed in Section 2.9.

If you find the chapter heavy going, you may prefer to skim it now and come back to it when the text refers to it.

Our aim in this chapter is to understand some key concepts in graph theory, such as *closure*, *reduction*, and *transitive root*. *Graphs* are used to visualise relations between sets. *Sets* and *relations* are often defined by *predicates*, and predicate calculus is easily understood following a discussion of *propositions*.

2.1 PROPOSITIONAL CALCULUS

Propositional calculus is the mathematics of Boolean variables, so the basic ideas should be familiar to most readers. We shall not make much *direct* use of it in this book, but it will help to define several ideas presented later in this chapter.

A **proposition** is a statement that is either true or false, such as, 'Peter is male', 'John is female', or 'Peter has parent Mary'.

2.1.1 LOGICAL OPERATORS

Propositions can be combined using **logical operators** to form **logical expressions**. The most frequently used operators are as follows (where the literals P and Q are arbitrary propositions):

$\neg P$	'**not** P'.
$P \wedge Q$	'P **and** Q'.
$P \vee Q$	'P **or** Q'.
$P \Rightarrow Q$	'P **implies** Q'.
$P \Leftarrow Q$	'P **if** Q', or 'P **is implied by** Q'.
$P \Leftrightarrow Q$	'P **if and only if** Q', or 'P **is equivalent to** Q'.

The results of the operators are defined by the following *truth table*,

P	Q	$\neg P$	$P \wedge Q$	$P \vee Q$	$P \Rightarrow Q$	$P \Leftarrow Q$	$P \Leftrightarrow Q$
False	*False*	*True*	*False*	*False*	*True*	*True*	*True*
False	*True*	*True*	*False*	*True*	*True*	*False*	*False*
True	*False*	*False*	*False*	*True*	*False*	*True*	*False*
True	*True*	*False*	*True*	*True*	*True*	*True*	*True*

Given two component expressions P and Q, a **truth table** shows the value of each expression for each combination of values of P and Q. *True* and *False* are constants. The operators are displayed from left to right in order of precedence, with 'not' ($\neg$) binding the tightest. That is to say, the expression $P \wedge Q \vee \neg P \wedge R \Rightarrow Q \vee R$ means the same as $(P \wedge Q) \vee ((\neg P) \wedge R) \Rightarrow (Q \vee R)$.

$\neg P$ is called the **logical complement** of P.[1]

[1] An electronics engineer would write $\bar{P}$.

The $\vee$ (or) operator has the meaning often written as 'and/or'. Its everyday meaning does not always match its logical meaning. This can confuse the uninitiated.[2]

Beginners often find the entries for $P \Rightarrow Q$ confusing, especially because *False* $\Rightarrow$ *True* has the value *True*. Consider the case 'Pluto is a dog' $\Rightarrow$ 'Pluto is an animal'. The table does not say that if 'Pluto is a dog' is *False* then 'Pluto is an animal' *must* be *True*, it merely says that there is nothing wrong with a universe in which Pluto is an animal because Pluto is a cat. Likewise, in the case when P and Q are both *False*, there is nothing wrong with a universe where Pluto is not an animal because Pluto is a minor planet. The main points to remember are that if $P \Rightarrow Q$, when P is *True*, Q must be *True*, and when Q is *False*, P must be *False*, but knowing that P is *False* tells us nothing about Q, and knowing that Q is *True* tells us nothing about P.

The word 'implies' can itself be a source of confusion. $P \Rightarrow Q$ does not mean 'P causes Q', but is best read as 'P is a special case of Q', as in, 'A dog is a special case of an animal'.

Equivalently, the operators may be defined by using the following axioms:

$$True = \neg False \tag{2.1.1}$$

$$False = \neg True \tag{2.1.2}$$

$$P \wedge True = P \tag{2.1.3}$$

$$P \wedge False = False \tag{2.1.4}$$

$$P \vee True = True \tag{2.1.5}$$

$$P \vee False = P \tag{2.1.6}$$

$$P \Rightarrow Q = \neg P \vee Q \tag{2.1.7}$$

$$P \Leftarrow Q = Q \Rightarrow P = P \vee \neg Q \tag{2.1.8}$$

$$P \Leftrightarrow Q = (P \wedge Q) \vee (\neg P \wedge \neg Q) \tag{2.1.9}$$

Axioms 2.1.1 and 2.1.2 define negation, Axioms 2.1.3 and 2.1.4 define the $\wedge$ operator, and Axioms 2.1.5 and 2.1.6 define the $\vee$ operator. Axioms 2.1.7, 2.1.8, and 2.1.9 define the implication operators in terms of more basic operators. From these axioms, we can easily prove these two well-known laws:

$$P \wedge \neg P = False, \qquad P \vee \neg P = True \tag{2.1.10}$$

$$(P \Rightarrow Q) \wedge (Q \Rightarrow R) \Rightarrow (P \Rightarrow R) \tag{2.1.11}$$

Theorem 2.1.10 is called the **law of the excluded middle**, and Theorem 2.1.11 demonstrates the transitivity of the $\Rightarrow$ operator.

[2] In an an old story, an army general is confronted by a computer intended to advise him on military strategy. The dialogue goes like this:
General: Shall I advance or retreat?
Computer: `Yes.`
General: Yes what?
Computer: `Yes, sir!`

2.1.2 PROPERTIES OF LOGICAL OPERATORS

We can always prove whether two propositional expressions are equivalent by writing out their truth tables. Unfortunately, this approach is inherently *intractable*, because an expression containing N propositions requires a table containing 2^N entries. It is often more practical to work algebraically. Some useful theorems are listed below.

$$P \wedge Q = Q \wedge P \tag{2.1.12}$$
$$P \vee Q = Q \vee P \tag{2.1.13}$$
$$(P \wedge Q) \wedge R = P \wedge (Q \wedge R) = P \wedge Q \wedge R \tag{2.1.14}$$
$$(P \vee Q) \vee R = P \vee (Q \vee R) = P \vee Q \vee R \tag{2.1.15}$$
$$P \wedge (Q \vee R) = (P \wedge Q) \vee (P \wedge R) \tag{2.1.16}$$
$$P \vee (Q \wedge R) = (P \vee Q) \wedge (P \vee R) \tag{2.1.17}$$
$$P \wedge P = P \tag{2.1.18}$$
$$P \vee P = P \tag{2.1.19}$$
$$\neg(\neg P) = P \tag{2.1.20}$$
$$\neg(P \wedge Q) = \neg P \vee \neg Q \tag{2.1.21}$$
$$\neg(P \vee Q) = \neg P \wedge \neg Q \tag{2.1.22}$$

Theorems 2.1.12 and 2.1.13 mean that the operands of $\wedge$ and $\vee$ *commute*, i.e., may appear in either order. Theorems 2.1.14 and 2.1.15 mean that $\wedge$ and $\vee$ are *associative*, i.e., that a string of $\wedge$ or $\vee$ operators may be evaluated in any order (and therefore the order needn't be spelled out). The distribution laws of 2.1.16 and 2.1.17 allow products to be expanded into simple terms. The idempotence laws of 2.1.18 and 2.1.19 allow multiple occurrences of the same variable to be simplified. Equation 2.1.20 expresses the idea that two negatives make a positive. Finally, **De Morgan's laws**, shown in 2.1.21 and 2.1.22, allow '$\wedge$' to be replaced by '$\vee$', or *vice versa*. A generalisation, known as **De Morgan's Theorem**, states that the logical complement of any proposition can be found by replacing '$\wedge$' by '$\vee$' and '$\vee$' by '$\wedge$' and complementing each basic proposition.

Informally, we often see $P \wedge Q$ written as $P.Q$, $P \vee Q$ written as $P + Q$, and $\neg P$ as $-P$. This is because, with such an interpretation, logical expressions obey some of the rules of ordinary school algebra — except that $P + P = P$, etc.[3] Thus, we often see $P \wedge Q$ referred to as a 'product', and $P \vee Q$ referred to as a 'sum'.

2.1.3 CONJUNCTIVE NORMAL FORM

To prove that two expressions are equal, a frequently used technique is to transform both expressions to a standard form. One such standard form is called **conjunctive normal form** or **CNF**. An expression in CNF is a 'product of sums'. The 'sums' are

---

[3] School algebra correctly models the distribution law of equation 2.1.16, but fails to express that of equation 2.1.17.

literals (simple propositions or negated propositions, e.g., P, or $\neg Q$) linked by $\vee$, which are then formed into a 'product' using $\wedge$.[4]

Consider the expression

$$(P \Leftrightarrow Q) \tag{2.1.23}$$

Its conjunctive normal form is

$$(\neg P \vee Q) \wedge (P \vee \neg Q) \tag{2.1.24}$$

To get this result, (using Axiom 2.1.9) we reduce all the operators to $\wedge$, $\vee$, and $\neg$

$$(P \wedge Q) \vee (\neg P \wedge \neg Q) \tag{2.1.25}$$

We then use the first distribution law, three times:

$$\begin{aligned}(P \wedge Q) \vee (\neg P \wedge \neg Q) &= (P \vee \neg P \wedge \neg Q) \wedge (Q \vee \neg P \wedge \neg Q)\\ &= (P \vee \neg P) \wedge (P \vee \neg Q) \wedge (Q \vee \neg P \wedge \neg Q)\\ &= (P \vee \neg P) \wedge (P \vee \neg Q) \wedge (Q \vee \neg P) \wedge (Q \vee \neg Q)\end{aligned}$$

Finally, we eliminate the sums $(P \vee \neg P)$ and $(Q \vee \neg Q)$, which are always *True*, leaving

$$(P \vee \neg Q) \wedge (Q \vee \neg P) \tag{2.1.26}$$

Normalisation is a purely mechanical process that a computer can do (although it is NP-hard).

We can prove the theorem

$$(P \Leftrightarrow Q) = (P \Rightarrow Q) \wedge (P \Leftarrow Q) \tag{2.1.27}$$

by converting both sides to CNF. We have already dealt with the left-hand side above. Normalising its right-hand side is left as a simple exercise for the reader.[5]

---

[4] An interesting connection exists between CNF and the notion of NP-completeness discussed in the previous chapter. Consider the question of whether any assignment of values exists that makes a CNF expression *False*. All we need to do is render any one sum *False* and the whole conjunction will be *False*. To do this, we merely need to make each literal in the sum *False*. For example, in the CNF, $(\neg P \vee Q) \wedge (P \vee \neg Q)$, we can either set P = *True* and Q = *False*, rendering the first sum *False*, or set P = *False* and Q = *True*, rendering the second sum *False*.

Sadly, the question of finding a set of values that makes a given CNF expression *True* isn't so easy, because we have to find a set of assignments that make *all* the sums *True* simultaneously. In fact (for three literals or more), this was the first problem to be considered to be NP-complete.

The astute reader will realise that, equivalently, we could find the converse (negation) of the expression using De Morgan's Theorem, and find an assignment for which it is *False*. But, needless to say, expressing the converse of the expression in CNF can itself can take time 2^N.

[5] There are several normal forms for Boolean expressions. For example, **DNF (Disjunctive Normal Form)** is expressed as a sum of products.

2.1.4 LOGICAL INFERENCE

In a **logical proof**, it is conventional for propositions to be stacked in columns. Stacked propositions are implicitly linked by $\wedge$. From the propositions already known, new propositions can be deduced using **rules of inference**. These are usually stacked below those already known as the proof progresses, with the final conclusion separated from the others by a horizontal line.

The best-known rule is called **modus ponens**: if P implies Q and P is true, then Q must be true.

$$\begin{array}{c} P \Rightarrow Q \\ P \\ \hline Q \end{array}$$

Its converse is called **modus tollens**: if P implies Q and Q is false, then P must be false.

$$\begin{array}{c} P \Rightarrow Q \\ \neg Q \\ \hline \neg P \end{array}$$

Some rules are trivial, such as **and elimination**, and **or introduction**:

$$\begin{array}{c} P \wedge Q \\ \hline P \end{array}$$

$$\begin{array}{c} P \\ \hline P \vee Q \end{array}$$

All these rules are special cases of one general rule, called **resolution**:

$$\begin{array}{c} P \vee Q \\ \neg P \vee R \\ \hline Q \vee R \end{array}$$

P must be either *True* or *False*, so either Q or R must be *True*. P is said to **cancel**, leaving $Q \vee R$ as the **resolvent**. Resolution is well-suited to automated proof, because it is merely necessary to find two propositions containing complementary terms. This is easiest if propositions are simple literals linked by $\vee$. Thus logical proof is straightforward if propositions are in CNF.

2.2 FIRST-ORDER PREDICATE CALCULUS

The (first-order) **predicate calculus** is the mathematics of Boolean functions, so, again, most readers should find the basic ideas familiar.[6]

[6] In *first-order* predicate calculus, the predicates themselves cannot be variables.

A **predicate** is a sentence that may be *True* or *False*, depending on the values of the variables it contains. For example, 'm is male', 'f is female', and 'c has parent p' are predicates whose truths depend on m, f, c, and p. Predicates are usually written the same way as Boolean *functions* are written, e.g., $Male(m)$, $Female(f)$, or *Has Parent*(c,p), with the convention that constants are capitalised, and variables are written in lower case.

2.2.1 QUANTIFIERS

First-order predicate calculus uses the same set of operators as propositional calculus, but it also employs **quantifiers**. We use an 'A' upside-down (∀) to mean 'for all', and an 'E' backwards (∃) to mean 'there exists'. For example, to represent the usual meaning of the sentence, 'Everybody has a parent', we would write

$$\forall c(\exists p(Parent(c,p))) \tag{2.2.1}$$

unless we thought it really meant, 'There is somebody who is everybody's parent', which would be written

$$\exists p(\forall c(Parent(c,p))) \tag{2.2.2}$$

Provided we are clear about the way we interpret predicates, predicate calculus is less ambiguous than natural language and is therefore widely used to specify computer programs.

Variables preceded by ∀ are said to be **universally quantified**. Variables preceded by ∃ are said to be **existentially quantified.**

If x can take values x_1, x_2, and x_3, then $\forall x P(x)$ means $P(x_1) \wedge P(x_2) \wedge P(x_3)$, and $\exists x P(x)$ means $P(x_1) \vee P(x_2) \vee P(x_3)$. In other words, ∀ functions like ∧, and ∃ functions like ∨. As a result, the following extensions of De Morgan's laws are sometimes useful.[7]

$$\neg(\forall x P(x)) \Leftrightarrow \exists x(\neg P(x))$$
$$\neg(\exists x P(x)) \Leftrightarrow \forall x(\neg P(x))$$

Because the names we choose for constants have no inherent meaning to predicate calculus, we must always reason formally. For example, from *Has Parent(Peter, Mary)*, we cannot deduce *Has Child(Mary, Peter)*. Although *Has Parent(Peter, Mary)* = *True* suggests to *us* that *Has Child(Mary, Peter)* = *True*, predicate calculus cannot deduce it, unless it has some additional rule with which to reason, such as *Has Child*$(p,c) \Leftrightarrow$ *Has Parent*(c,p).

As long as we deal with *finite* sets of variables, predicate calculus remains **complete**: in principle, any theorem can be proved by enumerating all possible cases. On the other hand, once we deal with infinite sets, such as the integers, predicate calculus becomes capable of expressing the deepest problems in mathematics, including unresolvable paradoxes.

---

[7] Since P here is a variable that stands for *any* unary predicate, these theorems are not first-order, they are 'higher-order'.

2.3 SETS

A **set** is any collection of objects, called its **elements**. Normally the elements have the same type (e.g., *integers* or *animals*), in which case the set is said to be **typed**. The set of all elements of that type is called the **universe of discourse**, often symbolised by **U**. (Sets of elements of mixed types are often referred to as **collections**.)

2.3.1 SET NOTATION

A set of three males may be written as $\{Peter, Paul, Mark\}$. The members of a set have no order, so $\{Paul, Mark, Peter\}$ is the same set. Nor does it matter if elements are duplicated, so $\{Peter, Paul, Mark, Peter\}$ is the same set too. Sets may be given names, such as $Males = \{Peter, Paul, Mark\}$.

A special case is an empty set (written as $\{\}$ or $\emptyset$), which has no elements. In a sense, all empty sets are the same, but if we are using **typed sets**, then the empty set of males would be considered distinct from the empty set of females.

We write $T \subseteq S$ to say that T is a **subset** of S. A subset of S is any set whose elements are all elements of S. The set *Males* has 8 subsets:

$$\{Peter, Paul, Mark\},$$
$$\{Paul, Mark\},$$
$$\{Peter, Mark\},$$
$$\{Peter, Paul\},$$
$$\{Peter\},$$
$$\{Paul\},$$
$$\{Mark\},$$
$$\{\}$$

Note that *Peter* isn't a *subset* of $\{Peter, Paul, Mark\}$; *Peter* is an *element*, not a set.

If we want to say that element x is a member of set S, we write $x \in S$. To say that x is *not* a member of S, we write $x \notin S$.

The number of elements of set S is called its **cardinality**, $|S|$. A set with cardinality $|S|$ has $2^{|S|}$ subsets. For this reason, the set of all subsets of S, called the **powerset** of S, may be written as 2^S.

Any set is a subset of itself. If we exclude this possibility, we refer to the subset as a **proper subset** and write $T \subset S$.

The elements of a set don't have to be simple. They may, for example, consist of pairs of values, such as the set

$$Has\ Parent = \{(Peter, Mary), (Peter, Mark), (Paul, Mary), (Paul, Mark)\}$$

Sets don't have to be finite, or even countable; we may speak of the set of integers, or the set of real numbers, although in such cases we cannot write down the set as a list of elements. Usually, we define such sets recursively, e.g.,

$$\begin{cases} 0 \in \mathit{Integers}, \\ \forall n(n \in \mathit{Integers} \Rightarrow (n+1) \in \mathit{Integers}), \\ \forall n(n \in \mathit{Integers} \Rightarrow (n-1) \in \mathit{Integers}). \end{cases} \tag{2.3.1}$$

We often define sets using the notation,

$$S = \{E(x, y, \ldots) \mid P(x, y, \ldots)\} \tag{2.3.2}$$

where S is the set being defined, $E(x, y, \ldots)$ is an expression in one or more variables, and $P(x, y, \ldots)$ is a predicate. (Out loud, we say, 'The set of all $E(x, y, \ldots)$ such that $P(x, y, \ldots)$.') For example, we can write,

$$\mathit{Squares} = \{n^2 \mid n \in \mathit{Integers}\} \tag{2.3.3}$$

to define the set of the squares of the integers. The expression is read as, '*Squares* is the set of all n^2 such that n is an integer'.

2.3.2 SET OPERATORS

The **union** ($\cup$) of sets S and T is the set of all elements that are members of *either* S *or* T. We write

$$S \cup T = \{x \mid x \in S \lor x \in T\} \tag{2.3.4}$$

which we read as, 'The union of S and T is the set of all elements x such that either x is a member of S or x is a member of T'.

If *Males* = {*Peter*, *Paul*, *Mark*} and *Females* = {*Mary*, *Jane*}, then

$$\begin{aligned} \mathit{Persons} &= \mathit{Males} \cup \mathit{Females} \\ &= \{\mathit{Peter}, \mathit{Paul}, \mathit{Mark}, \mathit{Mary}, \mathit{Jane}\} \end{aligned}$$

Similarly, if *Children* = {*Peter*, *Mark*, *Mary*}, and *Parents* = {*Paul*, *Jane*, *Mark*, *Mary*}, then *Children* $\cup$ *Parents* = {*Peter*, *Paul*, *Mark*, *Mary*, *Jane*}.

It is always the case that $S \subseteq S \cup T$ and $T \subseteq S \cup T$.

The **intersection** ($\cap$) of sets S and T is the set of all elements that are members of both S and T. We write

$$S \cap T = \{x \mid x \in S \land x \in T\} \tag{2.3.5}$$

If *Children* = {*Peter*, *Mark*, *Mary*}, and *Parents* = {*Paul*, *Jane*, *Mark*, *Mary*}, then *Children* $\cap$ *Parents* = {*Mark*, *Mary*}. Not surprisingly, *Males* $\cap$ *Females* = {}.

It is always the case that $S \cap T \subseteq S$ and $S \cap T \subseteq T$.

The **ordered difference**, or **asymmetric difference**, of S and T consists of all elements of S that are *not* members of T:

$$S \setminus T = \{x \mid x \in S \land x \notin T\} \tag{2.3.6}$$

If *Children* = {*Peter*, *Mark*, *Mary*}, and *Parents* = {*Paul*, *Jane*, *Mark*, *Mary*}, then *Children* \ *Parents* = {*Peter*}, but *Parents* \ *Children* = {*Paul*, *Jane*}.

2.3.3 PROPERTIES OF SET OPERATORS

Because the ∪ operator is defined using ∨, and the ∩ operator is defined using ∧, it isn't surprising that set theory contains analogues of the commutation, association, distribution, and idempotence rules of propositional calculus, given in Equations 2.1.12–2.1.19 on pages 27–27.

$$
\begin{aligned}
S \cap T &= T \cap S \\
S \cup T &= T \cup S \\
(S \cap T) \cap R &= S \cap (T \cap R) = S \cap T \cap R \\
(S \cup T) \cup R &= S \cup (T \cup R) = S \cup T \cup R \\
S \cap (T \cup R) &= (S \cap T) \cup (S \cap R) \\
S \cup (T \cap R) &= (S \cup T) \cap (S \cup R) \\
S \cup S &= S \\
S \cap S &= S
\end{aligned}
$$

To extend the analogy, the empty set ∅ corresponds to *False*, and the universal set **U** corresponds to *True*:

$$
\begin{aligned}
S \cap \mathbf{U} &= S \\
S \cap \emptyset &= \emptyset \\
S \cup \mathbf{U} &= \mathbf{U} \\
S \cup \emptyset &= S
\end{aligned}
$$

which match 2.1.3–2.1.6 above.

Although set theory has no ¬ operator, we can find a similar role for asymmetric difference in Axioms 2.1.1–2.1.2 and Theorem 2.1.10.

(Within the universe of discourse, **U**, $x \notin S$ if and only if $x \in (\mathbf{U} \setminus S)$.)

$$
\begin{aligned}
\mathbf{U} \setminus \mathbf{U} &= \emptyset \\
\mathbf{U} \setminus \emptyset &= \mathbf{U} \\
S \setminus (\mathbf{U} \setminus S) &= S \\
S \cap (\mathbf{U} \setminus S) &= \emptyset \\
S \cup (\mathbf{U} \setminus S) &= \mathbf{U}
\end{aligned}
$$

There are even analogues of De Morgan's laws 2.1.21 and 2.1.22:

$$
\begin{aligned}
\mathbf{U} \setminus (S \cap T) &= (\mathbf{U} \setminus S) \cup (\mathbf{U} \setminus T) \\
\mathbf{U} \setminus (S \cup T) &= (\mathbf{U} \setminus S) \cap (\mathbf{U} \setminus T)
\end{aligned}
$$

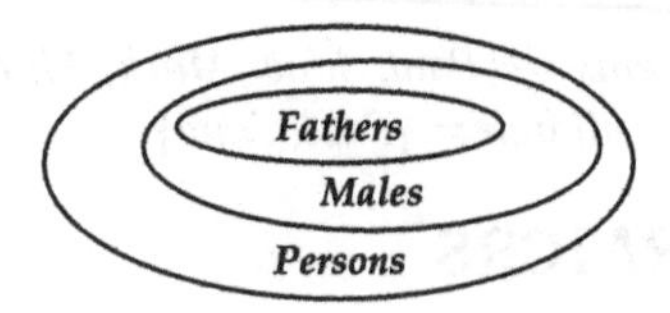

FIGURE 2.1

An Euler diagram showing subset relationships between *Fathers*, *Males*, and *Persons*.

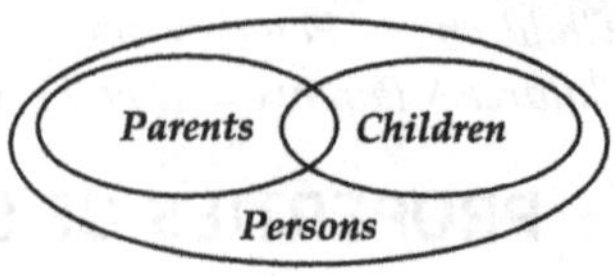

FIGURE 2.2

An Euler diagram showing that *Parents* and *Children* are overlapping subsets of *Persons*.

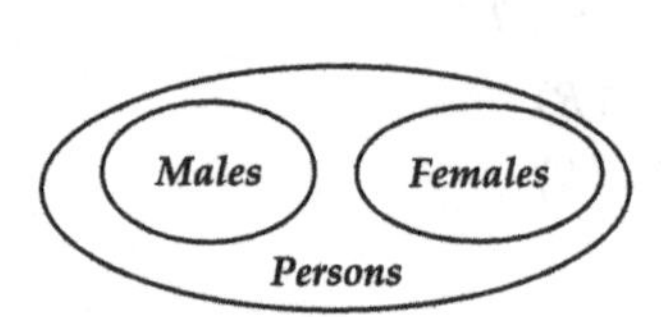

FIGURE 2.3

An Euler diagram showing that *Males* and *Females* are disjoint subsets of *Persons*.

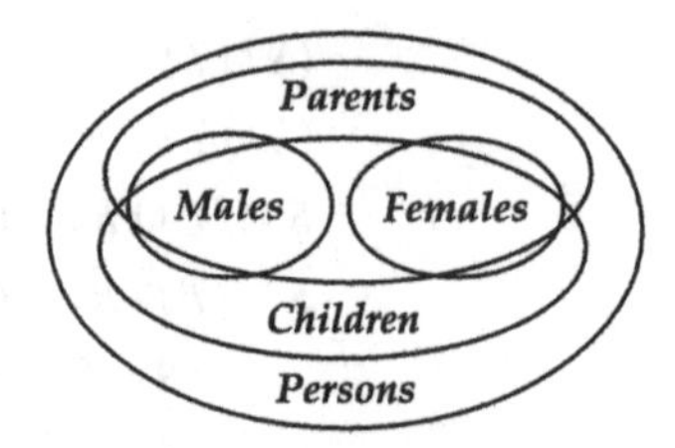

FIGURE 2.4

An Euler diagram that tries to show too much. As a result, it fails to show anything clearly.

2.3.4 EULER DIAGRAMS

We may visualise the relationships between sets by drawing **Euler diagrams**. Figure 2.1 shows that *Males* is a subset of *Persons*; every element of *Males* is also an element of *Persons*. Likewise, every member of *Fathers* is a member of *Males*. The outermost ellipse defines our universe of discourse, the set *Persons*. This gives us a visual way to understand the $\Rightarrow$ operator of predicate calculus: $x \in Males \Rightarrow x \in Persons$, and $x \notin Persons \Rightarrow x \notin Males$. On the other hand, $x \notin Males$ permits either $x \in Females$ (for example) or $x \notin Persons$, and $x \in Persons$ permits either $x \in Males$ or $x \notin Males$. Not surprisingly, like $\Rightarrow$ in Equation 2.1.11, the $\subseteq$ operator is transitive.

Figure 2.2 shows a situation where *Parents* and *Children* **overlap**; at least one element of *Parents* is also an element of *Children*, but not all of them.

Figure 2.3 shows a situation where *Males* and *Females* are **disjoint**; *Males* and *Females* have no common element.

If $S_1, S_2, \ldots S_N$ are all mutually disjoint and $S_1 \cup S_2 \cup \ldots S_N = S$, then $S_1, S_2, \ldots S_N$ are said to **partition** S. For example, if $Persons = Males \cup Females$, and *Males* and *Females* are disjoint, then *Males* and *Females* partition *Persons*.

Often, it is possible to combine several relationships into a single diagram, as in Figure 2.4, which shows that *Males* and *Females* are disjoint sets entirely contained within the union of the *Parents* and *Children* sets. There is a limit to how much information a single Euler diagram can express: for example, Figure 2.4 doesn't show that the union of *Males* and *Females* is the same as the set of *Persons*.

If we want to show the elements of a set on an Euler diagram, they must be drawn as points, not areas. Elements of a set are not subsets of it — although it is possible to define **singleton** sets, containing only one element, e.g., $\{Peter\}$.

2.4 RELATIONS AND FUNCTIONS

2.4.1 CARTESIAN PRODUCTS

The **Cartesian product** of two sets, X and Y, denoted by $X \times Y$, is the set of *all* ordered pairs (x, y), where x is an element of X and y is an element of Y:[8]

$$X \times Y = \{(x, y) \mid x \in X \wedge y \in Y\} \tag{2.4.1}$$

For example, if $Children = \{Peter, Mark, Mary\}$, and $Parents = \{Paul, Jane, Mark, Mary\}$, then

$$\begin{aligned} Children \times Parents = \{&(Peter,\ Paul), (Peter,\ Jane), (Peter,\ Mark), (Peter,\ Mary),\\ &(Mark,\ Paul), (Mark,\ Jane), (Mark,\ Mark), (Mark,\ Mary),\\ &(Mary,\ Paul), (Mary,\ Jane), (Mary,\ Mark), (Mary,\ Mary)\} \end{aligned}$$

The order of terms within a pair is important: $(Mary, Mark)$ is not the same ordered pair as $(Mark, Mary)$; $X \times Y = Y \times X$ if and only if $X = Y$.

2.4.2 BINARY RELATIONS

The equation $x^2 + y^2 = 1$ describes a *relationship* between x and y. A **binary relation** is a similar notion, but it has a sense of direction, for example, **from** x **to** y.

A **binary relation** R **from domain** X **to codomain** Y is a *selected* set of ordered (x, y) pairs. The first element of each pair is drawn from the domain, and the second element from the codomain. Formally, $R \subseteq X \times Y$, where $X = \text{dom}(R)$ and $Y = \text{codom}(R)$.

As an example, consider

$$Unit\ Circle = \{(x, y) \mid x \in Reals \wedge y \in Reals \wedge x^2 + y^2 = 1\} \tag{2.4.2}$$

This describes a binary relation **from** a real value x **to** (typically) two y values, i.e., $y = +\sqrt{1 - x^2}$ and $y = -\sqrt{1 - x^2}$. For example $(0, +1)$ and $(0, -1)$ are both

[8] The name 'Cartesian' derives from the Cartesian co-ordinate system used in co-ordinate geometry, in which points in a plane are represented as ordered pairs, such as (3.0, 2.5). As in geometry, it is possible to have Cartesian products of higher order, such as $X \times Y \times Z$.

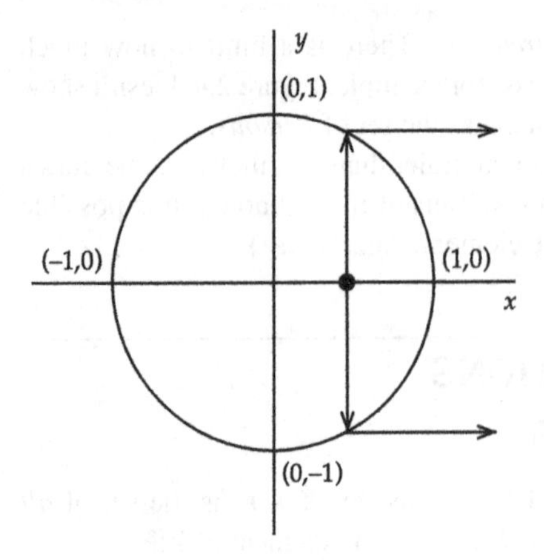

FIGURE 2.5

Unit Circle considered as a relation from x to y.

members of relation *Unit Circle*. (There is only one y value when $x = \pm 1$. See Figure 2.5.)

To emphasise the fact that a relation has direction, we shall write a pair such as (x, y) in the form $x \mapsto y$ (x **'maps to'** y), thus,

$$Unit\ Circle = \{x \mapsto y \mid x \in Reals \wedge y \in Reals \wedge x^2 + y^2 = 1\} \qquad (2.4.3)$$

An alternative way to express that $x \in X$ and $y \in Y$ is to include the domain and codomain as the type of the relation:

$$Unit\ Circle : Reals \rightarrow Reals = \{x \mapsto y \mid x^2 + y^2 = 1\} \qquad (2.4.4)$$

(We shall make particular use of this notation when we discuss schemas.) If $x \mapsto y \in R$, we may alternatively write $x\ R\ y$, for example, $x \mapsto y \in Unit\ Circle$ means the same as $x\ Unit\ Circle\ y$.

As in the case of *Unit Circle*, relations are often referred to by name.

For a second example, please forgive the somewhat circular definition, and consider

$$\leq\ = \{x \mapsto y \mid x \in Integers \wedge y \in Integers \wedge x \leq y\} \qquad (2.4.5)$$

In other words, '$\leq$' is the name of the relation that maps integer x to all integers y no less than itself.

When, as here, the domain and codomain of a relation are infinite sets, the mapping cannot be enumerated and must be given by a formula. On the other hand, relations may also be defined over discrete sets, in which case they can be defined by enumeration. For example, consider the relation *Has Parent*, whose domain is {*Peter*, *Mark*, *Mary*} and whose codomain is {*Paul*, *Jane*, *Mark*, *Mary*}, with the

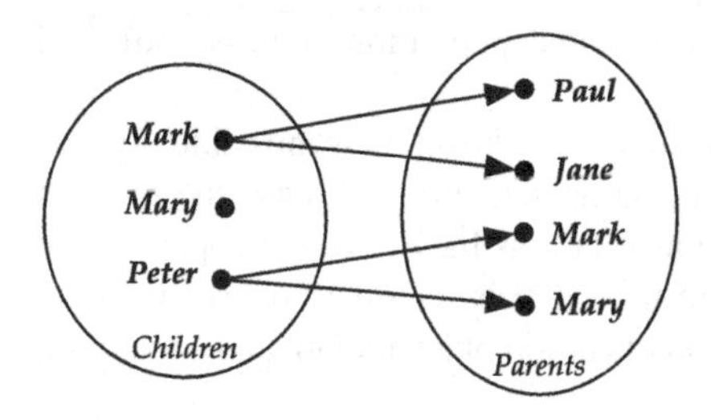

FIGURE 2.6

The *Has Parent* relation *from* the set {*Peter*, *Mark*, *Mary*} *to* the set {*Paul*, *Jane*, *Mark*, *Mary*}.

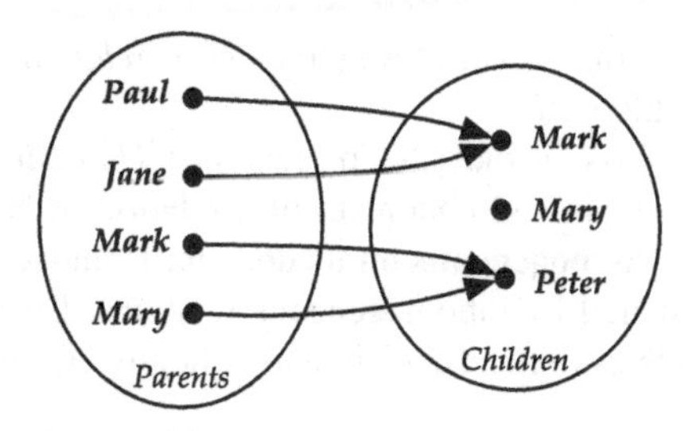

FIGURE 2.7

Has Child, the converse of the *Has Parent* relation.

mapping $\{Peter \mapsto Mary,\ Peter \mapsto Mark,\ Mary \mapsto Jane,\ Mary \mapsto Paul\}$. We would describe this particular relation as being **from** the set {*Peter*, *Mark*, *Mary*} **to** the set {*Paul*, *Jane*, *Mark*, *Mary*}.

We may visualise finite relations by drawing a mapping, as in Figure 2.6. The ellipses represent the domain and codomain; the black dots, elements; and the arrows, the mapping, i.e., the ordered pairs.

If every element of the domain appears in at least one ordered pair, the relation is **total**; otherwise it is **partial**. If every element of the codomain appears in at least one pair, the relation is **onto** its codomain; otherwise it is **into** its codomain. The relation in Figure 2.6 is partial and onto.

Every relation has a **converse**, which is obtained by exchanging the roles of the domain and codomain and reversing the mapping.

$$R^{-1} = \{y \mapsto x \mid x\ R\ y\} \tag{2.4.6}$$

The converse of the *Has Parent* relation, which we may call *Has Child*, is illustrated in Figure 2.7. In general, it is a relation, because parents can have any number of children, including zero, but in this particular example, each parent has exactly one child.

It is possible to have relations of orders higher than binary. A ternary relation is defined by

$$R \subseteq X \times Y \times Z \tag{2.4.7}$$

For our purposes, such relations will always be reduced to binary relations. We might regard $X \times Y$ as the domain and Z as the codomain, i.e.,

$$R : X \times Y \to Z = \{(x, y) \mapsto z \mid P(x, y, z)\} \tag{2.4.8}$$

Alternatively, we might make X the domain and $Y \times Z$ the codomain:

$$R : X \to Y \times Z = \{x \mapsto (y, z) \mid P(x, y, z)\} \tag{2.4.9}$$

In what follows, the term *relation* means binary relation, unless indicated otherwise.[9]

As in the case of relations *Unit Circle* and '≤' above, it is possible for the domain and codomain of a relation to be the same set, in which case the relation is **homogeneous on** its domain. Homogeneous relations have several properties not shared by inhomogeneous relations. Despite this, if it is useful to do so, we may always extend a relation to be homogeneous on the union of its domain and codomain.

2.4.3 SPECIAL RELATIONS

A *general* relation has no restrictions: elements of the domain and codomain may appear in ordered pairs once, several times, or not at all. If each element of the domain occurs at most once, then each element of the domain associates with a single value in the codomain. Such a relation is called a **function**, and its codomain is called its **range**. This is an extension of the familiar idea of a function, such as sine or cosine: given a value in the domain, it yields a *unique* value in the codomain. This notion of function also corresponds closely to the idea of function in a programming language. The *Has Parent* relation of Figure 2.6 is *not* a function, but its converse, *Has Child*, in Figure 2.7, happens to be a function.

We may use infix notation to show that elements are related, for example, we may write *Jane Has Parent Mary* in just the same way that we would write $1 \leq 2$. We may also **apply** a relation to an argument using the dot notation familiar to programmers, e.g., *Jane.Has Parent* = *Mary*. In the case of a function, such as *Has Mother*, we may still write *Jane.Has Mother* = *Mary*, but we may also use the more familiar functional notation *Has Mother*(*Jane*) = *Mary*.

A Cartesian product is always associated with two or more (trivial) **projection functions**. For example, given the domain $X \times Y$,

$$\pi_x : X \times Y \to X = \{(x, y) \mapsto x\}$$
$$\pi_y : X \times Y \to Y = \{(x, y) \mapsto y\}$$

We may extend this idea to relations. For example, given

$$\mathit{Has\ Parent} = \{\mathit{Peter} \mapsto \mathit{Mary}, \mathit{Peter} \mapsto \mathit{Mark},\ \mathit{Mary} \mapsto \mathit{Jane}, \mathit{Mary} \mapsto \mathit{Paul}\}$$

we have

$$\begin{aligned}\pi_{\mathit{Children}}(\mathit{Has\ Parent}) = \{&(\mathit{Peter},\ \mathit{Mary}) \mapsto \mathit{Peter}, (\mathit{Peter},\ \mathit{Mark}) \mapsto \mathit{Peter},\\ &(\mathit{Mary},\ \mathit{Jane}) \mapsto \mathit{Mary}, (\mathit{Mary},\ \mathit{Paul}) \mapsto \mathit{Mary}\}\\ \pi_{\mathit{Parents}}(\mathit{Has\ Parent}) = \{&(\mathit{Peter},\ \mathit{Mary}) \mapsto \mathit{Mary}, (\mathit{Peter},\ \mathit{Mark}) \mapsto \mathit{Mark},\\ &(\mathit{Mary},\ \mathit{Jane}) \mapsto \mathit{Jane}, (\mathit{Mary},\ \mathit{Paul}) \mapsto \mathit{Paul}\}\end{aligned}$$

The converse of a function isn't necessarily another function, but it is always a relation. For example, the relation that maps each integer onto its square is a

[9] The term 'relational database' refers to exactly the same notion of relation as described here.

function, $\{\ldots,-2 \mapsto 4,-1 \mapsto 1,0 \mapsto 0,1 \mapsto 1,2 \mapsto 4,\ldots\}$, but its inverse, $\{\ldots,0 \mapsto 0,1 \mapsto 1,1 \mapsto -1,4 \mapsto 2,4 \mapsto -2,\ldots\}$, is not. Similarly, although sin is a well-known trigonometric function, since $\sin^{-1}(0)$ has an infinite set of values, $\{\ldots,-2\pi,-\pi,0,\pi,2\pi,\ldots\}$, $\sin^{-1}$ isn't a function. In the case that the converse of a function *is* also a function, it is then called its **inverse**. If a function and its inverse are both total, they define a (one-to-one) **correspondence**. The function,

$$Successor : Integers \rightarrow Integers = \{x \mapsto y \mid y = x + 1\} \quad (2.4.10)$$

has the inverse,

$$Predecessor : Integers \rightarrow Integers = \{x \mapsto y \mid x = y + 1\} \quad (2.4.11)$$

and both are examples of correspondences; every integer has both a successor and a predecessor.

A **sequence** or **list**, such as $[a,b,a,c,b,d]$ (in which both the order and repetition of terms matter), is simply a function that maps positive integers onto its terms, i.e., the function,

$$\{1 \mapsto a,2 \mapsto b,3 \mapsto a,4 \mapsto c,5 \mapsto b,6 \mapsto d\} \quad (2.4.12)$$

A **bag** is like a sequence in that each element can appear more than once, but like a set in that the order of the elements is unimportant. For an example, think of your loose change. It is not only the set of coins that matters, but how many you have of each denomination. The bag {5¢, 5¢, 10¢, 20¢, 20¢, 20¢} can be expressed as the function {5¢ ↦ 2, 10¢ ↦ 1, 20¢ ↦ 3}.

A relation R such that $\forall x \forall y(x\ R\ y \Rightarrow y\ R\ x)$ is said to be **symmetric**. The relations *Sibling Of* and *Married To* (with their everyday meanings) are symmetric. So is '=', because $\forall x \forall y(x = y \Rightarrow y = x)$.

If $(x\ R\ y \wedge y\ R\ x) \Rightarrow x = y$, then R is said to be **anti-symmetric**. The relation '≤' is anti-symmetric, because $x \leq y \wedge y \leq x \Rightarrow x = y$. The relation '<' is vacuously anti-symmetric, because there is no case where $x < y \wedge y < x$.

A relation R that (among other things) maps each element to itself, so that $x\ R\ x$, for all x, is said to be **reflexive.** The relation '≤' is reflexive, because $\forall x(x \leq x)$. The relation '<' is **irreflexive** because there is no case where $x < x$.

A relation R such that $\forall x \forall y \forall z(x\ R\ y \wedge y\ R\ z \Rightarrow x\ R\ z)$ is said to be **transitive**. The relation '≤' is transitive, because $x \leq y \wedge y \leq z \Rightarrow x \leq z$. The relation '*Part Of*' is transitive: if a spoke is part of a wheel and the wheel is part of a bicycle, then the spoke is part of the bicycle.

A **partial ordering** is a relation that is reflexive, anti-symmetric, and transitive. The subset relation is a paradigm of a partial order: $S \subseteq S$, $(S \subseteq T \wedge T \subseteq S) \Rightarrow S = T$, and $(S \subseteq T \wedge T \subseteq U) \Rightarrow S \subseteq U$. For example, the subset relation on the subsets of {*Peter*, *Paul*, *Mark*} is clearly transitive:

$$\begin{aligned}\{Peter\} &\subseteq \{Peter,\ Paul\} \subseteq \{Peter,\ Paul,\ Mark\}\\ &\Rightarrow \{Peter\} \subseteq \{Peter,\ Paul,\ Mark\}\end{aligned}$$

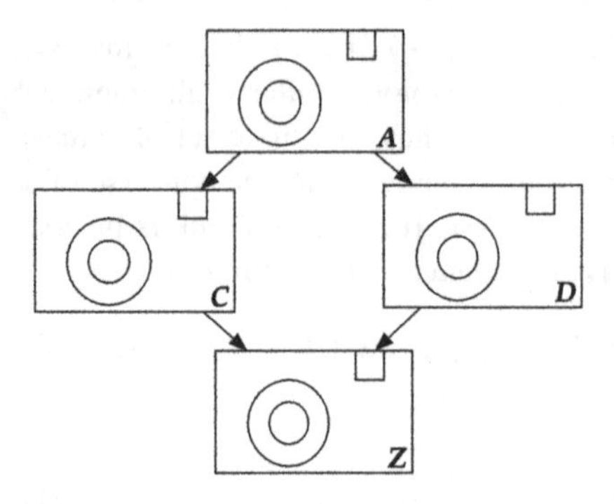

FIGURE 2.8

A comparison between cameras. Camera C has more pixels than D, but D has zoom and flash. C and D aren't comparable. Camera A has as many pixels as C and has zoom and flash like D. It is superior to both. Z has only as many pixels as D and like C, no zoom or flash. It is inferior to both.

Not every pair of elements in a partial order needs to be comparable. For example, neither $\{Peter, Mark\} \subseteq \{Paul, Mark\}$ nor $\{Paul, Mark\} \subseteq \{Peter, Mark\}$ is true. As a further example, consider a comparison between cameras. Camera C may have more pixels than camera D, but may not have as many features, so we cannot (in general) say that one is better than the other. In contrast, camera A may be superior to both C and D in both pixels and features, while camera Z may be inferior to both. (See Figure 2.8.)

2.4.4 OPERATIONS ON RELATIONS

It is possible to define many useful operations that can combine binary relations. These form an **algebra of relations**. The reader should note that the *algebra of relations* is distinct from the relational algebra associated with databases — which typically involves higher-order relations.

Since relations are sets, we can perform all the usual set operations. In addition, given a relation from X to Y, we can find the subset of y values associated with a given value of x, the converse relation from Y to X, etc.

If two relations share the same domain and codomain, we can find their union, intersection, or difference simply by applying these operations to their sets of pairs. Similarly, we can speak of one relation being a subset of another, or of two relations overlapping, or being disjoint.

If the codomain of relation Q is also the domain of a second relation R, we may form their **relational product** (denoted by ;) defined as follows[10]:

$$Q\,;R = \{x \mapsto z \mid x\ Q\ y \wedge y\ R\ z\} \tag{2.4.13}$$

Consider the following two relations:

$$Sqrt = \{\ldots, 0 \mapsto 0, 1 \mapsto 1, 1 \mapsto -1, 4 \mapsto 2, 4 \mapsto -2, \ldots\}$$
$$Seq = \{1 \mapsto a, 2 \mapsto b, 3 \mapsto a, 4 \mapsto c, 5 \mapsto b, 6 \mapsto d\}$$

Since the codomain of *Sqrt* and the domain of *Seq* below are both integers, it is possible to form their product:

$$Sqrt\,;Seq = \{1 \mapsto a, 4 \mapsto b, 9 \mapsto a, 16 \mapsto c, 25 \mapsto b, 36 \mapsto d\}$$

(Values in the codomain of *Sqrt*, such as −2, that have no corresponding element in the domain of *Seq*, do not contribute to the product.)

A **homogeneous relation** has the same domain and codomain ($dom(R) = codom(R)$), so it is possible for it to form a product with itself. Such products are written as powers:

$$R^2 = R\,;R$$
$$R^3 = R\,;R\,;R$$

By extension, we write

$$R^1 = R \tag{2.4.14}$$

and even

$$R^0 = \{x \mapsto x \mid x \in dom(R)\} \tag{2.4.15}$$

to denote the **identity relation**, $I_{dom(R)}$, **on** the domain of R.

A **closure of a relation** is formed by adding just enough pairs to the relation to give it some desired property:

- The **symmetric closure** of homogeneous relation R is the union of R with its converse, i.e., $R \cup R^{-1}$. (If $x \mapsto y$ is a pair in R, then $y \mapsto x$ is a pair in R^{-1}.)
- The **reflexive closure** of homogeneous relation R is the union of R with its corresponding identity, i.e., $R \cup R^0$. (Since R^0 is reflexive, so is the union.)
- The **transitive closure**, R^+, of homogeneous relation R is defined by the infinite union

$$R^+ = R^1 \cup R^2 \cup R^3 \cup \cdots \tag{2.4.16}$$

(If $x \mapsto y$ and $y \mapsto z$ are pairs in R, then $x \mapsto z$ is a pair in R^2, and so on for R^3, etc.)

---

[10] The reader may wish to pronounce ' ; ' as 'followed by.'

- Finally, the **reflexive transitive closure**, R^*, of homogeneous relation R is the union $R^+ \cup R^0$, alternatively defined by the infinite union

$$R^* = R^0 \cup R^1 \cup R^2 \cup R^3 \cup \cdots \tag{2.4.17}$$

Even though transitive closures are defined by infinite unions, if a relation has a finite domain, X, its closure must also be finite, being at most $X \times X$.[11]

To illustrate various operations on relations, consider the following relation[12]:

$$\textit{Daughter Of} = \{c \mapsto p \mid c \textit{ Has Parent } p \wedge Female(c)\} \tag{2.4.18}$$

This is rather an awkward definition because it mixes relational notation and predicate calculus. In using relations, we prefer to use an identity relation, as follows:

$$\textit{Is Female} = I_{Females} \tag{2.4.19}$$

where $I_{Females}$ is the identity relation on the set *Females*.

Relations 2.4.20 and 2.4.21 are subsets of the *Has Parent* relation, obtained by taking a subset of its domain or codomain, and are called **left-restrictions** and **right-restrictions**, respectively.

$$\textit{Daughter Of} = \textit{Is Female} \,;\, \textit{Has Parent} \tag{2.4.20}$$

$$\textit{Has Mother} = \textit{Has Parent} \,;\, \textit{Is Female} \tag{2.4.21}$$

With a similar definition, *Is Male* = I_{Males}, we can then proceed as follows:

$$\begin{aligned}
\textit{Has Child} &= \textit{Has Parent}^{-1} \\
\textit{Has Father} &= \textit{Has Parent} \,;\, \textit{Is Male} \\
\textit{Son Of} &= \textit{Is Male} \,;\, \textit{Has Parent} \\
\textit{Has Daughter} &= \textit{Daughter Of}^{-1} \\
\textit{Mother Of} &= \textit{Has Mother}^{-1} \\
\textit{Has Son} &= \textit{Son Of}^{-1} \\
\textit{Father Of} &= \textit{Has Father}^{-1} \\
\textit{Is Person} &= \textit{Is Male} \cup \textit{Is Female} \\
\textit{Has Grandparent} &= \textit{Has Parent}^2 \\
\textit{Has Grandchild} &= \textit{Has Child}^2 = \textit{Has Grandparent}^{-1} \\
\textit{Has Ancestor} &= \textit{Has Parent}^+ \\
\textit{Has Descendant} &= \textit{Has Child}^+ = \textit{Has Ancestor}^{-1}
\end{aligned}$$

---

[11] Not all closures are finite. The $\leq$ relation is the infinite reflexive transitive closure of the *Successor* relation.

[12] The names of the relations in this section have self-explanatory meanings equivalent to the normal family relationships recognised by English speakers.

$$\textit{Has Aunt} = (\textit{Has Grandparent}\,;\textit{Has Daughter}) \setminus \textit{Has Mother} \qquad (2.4.22)$$

(In Relation 2.4.22, your aunt is any daughter of your grandparents except your own mother.)

Since relations are special kinds of sets, all the axioms and theorems of set theory apply to relations. In addition, they have properties concerning products and converses:

$$(R\,;S)\,;T = R\,;(S\,;T) = R\,;S\,;T$$

$$R\,;(S \cup T) = (R\,;S) \cup (R\,;T)$$

$$(S \cup T)\,;R = (S\,;R) \cup (T\,;R) \qquad (2.4.23)$$

$$(R^{-1})^{-1} = R \qquad (2.4.24)$$

$$(R\,;S)^{-1} = S^{-1}\,;R^{-1}$$

However, the following are inequalities, except in special cases:

$$R\,;S \neq S\,;R \qquad (2.4.25)$$

$$R^0 \subseteq R^1\,;R^{-1} \qquad (2.4.26)$$

$$(S \cap T)\,;R \subseteq (S\,;R) \cap (T\,;R) \qquad (2.4.27)$$

$$R\,;(S \cap T) \subseteq (R\,;S) \cap (R\,;T) \qquad (2.4.28)$$

Inequality 2.4.25 reflects the fact that your mother's father isn't your father's mother. To illustrate inequality 2.4.26, your parent's child may be yourself, but it may also be your brother or sister.[13]

In the case of 2.4.27, consider

$$(\textit{Has Father}\,;\textit{Has Child}) \cap (\textit{Has Mother}\,;\textit{Has Child}),$$

which, when applied to you, yields yourself and your full brothers and sisters. In contrast, the expression (*Has Father* ∩ *Has Mother*) is an empty relation, and therefore so is the equivalent left-hand side of 2.4.27.

In the case of 2.4.28, consider

$$(\textit{Has Ancestor}\,;\textit{Has Child}) \cap (\textit{Has Ancestor}\,;\textit{Has Grandchild}),$$

which is certainly not empty, and when applied to you, yields yourself and your ancestors. But in a well-ordered society, the relation (*Has Child* ∩ *Has Grandchild*) — and thus the equivalent left-hand side of 2.4.28 — should be empty.

Finally, another potentially useful way to combine relations is by **parallel composition:**

$$Q \| R = \{x \mapsto (y,z) \mid x\ Q\ y \wedge x\ R\ z\} \qquad (2.4.29)$$

Contrast *Has Parent* with *Has Father*||*Has Mother*: The first is a relation from a child to both its father and mother; the second is a relation (which happens to be a function) that maps a child to a single (*Father*, *Mother*) pair.

---

[13] For this reason, and despite Equation 2.4.24, not everyone is happy with the notation R^{-1} to denote the inverse of R. Many prefer $R^{\leftarrow}$. Even so, such usage is well-established for $\sin^{-1} x$, $\cos^{-1} x$, etc.

2.5 GRAPHS AND SCHEMAS

Systems analysts love to draw diagrams. The virtue of diagrams is that they appeal to the eye. Because of the way the brain processes visual information, it is usually easier to assimilate the information they display than the same information given in algebraic or tabular form. Even so, diagrams are only useful when they are simple. An attempt to make a diagram useful for more than one purpose can sometimes make it useful for no purpose at all.

Many diagrams are actually mathematical objects called **graphs** or schemas.[14] A graph is a diagram that defines a relation. A *schema* shows a **meta-relation**: *a relationship between relations.*

2.5.1 GRAPHS

Most of us should be familiar with the idea of a graph that shows a relation between real numbers. Figure 2.5 shows the graph of the relation $x^2 + y^2 = 1$, describing the unit circle. We may use a similar approach to show a relation between integers. On the other hand, when we want to show the relation between *Parents* and *Children*, axes are less appropriate, and a different kind of graph is used. Such a graph consists of labelled **vertices** (usually drawn as small circles or rectangles) connected by **edges** (usually drawn as arrows). (See Figure 2.9.)[15] The edges are **directed** from the domain to the codomain. In other words, an edge **from** u **to** v means that the pair $u \mapsto v$ is a member of the relation. As a result, there may be at most one edge from u to v. The edge has **initial vertex** u and **terminal vertex** v. We denote an edge from u to v by $u \to v$. The edge $u \to v$ is said to be an **out-edge** of u and an **in-edge** of v. The number of edges entering v is called its **in-degree**, and the number of edges leaving v is called its **out-degree**. If we want to emphasise that the edge $u \to v$ belongs to a particular graph (or relation) G, we write $u \underset{G}{\to} v$. Because their edges are directed, such graphs are called directed graphs, or **digraphs**.

A **subgraph** H of graph G comprises some subset of the vertices of G and some subset of the edges of G, subject to the sensible restriction that the edges of H must link vertices that belong to H. (See Figure 2.10.) H is therefore a graph of a subset of the pairs of the relation represented by G.

Sometimes a diagram forms a **bi-partite graph**. This is a graph with two kinds of vertices and two kinds of edges. The edges link vertices of different kinds, but never of the same kind. For example, Figure 2.11 is a bi-partite graph showing the relationship between some processes (rectangles) and files (circles). It contains *two* subgraphs that define *two* relations, an *Output* relation from processes to files, and an

[14] Strictly speaking, since I consistently use the Latin-derived plurals, 'matrices', 'vertices', and 'indices', rather than 'matrixes', 'vertexes', and 'indexes', I should say 'schemata' rather than 'schemas'. But I don't. 'Schema' is derived from *Greek*, and there I draw the line.

[15] We are distinguishing here between Mark and Mary as *Parents* and Mark and Mary as *Children*. This isn't perhaps the best way to express their family relationships. Figure 2.12 expresses it better.

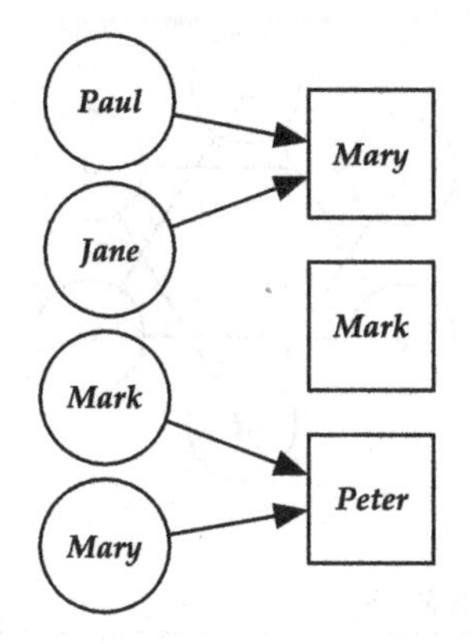

FIGURE 2.9

Graph of the *Has Child* relation from the domain *Parents* (circles) to the codomain *Children* (squares). The edges represent ordered pairs.

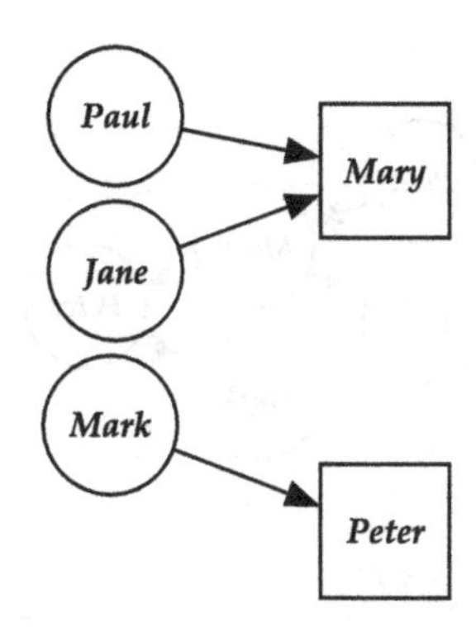

FIGURE 2.10

A subgraph of the *Has Child* relation of Figure 2.9.

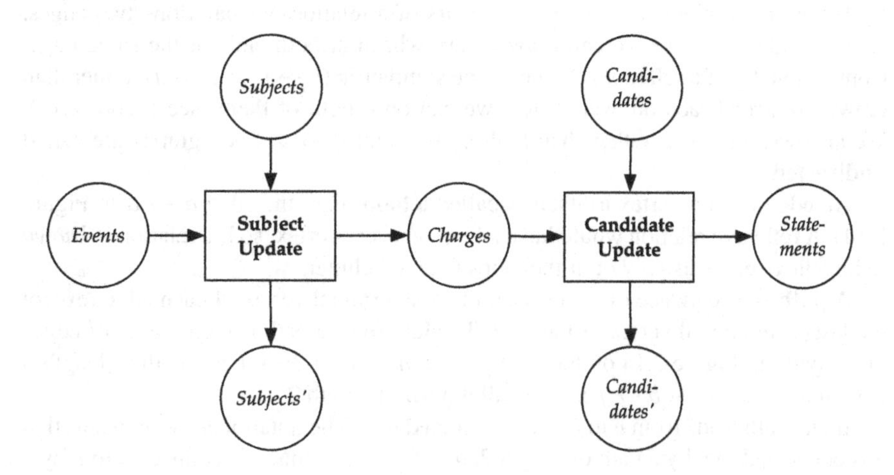

FIGURE 2.11

A bi-partite graph, showing both an *Output* relation (rectangles to circles) and an *Input* relation (circles to rectangles).

Input relation from files to processes. The *Output* relation is given by the edges from rectangles to circles, and the *Input* relation by the edges from circles to rectangles.

Similar conventions are used to draw homogeneous relations, where the domain and codomain are identical, although there now is only one kind of vertex. If we

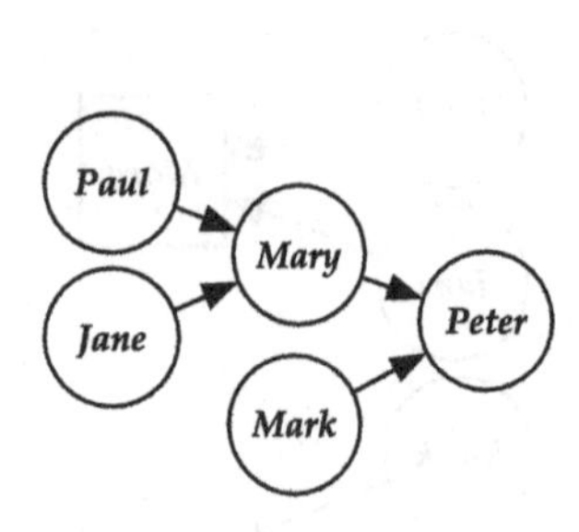

FIGURE 2.12

The *Has Child* relation on *Persons*. Compare this with Figure 2.9, where the *Mary* vertex appears twice.

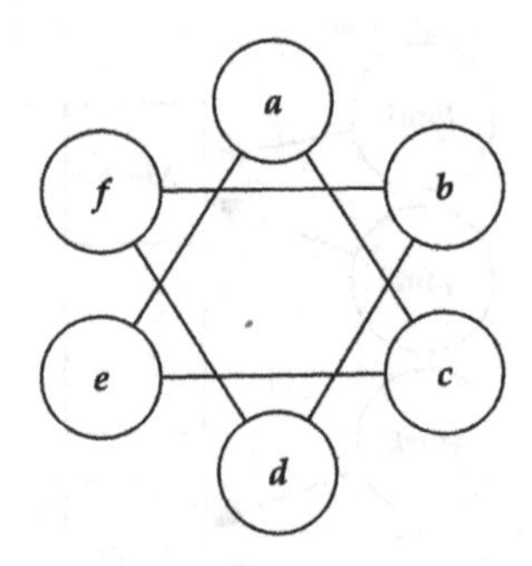

FIGURE 2.13

An undirected graph. Each edge is equivalent to a pair of directed edges with opposite directions.

consider the *Has Child* relation on *Persons* (the union of *Parents* and *Children*), we obtain the homogeneous graph of Figure 2.12.

If both $u \mapsto v$ and $v \mapsto u$ are both pairs of a relation, we can draw two edges, $u \rightarrow v$ and $v \rightarrow u$, or we can draw an arrowhead at both ends of the same edge. Conventionally, if a relation is *known* to be symmetric ($u \rightarrow v \Rightarrow v \rightarrow u$), rather than draw two arrowheads on every edge, we can omit both of them (see Figure 2.13). We also write $v \leftrightarrow u$, rather than both $u \rightarrow v$ and $v \rightarrow u$. Such graphs are called **undirected**.

An edge from a vertex to itself is called a **loop** (e.g., the edge $a \rightarrow a$ in Figure 2.14). A reflexive relation would have a loop on every vertex, so if a relation is *known* to be reflexive, we usually omit the loops to avoid clutter.

A **path** is a sequence of edges such that the terminal vertex of each edge (except the last) is the initial vertex of the next. The **length** of a path is the number of edges in the path. In Figure 2.14 on the facing page, $a \rightarrow c \rightarrow e \rightarrow f$ is a path of length 3 from a to f, and $a \rightarrow b \rightarrow f$ is a **parallel path** of length 2.

If any path leads from u to v, u is **connected** to v. The notation $u \rightarrow^2 v$ means that u is connected to v by a path of length 2, $u \rightarrow^+ v$ means that u is connected to v by a path of at least one edge, and $u \rightarrow^* v$ means that u is connected to v by a path of any length, including zero. Clearly, $v \rightarrow^* v$ is true for every vertex v of a homogeneous graph.

A path that has the same initial and terminal vertex and at least two edges is called a **cycle**.[16] If $u \rightarrow^+ v \rightarrow^+ u$, where $v \neq u$, then a cycle passes through vertices v and u. (In Figure 2.14 the path $a \rightarrow c \rightarrow e \rightarrow a$ is a cycle passing through a, c, and e.) A graph that contains at least one cycle is said to be **cyclic**, otherwise it is said to be **acyclic**.

[16] Some authors count loops as cycles, too.

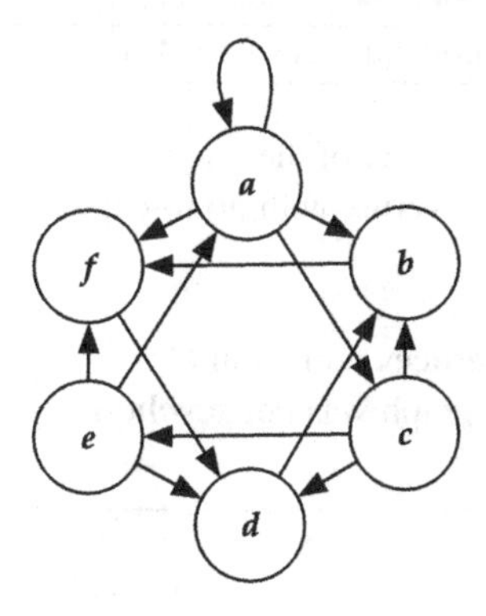

FIGURE 2.14

A homogeneous directed graph. The edge $a \to a$ is a loop. The path $a \to c \to e \to a$ is a cycle.

A **directed acyclic graph** is often referred to as a **DAG**. If $u \to^* v$ then v is a **successor** of u, and u is an **antecedent** of v. Any finite DAG must contain at least one vertex that has no antecedent. Such a vertex is called an **initial vertex** of the graph.[17] Likewise, a finite DAG must contain at least one vertex with no successor. Such a vertex is called a **final vertex**.

Every DAG represents a partial ordering. Every DAG and every partial order has at least one topological sort. A **topological sort** of a graph is a sequence of its vertices, such that, for all edges $u \to v$, u precedes v in the sequence. Consider the $\subset$ relation on $2^{\{a,b,c\}}$, i.e., the domain

$$\{\{a,b,c\},\{a,b\},\{b,c\},\{a,c\},\{a\},\{b\},\{c\},\{\}\}.$$

(See Figure 2.15 on the next page.) Then the sequences

$$[\{\},\{a\},\{b\},\{a,b\},\{c\},\{b,c\},\{a,c\},\{a,b,c\}]$$
$$[\{\},\{c\},\{b\},\{a\},\{a,c\},\{b,c\},\{a,b\},\{a,b,c\}]$$

are two of its 48 possible topological sorts.

A simple method for finding a topological sort is given by Algorithm 2.1.

The **transitive closure** G^+ of graph G is the graph of the transitive closure R^+ of its underlying relation R. Therefore, for example, if $u \to v \to w \to x$ is a *path* in G, then $u \to x$ is an *edge* in G^+. (See Figure 2.16.) (The symmetric, reflexive, and reflexive transitive closures of G are defined in similar ways.)

[17] The *Successor* relation on the integers forms an infinite DAG, but it has no initial vertex because there is no smallest number.

Algorithm 2.1 A simple method for finding a topological sort of a DAG.

1. Begin with an empty sequence, S, and a copy G of the graph.
2. Choose any initial vertex, V of G, i.e., any vertex with no in-edges.
3. Append V to S.
4. Delete V and all its out-edges from G.
5. Repeat Steps 2–4, until no more initial vertices remain in G.
6. (If any vertices remain in G, the original graph was not acyclic.)

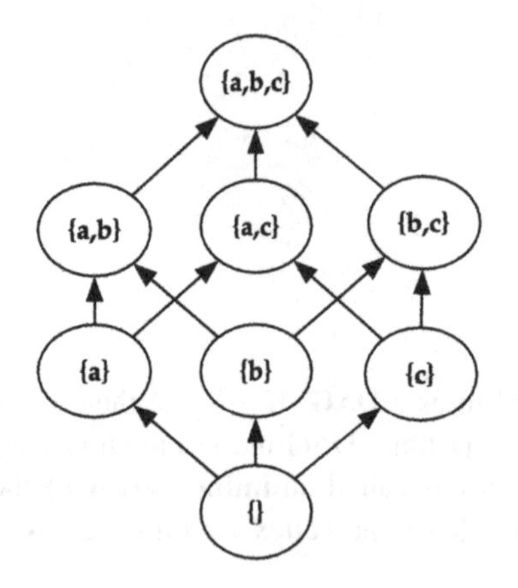

FIGURE 2.15

A DAG representing the $\subset$ relation on the powerset of $\{a,b,c\}$.

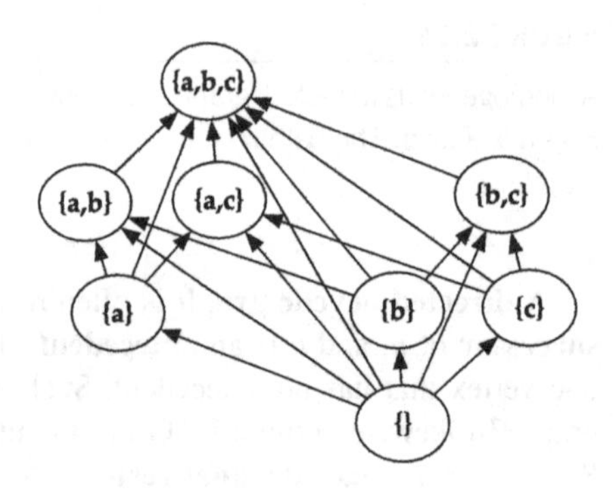

FIGURE 2.16

The transitive closure of the $\subset$ relation on the powerset of $\{a,b,c\}$.

A **transitive reduction** of graph G is any subgraph $H \subseteq G$ such that H and G have the same transitive closure ($H^+ = G^+$). If no edge can be removed from H without destroying the relation $H^+ = G^+$, then H is a **minimal transitive reduction**. If G is acyclic, its minimal transitive reduction is unique and is called its **transitive root**.

Transitive reductions are useful when we want to present the graph of a transitive relation with minimum clutter. Contrast Figure 2.16 (with 19 edges), which shows the transitive closure of the $\subset$ relation on the subsets of $\{a,b,c\}$, with its transitive root (with 12 edges), which is shown in Figure 2.15.

2.5.2 CONNECTED COMPONENTS

In a cyclic graph, if both $u \rightarrow^* v$ and $v \rightarrow^* u$, then u and v are said to be **strongly connected**. Clearly, every vertex is strongly connected to itself, and every vertex that is part of a cycle is strongly connected to every other vertex in the cycle.

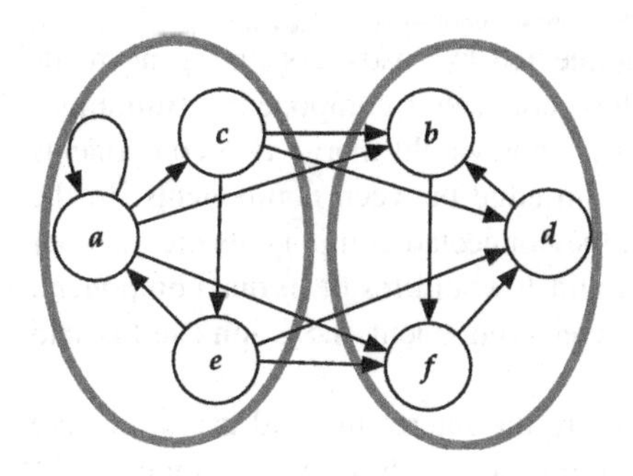

FIGURE 2.17

The graph of Figure 2.14 rearranged to reveal its strongly connected components. Edges lead from the left to the right component, but no edges lead from the right to the left component.

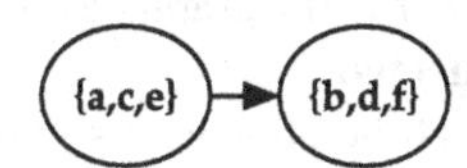

FIGURE 2.18

The reduced graph of the strongly connected components of the graph of Figure 2.14.

A **strongly connected component** of a graph is a set of vertices C, such that,

$$\forall u \forall v (u \in C \wedge v \in C \Rightarrow u \rightarrow^* v \wedge v \rightarrow^* u) \tag{2.5.1}$$

In other words, at least one cycle passes through every vertex within a strongly connected component. From now on, when we talk about a strongly connected component, we mean a *maximal* strongly connected component, one to which no more strongly connected vertices could be added.

The strongly connected components of a graph partition the vertices of the graph; every vertex belongs to exactly one strongly connected component. Figure 2.17 shows the same graph as Figure 2.14, but the vertices have been rearranged. The grey ellipses enclose its strongly connected components, $\{a, c, e\}$ and $\{b, d, f\}$.

Every directed graph G has a corresponding **reduced graph** H. Each vertex of H is a *set* of vertices of G that belong to the same (maximal) strongly connected component. The edges of H are the sets of edges of G that link different components. Clearly, a reduced graph must be acyclic; if a cycle existed between two strongly connected components, they could not be maximal. Figure 2.18 shows the reduced graph of Figure 2.14.

A cyclic graph has several minimal transitive reductions, because the vertices within its strongly connected components may be connected in any order while still preserving the same closure. In contrast, its reduced graph, being acyclic, has a unique transitive root.

In an *undirected* graph, if u is connected to v, then v is connected to u, so u and v must be strongly connected. The strongly connected components of an undirected graph are known simply as **connected components**. There can be no edges between its connected components, which therefore form disjoint subgraphs. As an example, think of the towns in a group of islands, connected by roads. Assuming no roads cross between the islands, the islands enclose connected components. Building a road bridge between towns on two different islands would merge two components into one; it would *not* (by definition) create an edge between components. In the case of the undirected graph of Figure 2.13, the connected components are $\{a,c,e\}$ and $\{b,d,f\}$. A single new edge between a and b would result in one component, $\{a,b,c,d,e,f\}$. If a graph has only one connected component, the graph itself is said to be **connected**.

Although heterogeneous graphs do not share the interesting and useful closure properties of homogeneous graphs, a bi-partite graph such as that of Figure 2.11 (page 45) — which displays an *Input* relation from files to processes and an *Output* relation from processes to files — can be expressed as two homogeneous products: *Feeds* = *Output* ; *Input* on processes, and *Determines* = *Input* ; *Output* on files. Thus, we can say that the *Subject Update* process *Feeds* the *Candidate Update* process, and that the data in the *Events* and *Subjects* files *Determines* the values in the *Charges* and *Subjects'* files and transitively *Determines* the *Statements* and *Candidates'* files. We shall make good use of these relations in Chapter 7.

2.5.3 ROOTED TREES

A **tree** is a connected undirected acyclic graph. It follows that a tree has the minimum number of edges to connect its vertices; if the tree has V vertices, it has exactly $(V-1)$ edges.

In contrast, a **rooted tree** is directed. It has a special vertex called the **root**. The vertices of a tree are called **nodes**; its edges are called **branches**. Implicitly, the branches are normally considered to be directed *away* from the root.

Usually, a tree is drawn with its root at the top, with the nodes ordered downwards according to their distances from the root. (See Figure 2.19 on the facing page.) Given any branch, the node closer to the root is called the **parent node**, and the node further from the root is called the **child node**. It is a property of rooted trees that every child node has exactly one parent, but a parent node may have any number of children. The number of its children is called its **degree**. The mapping from child to parent defines the *Parent* function, but *Child* is a one-to-many relation, where $Child = Parent^{-1}$. A **descendant** is any node in the transitive closure of the *Child* relation; an **ancestor** is any node in the closure of the *Parent* function. A node with no children is called a **leaf**, or **external node**. All other nodes are called **internal nodes**.

Any node V of a tree, its descendants, and their associated edges form a **sub-tree** of the tree. V is called the root of the sub-tree.

The **height** of a tree or sub-tree is 1 plus the length of the longest path from its root to any descendant.[18] The total number of nodes in the tree or sub-tree is called its **weight**.

---

[18] We say that a degenerate tree consisting only of a root node has height 1. Some authors say that such a tree has height zero, a privilege we reserve for the empty tree. The distinction is immaterial here.

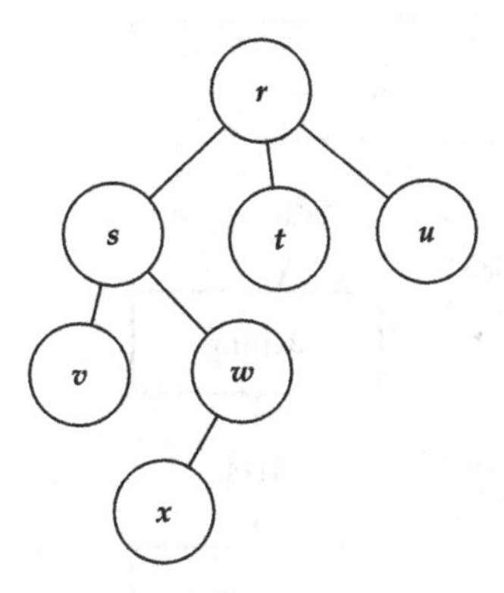

FIGURE 2.19

A rooted tree. Node r is the root; t, u, v, and x are leaves; s and w are internal nodes. The sub-tree rooted at s contains s, v, w, and x.

2.5.4 LABELLED GRAPHS

The vertices or the edges of a graph may be **labelled** with numbers or other symbols.

Although the vertices of a graph are normally labelled with domain values, it is possible to add further labels to them, representing the values of one or more functions of the vertices. For example, a graph of airline connections might show the time zone of each airport.

When the edges are labelled, the graph then represents a ternary relation. In a labelled graph, it isn't unusual to have more than one edge from vertex u to vertex v, provided they have different labels. For example, a labelled graph may contain both the edges $u \underset{A}{\rightarrow} v$ and $u \underset{B}{\rightarrow} v$. Labelled graphs may be used for three different purposes.

First, a labelled graph might represent a function from *pairs* of vertices to labels, for example, the vertices might represent airports, and the labels might represent the great circle distances between them. (See Figure 2.20 on the next page.) In such a case, there can be at most one edge in each direction between any two vertices.

Second, a labelled graph might represent a function where each edge maps its label and source vertex to its target vertex. In such a case, at most one edge may leave each vertex with a given label, although several (differently labelled) edges could connect the same pair of vertices. (See Figure 2.21.)

Third, and equivalently, a labelled graph may represent a *set of relations*, whose names are given by the labels on the edges. For example, the edge $u \underset{A}{\rightarrow} u$ means that (u,v) is an element of relation A, and the edge $v \underset{B}{\rightarrow} w$ means that (v,w) is an element of relation B. The graphs of these component relations are therefore simply the subgraphs we obtain by considering only the correspondingly labelled edges of

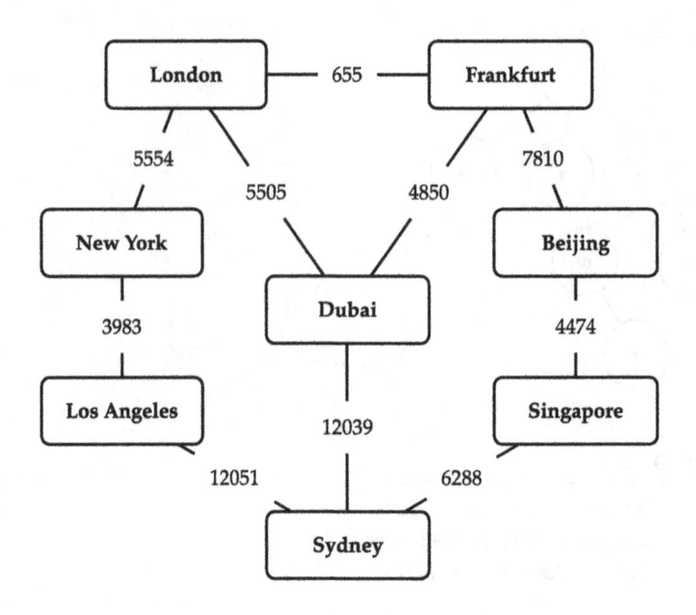

FIGURE 2.20

A labelled graph, representing a function from pairs of vertices to labels. It gives the distances in kilometres between pairs of international airports.

the graph. Conversely, we may superimpose the graphs of several relations, provided we label the edges properly. (See Figure 2.22 on the facing page.)

(A bi-partite graph, such as Figure 2.11, is a special instance of this practice, where the two sets of edges (*Input* and *Output*) can be distinguished by their direction, so they need no labels.)

2.5.5 MATRIX REPRESENTATION OF GRAPHS

It can be useful to represent a graph (and therefore its underlying relation) as an **adjacency matrix**. Each row of the matrix represents a source vertex, and each column represents a destination vertex. Each cell of the matrix therefore represents a potential edge, from the row to the column. If the graph is labelled, we can write the label of each edge into its cell; if it is unlabelled, we can mark the edges with any symbol we choose.

Although graphs are useful for tutorial purposes, the graphs of real-world problems are often too complex to be easily understood, and finding their products and closures (for example) by eye can be almost impossible. Matrices, on the other hand, are easy to manipulate within a computer.

The adjacency matrix of the graph of Figure 2.14 is given in Table 2.1. The entries in row a represent the edges $a \to a$, $a \to b$, $a \to c$, and $a \to f$.

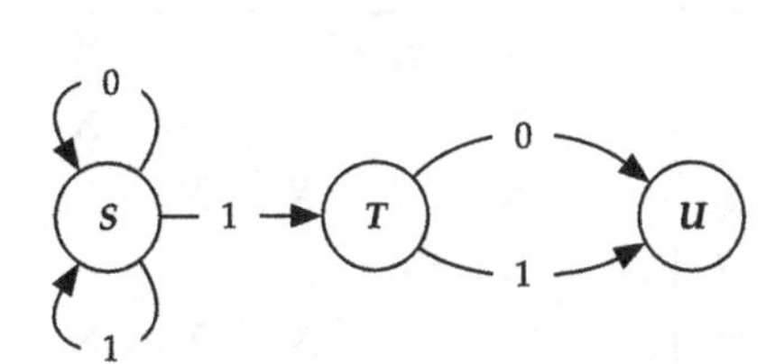

FIGURE 2.21

A graph whose edges are labelled 0 and 1. There are pairs of differently labelled edges from T to U and from S to S.

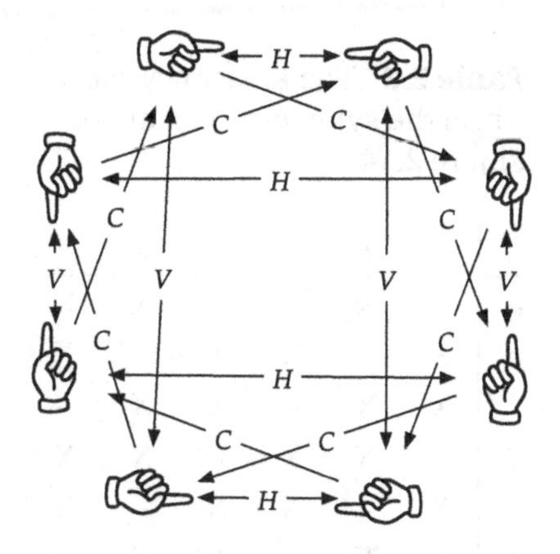

FIGURE 2.22

A labelled graph representing some transformations of a simple figure. The C relation is clockwise rotation, V is vertical reflection, and H is horizontal reflection. V and H are reflexive. C^{-1} is anticlockwise rotation. Cycles and parallel paths reveal interesting properties, such as $H\,;C = C\,;V$.

Table 2.1 The adjacency matrix of the graph of Figure 2.14

	a	b	c	d	e	f
a	X	X	X	–	–	X
b	–	–	–	–	–	X
c	–	X	–	X	X	–
d	–	X	–	–	–	–
e	X	–	–	X	–	X
f	–	–	–	X	–	–

Table 2.2 The adjacency matrix of the reverse of Figure 2.14

	a	b	c	d	e	f
a	X	–	–	–	X	–
b	X	–	X	X	–	–
c	X	–	–	–	–	–
d	–	–	X	–	X	X
e	–	–	X	–	–	–
f	X	X	–	–	X	–

To reverse all the edges of the graph of Figure 2.14 and thus find its converse, we exchange rows and columns in Table 2.1, giving Table 2.2.

To find the adjacency matrix of the union of two relations, we mark each cell that is marked in *either* of their matrices. To find their intersection, we mark each cell that is marked in *both* their matrices. In similar style, we can find the asymmetric difference $R \setminus S$ by deleting all entries from R that are marked in S.

Table 2.3 The adjacency matrix of the symmetric closure of Figure 2.14

	a	b	c	d	e	f
a	X	X	X	–	X	X
b	X	–	X	X	–	X
c	X	X	–	X	X	–
d	–	X	X	–	X	X
e	X	–	X	X	–	X
f	X	X	–	X	X	–

Table 2.4 The adjacency matrix of the reflexive closure of Figure 2.14

	a	b	c	d	e	f
a	X	X	X	–	–	X
b	–	X	–	–	–	X
c	–	X	X	X	X	–
d	–	X	–	X	–	–
e	X	–	–	X	X	X
f	–	–	–	X	–	X

The union of Table 2.1 and Table 2.2 is shown in Table 2.3. It represents the symmetric closure of Figure 2.14, i.e., the undirected graph with the same edges. (Not surprisingly, the matrix is symmetrical about its leading diagonal.)

To create the matrix of the reflexive closure, we simply mark each cell along the leading diagonal of the matrix, creating a loop on each vertex, as in Table 2.4.

We can find the relational product $(R\,;S)$ of relations $R : U \to V$ and $S : V \to W$ using adjacency matrices. Suppose $u \to v$ is an edge in R and $v \to w$ is an edge in S. Then $u \to w$ is an edge in $R\,;S : U \to W$. Suppose we want to discover if edge $u \to w$ is in $R\,;S$. If it is, then $u \to v \to w$, for some v. We therefore scan row u of R and column w of S for each potential value of V. If both (u,v) and (v,w) are marked, then we mark (v,w) in the matrix of the product. Clearly, this procedure has complexity $O(|\,U\,| \times |\,V\,| \times |\,W\,|)$.

R and S don't need to be distinct. Suppose we want to find all paths of length 2 in Figure 2.14 on page 47. Beginning with the adjacency matrix of Table 2.1, we construct Table 2.5, as just described. To help the reader understand the method, the cells of Table 2.5 are marked with the vertex or vertices that complete the path. For example, there are two paths from c to f: $c \to b \to f$ and $c \to e \to f$.

Finally, consider the problem of finding the strongly connected components of Figure 2.14. Although rearranging the graph as in Figure 2.17 makes the solution obvious, not everyone finds it easy to visualise such a rearrangement. Since every vertex in a strongly-connected component has a path to every other vertex in the component, the reflexive transitive closure of the graph highlights the components. A drawing of the closure would contain many edges and would confuse the eye. Instead, we manipulate the matrix.

One way to find the closure would be to find the union of all paths of lengths $1, 2, \ldots, (V-1)$ where V is the number of vertices. (No path can have more than $(V-1)$ edges without doubling back on itself.) This would have complexity $O(V^4)$. Fortunately, a more efficient method, called **Warshall's Algorithm**, has complexity $O(V^3)$. It works by modifying the matrix *in situ*.

Table 2.5 The matrix of all paths of length 2 in Figure 2.14

	a	b	c	d	e	f
a	a	a, c	a	c, f	c	b
b	–	–	–	f	–	–
c	e	d	–	e	–	b, e
d	–	–	–	–	–	b
e	a	a, d	a	f	–	a
f	–	d	–	–	–	–

Table 2.6 The effect of considering paths through vertex *a*

	a	b	c	d	e	f
a	X	X	X	–	–	X
b	–	X	–	–	–	X
c	–	X	X	X	X	–
d	–	X	–	X	–	–
e	X	X	X	X	X	X
f	–	–	–	X	–	X

Table 2.7 The effect of considering paths through vertices *a* and *b*

	a	b	c	d	e	f
a	X	X	X	–	–	X
b	–	X	–	–	–	X
c	–	X	X	X	X	X
d	–	X	–	X	–	X
e	X	X	X	X	X	X
f	–	–	–	X	–	X

We begin by marking the diagonal of the matrix, forming its reflexive closure. The result was already shown in Table 2.4. (If we were forming the *transitive closure* rather than the reflexive transitive closure, we would omit this step.)

We now reason as follows: any vertex that has a path *to* vertex a must also have a path to every vertex that has a path *from a*. Therefore, we copy the non-blank entries in *row a* into every row that has an entry in *column a*. In fact, e is the only vertex other than a that has an edge leading to a. The result is shown in Table 2.6.

We continue to reason in the same way: any vertex that has a path to vertex b must have a path to every vertex that can be reached from b. Therefore, we copy the entries in *row b* into every row that now has an entry in *column b*. In this case, f is the only vertex that doesn't. The result is shown in Table 2.7.

We continue the iteration through vertices c, d, e, and f in turn. The result is shown in Table 2.8.

The procedure we have just described is guaranteed to find such a **connectivity matrix** in just one iteration through intermediate vertices. The reader should pay attention to the order of the iterations. It is essential that the intermediate vertices (the ones the paths pass through) are considered in the outermost iteration, otherwise

Table 2.8 The effect of considering paths through all vertices *a* to *f*

	a	b	c	d	e	f
a	X	X	X	X	X	X
b	–	X	–	X	–	X
c	X	X	X	X	X	X
d	–	X	–	X	–	X
e	X	X	X	X	X	X
f	–	X	–	X	–	X

Table 2.9 The effect of grouping vertices with equal closures

	a	c	e	b	d	f
a	X	X	X	X	X	X
c	X	X	X	X	X	X
e	X	X	X	X	X	X
b	–	–	–	X	X	X
d	–	–	–	X	X	X
f	–	–	–	X	X	X

the algorithm will need to be applied repeatedly until the result is stable. Warshall's Algorithm can also be expressed independently of the matrix representation, as in Algorithm 2.2 below.

Algorithm 2.2 Warshall's Algorithm for finding the transitive closure of graph G.

1. Initialise a set of edges E to equal the set of edges in G.
2. If the *reflexive* transitive closure is required, add the edge $V \rightarrow V$ to E, for each vertex V of the graph.
3. For each vertex V of G,
   - For each pair of vertices U and W (not necessarily distinct),
     - If $U \rightarrow V$ and $V \rightarrow W$ are both members of E, add the edge $U \rightarrow W$ to E.
4. E is the set of edges in the transitive closure of G.

Notice that Table 2.8 contains only two different kinds of row: row a, c, or e, and row b, d, or f. Once we group like rows together, we have found the strongly connected components. Every vertex in a strongly connected component has a path to every other vertex in it and therefore to every vertex that can be reached from it. As a result, they have identical matrix entries. Reordering the rows and columns representing the vertices, as in Table 2.9, makes this clear.

Finally, by grouping vertices a, c, and e as the set $\{a,c,e\}$ and vertices b, d, and f as the set $\{b,d,f\}$, we obtain, in Table 2.10, the adjacency matrix of the closure of the reduced graph, shown in Figure 2.18. We can therefore find the reduced graph H of the strongly connected components of a directed graph G using Algorithm 2.3.

Finally, we can use matrix methods to find the transitive root of an acyclic graph G: we first find its transitive closure G^+ and then the product $G\,;G^+$, using methods

Algorithm 2.3 Finding the reduced graph H of the strongly connected components of G.

1. Find the transitive closure G^* of G (e.g., using Warshall's Algorithm).
2. Form the maximal sets of vertices in G^* that share the same closure. These sets are the strongly connected components of G. Create a vertex of the reduced graph H to correspond to each component of G.
3. If there is any edge from vertex V in component S to vertex U in component T in graph G, where S and T are different strongly connected components of G, then create an edge from the corresponding vertex S to vertex T of H.

Table 2.10 The connectivity matrix of the closure of the reduced graph of Figure 2.18

	$\{a,c,e\}$	$\{b,d,f\}$
$\{a,c,e\}$	X	X
$\{b,d,f\}$	–	X

just explained. The expression $G\,;G^+$ gives the set of all paths of length 2 or more in G and therefore includes any compound paths already present in G. Finding the asymmetric difference $G \setminus (G\,;G^+)$ eliminates all but the simple paths in G.

2.5.6 SCHEMAS

A **schema**, as we use the term here, is a collection of domains and the relations between them. There are three kinds of schema: an abstract schema, which is a mathematical object, a conceptual schema, which is used to design the structure of a database, and a physical schema, which is the embodiment of a schema in software.

A schema may be drawn as a labelled graph that shows, not a relation between elements of sets, but a meta-relation between whole sets. The vertices of a schema represent the domains or codomains of relations. The edges of a schema are labelled with the names of relations. A labelled directed edge $S \underset{R}{\rightarrow} T$ means that R is a relation with domain S and codomain T.

It is often important to distinguish functions and correspondences from more general relations. Different authors have used just about every conceivable convention

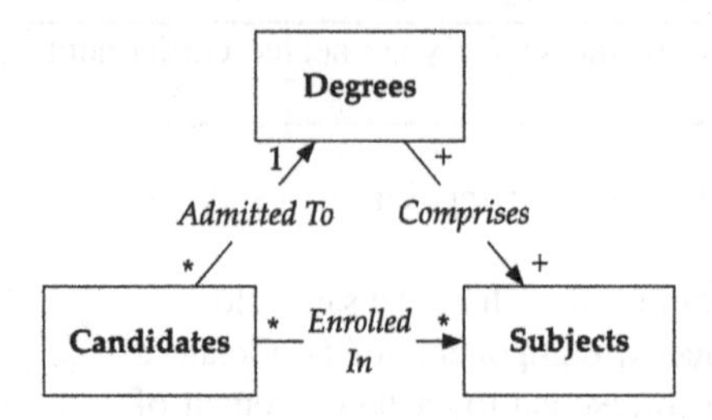

FIGURE 2.23

An example of a schema. *Candidates* are *Admitted To* exactly one *Degree* and may be *Enrolled In* any number of *Subjects*. Each *Degree Comprises* at least one *Subject*, and every *Subject* must be part of at least one *Degree*. *Admitted To* is a **total into** function, *Enrolled In* is a **partial into** relation, and *Comprises* is a **total onto** relation.

to show these distinctions.[19] Here, we label the ends of the edges according to the following conventions:

?: at most one element of the domain or codomain,

1: exactly one element of the domain or codomain,

$*$: zero or more elements of the domain or codomain,

+: one or more elements of the domain or codomain.

For example, Figure 2.23 represents the schema

$$\mathit{Candidates} \xrightarrow[\textit{Admitted To}]{*\qquad\qquad 1} \mathit{Degrees}$$

$$\mathit{Candidates} \xrightarrow[\textit{Enrolled In}]{*\qquad\qquad *} \mathit{Subjects}$$

$$\mathit{Degrees} \xrightarrow[\textit{Comprises}]{+\qquad\qquad +} \mathit{Subjects}$$

If the *destination* of an edge (the arrowhead) is labelled '1' or '+', this implies that the relation is total, because every element of its *source* maps to at least one element in the codomain. On the other hand, if the destination is labelled '?' or '$*$',

[19] Except, perhaps, the one used here. For example, the Z notation for program specification uses a variety of arrows that seem — to me at least — rather arbitrary and that still fail to cover all 16 possibilities.

Table 2.11 All 16 possible ways of labelling the edges of a schema

$S \xrightarrow[F]{?\ \ ?} T$	$F : S \rightarrow T$ is a partial correspondence from S into T.
$S \xrightarrow[F]{?\ \ 1} T$	$F : S \rightarrow T$ is a total correspondence from S into T.
$S \xrightarrow[R]{?\ \ *} T$	$R : S \rightarrow T$ is the converse of a partial function from T into S.
$S \xrightarrow[R]{?\ \ +} T$	$R : S \rightarrow T$ is the converse of a partial function from T onto S.
$S \xrightarrow[F]{1\ \ ?} T$	$F : S \rightarrow T$ is a partial correspondence from S onto T.
$S \xrightarrow[F]{1\ \ 1} T$	$F : S \rightarrow T$ is a total correspondence from S onto T.
$S \xrightarrow[R]{1\ \ *} T$	$R : S \rightarrow T$ is the converse of a total function from T into S.
$S \xrightarrow[R]{1\ \ +} T$	$R : S \rightarrow T$ is the converse of a total function from T onto S.
$S \xrightarrow[F]{*\ \ ?} T$	$F : S \rightarrow T$ is a partial function from S into T.
$S \xrightarrow[F]{*\ \ 1} T$	$F : S \rightarrow T$ is a total function from S into T.
$S \xrightarrow[R]{*\ \ *} T$	$R : S \rightarrow T$ is a partial relation from S into T.
$S \xrightarrow[R]{*\ \ +} T$	$R : S \rightarrow T$ is a total relation from S into T.
$S \xrightarrow[F]{+\ \ ?} T$	$F : S \rightarrow T$ is a partial function from S onto T.
$S \xrightarrow[F]{+\ \ 1} T$	$F : S \rightarrow T$ is a total function from S onto T.
$S \xrightarrow[R]{+\ \ *} T$	$R : S \rightarrow T$ is a partial relation from S onto T.
$S \xrightarrow[R]{+\ \ +} T$	$R : S \rightarrow T$ is a total relation from S onto T.

the relation is partial. By similar reasoning, if the *source* of the edge is labelled '1' or '+', this implies that the relation is *onto* its *destination*, because every element of the codomain is associated with at least one element of the domain. On the other hand, '?' or '*' implies that it is *into* its codomain. The rule — which is logical, if a little confusing — is to examine the *codomain* labelling to see if the relation is total or partial and to examine the *domain* labelling to see if it is *into* or *onto*. All sixteen possible labellings are explained in Table 2.11.

If the destination of an edge is labelled '1' or '?', this implies that the relation is a function, because no element of the domain maps to more than one element of the

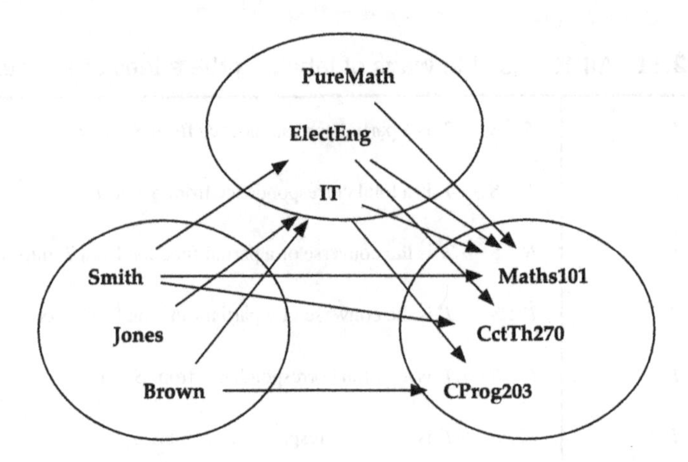

FIGURE 2.24

An *instance* of Figure 2.23. *Smith* is admitted to *ElectEng* and enrolled in *Maths101* and *CctTh270*. *Jones* is admitted to *IT* but is not currently enrolled in any subject. *Brown* is admitted to *IT* and is enrolled in *CProg203*. No one is currently admitted to *PureMath*. *PureMath* comprises *Maths101*, *ElectEng* comprises *CctTh270* and *Maths101*, and *IT* comprises *CProg203* and *Maths101*.

codomain. Conversely, if the destination is labelled '∗' or '+', the relation is *not* a function.

Correspondences introduce a special case. An edge such as $S \xrightarrow[C]{?\quad 1} T$ means that C maps an element of S to exactly one element of T, but its converse maps an element of T to at most one element of S. In the case that C is an identity relation, this means that S is actually a subset of T. In such a case, we omit the name of the relation, and label the edge $S \xleftrightarrow{?\quad 1} T$. The edge is given two arrowheads.

Projection functions introduce a second special case. A **composite domain**, such as $X \times Y \times Z$, has nine total projection functions: into X, Y, Z, $X \times Y$, $X \times Z$, $Y \times Z$, $Y \times X$, $Z \times X$, and $Z \times Y$. Such projection functions are normally labelled π_X, π_Y, etc.

It is important to distinguish a schema from the graphs that it summarises. Vertices of the schema represent domains, i.e., *sets* of vertices. Edges of the schema represent *sets* of pairs in a relation. Since many graphs can map to the same schema, any particular graph that does so is called an **instance** of the schema.[20] Compare the instance of Figure 2.24 with the schema of Figure 2.23.

[20] This is the terminology used in database theory. In the jargon of artificial intelligence, such a graph is called a semantic network.

2.6 REPRESENTING SETS

Most readers will be familiar with the programming techniques described below; they are included here to reinforce the notion of 'set' in a programming context, thereby integrating many different data structures into a single concept.

In computer languages that support multiple assignments to variables,[21] a set is stored as a dynamic, or modifiable, object. Set representations should support several operations, including adding an element to a set, removing an existing member from a set, testing if a given element is a member of a set, enumerating the members of a set, finding the union, intersection, etc., of two sets, and testing if two sets overlap, are disjoint, etc. There is no universal best way to represent a set; the best choice depends, among other things, on which operations occur most often.

As a running example, we shall use the set obtained by adding the sequence of alphabetic character strings [**'first'**, **'second'**, **'third'**, **'fourth'**, **'fifth'**, **'sixth'**] to an empty set.

2.6.1 ARRAYS

A set S may be represented by an array, A, of length N, where $|S| = n \leq N$. Typically, entries $1 \ldots n$ contain representations of the elements of S, the remaining entries being unused; if $A(i) = x$, $(1 \leq i \leq n)$, then $x \in S$.[22] The entries of A may be ordered or unordered. In an **unordered array**, a new element is added at position $n + 1$ (n is then increased by 1). (See Figure 2.25.)

In an **ordered array**, the entries are stored in ascending (or, more rarely, descending) order of *value* (e.g., **fifth**, **first**, **fourth**, **second**, **sixth**, **third**). (See Figure 2.26.) The representation of the set is independent of the order in which the values were added to it. On the other hand, if a new element x is added, where $A(j-1) < x < A(j)$, entries $A(j) \ldots A(n)$ must be moved to higher locations so that x may be inserted at location $A(j)$. In Figure 2.26, the element **seventh** should occupy the 5th array entry, so **sixth** and **third** need to be moved up into the 6th and 7th entries.

Adding an element to an unordered array takes constant time, but adding an element to an ordered array takes time $O(|S|)$. On the other hand, it takes time $O(|S|)$ to test if $x \in S$ in an unordered array, but only time $O(\log |S|)$ in an ordered array, using **binary search**.[23] Similarly, the ordered representation allows the union,

---

[21] That is to say, languages such as *C*, *Java*, etc. In single-assignment languages, such as *Prolog* or *Haskell*, a *new* set must be formed, rather than an existing set being modified. This can make some of the techniques described here inefficient.

[22] Strictly, if we want to argue *formally* that data structure D represents set S, we first need to define a **representation function** *Rep* such that $S = Rep(D)$. A representation function needn't be a one-to-one correspondence. Many *unordered* arrays can represent a given set S. (We leave things informal here.)

[23] The middle element of the array is tested first; if this isn't the element that is sought, the search is repeated using either the smaller entries or the larger entries, as appropriate. To locate **second** in Figure 2.26, **fourth**, **sixth**, and **second** would be inspected, in that order.

1	*first*
2	*second*
3	*third*
4	*fourth*
5	*fifth*
6	*sixth*
7	
8	

FIGURE 2.25

An Unordered Array. The element '**seventh**' would be added as its 7th entry.

1	*fifth*
2	*first*
3	*fourth*
4	*second*
5	*sixth*
6	*third*
7	
8	

FIGURE 2.26

An Ordered Array. The element '**seventh**' would be added as its 5th entry, requiring the existing 5th and 6th entries to be moved.

intersection, etc., of sets S_1 and S_2 to be found in time $O(|S_1| + |S_2|)$ by merging,[24] whereas the unordered representation takes time $O(|S_1| \times |S_2|)$. Finally, enumerating the elements of an ordered array in ascending or descending order takes time $O(|S|)$, whereas it takes time $O(|S| \log |S|)$ for an unordered array.

Therefore, if many elements are stored but few retrieved, an unordered representation may prove better; if many elements are retrieved, the ordered representation may be better.[25] *This illustrates a general point about all data representations: the time spent in maintaining a higher degree of organisation may be more than recouped in other operations.*

The chief problem with array-based representations is that once N has been chosen, it cannot be increased.[26] Fortunately, several representations exist in which additional storage can be allocated dynamically.

2.6.2 LINKED LISTS

The simplest dynamic representation is a **linked list**. Such a list consists of nodes,[27] each of which is labelled with the value of a member of the set, and which has a directed branch to the remaining nodes of the list. The list must be accessed starting

[24] Pairs of entries from each array are compared, starting with the first ones. If both are equal, comparison continues with the next pair of entries from each array. If they are unequal, the lesser entry is processed, and the greater entry and the entry *following* the lesser entry form the next pair to be compared.

[25] It may sometimes be best to add elements to an unordered array, then sort the array into order.

[26] This isn't entirely true. It may be possible, for example, to dynamically create a larger array, and move the existing entries into it.

[27] We use the word 'node' because a list is technically a unary tree, whose root is the head of the list.

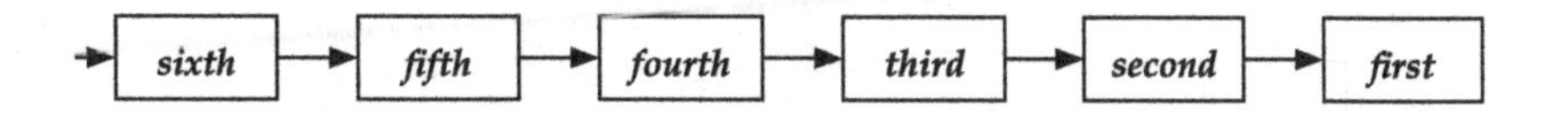

FIGURE 2.27

An Unordered Linked List. Elements are added to the head of the list in order of arrival. The element '**seventh**' would be added as its head.

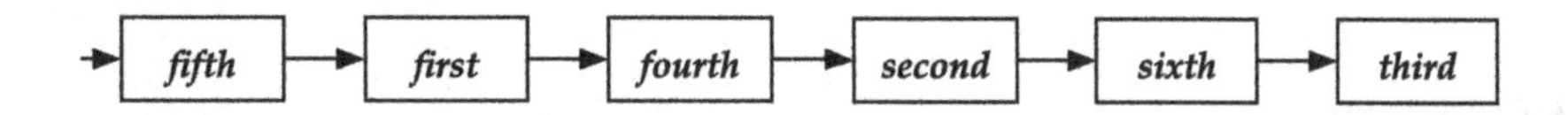

FIGURE 2.28

An Ordered Linked List. Elements are inserted in alphabetical order. The element '**seventh**' would be added between '**second**' and '**sixth**'.

with its first node, or **head**. The last node contains a special **null link**. When a new element is added to the set, a new node is created and added to the list.

As with arrays, the elements may be unordered (Figure 2.27), or ordered (Figure 2.28). When a node is added to an unordered list, it is most conveniently added at its head. Thus, the head of the list in Figure 2.27 is **sixth**, the most recent element to have been added. When a new node is added to an ordered list, it is added in the correct position, which merely needs two links to be adjusted.

As with arrays, the ordered representation allows the union, intersection, etc., of sets S_1 and S_2 to be found in time $O(|S_1|+|S_2|)$, whereas the unordered representation takes time $O(|S_1| \times |S_2|)$. Unfortunately, it takes time $O(|S|)$ to test if $x \in S$ *whether the list is ordered or not*; the nodes of a list can only be accessed in order from first to last. Search trees overcome this problem.

2.6.3 SEARCH TREES

A **binary search tree** is a rooted binary tree with labelled nodes. The nodes are labelled with representations of the members of the set. Each node except the root has an incoming branch from its parent. For uniformity, each node is considered to have two outgoing branches, called *left* and *right*. Each branch is either directed to a sub-tree or is **null**. (If both its branches are **null**, the node is a *leaf*.) In Figure 2.29 the root of the tree contains the value **first**. The nodes labelled **fourth**, **fifth**, and **sixth** are leaves. The set of elements with values less than **first**, {**fifth**}, is found by following the *left* branch, the set with greater values, {**fourth**, **second**, **sixth**, **third**}, by following the *right* branch. Such a tree is said to be ordered from left to right, and algorithms exist for listing its elements in ascending or descending order in time $O(|S|)$.

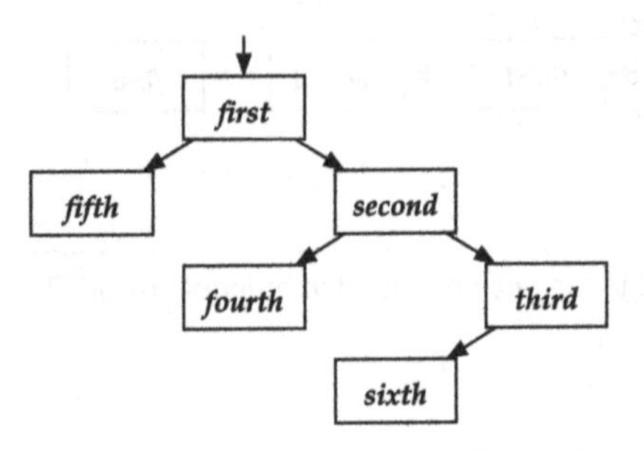

FIGURE 2.29

A Binary Search Tree. Elements are in order from left to right. Existing nodes aren't moved. The element '**seventh**' would be added to the left branch of '**sixth**'.

To test if an element $x \in S$, and therefore in the tree representing S, first the root is tested. If the node contains x, then $x \in S$. Otherwise, if x is less than the root, the *left* sub-tree is searched (recursively); if it is greater than the root, the *right* sub-tree is searched. If the sub-tree is **null**, then $x \notin S$.

When a new element x must be added to a set, in the simplest algorithm, the tree is first searched for x, until a **null** branch is found. A new leaf node labelled with x is then created, which replaces the **null** branch. To add **seventh** to the tree of Figure 2.29, the nodes **first**, **second**, **third**, and **sixth** would be inspected, and **seventh** would be added as the left branch of **sixth**. Unfortunately, if elements are added in certain sequences, the tree becomes tall and spindly. For example, the sequence, **fifth**, **first**, **fourth**, **second**, **sixth**, **third**, results in a tree like the ordered linked list of Figure 2.28.

If both sub-trees of every node in a binary tree have equal weight and height, the tree is said to be **perfectly balanced**. A perfectly balanced binary tree of height h contains $2^h - 1$ nodes. (See Figure 2.30.) Conversely, the height of a balanced binary tree is proportional to $\log_2 |S|$. For example, a balanced tree of height 3 contains 7 nodes; any member of the set it represents can be found by inspecting at most 3 nodes, i.e., in time $O(\log |S|)$. Searching is most efficient when a tree is perfectly balanced. Accordingly, search trees (and their sub-trees) are often **balanced** by rotating them, i.e., choosing a new root.

In practice, maintaining *perfect* balance proves too costly: additions to the tree can take time $O(|S|)$. Fortunately, several approximate balancing methods have been developed for which balancing only takes time $O(\log |S|)$. As a result, inserting a new element into such trees takes total time $O(\log |S|)$, and trees representing the union, intersection, etc., of sets S_1 and S_2 can be constructed in time $O(|S_1| + |S_2|)$.

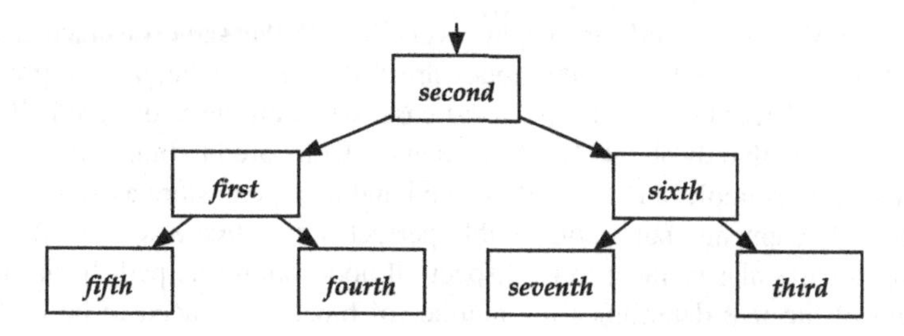

FIGURE 2.30

A perfectly balanced binary search tree.

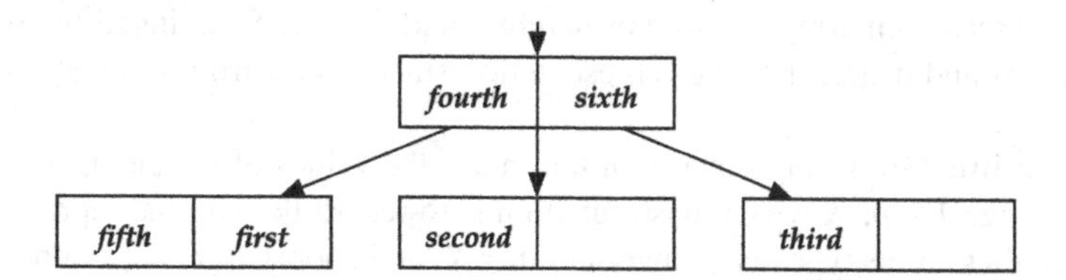

FIGURE 2.31

A B-tree with $b = 2$. The element '**seventh**' would be inserted into the empty cell adjacent to '**second**'.

2.6.4 B-TREES

It is possible to construct trees with orders higher than binary: in an n-ary tree, each node contains $(n - 1)$ elements and n branches to other nodes. For example, a ternary (3-ary) search tree has nodes containing 2 elements and 3 branches.

Of particular interest to us here are **B-trees**, which are widely used to implement large databases stored in secondary storage. For efficiency, information is transferred between secondary storage and RAM in blocks. Typically, the nodes of the tree correspond to blocks and contain many elements and links.

A B-tree is balanced by controlling the number of elements in each node. Each node of a B-tree except the root contains at most b and at least $b/2$ elements, and at most $b + 1$ links. If a node contains b elements, it is said to be **full**. If an attempt is made to add an element to a full node, it is said to **overflow**, and a new node is allocated to the tree.[28] Conversely, if removing an existing member of the set would reduce the number of elements in a node below $b/2$, a node may be removed.[29] Figure 2.31 illustrates the simplest possible B-tree, with $b = 2$ and maximum degree 3.

[28] This may cause the parent node to overflow, and so on. If the root overflows, a new parent root is created, and the height of the tree increases by 1.

[29] Which may cause the parent to underflow, and so on, possibly reducing the height of the tree by 1.

B-trees always have uniform height: every leaf is the same distance from the root. In the best case, when all the nodes are full, a tree of height h represents a set with $|S| \approx b^h$ (for large b). In the worst case, when all the nodes are half-full, it represents a set with only about $(b/2)^h$ members.[30] Compared to binary trees, B-trees are shallow. For example, a B-tree with $b = 64$ and $h = 2$ can store as many as 4,224 ($64 + 65 \times 64$) elements, but a comparable perfect binary tree has $h = 12$. Since a search for a particular element must inspect all nodes along its path from the root, the height of the tree determines the number of blocks that need to be read from secondary storage.

2.6.5 HASH TABLES

All the representations we have discussed so far may be considered to be trees of various degrees. (An array is a tree of height 1 and degree $|S|$, a linked list is a tree of height $|S|$ and degree 1.) One representation stands apart from the rest: the hash table.

A **hash function** is any function h that maps the values of the elements of a set onto the range $1 - p$. A typical hash function is to regard the internal representation of x (regardless of its type) as a binary number X, and choose $h(x) = (X \bmod p) + 1$, where p is a prime number roughly equal to $|S|$. The effect is to partition the elements of the set into p subsets. Each element x is a member of subset $h(x)$.

The subsets are located using an array called a **hash table**. Given an element x, its corresponding subset is found via entry $h(x)$ of the array. The average size of a subset is $|S|/p$. If $p \simeq |S|$, the subsets will contain an *average* of about 1 element, and take little time to search. Since it also takes little time to compute a hash function, hashing can greatly speed access to large sets. Sadly, in the (unlikely) worst case, all the elements might be assigned to the same subset, and all the other subsets might be empty. If no subset contains more than one element, the hash function is said to be **perfect**. The subsets themselves may be represented in any of the ways we have discussed above (including structures that utilise unused entries within the hash table). When the subsets are represented in a way that takes time $O(|S|)$ to access (e.g., a linked list), hashing *divides* the access time by a factor of p. On the other hand, if the subsets can be accessed in time $O(\log |S|)$ (e.g., a tree), hashing merely *subtracts* a constant time $O(\log p)$.

Although hash tables speed up the search for a given element, they make it less easy to enumerate the elements of a set in ascending or descending order.

2.6.6 BIT MAPS

An array of n bits (a **bit map**) may be used to represent a finite set of integers. Assume bits $1 \ldots n$ represent the integers $1 \ldots n$. If bit n is **on** (1), n is a member of the set; if bit n is **off** (0), n is not a member of the set. (By extension, a bit map can represent any collection of elements that corresponds to a finite range of

[30] Various refinements of the basic idea are used to avoid the worst case.

integers.) Such a bit map can occupy one or more words in memory. For example, a computer with 64-bit words can represent the integers 1–128 using two words. This representation is actually a *function* from each element to a Boolean value, and is called the **characteristic function** of the set.

To add an element to the set, it is merely necessary to turn the corresponding bit on; to remove it, the bit is turned off. The union of two sets is found using logical **or**, their intersection by **and**, and so on. Most computers have built-in operations that can manipulate bit maps in these ways.

2.6.7 DECISION TREES AND DECISION DIAGRAMS

A **decision tree** is a rooted tree with labelled *edges*, in which a path from the root to a node *spells* the value of an element. By 'spells', it is meant that its value is given by the *sequence* of the labels on the edges along the path.

A **binary decision tree** has edges labelled 0 or 1. To add a new element, its representation is considered as a binary number, and nodes are added as necessary to create a path that spells the number.

An **alphabetic decision tree** (or **dictionary tree**) has edges labelled by alphabetic characters, so that paths literally spell character strings. To represent a set of strings of unequal length, a path may need to terminate at an internal node, which must therefore be marked to indicate that it ends a possible path.[31]

A **decision diagram** is an optimised form of decision tree. If two sub-trees anywhere within the tree are identical, only one instance is stored. Both parent nodes then branch to the remaining instance.[32] The resulting directed acyclic graph is *not* a tree. Because identical sub-trees are merged, decision diagrams can be a very compact way of representing sets. Unfortunately, finding identical pairs of sub-trees is costly.

2.6.8 OTHER STRUCTURES

Many other data structures can represent sets. Some of them are mixtures of those we have described, for example, a linked list of arrays and an array of linked lists are both possible structures. Certain special representations are suited to particular problems. For example, a **union-find tree** has its branches directed from children to parents rather than from parents to children. It makes it easy to merge sets and to find of which set a given element is a member, but makes it less easy to enumerate the members of any given set.

---

[31] If, for example, the tree contained both the word 'boring' and the word 'boringly', they would share the same path, but the 'g' and 'y' in 'b-o-r-i-n-g-l-y' would both be marked to show that they end words.

[32] Consider, for example, that many English words share common endings. In an alphabetic decision diagram, the words 'boring' and 'exciting' would both share the sub-tree for 'ing'.

2.7 REPRESENTING FUNCTIONS, RELATIONS, AND GRAPHS

2.7.1 FUNCTIONS

A function from X to Y associates one y value with each x value. A simple set of $x \mapsto y$ pairs is an adequate way to represent a function, there being at most one pair for any given value of x. Therefore any structure that can represent the set X can be extended to represent a function. For example, we might choose to use a binary search tree whose nodes contain $x \mapsto y$ pairs to represent a function from X to Y. In addition, when X is, or can be mapped onto, a small range of integers, a function from X to Y can conveniently be represented by an array in which element x contains the corresponding value of y.

In practice, a system often needs to represent several functions with a common domain X. Suppose we have $f : X \rightarrow Y$ and $g : X \rightarrow Z$. Then it will usually prove wise to combine them into a single function $f \| g : X \rightarrow Y \times Z$, by storing the set of triples $x \mapsto (y, z)$. This has the advantage that the domain X is stored only once, saving space, and also saving time whenever $f(x)$ and $g(x)$ are both needed.

A small complication arises when f or g are *partial* functions. There may be some values of x for which there is a corresponding value of y but no corresponding value of z, or *vice versa.* In this case, some means must be used to show when there is no y (or z) value; y (or z) is said to be **null**. This can be implemented by reserving a special value of y (or z) that is not a true member of Y (or Z) or by storing additional bits to indicate which values are present.

2.7.2 CORRESPONDENCES

A correspondence is a function whose inverse is also a function. A correspondence $f : X \rightarrow Y$ is a set of $x \mapsto y$ pairs such that, not only is each x value unique, but also each y value is unique. This causes a difficulty: an ordered data structure of $x \mapsto y$ pairs, such as a search tree, offers a fast way to check if a new $x \mapsto y$ pair duplicates an existing x value, but no fast way to check if the y value is duplicated. The usual solution to this problem is to store the set of y values redundantly in a second data structure. In many applications, it may even pay to store the set of $y \mapsto x$ pairs, forming the inverse function f^{-1}.

If there are other functions of X, say $g : X \rightarrow Z$, then, as just mentioned, a single structure can be used, holding triples of the form $x \mapsto (y, z)$. If f is a correspondence, it is implicit that a second function $h : Y \rightarrow Z = f^{-1}; g$ must also exist, but it is pointless to store triples of the form $y \mapsto (x, z)$, as h can be found from f^{-1} and g by composition.

In the case that the correspondence is a subset relation, e.g., $Y \subseteq X$, then it is sufficient to store a function $X \rightarrow \{True, False\}$, where the pair $x \mapsto True$ indicates $x \in Y$, but the pair $x \mapsto False$ indicates $x \notin Y$.

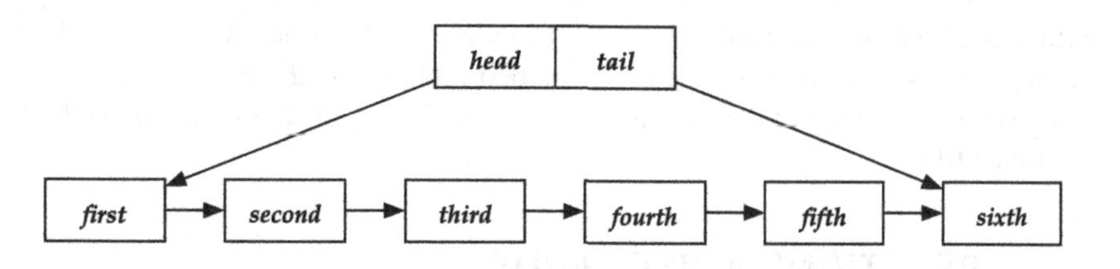

FIGURE 2.32

A linked list with pointers to both its head and its tail. Such lists can be concatenated in constant time.

2.7.3 SEQUENCES

A sequence of length n is, strictly speaking, a special case of a function whose domain is $1 - n$. In practice, certain operations commonly performed on sequences mean that storing the domain values explicitly would be a poor choice. For example, adding a new element to the start of a sequence means that all the existing elements must be renumbered. Likewise, concatenating a second sequence onto the end of a first (e.g., concatenating $[a,b,c]$ and $[d,e,f]$ to give $[a,b,c,d,e,f]$) means that the elements of the second sequence must be renumbered. To avoid renumbering, the numbering should be implicit. This is easy in the case of a linked list; the first element is implicitly numbered 1, the second numbered 2, and so on. To concatenate two lists, the last node of the first list simply has to be redirected to point to the head of the second list. To speed this process, a pointer may be maintained to point directly to the last element of the list. (See Figure 2.32.)

A common use of such a linked list is to represent a first-in, first-out (FIFO) **queue**, in which elements are added one-by-one to the **tail** of the queue and removed one-by-one from the head. A similar list structure is used for a last-in, first-out (LIFO) queue or **stack**, where elements are both added to and removed from the head. In this case, the tail pointer is unnecessary.

The problem with a linked list is that it doesn't offer a fast way of accessing its ith member. For very long sequences, a better way is to use a binary tree. Each node of the tree contains a member of the list, pointers to sub-trees containing the elements that precede it and follow it, and the weight of its sub-tree. (In the case of the root, its weight equals the length of the sequence.) The position of any node in the sequence is one plus the weight of its left sub-tree. Typically the tree is approximately balanced by weight. Any element can be found in time $\mathbf{O}(\log n)$. Also, adding or deleting an element requires only $\mathbf{O}(\log n)$ nodes to be rebalanced.

A **heap** is a special kind of balanced binary tree used to represent **priority queues**. A priority queue is a sequence in which nodes are ordered by priority — earliest priority first. The root of any sub-tree has an earlier priority than any of its descendants, so the root (top) of the heap has the earliest priority of all. The tree is filled strictly from top to bottom, left to right. A node added to the tree starts at the bottom right and is then made to sift up to its correct position. When the earliest priority node is removed from the root, it is first replaced by the bottom right node, which is then made to sift down to its proper level. Sifting up and sifting down both take time $\mathbf{O}(\log n)$.

2.7.4 RELATIONS AND GRAPHS

All the set representations we described earlier may be adapted to represent relations. There are two ways to proceed: we either represent the set of pairs in the relation, or we associate each value in the domain with its corresponding *set* of values in the codomain, called its **image**.

In a linked list consisting of $x \mapsto y$ pairs, it takes time $O(|S|)$ to test for the presence of a given value of $x \mapsto y$ or to enumerate all pairs with a given value of x or a given value of y. In contrast, an ordered array allows a given value of $x \mapsto y$ to be found in time $O(\log |S|)$ using binary search, and, since all pairs with a given value of x will be stored adjacently, binary search can be adapted to allow all pairs with a given x value to be enumerated efficiently. Unfortunately, a structure that is ordered by $x \mapsto y$ is *not* ordered by $y \mapsto x$, so the whole array must be scanned to enumerate all pairs with a given value of y. Other ordered structures, such as search trees, or decision diagrams, have a similar behaviour.

The second approach stores pairs of the form $x \mapsto \{y_1, y_2, \ldots\}$.[33] It has the advantage that it can better represent partial relations. A pair of the form $x \mapsto \{\}$ means that x has no corresponding y value. Such a representation is usually called a **graph data structure**, where the x values correspond to vertices and the y values correspond to out-edges. Any of the set representations may be used to represent the set of vertices, and any representation (not necessarily the same) may be used to represent the sets of edges. The structure is easily adapted to represent a labelled graph.

There are three commonly used structures of this type.

2.7.4.1 *Adjacency Matrices*

The first structure is a two-dimensional array, or **adjacency matrix**.[34] The rows of the matrix represent x values, and the columns represent y values. The entry in cell (x, y) is *true* if $x \mapsto y$ is a member of the relation, and is *false* otherwise. By extension, a matrix can also represent a labelled graph, the entry in cell (x, y) containing the label of the edge $x \to y$. The matrix representation allows all (x, y) pairs with a given x value or a given y value to be easily enumerated (by scanning row x or column y), but it usually makes poor use of memory: the number of cells is $X \times Y$, which may be many times more than the number of pairs in the relation.

---

[33] Strictly, this is not a relation from X to Y, but a *function* from X to 2^Y, the set of all subsets of Y.

[34] We described adjacency matrices in Section 2.5.5.

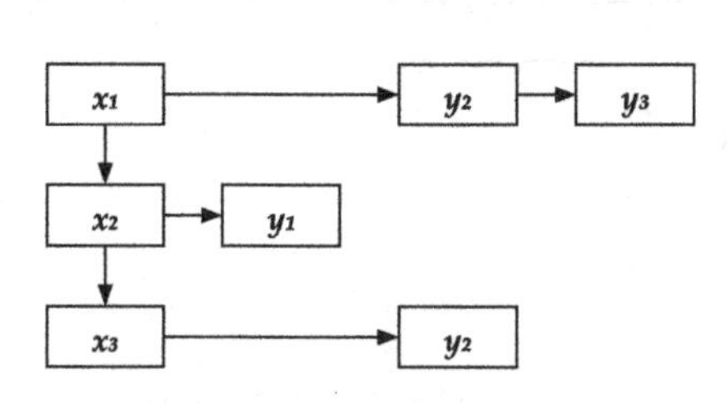

FIGURE 2.33

An adjacency list. The domain contains the values x_1, x_2, and x_3; the codomain contains y_1, y_2, and y_3.

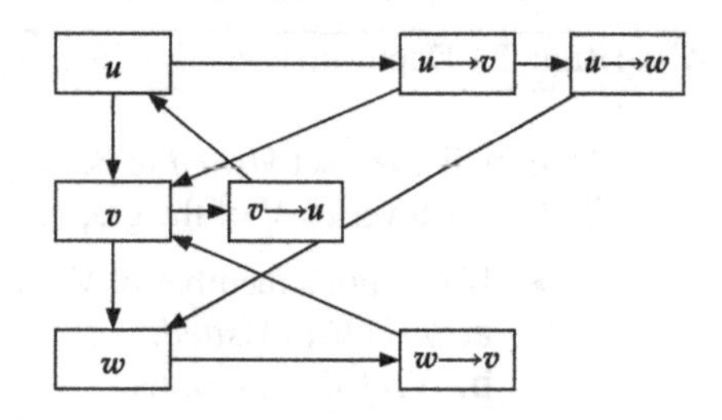

FIGURE 2.34

A graph data structure. Paths in the graph can be traced following the links between cells.

2.7.4.2 Adjacency Lists

The second structure is called an **adjacency list**. (See Figure 2.33.) It consists of a list (or possibly an array) of domain values, each cell of which has a list of out-edges labelled with values in the codomain.

In the case of a homogeneous relation, where the domain and codomain are the same, each edge cell is often a pointer to a vertex, as in Figure 2.34. This graph data structure is the basis of some very fast algorithms, based on a procedure known as **depth-first search (DFS)**.

In a depth-first search of a *vertex*, the vertex is visited, all its successors are visited and searched *recursively*, and then the vertex is visited again. The first visit to the vertex is called its **preorder** visit, and the second visit is called its **postorder** visit. Any vertex that has already been visited during the search is ignored, which prevents an infinite recursion in the case of a cycle. The recursion means that *all* the vertex's successors are visited before its postorder visit.

For example, a DFS of u in Figure 2.34 would make a preorder visit to u, then follow the edge $u \to v$ to v. Continuing recursively, DFS would then follow the edge $v \to u$, but finding that u had already been visited, the recursion would unwind, and v would be visited in postorder. The DFS would continue to unwind to u, then follow its second edge to w. Since w would not yet have been visited, the edge $w \to v$ would then be considered, but since v had been visited, the recursion would unwind twice, making a postorder visit to w, and finally a postorder visit to u.

A depth-first search of a *graph*, shown in Algorithm 2.4, is a DFS of all its (as yet unvisited) vertices. A DFS of the graph of Figure 2.34 would begin with a DFS of vertex u, then consider v and w, but they would not become the starts of new searches because they were already visited during the search from u. A DFS of a graph visits each vertex and each edge exactly once and has complexity $O(|V| + |E|)$, where $|V|$ and $|E|$ are the numbers of vertices and edges of the graph.[35]

[35] $|E|$ is at most $|V|^2$.

Algorithm 2.4 Depth-first search of a graph.

1. Initialise the set *Visited* to be empty.
2. For each vertex V of the graph,
   - If V is not a member of *Visited*
     a. Add V to *Visited*.
     b. Visit V in preorder.
     c. For each successor U of V, make a depth-first search of U recursively, as in Step 2.
     d. Visit V in postorder.

Depth-first search has the useful property that the postorder visits to the vertices of an acyclic graph occur in reverse topological order.[36] In the case of a cyclic graph, the postorder visit to the first vertex visited in a strongly connected component doesn't occur until every other vertex in the component has been visited. Depth-first search is therefore the basis of algorithms that can find the transitive closure or the strongly connected components of a graph in time $O(|V| + |E|)$.

The graph data structure doesn't offer any convenient way of finding the set of *in*-edges of a vertex, although there is a fast algorithm for reversing *all* the edges to find the converse relation.

2.7.4.3 Sparse Matrices

The third structure is an elaboration of the adjacency list: the **sparse matrix** representation. Imagine an adjacency list representation of a relation and another adjacency list representation of its converse. Now share both sets of edge cells, as in Figure 2.35.

2.7.5 FILES

A **file** is a data structure stored on a persistent secondary storage medium. Physically, the medium is arranged in blocks of fixed size, e.g., 4 KB. The computer's operating system will usually try to arrange that a file occupies a contiguous series of blocks, or if that isn't possible, a few contiguous areas, called **extents**.

From the programmer's point of view, a file consists of **records**. There are typically several records per block. Usually, the operating system unpacks the blocks into individual records when they are read and packs records into blocks when they are written. Each record typically represents one element of a set, one pair of a relation, or one pair, triple, etc., of one or more functions. The components of records are called **attributes**. For example, a function from X to Y would have attributes named X and Y.

[36] Because their successors are first visited recursively.

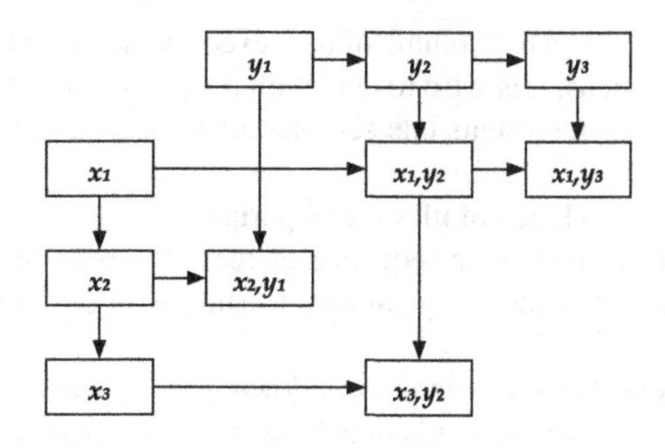

FIGURE 2.35

A sparse matrix. Each edge cell has two links to other edge cells. In mathematical applications involving matrices, the domain and codomain arrays replace the lists, and only non-zero values are explicitly represented.

As a rule, each record has a unique identifier: each pair in a *function* from X to Y must have a different value from the domain X. X is called a **key** of the file; Y is a **non-key attribute**. In the case of a *relation* from X to Y, only the $x \mapsto y$ *pairs* are unique, so the file has the **composite key** (X,Y). In the case of a sequence, records are physically stored in order of the sequence.

Records don't have to be fixed in length. For example, a file could represent a graph, in which each record could contain a vertex followed by a list of edges of arbitrary length.

A file can represent any data structure. Records can be added to a file dynamically, so that dynamic data structures can be created.

A record has both an absolute and a relative address. The **absolute address** specifies the physical storage location (e.g., surface, track, sector and byte) of the record. The **relative address** specifies the number of bytes from the start of the file. The computer's operating system translates relative addresses into physical addresses by reference to the file's directory entry, which lists the physical locations of its extents. When a file is copied, all its physical addresses change, but the relative addresses stay the same. Therefore, records can be linked into list or tree structures using their relative addresses.

Most modern computer systems allow an entire file to be read into and processed entirely within RAM, the file being saved periodically and, again, when processing is complete. The file can therefore contain whatever data structures are convenient to the application. There is one restriction: it cannot be guaranteed that a saved file will occupy the same memory locations when it is later reloaded into RAM. Therefore, the file cannot contain absolute memory addresses. It may, on the other hand, contain relative addresses (i.e., relative to the start of the file), or it may be saved in a form that is free of addresses. For example, it can be reduced to a textual form with mark-up, such as *XML*.

On the other hand, when the volume of data exceeds the available RAM, or when saving must be continuous, it is wise to deal with the persistent data directly. In such a case, each file usually represents one set, one relation, a cluster of functions, or a sequence.

For our purposes, two kinds of files are important.

A **sequential file** consists of a sequence of records stored one after another. A sequential file is read or written sequentially, starting with the first record, then the second record, and so on.

An **indexed-sequential file** is indexed, allowing the records to be located either by key or in sequential order. Records may be added or deleted anywhere within the file. The file is structured to make sequential access efficient: blocks contain records with contiguous key values, but the blocks may only be partly filled, and the blocks themselves may not always be stored in key order. A B-tree structure is often used, with the nodes of the tree mapped directly onto blocks. Often, it is possible to store the entire index of a file in RAM, so that, after consulting the index, a record can be located by direct access to the block that contains it.

An indexed file can have several indices, allowing access via more than one attribute. It is also possible for several records to share the same value of a secondary key, in which case the index will need to store the *set* of records with each key value, perhaps as a linked list.

Finally, it is perhaps worth mentioning that the directories or folders that a hierarchical file store uses to contain files form a kind of decision tree, in which the directory names and filename spell the path from the root to the file. For example, the path '/Users/barry/Documents/Synthesis/Maths.tex' describes a path from the root of the file system to the document I am currently editing. Each step in the path translates a name into the storage address of the next node.

2.8 SUMMARY

It should be obvious to the reader that there are strong connections between the various ideas presented in this chapter. *Relations* are just special kinds of *sets*. Although some relations are best defined by enumeration (such as who is the parent of whom), others are better defined by *predicate calculus* expressions. *Graphs* are a way of representing relations that is better suited to the human visual system than matrices or lists of pairs. *Labelled graphs* are often a superimposition of several simpler graphs.

For example, if $R = \{x \mapsto y \mid P(x, y)\}$, the ideas of relation, graph, and predicate are connected as follows:

$$x\, R\, y \Leftrightarrow x \underset{R}{\mapsto} y \Leftrightarrow P(x, y)$$

Finally, *schemas* are special graphs that describe relations between relations.

Table 2.12 on the next page lists the mathematical symbols explained in this chapter.

Table 2.12 The mathematical symbols explained in this chapter

Symbol	Name	Meaning
$\neg$	not	$\neg P$ is false if P is true and true if P is false.
$\wedge$	and	$P \wedge Q$ is true only if both P and Q are true.
$\vee$	or	$P \vee Q$ is false only if both P and Q are false.
$\Rightarrow$	implies	$P \Rightarrow Q$ is true unless P is true and Q is false.
$\Leftarrow$	if	$P \Leftarrow Q$ is true unless P is false and Q is true.
$\Leftrightarrow$	if and only if	$P \Leftrightarrow Q$ is true when P and Q have the same value.
$\forall$	for all	$\forall x(P(x))$ is true only if $P(x)$ is true for all x.
$\exists$	there exists	$\exists x(P(x))$ is false only if $P(x)$ is false for all x.
$\in$	is a member of	$X \in S$ is true only if x is a member of set S.
$\notin$	is not a member of	$X \notin S$ is true only if x is not a member of set S.
$\mid$	such that	$\{X \mid P\}$ is the set of all X such that P is true.
$\vert\ \vert$	cardinality	$\vert S \vert$ denotes the number of elements in S.
$\mathbf{U}$	universe of discourse	The set of all elements x under discussion.
$\emptyset$	empty set	The empty set, alternatively written as $\{\}$.
$\subseteq$	subset	$S \subseteq T$ if and only if $x \in S \Rightarrow x \in T$.
$\subset$	proper subset	$S \subset T$ if and only if $S \subseteq T \wedge S \neq T$.
$\cup$	union	$x \in S \cup T$ if and only if $x \in S \vee x \in T$.
$\cap$	intersection	$x \in S \cap T$ if and only if $x \in S \wedge x \in T$.
$\setminus$	set minus	$x \in S \setminus T$ if and only if $x \in S \wedge x \notin T$.
$\times$	Cartesian product	$(x, y) \in S \times T$ if and only if $x \in S \wedge y \in T$.
$\underset{R}{\mapsto}$	maps to under R	$x \underset{R}{\mapsto} y$ if $x \; R \; y$. (Alternatively, $x \underset{R}{\mapsto} y$ if $(x, y) \in R$.)
$;$	relational product	$x \; R \,;S \; z$ if and only if $\exists y(x \; R \; y \wedge y \; S \; z)$.
$\Vert$	parallel composition	$x \; R \Vert S \; (x, y)$ if and only if $x \; R \; y \wedge x \; S \; z$.
R^n	relational power	$R^n = R\,;R\,;R\ldots$, where R occurs n times.
R^{-1}	relational converse	$x \; R^{-1} \; y$ if and only if $y \; R \; x$.
R^0	relational identity	The identity function on R; $x \; R^0 \; x$ if $\exists y(x \; R \; y)$.
R^+	transitive closure	$R^+ = R \cup R^2 \cup R^3 \cdots$.
R^*	reflexive transitive closure	$R^* = R^0 \cup R^+$.
$u \underset{G}{\rightarrow} v$	labelled directed edge	There is an edge from u to v in G.
$u \rightarrow v$	directed edge	$u \underset{G}{\rightarrow} v$, where G is understood.
$u \rightarrow^+ v$	directed compound path	There is a directed path of length ≥ 1 from u to v.
$u \rightarrow^* v$	directed path	There is a directed path of length ≥ 0 from u to v.
$u \leftrightarrow v$	undirected edge	$u \rightarrow v \wedge v \rightarrow u$.
$u \leftrightarrow^+ v$	undirected compound path	$u \rightarrow^+ v \wedge v \rightarrow^+ u$.
$u \leftrightarrow^* v$	undirected path	$u \rightarrow^* v \wedge v \rightarrow^* u$.

Provided we use graphs formally, as representations of relations, being clear what the underlying domains and relations are, we shall keep out of trouble. There are two dangers we must avoid: being sloppy about what relations are being represented, or piling so much information onto a single graph that our visual systems can no longer make sense of it.

We reluctantly admit that graphs are mainly useful for tutorial purposes, because real-world relations are typically so complex that their graphs overwhelm the eye. They are also time-consuming to draw, especially if care is taken to minimise the crossing of edges. In practice, a list-of-pairs representation or a matrix representation is often better.

We observe that there are many data structures that can be used to represent sets and relations within a computer system, both within RAM and in secondary storage. This chapter has only listed the most common possibilities. As a general rule, the more time is devoted to storing an element, the less is spent in retrieving it. Structures and algorithms exist that are able both to store and to retrieve the elements of a set S in time $O(\log |S|)$.

Conversely, all the various data structures that programmers have devised can be seen as realisations of a few mathematical concepts. It is better to work first at a conceptual level, and defer implementation decisions by naming functions and procedures abstractly — as in $Member(x, S)$, rather than $SearchTree.Find(x, S)$. Whether to use a search tree or an ordered list, etc., can be decided later.

A theme that will occur several times in the following chapters is the usefulness of finding the strongly connected components of directed graphs. Consider the graph of Figure 2.14, and imagine that it was derived from the following set of six simultaneous equations:

$$
\begin{aligned}
a &= e - a - 12 \\
b &= a + 2c - d \\
c &= 3a + 1 \\
d &= 2c - e + 3f - 7 \\
e &= 2c + 2 \\
f &= 2a - 3b + e
\end{aligned}
$$

where there is an edge in Figure 2.14 from variable x to variable y if x is on the right-hand side of the equation for y.

In Figure 2.18, we saw that this graph has two strongly connected components, $\{a, c, e\}$ and $\{b, d, f\}$. The implication is that we have two separable sub-problems, one for each component:

The set of three equations

$$
\begin{aligned}
a &= e - a - 12 \\
c &= 3a + 1 \\
e &= 2c + 2
\end{aligned}
$$

is independent of b, d, and f and has the solution $a = 2$, $c = 7$, $e = 16$. We can then substitute these values into the remaining equations to obtain

$$b = 2 + 14 - d$$
$$d = 14 - 16 + 3f - 7$$
$$f = 4 - 3b + 16$$

Thus, we have two small problems rather than one big one.

In general, finding the strongly connected components of a graph breaks it into more manageable subgraphs. Since the resulting reduced graph is acyclic, it has a topological sort, which lets each sub-problem be tackled in turn. Unfortunately, the cycles *within* a sub-problem may make it hard to understand or hard to solve. In the following chapters, we shall see over and again that it is cyclic relationships that cause problems for the systems analyst and the implementor.

2.9 FURTHER READING

I have yet to find a single textbook that includes all the material in this chapter, although chapters 5–9, 12, and 14 of 'Foundations of Computer Science: C Edition' by Alfred V. Aho and Jeffrey D. Ullman (1994, ISBN: 0-7167-8284-7) come close.

Most of the mathematical ideas can be found in Kenneth A. Ross and Charles R. Wright, 'Discrete Mathematics' (ISBN: 0-13-065247-4), but it doesn't cover some of the material on graphs and relations. This material is covered in 'Relations and Graphs' by Gunther Schmidt and Thomas Ströhlein (ISBN: 3642779700), but in far more depth than is needed in this book. Representations of sets, relations, and graphs can be found in 'Data Structures and Algorithms' by Alfred V. Aho, Jeffrey D. Ullman, and John E. Hopcroft (ISBN: 0-201-00023-7). There are many other good books that contain similar material, but before borrowing one, check that its index contains a reference to 'strongly connected components of a graph'.

Warshall's Algorithm for finding the transitive closure of a relation was published in January 1962 under the rather obscure title, Warshall, Stephen, 'A theorem on Boolean matrices', *Journal of the ACM* 9 (1): 11–12. The article by James Eve and Reino Kurki-Suonio, 'On computing the transitive closure of a relation', *Acta Informatica*, 8(4), 303–314 (1977), describes a much faster graphical method, based on Robert Endre Tarjan's 1972 article 'Depth-first search and linear graph algorithms', *SIAM Journal on Computing* 1 (2): 146–160.

Finally, the paper, 'The transitive reduction of a directed graph', *SIAM Journal of Computing*, 1 (2): 131–137, (1972) by Alfred V. Aho, Michael Garey, and Jeffrey D. Ullman, describes an efficient way to compute a transitive reduction.[37]

---

[37] Yes, both these final two articles appeared in the same issue.

2.10 EXERCISES

1. Complete the proof of Equation 2.1.27.
2. Prove Equation 2.1.11. (Hint: Use Axiom 2.1.7 and the resolution theorem (page 29). Set out your proof as a vertical stack of propositions.)
3. Use Axiom 2.1.7 to reduce $P \Rightarrow Q$ to CNF, then use De Morgan's Theorem to find its logical complement. How do you interpret your result?
4. Why does a set of cardinality n have 2^n possible subsets? (See 2.3.1.)
5. In what special circumstances does $R^1 ; R^{-1} = R^0$? (See inequality 2.4.26.) (Hint: Consider the difference between the *Child* relation in connection with family relationships, and the *Child* relation in rooted trees such as Figure 2.19.)
6. By analogy with Equation 2.4.22, define the *Has Uncle* relation.
7. Represent the labelled graph of Figure 2.21 as a matrix, with entries '0', '1', '0,1' or blank.
8. Sketch a graph data structure corresponding to Figure 2.21 in the style of Figure 2.34, labelling its edges '0' or '1'.
9. Graph G has vertices V, W, X, Y, and Z, and edges $V \rightarrow W$, $V \rightarrow X$, $W \rightarrow V$, $W \rightarrow X$, $X \rightarrow Y$, $X \rightarrow Z$, $X \rightarrow Y$, $Y \rightarrow Z$, and $Z \rightarrow Y$. Find the strongly connected components of G and sketch its reduced graph. (See Section 2.5.5.)

CHAPTER

Atoms

3

CHAPTER CONTENTS

INTRODUCTION

We normally regard an information system as a model, or analogue, of part of the real world. The real world contains things, usually many things, that we wish to model. We shall call the real-world things we model **atoms**. We use the word 'atom' because they are the smallest things that we model. Depending on the nature of the system, the number of atoms to be modelled will vary. A flight control system has to model only one aircraft, with two wings, four engines, and so on, but a bank has to model thousands of customers.

We don't just want to model the existence of atoms, we also want to model their behaviour.

Of course, the analyst must choose what things in the real world need to be modelled by atoms. Further, the analyst should not, and cannot, model the chosen atoms in their entirety, but only certain aspects of them. For example, a bank is not much interested in what its customers do at home, but only in the transactions that they conduct with the bank. The flight control system of an aircraft is concerned with the position and velocity of the aircraft, but not with which passengers have asked for a cup of coffee.

The things that happen in the real world that cause us to inspect or update our model, we shall call **events**. **Update events** cause the **states** of atoms to change. Update events are **atomic**, or indivisible; an atom has a state *before* an event and a state *after* an event, but its state *during* an event is undefined. The states of atoms between updates, we shall call **snapshots**. Snapshots are represented by records in files, rows in database tables, or other data structures. Update events invoke **event procedures** (or methods) that transform a **before snapshot** into an **after snapshot**.[1] **Inspection events** simply analyse and report the states of atoms. They leave the states of atoms unaltered. Even so, it isn't unusual for events both to inspect and to update atoms, perhaps reporting the before snapshot or the after snapshot. When it is necessary to keep a permanent record of *events*, it is typically stored in sequential files.

Events of interest to a banking system might be the opening of a new account, depositing money, withdrawing money, a balance enquiry, printing a statement, and so on. We only need to inspect or update the snapshot of the account when these events happen in the real world. On the other hand, an aircraft control system must monitor the state of an airframe *continuously*. In a digital control system, this is approximated by updating the state at frequent intervals. The passage of time in the real world generates a series of internal clock tick events. The succession of resulting snapshots is like a movie.[2]

What we consider to be an event depends on our perspective. A person applying for a job might consider that completing a computer course was an event worth including in a curriculum vitae. On the other hand, the institution that offered the course would not only consider each subject completed as one event, but even more detail: enrolling, sitting an examination, etc. In this way, the course and the subject enrolments within it, although events at one level, become atoms at a greater level of detail. Think of a bank account: it is not enough for the bank to keep track of the balance of an account. A customer needs to see the history of transactions that have led to that balance. For that reason, records of deposits and withdrawals need to be kept, which need to be reported. In the context of updating a balance, a deposit is an event. In the context of being reported, the deposit is an atom.

The state of an aircraft's airframe may be described by 12 variables: its position relative to the earth is given by its latitude, longitude, and altitude. Each of these variables has a corresponding linear velocity. Its attitude is given by three angles: roll, pitch, and yaw. Each of these has a corresponding angular velocity. All 12 variables are *continuous* quantities that can be represented by real numbers, so the possible states of an airframe are therefore infinite in number.

---

[1] An object (in the sense used in object-oriented programming) might therefore correspond to an atom and some events that affect it.

[2] We shan't discuss continuous-time controllers in this book, only sampled-data systems.

The state of a customer's bank account is given primarily by the balance of the money owed to the customer. This can be described by an integer. If a customer may deposit any amount, without limit, the states of an account are *discrete*, but are also effectively infinite in number.[3]

We begin by considering discrete finite-state systems.[4] Our motive is that, by identifying the states of atoms, we discover what kinds of events a system needs to consider. Conversely, learning about the existence of a certain kind of event may help us to identify the need for a new state. This interaction between events and states can be very useful in eliciting information from a **domain expert**, a person familiar with the real-world situation that the system is meant to model. Indeed, as we shall learn, such analysis can help us discover additional atoms that we need to model.

Finite-state systems can even be used to model certain aspects of continuous or infinite-state systems. For example, a *Bank Account* might be considered to have one of three states: *Null*, *Open*, or *Closed*; and from the point of view of an airport, an *Aircraft* might be in one of four states: *Arriving*, *Landed*, *Boarding*, or *Departed*.

Our finite-state models will have only a few states: *we are concerned with states only in so far as they determine what events are possible*. When an *Aircraft* has *Departed*, it may climb, dive, bank, or yaw; when it has *Landed*, diving isn't really an option. When a *Bank Account* is *Open*, it may be credited or debited by any reasonable amount. When it is *Closed*, it may not.

When an aircraft has departed, it may have many possible spatial positions. When a bank account is open, it may have any reasonable balance. We therefore consider *Departed* or *Open* as **families of states**, which are related in that they allow the same set of events.

3.1 FINITE-STATE AUTOMATA

Systems that have a finite number of states are called **finite-state automata** (**FSA**s). We can model them by labelled directed graphs, called **state-transition diagrams** (see Figure 3.1). Vertices are labelled with the names of states, and edges are labelled with **transitions**: changes of state. In the present context, transitions are caused by events.

The vertex containing a dot represents the **initial state** of a finite-state automaton. This is the state of the automaton before any transitions have occurred. A vertex with a bold outline marks a **final state**. An automaton in a final state has done its job and may often cease to be of interest.

---

[3] An alternative view, discussed below, is that being in the customer's account is one of a finite number of states of some coins.

[4] Whenever a system is modelled by a digital computer, there is a trivial sense in which it *must* be finite-state, because the computer's memory has only a finite number of states. One bit has 2 possible states, one byte has $2^8 = 256$ possible states, and so on.

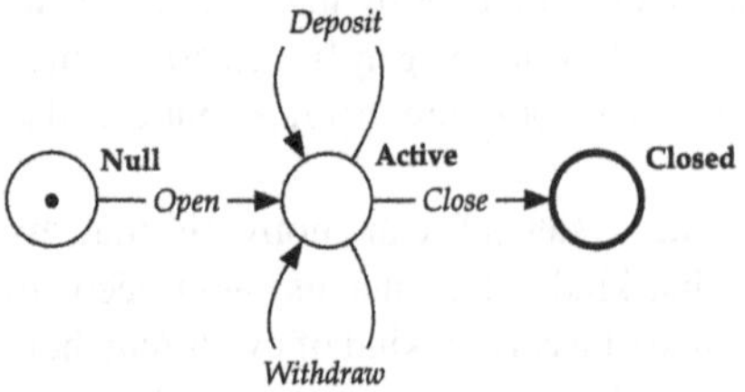

FIGURE 3.1

A state-transition diagram for a bank account atom. Once the account has been opened, deposits and withdrawals can be made in any order, until it is closed.

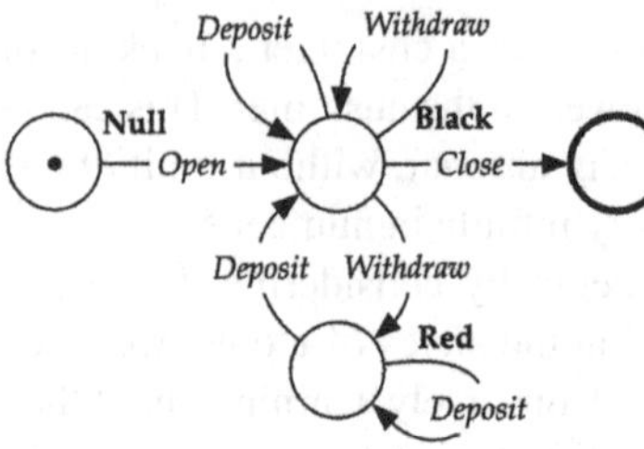

FIGURE 3.2

An alternative non-deterministic state-transition diagram for a bank account atom. If the account is in the *Red* state, withdrawals are not allowed.

It may help the reader to think of the dot as a **token** that is passed from state to state, e.g., when an *Open* event occurs, the token is passed from the *Null* state to the *Active* state; the *Null* state loses the dot, and the *Active* state gains it. The state containing the token is called the **current state**.

A path through a labelled graph is said to **spell** the sequence of the labels on its edges. One path from the initial state to the final state of Figure 3.1 spells the sequence [*Open*, *Deposit*, *Withdraw*, *Withdraw*, *Deposit*, *Withdraw*, *Close*]. There are many other paths from the initial to the final state, but they all begin with *Open* and end with *Close*. Between them can come any number of occurrences of *Deposit* and *Withdraw*, in any order. The possible sequences are called the **sentences** accepted by the automaton, and in general, the labels on the edges are called **letters**. For our purposes, letters and events are synonymous.

An FSA is an abstract notion, whose primary purpose is to define a set of sentences, called a **language**. A state-transition diagram may be used to describe an **acceptor**, an automaton that accepts or classifies input. If the input spells a sequence of transitions that takes the automaton from its initial state to one of its final states, it is said to **accept** the input, and the input is classified as a **valid sentence**. If the input does not spell a path from the initial state to a final state, it is **invalid**. (In a systems context, this usually indicates a human error.) An FSA may be used to classify sentences according to which final state it reaches — if any.

Alternatively, a transition diagram may describe a **generator**, an automaton that generates output. If the automaton moves from its initial state to a final state, then the path it follows generates a valid sentence. In the systems we are considering, the models will be acceptors; the real world will generate the sentences they accept.

In the case of an acceptor, if, in any given state, the input letter uniquely determines the following state, the diagram and its corresponding automaton are said to be **deterministic**. If more than one following state is possible, the diagram is said to be

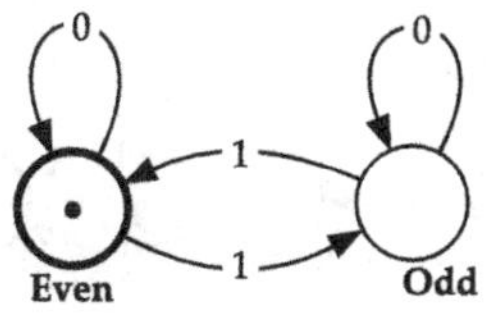

FIGURE 3.3

A state-transition diagram of an FSA to check the parity of a stream of bits. When it is in the *Even* state the parity of the stream received so far is even.

non-deterministic. In other words, it is deterministic if and only if the state-transition diagram defines a function from the input letter and the current state to the next state.

It is the systems analyst's task to decide what kinds of events and what states are relevant. In the case of the banking example, the analyst might, or might not, wish to distinguish between withdrawals that are refused because they would leave an account overdrawn, and those that don't. (See Figure 3.2.)

In the *Black* state, both deposits and withdrawals are possible, but if the amount of a withdrawal puts the account into the *Red* state, only deposits are allowed. In the case of Figure 3.1, a withdrawal might be dishonoured, but in *this* case, a withdrawal is not even a valid input. The amount of a deposit might or might not be sufficient to put the account back into the *Black* state.

3.2 DETERMINISTIC AND NON-DETERMINISTIC AUTOMATA

To illustrate some basic properties of finite-state automata, let us consider a simple problem: to determine whether the parity of a stream of binary digits is even (i.e., contains an even number of 1's). Figure 3.3 is the state-transition diagram of such an automaton. It has two states, labelled *Even* and *Odd*. It starts in the *Even* state. On an input of 1 it always changes state, but on an input of 0 its state doesn't change. If at the end of the input sequence it is in the *Even* state, the parity is even; if it is in the *Odd* state, the parity is odd. Since the goal here is to detect if the input has even parity, the *Even* state is its final state.[5]

As a more complex example, consider an FSA that accepts all sequences of binary digits whose *second* bit is a 1. State U is its initial state. (See Figure 3.4 on the next page.) The automaton moves to state T on the first bit of the input, irrespective of

[5] If, instead, both *Odd* and *Even* were final states, the automaton could be used to *classify* the input. The final state would indicate the parity.

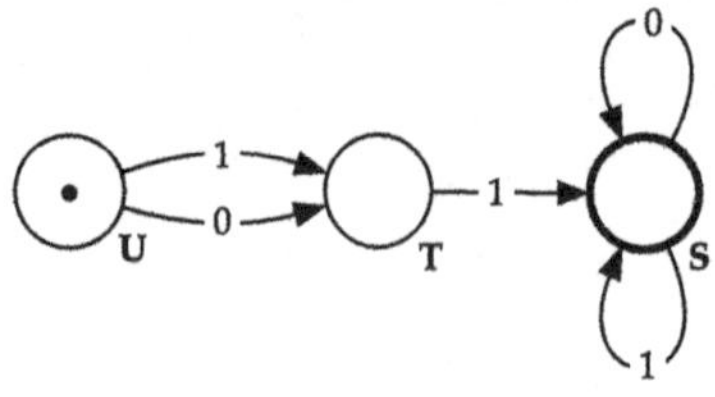

FIGURE 3.4

An FSA that accepts all sequences of bits whose second bit is a '1'. After the first bit, the FSA enters the state T. In state T, only a '1' bit is acceptable. If the bit is accepted, the FSA then remains in state S, its final state.

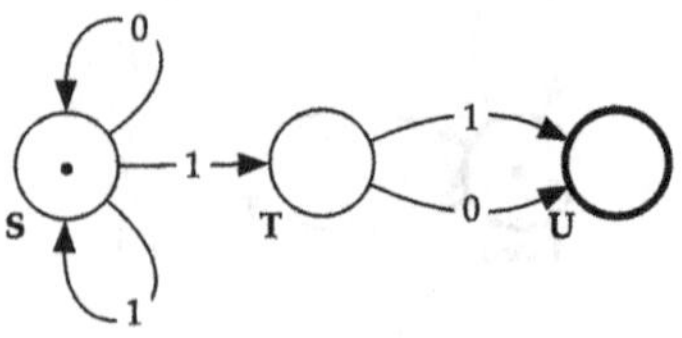

FIGURE 3.5

An FSA that accepts all sequences of bits whose second-last bit is a '1'. It is the reverse of the FSA of Figure 3.4. There are two edges leaving state S labelled 1. Therefore the FSA is non-deterministic.

whether the input is 1 or 0. On the second bit of the input, it moves to state S, but only if the input is a 1. If the input is a 0, the FSA rejects the sentence. S is its final state. Once in state S, the FSA remains in state S, irrespective of the sequence of bits that follows.

If we reverse all the edges of Figure 3.4 and exchange its initial and final states, we have an FSA that accepts the *reverse* of the original sentence, i.e., all binary sequences whose penultimate (second-to-last) bit is 1. (See Figure 3.5.) In this case, there is a difficulty: in state S, given an input of 1, it is impossible to know whether this will prove to be the second-to-last element of the input, so it is impossible to guess whether the automaton should move to state S or T.

This is an example of a non-deterministic finite-state automaton (**NFA**). Usually, an NFA must be converted into a deterministic finite-state automaton (**DFA**) before it can be useful. Even so, it remains possible to use the NFA itself, by keeping track of the *sets* of states that it can be in. In this case, in state S, on an input of 1, the NFA will move into a state that is a member of $\{S,T\}$. Only time will tell which is correct, so we must keep track of the set of possible states.

This idea also provides the key to converting an NFA into a DFA. We do this by mapping each distinct set of possible states of the NFA onto a *single state* of the DFA. The sets of states we consider are those that the NFA can reach after accepting the start of a valid sentence. We now demonstrate this idea by converting the NFA of Figure 3.5 into a DFA.

Before any input has been accepted, the NFA can only be in state S, so $\{S\}$ is its initial set of states, which will map to the initial state of the DFA.

Now consider the situation following the first digit of the input. If it is a 0, the NFA could only move from state S back to state S. Ambiguously, on an input of 1, the NFA can either move back to S or on to state T. Therefore, we have two transitions that are possible between *sets* of states,

$$\{S\} \underset{0}{\rightarrow} \{S\}$$
$$\{S\} \underset{1}{\rightarrow} \{S,T\}$$

We have now found a new set of states to think about, $\{S,T\}$. On input 0, the NFA can move from state S to state S or from state T to state U. On input 1, it can move from state S to either S or T, or it can move from state T to state U.

$$\{S,T\} \underset{0}{\rightarrow} \{S,U\}$$
$$\{S,T\} \underset{1}{\rightarrow} \{S,T,U\}$$

This time, we have discovered *two* further sets of states, $\{S,U\}$ and $\{S,T,U\}$.

In the set $\{S,U\}$, on input 0, the NFA can move only from state S to state S. On input 1, it can move from state S to either S or T. (In state U, it cannot move at all.)

$$\{S,U\} \underset{0}{\rightarrow} \{S\}$$
$$\{S,U\} \underset{1}{\rightarrow} \{S,T\}$$

In the set $\{S,T,U\}$, on input 0, the NFA can move from state S to state S or from state T to state U. On input 1, it can move from state S to either S or T, or it can move from state T to state U.

$$\{S,T,U\} \underset{0}{\rightarrow} \{S,U\}$$
$$\{S,T,U\} \underset{1}{\rightarrow} \{S,T,U\}$$

Fortunately, no new sets of states have appeared, so the above states are the only ones that can be reached.

We now construct a DFA by mapping each *set* of states of the NFA onto a corresponding state of the DFA. We can draw the state-transition diagram of the resulting automaton, which has four states in all. (See Figure 3.6.) Any set of states that includes U, the final state of the NFA, becomes a final state of the DFA.

Because we construct only one transition from each set of states for each input letter, we can be confident that the resulting automaton will always be deterministic, but can we be confident that the construction process will terminate? The NFA of Figure 3.5 has three states. There are only $2^3 - 1$ non-empty subsets of these states, so we can see that its corresponding DFA can have at most seven states. Therefore, the construction process must end after at most seven iterations.

This example clearly illustrates that, in pursuing an efficient implementation, synthesis (in this case from an NFA to a DFA) can lead to a result that is only obscurely related to its origin. The purpose of the NFA of Figure 3.5 is intuitive; the purpose of the DFA of Figure 3.6 is not.

In this example, it is easy to see why the resulting DFA needs at least four states: an obvious way to detect if the penultimate bit of a sequence is a 1 would be to store

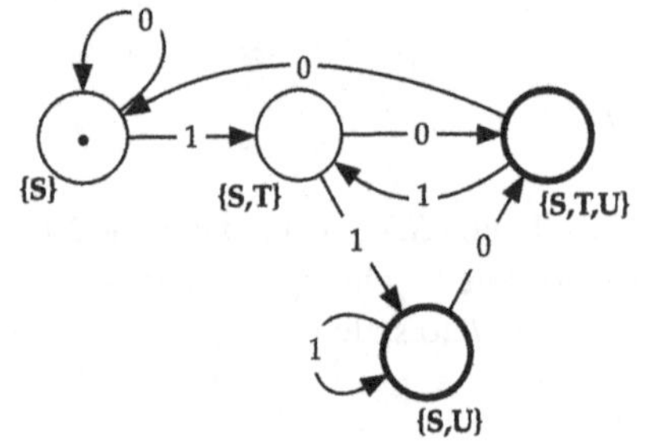

FIGURE 3.6

A deterministic state-transition diagram of an automaton that detects if the last but one bit of a binary sequence is a 1. Its states are labelled with sets of the states of the NFA of Figure 3.5.

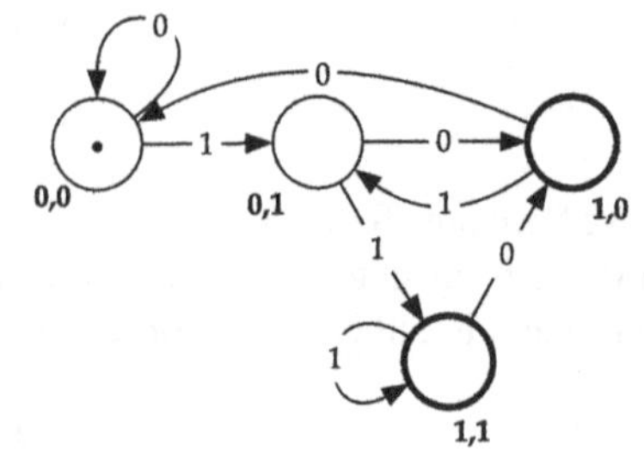

FIGURE 3.7

The unique minimal deterministic state-transition diagram of an automaton that detects if the second-to-last letter of a binary sequence is a 1. Its states are labelled with the most recent two bits of the input.

the most recent 2 bits of the input in a shift register (initially 00), and check its leading bit once the input is complete. A 2-bit shift register has precisely four states. Figure 3.7 shows the result of relabelling the states of Figure 3.6 as follows:

$$
\begin{aligned}
00 &= \{S\} \\
01 &= \{S,T\} \\
10 &= \{S,U\} \\
11 &= \{S,T,U\}
\end{aligned}
$$

In systems analysis, we rarely meet situations where an automaton is non-deterministic, but if we do, Algorithm 3.1 always lets us convert the resulting NFA to a DFA.[6]

In some cases, it may be more convenient to draw a state-transition diagram that appears to be non-deterministic, when actually it is not. Consider Figure 3.2, which shows the more detailed view of a bank account. The state *Red* indicates that the account is overdrawn, and the state *Black* indicates that the account is solvent. The event *Withdraw* can move the account from the *Black* state to either the *Black* or the *Red* state. In this example, it happens deterministically: if the amount of the withdrawal exceeds the current balance, the account becomes *Red*, otherwise it remains *Black*. The analyst has distinguished the *Black* and *Red* states to show that *Withdraw* is forbidden in the *Red* state. The deterministic nature of the FSA could have been shown by considering withdrawals that do or do not overdraw the account as two different kinds of event. Alternatively, the analyst could have combined the

[6] If the NFA has N states, the resulting DFA may have as many as 2^{N-1} states, so the problem is intractable in theory, but this is most unlikely to be an issue in practice.

Algorithm 3.1 Converting an NFA into an equivalent DFA.

1. Begin with an NFA N and a set of sets of states D initially containing a singleton set containing the initial state of N.
2. For each newly added set of states S in D, and each letter L in N,
   - Find the maximal set of states T such that an edge $V \underset{L}{\rightarrow} U$ is present in N, where $V \in S$ and $U \in T$.
   - Add the edge $S \underset{L}{\rightarrow} T$ to D.
3. Repeat Step 2 until no new sets of states can be added to D.
4. The sets of states in D are the states, and the edges in D are the transitions of the required DFA.

Red and *Black* states into one (as in Figure 3.1) and made a note that there are further conditions on whether a *Withdraw* event will be honoured. We cannot expect an FSA to express all the rules of a business.

3.3 REGULAR EXPRESSIONS

We can model an atom in two distinct ways: we can describe its behaviour, or we can describe how it changes its state. In a **behavioural description** we focus on what *sequences* of events are possible. In a **state-based description**, we take the view that the current state of the system is determined by prior events and that the current state of the system then determines what future events can occur. These descriptions are equivalent; we can derive behaviour from states or derive states from behaviour.

We can describe the possible sentences accepted by an automaton using a formula, called a regular expression, which uses three familiar constructs: *sequence*, *choice*, and *iteration*. We may have one thing followed by another, we may choose between alternatives, and we may repeat things any number of times. There is one limitation: regular expressions do *not* permit recursion. This isn't a problem, because real things are not recursive.[7] Although a component of a bicycle can be a part of a part of a part of a bicycle, etc., it can't be a part of *itself*.

We may define **regular expressions** formally as follows:

- If x is any letter (a transition), it is a regular expression.
- If R and S are regular expressions, $R;S$ denotes expression R *followed by* expression S.
- If R and S are regular expressions, $R \cup S$ denotes *either* expression R *or* expression S.

[7] Fiction is different: Dr Who once landed the TARDIS (which is bigger on the inside than on the outside) inside itself, creating an infinite regress.

- If R is a regular expression, R^* denotes a sequence of zero or more occurrences of R.
- ε (epsilon) denotes an empty expression, a path of length zero

where '$*$', binds tightest, '$;$' next, and '$\cup$' binds loosest.[8]

The language accepted by the bank account automaton of Figure 3.1 (page 82) is given by the regular expression

$$open\,;(deposit \cup withdraw)^*\,;close \tag{3.3.1}$$

A regular expression is a behavioural description: it says what sequences of letters can occur, without the need to invoke the idea of state. We might therefore prefer to use behavioural descriptions, because different automata can accept the same language. For example, the state-transition diagram of Figure 3.2 (page 82) defines an FSA that accepts exactly the same language as that of Figure 3.1 (page 82).

On the other hand, the expression

$$open\,;(deposit^*\,;withdraw^*)^*\,;close \tag{3.3.2}$$

describes the same language as expression 3.3.1, so neither a regular expression nor an FSA can claim to be a canonical way to describe a language.[9]

What we can say, in general, is that a given language can often be described by many possible regular expressions or by many possible transition diagrams. The good news is that nothing can be described using regular expressions that cannot be described using state-transition diagrams, and *vice versa*. Indeed, algorithms exist that enable each to be synthesised from the other.

The attentive reader will have noticed that the operators we have used here to construct regular expressions are the same relational operators we defined in Section 2.4.4. This is because each input letter defines a relation between states. Consequently, any theorem or transformation that is valid for binary relations is also valid for regular expressions.[10]

In the case of the shift-register DFA of Figure 3.7 (page 86), **0** and **1** are the following functions:

$$\mathbf{0} : States \rightarrow States = \{00 \mapsto 00, 01 \mapsto 10, 10 \mapsto 00, 11 \mapsto 10\}$$
$$\mathbf{1} : States \rightarrow States = \{00 \mapsto 01, 01 \mapsto 11, 10 \mapsto 01, 11 \mapsto 11\}$$

In theory, no other operators are required; any language can be described solely by the operators listed above. Even so, a language can often be described more concisely using additional operators:

---

8 Yes, these are the same regular expressions used by utility programs such as *grep*. The difference is that *grep* uses a more convenient syntax. Nonetheless, any *grep* expression can be reduced to the form used here.

9 But see the discussion of *minimal DFAs* in Appendix A.

10 With this interpretation ε is the identity function.

- R^+ means the same as $R\,;R^*$ and denotes at least one occurrence of R.
- $R^?$ means the same as $R \cup \varepsilon$ and denotes an optional occurrence of R.
- $R \cap S$ is an expression that describes all sentences accepted by R that are also accepted by S.
- $R \setminus S$ is an expression that describes all sentences accepted by R that are *not* also accepted by S.

States and behaviours are closely connected. A state is both a summary of previous behaviour, and a restriction on future behaviour. If a bank account automaton is in the *Active* state (in Figure 3.1), it has accepted an *Open* event and zero or more *Deposit* or *Withdraw* events, but has never accepted a *Close* event. Likewise, it is ready to accept one or more *Deposit* or *Withdraw* events, or one *Close* event, but not an *Open* event. It has accepted the **phrase** $Open\,;(Deposit \cup Withdraw)^*$ and is ready to accept the phrase $(Deposit \cup Withdraw)^*\,;Close$. In the *Null* state, it has accepted the empty phrase ε and is ready to accept a complete sentence; in the *Closed* state, it has accepted a complete sentence and is ready to accept ε.

The argument in favour of describing behaviour is that it can be done without the need for states. The argument *against* describing behaviour is that before we can actually implement a model, we usually have to derive states from behaviour. Here, we consider it more convenient to deal with states directly, and let the behaviour emerge. The correspondence between the behavioural and state-based approaches is therefore peripheral to our purposes, so it has been relegated to Appendix A, where we present well-known algorithms for deriving a state-based description from a behavioural description and *vice versa*. Even so, the reader is strongly advised to read Appendix A, because it illustrates several important points about synthesis methods in general.

3.4 FINITE MEANS FINITE

Before looking at some practical uses of FSAs, let us consider a simple problem that they cannot deal with: checking to see if the left and right parentheses in a text are correctly nested, i.e., that the numbers of left and right parentheses are equal, and that the number of right parentheses never exceeds the number of left parentheses. For example, '$((a,b),(3+4))$' is a valid sentence, but '$(2*3)+)9($' and '$(f(x)$' are not.

Figure 3.8 shows part of the state-transition diagram of such an automaton. When '(' is scanned, the machine moves one state to the right; when ')' is scanned, it moves one state to the left. When any other character is scanned, it remains in the same state. The states are labelled, logically enough, with the integers 0, 1, 2, 3, and so on. If at the end of the input, the numbers of left and right parentheses are equal, the machine should be in State 0. If the number of right parentheses ever exceeds the number of left parentheses, the machine has no possible transition, so the error is detected immediately.

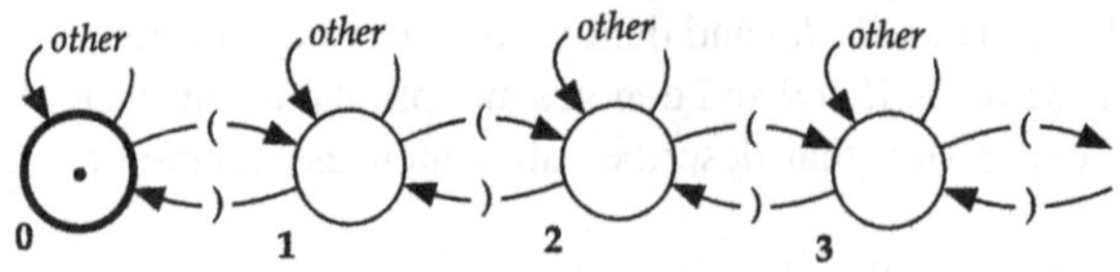

FIGURE 3.8

Part of the state-transition diagram of an automaton that detects if parentheses are correctly nested.

Unfortunately, it should be clear that if the automaton has a *finite* number of states, however large, an input sequence can start with so many left parentheses that the automaton runs out of states.

The reader will realise that such an automaton can easily be implemented by a counter, incremented for each '(', and decremented for each ')'. Nonetheless, if the representation of the counter is finite, eventually the counter will overflow, so the situation is like the state machine. Indeed, a counter *is* a state machine. We sum up this situation with the statement, 'FSA's can't count to infinity'.[11]

In practice, the FSAs we shall discuss will need only a handful of states. If we find a need for many states, we should think again. The sole purpose of our FSAs will be to keep track of what sequences of events make sense.

3.5 ANALYSING STATES

Consider the states of a copy of a *Book* from the point of view of a librarian. The important states of a copy might be: *On Order*, *Shelved*, *On Loan*, *Archived*, and *Scrapped*. From this point of view, although much detail has been omitted, the states of the book are discrete and finite in number. In contrast, a book retailer might consider a different set of states to be important: *On Order*, *In Stock*, *Sold*, and *Damaged*. Let us focus first on the librarian's view.

Libraries keep track of individual copies of books. The first action of the librarian would be to *Order* a copy from a supplier. After this, the copy would be *On Order*. Prior to this, the copy would be in its *Null* state, one that is of no particular interest to the library, but one that the state-transition diagram must include in order to show the *Order* event. Once a copy is *On Order*, the next event should be to *Receive* the copy, after which it is in the *Uncatalogued* state. The next action should be to *Catalogue*

[11] The reader will recognise that this task is similar to one performed by a compiler for a structured programming language in which statements and expressions can be nested to arbitrary depth. To overcome this limitation of FSAs, compilers use a *stack* of states, which can be infinitely deep in principle (although not in reality).

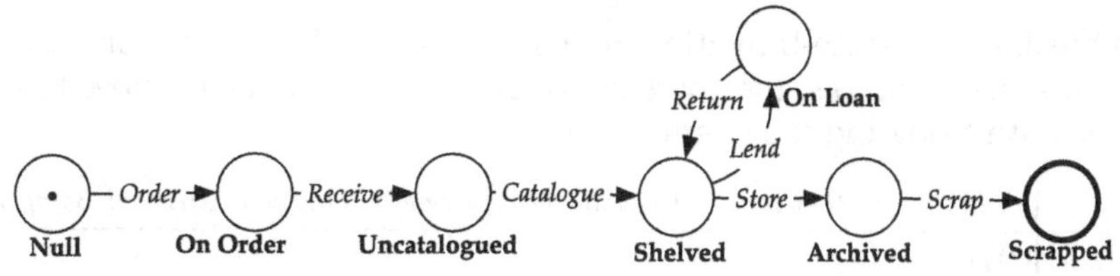

FIGURE 3.9

The state-transition diagram of a book, as seen by a librarian. Until a book is ordered, it is of no interest. Once it has been received, it is catalogued and made available for loan. Eventually, it may be archived or even scrapped. While the book is on loan, the only possibility is for it to be returned to the library.

the copy, after which it is said to be *Shelved*.[12] Once a copy is *Shelved*, a *Lend* event may occur, causing the state of the copy to become *On Loan*. In this state, a *Return* event will put the copy back into the *Shelved* state again. The resulting cycle means that a copy may be lent and returned any number of times. It is also possible for a *Store* event to change the state of a copy from *Shelved* to *Archived*. Eventually, a *Scrap* event may put the copy into the *Scrapped* state, which is a final state. The state-transition diagram of a library book is shown in Figure 3.9.

We shall have more to say about how states are represented later, but for the present, assume that the state of a copy is indicated by the value of a variable, a small integer, for example. Actually, *On Loan* is really a family of states, because a librarian cares who has borrowed each copy. Therefore, the representation of the *On Loan* state would need to be supplemented with the borrower's identity.

We may make a distinction between FSAs implemented by hardware and those implemented by software. Hardware FSAs have a permanent physical existence. The FSA to check the parity of a stream of bits can be implemented by a simple electrical circuit. On the other hand, software FSAs usually come into existence by being **instantiated**, i.e., memory is allocated to them as needed. Typically, when a library places a book on order, a record will be created to represent it. Prior to this, when the book is in the *Null* state, no record will exist.

It is wise to develop the transition diagram of an FSA further using a **state-transition matrix**. The DFA for a library book, as established so far, has the matrix shown in Table 3.1.

First, notice that if an FSA is deterministic, its current state (column) and the event (row) uniquely determine its next state. If this isn't the case, then the analyst should consider whether what appears to be one state or one kind of event is actually two or more.[13] For example, in an academic record system, a single event such as

---

[12] Libraries tend to claim that books are shelved even when they are sitting around *waiting* to be shelved. It is usually too much trouble to do anything else.

[13] As in the case of *Red* and *Black* bank withdrawals considered earlier.

Table 3.1 The state-transition matrix for a library book. The columns show existing states, the rows show events, and the cells show the new state following the event

	Null	*On Order*	*Uncatalogued*	*Shelved*	*On Loan*	*Archived*	*Scrapped*
Order	*On Order*	–	–	–	–	–	–
Receive	–	*Uncatalogued*	–	–	–	–	–
Catalogue	–	–	*Shelved*	–	–	–	–
Borrow	–	–	–	*On Loan*	–	–	–
Return	–	–	–	–	*Shelved*	–	–
Store	–	–	–	*Archived*	–	–	–
Scrap	–	–	–	–	–	*Scrapped*	–

Assess leading non-deterministically to the *Passed* state or *Failed* state could be subdivided into events, *Pass* and *Fail*, each leading deterministically to *Passed* and *Failed*. Alternatively, it may be more convenient to combine the *Passed* and *Failed* states into a single *Assessed* state. A third option is to allow the non-determinism to stand, with a footnote to the effect that the mark associated with the *Assess* event determines whether the new state is *Passed* or *Failed*. There is no right or wrong answer. The analyst will have to make a choice, probably based on requirements that will emerge later. For example, if the system must remember the marks that candidates obtain, it can then easily be established whether an *Assess* event is a *Pass* or *Fail*.

The most notable feature of the matrix of Table 3.1 is its predominance of empty cells. Most of these represent errors: for example, it makes no sense to borrow a copy that is already *On Loan*.[14] Even so, the analyst shouldn't rest until every cell has been accounted for. Can a copy be removed from the store and be re-shelved? Can a shelved copy be re-catalogued? What happens when a copy is lost by its borrower, stolen from the shelves, or damaged beyond repair? These issues can only be clarified by talking to a librarian, a domain expert. We call this process of discovery **event-state analysis**. Likewise, we may refer to a state-transition diagram as an **event-state diagram**, or a state-transition matrix as an **event-state matrix**. Event-state analysis is a two-way process: the model generates the questions, and the answers generate the model.

Apart from ensuring that all the cells are accounted for, it is important to ensure that no events or states have been missed. New events can often be discovered by asking questions such as, "When a copy is on the shelves, can anything happen to it apart from it being borrowed or archived?" New states can be discovered by questions like, "Are there different ways of borrowing a copy?" or, "Are there different kinds of

[14] This might result from the failure to record a *Return* event.

borrowers?" This might lead to the analyst's discovery of inter-library loans, resulting in a whole list of hitherto undiscovered events. In the case of physical objects, such as books, states often correspond to places. In this context, it is worth asking, "Apart from the shelves, the store, or the hands of a borrower, are there any other places where a copy can be?"

Finally, it is worth checking whether the final and initial states of atoms need to be distinct, since both states might correspond to a complete absence of data. For example, no record for a copy may exist at all until it is ordered. Perhaps, when it is scrapped, its record should be deleted, or maybe it should be retained for historical or statistical purposes. If its record *is* deleted, a scrapped copy looks no different from a copy that has never been ordered, and the record can be returned to the *Null* state. The record can be **destroyed**, i.e., the memory it occupies can be freed for reuse.

People make mistakes. A *Scrap* event might specify the wrong book, causing it to become *Scrapped*. No event we have considered can undo this action. To be able to reverse an erroneous event, a special, rarely used, **reversal event** may be introduced.

The analyst should also check that the state-transition diagram is strongly connected. This means that it has a path from every state to every other state, so *some* sequence of events can always restore an atom to any given state. For this purpose, the analyst may find an event-state diagram more useful than a matrix, as it is usually easy to see where a graph contains cycles. Alternatively, the analyst can reduce the event-state matrix to an adjacency matrix by ignoring the names of events and, from that, find its strongly connected components. (See Section 2.5.5.) One strategy (often used with hardware devices) is to provide a *Reset* event that restores an atom to its initial state. From there, a suitable sequence of events can drive it to any state that it can ever reach.

Care is needed: reversing an erroneous *Scrap* event by faking an *Order* and *Catalogue* event is likely to mean that the copy is paid for twice.[15] Reversing an event is typically more difficult, for example, than implementing *undo* in a word-processor. Erroneously recorded events may remain undiscovered for a long time, and it may be impractical to reverse every action that has occurred since the error was made. Sometimes too, an error can cascade, causing further errors that affect the snapshots of other atoms in the system. For example, erroneously recording that a book was loaned to one patron may not be detected until a second patron of the library actually borrows it. At this time, the system should baulk because the copy is already in the *On Loan* state. But by then, the first patron, unable to return the copy, may have been asked to pay a fine. Such errors can be hard to fix. As a result, it is tempting to allow an operator of a system to make arbitrary changes to states, but this is dangerous to do, especially if the changes they make go undocumented.

You are unlikely to hear about reversal events from a domain expert. Often, no formal system exists for dealing with mistakes, and each error is handled on an *ad hoc* basis. Nonetheless, persistent questioning can reveal some of the more common error-correcting strategies.

---

[15] Even so, this might be attractive because it is simple to implement. For example, the database could include an imaginary vendor who never charges for books.

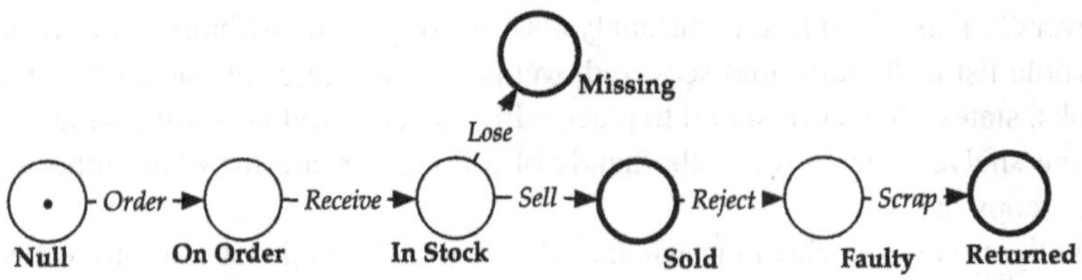

FIGURE 3.10

The event-state diagram of a book, as seen by a bookseller. There are three possible final states. A book may be sold, it may be missing (presumed stolen), or it may be returned to the publisher.

3.6 COUNTING

A bookseller's FSA for a book will be somewhat different from a librarian's. For a bookseller, a book's states might include *On Order*, *In Stock*, *Sold*, *Missing*, *Faulty*, and *Returned* (to the publisher). Its events might include *Order*, *Receive*, *Sell*, *Reject*, *Return*, and *Lose* (see Figure 3.10). However, the most important difference is that, unlike a librarian, a bookseller isn't interested in individual *copies* of books. A bookseller is interested in *titles*, and also whether the book is hardback or paperback, and so on. All books belonging to a given edition are alike. It makes no sense for a bookseller to track the progress of each copy of a book. Instead, it is the *number* of books in each state that matters.

This difference is reflected in the way states are represented. Instead of a variable that represents the state of an individual copy, a set of variables is needed, one integer to count the number of copies of the book in each state. When a copy moves from one state to another, for example, when a *Sell* event occurs, the number of copies *In Stock* decreases by 1, and the number *Sold* increases by 1. In other words, copies of books are *conserved*. If a copy is stolen, it is usually better practice to increment the number of books *Missing* than to merely decrement the number *In Stock*.[16] Similar conservation laws apply to virtually any system.

From this, we see that integers in a system often record the numbers of similar atoms in given states. By an extension of this idea, currency amounts represent the numbers of coins of some basic denomination: cents, pence, etc. Indeed, for many purposes, coins themselves have already become obsolete, and numbers have taken their place.

In an accounting system, coins (or their representations) can be in many states, one state for each account to which they might be credited. Each such account is

[16] Counting does not imply that we are no longer dealing with FSAs; the number of states of each atom (e.g., a book) is still quite small.

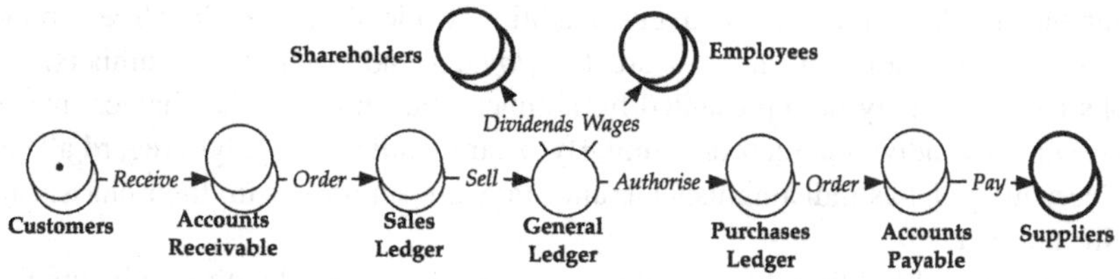

FIGURE 3.11

States of coins in a hypothetical accounting system. Only the main flows are shown.

typically represented by a separate record. Within the system, money is conserved. If the amount in one account increases, that in another decreases. That is the basis of double-entry bookkeeping. Indeed, because accounts can contain any amount of money, they are not FSAs and cannot be modelled by them. Consequently, although a coin's states are finite, a coin might be in so many states (one for each account) that drawing its event-state diagram might seem an impossible task. Fortunately, accounts fall into families that share similar behaviour, usually corresponding to ledgers. In drawing the event-state diagram for money, we should show families of states such as *Sales*, *Customers*, and so on. The *Accounts Receivable* ledger would be subdivided by customer, whereas the *Sales Ledger* might be subdivided by product. Figure 3.11 sketches part of an accounting system, in which money flows from customers to *Accounts Receivable*, and from there to a *Sales Ledger*, and so on. The states labelled *Customers* and *Suppliers* really lie outside the accounting system. Money is conserved *within* the system. When an *Order* event occurs, through the miracle of double-entry bookkeeping, money is transferred from *Accounts Receivable* to a *Sales Ledger*. On the other hand, when a *Receive* or *Pay* event occurs, marking a payment by a customer or a payment to a supplier, money crosses the system boundary and appears or disappears.

In an **open system**, conservation laws break down at its boundaries. Typically, no estimate is made of the number of atoms in a *Null* state. A bookseller doesn't care how many books have *not* been ordered. Numbers become important only when an order is placed.

A further extension of the conservation idea applies to **stuff**. Stuff cannot be counted; it has to be measured. A company that sells oil cannot be expected to count every molecule. Instead, it estimates the number of molecules it sells by measuring the volume sold.[17] Again, *stuff* is usually conserved, and the oil company will need to have a measure of the volume of fluid in each state, including, perhaps, the amount that has leaked or spilled.

Sometimes, a system's view of atoms changes; it may view the same atoms as sometimes interchangeable, and as sometimes distinct. A librarian might merely

[17] Because of thermal expansion, fluids really ought to be sold by mass rather than by volume.

count the number of books of a given edition while they are *On Order*, books only becoming distinct when they are *Catalogued* and given copy numbers. The books *On Order* may be represented by numbers, but those on the shelves may be represented by individual records. Similarly, a car manufacturer might regard all cars of a certain model as indistinguishable until they are stamped with their chassis and engine numbers.

Irrespective of whether the system keeps track of individual atoms, indistinguishable sets of atoms, money, or stuff, the story of every atom, coin, or molecule can be modelled by an event-state diagram. Despite this, ***and this is important***, the atoms (e.g., coins) that are modelled may not correspond one-to-one with objects or records in an information system; it may be their states (e.g., accounts) that are represented.[18] Trying to draw an event-state diagram for a *state* is an impossible, meaningless, task.

3.7 CONTINUOUS SYSTEMS

Although continuous systems aren't the focus of this book, it is worth giving some space to them. As we shall see in a later chapter, both discrete and continuous systems exhibit **dynamic behaviour**. Such behaviour can be **autonomous**, that is, it can occur without external input. Sometimes autonomous behaviour is desirable, and sometimes it is not.

We cannot apply the notion of an FSA to the continuous behaviour of an aircraft, so what alternative do we have? Such a subject would fill a book, so we turn instead to a simple pendulum.

Consider an ideal pendulum consisting of a point mass m hanging by a weightless string of length l. Let us assume that the mass is displaced a distance x from the vertical. Intuitively, we can see that a force acts on the mass to restore it to the vertical, and that the mass will therefore accelerate. Let us assume that it moves with velocity v. We assume further that the pendulum is *perfect*, i.e., that it doesn't lose energy through friction.

We suggested earlier that systems obey conservation laws. What is conserved here is energy. The point mass has two forms of energy, kinetic energy and potential energy, and their sum is constant. The pendulum's kinetic energy is given by the formula $\frac{1}{2}mv^2$. The pendulum's potential energy is given by mgh, where g is the acceleration due to gravity, and h is the vertical displacement of the mass above its lowest position. For small displacements $h = \frac{1}{2}x^2/l$. Therefore $\frac{1}{2}mv^2 + \frac{1}{2}mgx^2/l = E$, where E is the (constant) energy in the system.

The state of the pendulum is defined by its position and velocity. If we plot position, x, horizontally and velocity, v, vertically, then with suitable scaling, we see that all the points with energy E must lie on a circle whose area is proportional to E, as shown in Figure 3.12. This is actually a form of state-transition diagram. The

[18] This is why we don't call atoms 'objects'. This term is used in object-oriented programming to represent some data and its associated methods. Often, an atom can be represented by an object, but not always. What is more, an object often represents a set of atoms, rather than an individual atom.

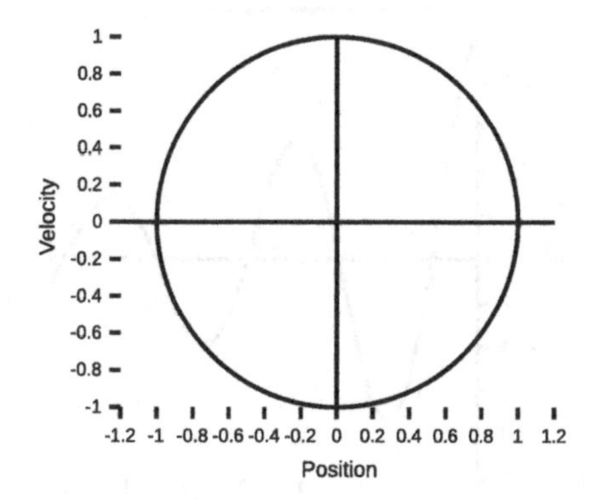

FIGURE 3.12

The state trajectory of a perfect undamped pendulum. The scaling is normalised so that the initial position of the pendulum is 1.0, and its maximum velocity is 1.0. Its state moves from the point (1.0, 0.0) clockwise around the circle, once for each complete swing.

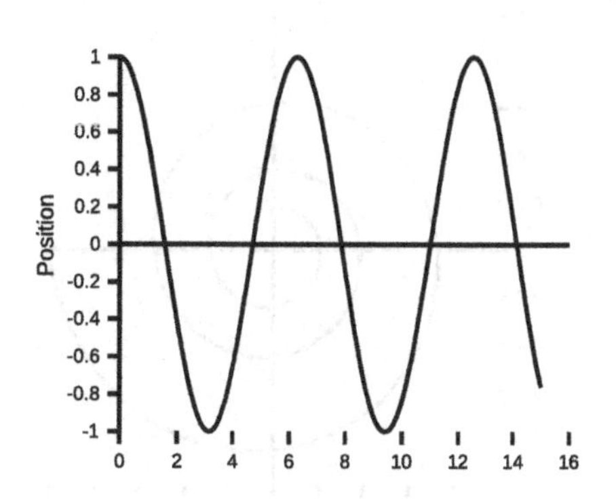

FIGURE 3.13

The position of a perfect pendulum as a function of time. Its initial position is normalised to +1.0. Each horizontal unit is normalised to represent $\sqrt{(l/g)}$. The pattern repeats at intervals of 2π.

state of the pendulum moves around the circle. The states lie infinitesimally close to one another, linked by infinitesimally small transitions. The circle is the pendulum's **state trajectory**. Figure 3.13 shows the oscillatory displacement of the pendulum as a function of time.[19]

[19] An analysis using differential calculus will tell us a little more: to counter the weight of the mass, the string must exert a vertical force equal to mg. For small displacements, this will have a horizontal component $f = -mgx/l$. The horizontal acceleration of the mass is therefore $a = f/m = -gx/l$. Mathematically, acceleration is the second differential of position, so that

$$\frac{d^2x}{dt^2} = -\frac{g}{l}x \tag{3.7.1}$$

In the case that the pendulum has zero velocity and initial displacement x_0, this differential equation has the well-known solution

$$x = x_0 \cos\left(\sqrt{\frac{g}{l}}t\right) \tag{3.7.2}$$

Differentiating with respect to t, we see that

$$v = \sqrt{\frac{g}{l}}x_0 \sin\left(\sqrt{\frac{g}{l}}t\right) \quad \text{and} \quad a = -\frac{g}{l}x_0 \cos\left(\sqrt{\frac{g}{l}}t\right) \tag{3.7.3}$$

So $a = -(g/l)x$, as required. Since the cosine function repeats after an interval of 2π, the oscillation has period $2\pi\sqrt{g/l}$.

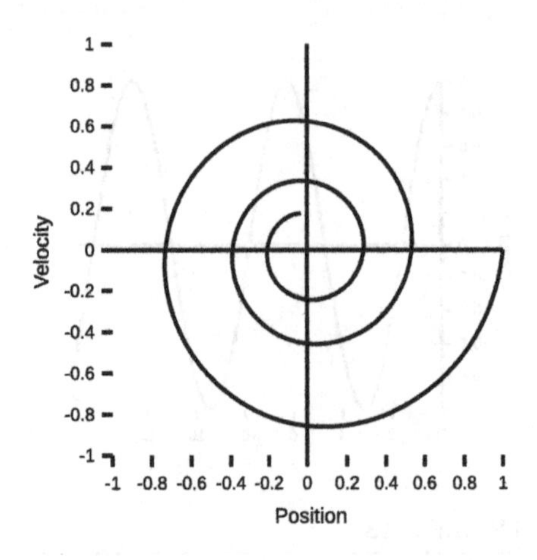

FIGURE 3.14

The state-transition diagram of a damped pendulum. Because energy is lost, the oscillations decay.

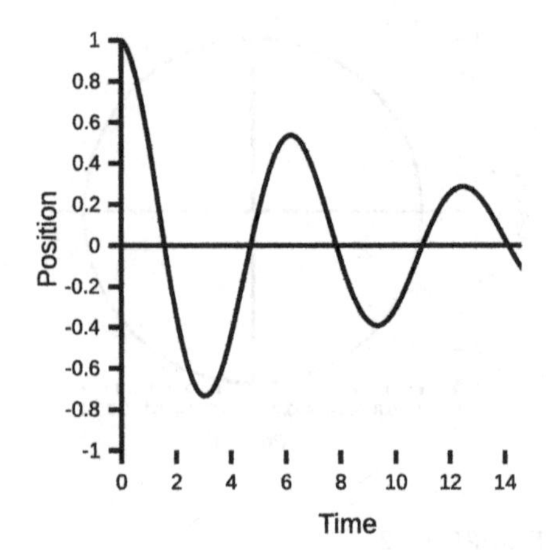

FIGURE 3.15

The damped oscillation of a pendulum. The amplitude of the oscillations decays with time.

In practice, since any real pendulum will lose energy to its environment, its state trajectory will be a spiral converging (asymptotically) towards the origin, as in Figure 3.14, and its oscillation will decay exponentially, as in Figure 3.15.

To extend these ideas to the analysis of an aircraft, we have to remember that its state is described by three linear positions and velocities, and three rotational positions and velocities. Therefore we would need to visualise its state trajectory in 12 dimensions. The equations that describe its motion are complex. Even so, an aircraft does have one mode of behaviour, the **phugoid oscillation**, which is roughly similar to the case of the simple pendulum.

Imagine an aircraft that is flying straight and level, but a little too slowly. Because of this, its wings won't generate enough lift to support its weight. The aircraft will start to descend, gaining speed as it does so. The additional speed will generate more lift, so the aircraft will climb again. At the top of the climb, it will again be flying too slowly, and start to descend again. Like the pendulum, the aircraft exchanges potential energy (height) for kinetic energy (speed).[20] Fortunately, this kind of oscillation is usually slow, easily controlled by the pilot, and even without intervention, like the swing of a pendulum, it gradually decays.[21]

[20] In theory, the height of an aircraft should be described by a real number of infinite precision. In practice, continuous systems are usually modelled digitally to finite precision. A 64-bit word has 2^{64} (almost 10^{20}) states, enough to represent the height of an aircraft far more precisely than it can be measured. Likewise, we must approximate the continuous flow of time by a series of sufficiently frequent snapshots. An airliner travels less than 0.3 millimetres per microsecond, so it isn't hard to compute sufficiently frequent approximations for practical purposes.

[21] In highly manoeuvrable fighter aircraft, the pilot's intervention can make things worse.

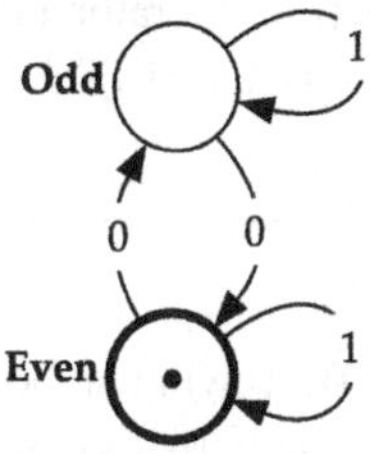

FIGURE 3.16

An FSA to check for an even number of 0s.

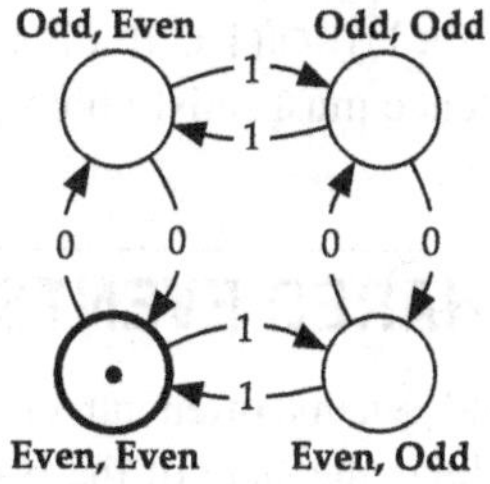

FIGURE 3.17

An FSA to check for an even number of 0s and an even number of 1s.

Similar dynamic behaviour is found in business systems and national economies: suppose you go to the supermarket only to find that your favourite blend of coffee is out of stock. Realising that its supply is unreliable, the next time you visit, you take twice your normal amount. Multiply this by a dozen like-minded individuals, and the supermarket manager quickly notices how fast the coffee is selling, and therefore orders more than usual. Shortly, the shelves are full of your favourite coffee, but you still have a spare supply at home and don't need more. At this point, the supermarket manager notices that the coffee isn't selling after all — and you can continue the story yourself.

3.8 PRODUCTS OF FSAS

As a preparation for understanding shared events, recall the FSA of Figure 3.3, which checks that a sequence of bits contains an even number of 1s. Figure 3.16 describes a similar FSA that checks whether such a sequence contains an even number of 0s. If both FSAs share the same events, together they will check that the input contains even numbers of 1s and 0s. We can describe their combination in a single FSA that has states $(Even, Even)$, $(Even, Odd)$, $(Odd, Even)$, and (Odd, Odd); the set of states is the Cartesian product of the states of the FSA of Figure 3.3 and the FSA of Figure 3.16. (See Figure 3.17.)

Clearly, the number of states of the FSA of Figure 3.17 is the product of the numbers of states of its component FSAs. If we collapse the diagram vertically, merging $(Even, Even)$ with $(Odd, Even)$ and $(Even, Odd)$ with (Odd, Odd), we recover the FSA of Figure 3.3, and if we collapse it horizontally, we recover the FSA of Figure 3.16.

As we discussed earlier, we may express the behaviour of an FSA using a regular expression. An expression that describes any sequence containing an even number of 1s is $(0 \cup (1\,;0^*\,;1))^*$. Similarly, $(1 \cup (0\,;1^*\,;0))^*$ describes any sequence containing an even number of 0s. The FSA of Figure 3.17 therefore accepts the expression

$(0 \cup (1 ; 0^* ; 1))^* \cap (1 \cup (0 ; 1^* ; 0))^*$, where the intersection operator means that a valid sentence must satisfy both $(0 \cup (1 ; 0^* ; 1))^*$ and $(1 \cup (0 ; 1^* ; 0))^*$.[22]

3.9 SHARED EVENTS

Real-world events often affect more than one atom. Borrowing a book from a library affects the state of the book and also the state of the borrower. We call such interactions **shared events.**

Often, a shared event involves a **rate of exchange**. Selling a book results in a loss of stock to the bookseller, but also an increase in money. The rates of exchange between books and money are prices. Rates of exchange are a second way in which numbers are important in a system.

The main point about shared events is that all parties to the event have to be in states where the event is permitted. That is so important, I shall say it again:

> ***Each atom involved in a shared event must have an event-state diagram that includes at least one edge labelled with that event, and the event itself is only valid if* all *the atoms concerned are in states that have such edges leaving them.***

Let's look at a simple student enrolment system. When a candidate enrols in a subject, two atoms participate in the event: the *Candidate* and the *Subject*.[23] For this to happen, the candidate must be *Admitted* and the subject must be *Offered*. Figure 3.18 shows the event-state diagram for a *Candidate*, and Figure 3.19 shows the diagram for a *Subject*. We see that once a candidate has been admitted to a degree, he or she may *Enrol* in subjects or *Withdraw* from them, until the candidate *Graduates*. Likewise, once a subject is *Offered*, candidates may *Enrol* in it or *Withdraw* from it, until some date at which enrolments become *Frozen*, after which candidates enrolled in it will be *Assessed*. As a result, a candidate cannot, for example, enrol in a subject that has been frozen, and a candidate who has already graduated cannot enrol in any subject.

Even so, these constraints aren't strong enough: the diagrams allow a candidate to *Withdraw* from a subject before ever enrolling in it. Neither the *Candidate* DFA nor the *Subject* DFA prevents this.

We now consider a corollary to the point we just made: the *only* things that determine whether an event can occur are the states of the atoms that share it. This is again so important I shall say it again:

> ***Unless the state of something prevents it, an event is valid.***

---

[22] Regular expression syntax doesn't usually include the $\cap$ operator, because it can always be eliminated. Unfortunately, eliminating it is an NP-complete problem, and the resulting expression can be ridiculously long.

[23] Actually, *three* atoms participate, as we shall soon see.

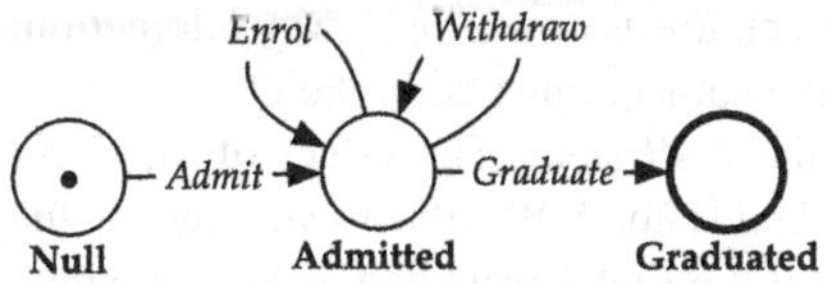

FIGURE 3.18

The event-state diagram for a *Candidate*.

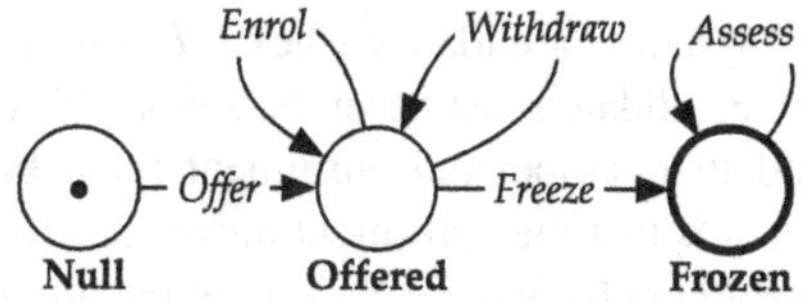

FIGURE 3.19

The event-state diagram for a *Subject*.

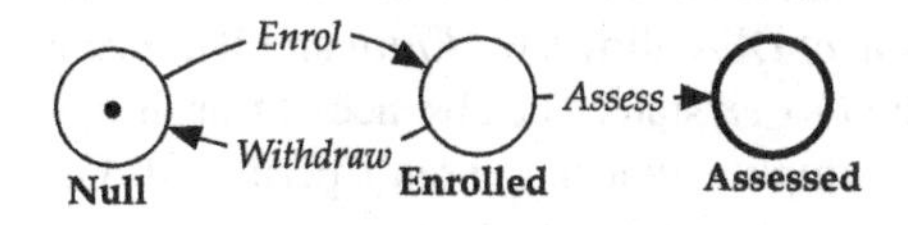

FIGURE 3.20

The event-state diagram of an *Enrolment*.

Can we adjust the *Candidate* and *Subject* DFAs to prevent silly sequences of events being possible? One approach would be to count the number of subjects in which a candidate is enrolled and to count the number of candidates enrolled in a subject. Clearly, this will prevent a candidate withdrawing before enrolling and will stop the number of candidates in a subject becoming negative. Despite this, other errors are still possible. Candidate *C* could enrol in subject *S* and candidate *D* could enrol in subject *T*, then *C* could withdraw from *T*, despite never having enrolled in it. Clearly, we must prevent this.

That we are discussing *counters* suggests that we are regarding some set of atoms as indistinguishable. However, the *number* of subjects in which a candidate is enrolled is secondary to *which* subjects they are. We would be wrong to regard subjects as indistinguishable. Similarly, to enable a candidate to withdraw from a subject, we need to know the *set* of candidates in the subject, not just the number of candidates.[24] We would be equally wrong to regard candidates as indistinguishable.[25] *What we need is a new kind of atom to constrain the order in which events can occur.*

The atom concerned is already known as an *Enrolment*. An *Enrolment* is a record of the relationship between a particular candidate and a particular subject. Figure 3.20 shows its event-state diagram.

[24] It would be possible, but unwise, to extend the set of *Candidate* states to include a state for every combination of classes the candidate enrolled in or to extend the set of *Subject* states to include a state for every possible class list. Even as few as 10 possible classes would require a candidate to have over 1,000 states, and 100 candidates would need a subject to have over 10^{30} states. Neither possibility really respects the notion of a DFA.

[25] Despite this, we shall later learn that it remains convenient to keep track of the number of candidates enrolled in a subject (*Filled*), because there is a limit on the number of places available (*Quota*).

When a candidate wishes to *Enrol* in a subject, this is now a three-way interaction: the candidate must be in the *Admitted* state, the subject must be in the *Offered* state, and their associated enrolment must be in the *Null* state. The effect of an *Enrol* event is that the enrolment moves to the *Enrolled* state. A *Withdraw* event moves the enrolment back to its *Null* state; its effect is as if the *Enrol* event had never happened. In the *Enrolled* state, an *Assess* event moves the *Enrolment* DFA to the *Assessed* state. Whether *Withdraw* or *Assess* events are allowed is controlled not by the *Enrolment* DFA itself, but by the state of the *Subject* DFA.

A convenient, but perhaps confusing, convention has been observed in labelling the DFAs. If an event is shared with other DFAs, it must be taken into account when considering *each* DFA. For example, the *Subject* DFA allows for *Enrol* and *Withdraw* events only in the *Offered* state and not in the *Frozen* state. The absence of transitions for *Enrol* and *Withdraw* in the *Frozen* state implies that they aren't permitted. On the other hand, the absence of any transition for *Admit* events doesn't mean they are forbidden, because this isn't an event that *Enrolment* shares with any other DFA.

We must remember that each DFA is really one of a family of similar DFAs. There is a *Candidate* DFA for every candidate, a *Subject* DFA for every subject, and several *Enrolment* DFAs for every candidate or every subject.

Admittedly, a real student enrolment system is much more complex than this discussion suggests. For example, a candidate may be forbidden to enrol in certain subjects until having passed examinations in certain others. Likewise, complex rules may determine when a candidate may graduate. Although such rules are not theoretically beyond the scope of DFAs, there are better ways to deal with them. Such rules are typically embedded in computer procedures, in (human) experts, or in expert systems.

If we were to try to draw the state diagram of the product of the *Candidate* FSA of Figure 3.18, the *Subject* FSA of Figure 3.19, and the *Enrolment* FSA of Figure 3.20, we would obtain a three-dimensional FSA that might have as many as $3 \times 3 \times 3 = 27$ states. Mathematically speaking, drawing such a diagram is intractable, the number of its states being exponential in the number of its component FSAs. Fortunately, interactions between more than three FSAs are rare, and because of shared events, the number of reachable states is usually only a subset of the Cartesian product. In this example, only fifteen states can be reached from the initial state (see Table 3.2). Each state of the product consists of a triple: (*Candidate*, *Subject*, *Enrolment*). The first three columns show the value of the triple before an event, the fourth column shows a possible event, and the final three columns show the value of the triple (the new state) after the event.

Table 3.2 was derived as follows: the first three columns of the first two rows show the initial state. Only *Admit* and *Offer* events are possible. The *Enrolment* DFA cannot move from its initial state because *Enrol* events are blocked. The effect of an *Admit* event is considered in the first row, and the effect of an *Offer* event is considered in the second. The new state resulting from the second row is considered in the third and fourth rows. The new state resulting from the first row is considered in the sixth and seventh rows. Each new state generated in the last three columns is transferred to

Table 3.2 The 21 possible transitions of interacting *Candidate*, *Subject*, and *Enrolment* FSAs. Only 15 of their 27 possible product states are actually reachable. (Two of the states are final, and have no outgoing transitions.)

Candidate	*Subject*	*Enrolment*	Event	*Candidate*	*Subject*	*Enrolment*
Null	*Null*	*Null*	*Admit*	*Admitted*	*Null*	*Null*
”	”	”	*Offer*	*Null*	*Offered*	*Null*
Null	*Offered*	*Null*	*Admit*	*Admitted*	*Offered*	*Null*
”	”	”	*Freeze*	*Null*	*Frozen*	*Null*
Null	*Frozen*	*Null*	*Admit*	*Admitted*	*Frozen*	*Null*
Admitted	*Null*	*Null*	*Offer*	*Admitted*	*Offered*	*Null*
”	”	”	*Graduate*	*Graduated*	*Null*	*Null*
Admitted	*Offered*	*Null*	*Enrol*	*Admitted*	*Offered*	*Enrolled*
”	”	”	*Freeze*	*Admitted*	*Frozen*	*Null*
”	”	”	*Graduate*	*Graduated*	*Offered*	*Null*
Admitted	*Offered*	*Enrolled*	*Withdraw*	*Admitted*	*Offered*	*Null*
”	”	”	*Freeze*	*Admitted*	*Frozen*	*Enrolled*
”	”	”	*Graduate*	*Graduated*	*Offered*	*Enrolled*
Admitted	*Frozen*	*Null*	*Graduate*	*Graduated*	*Frozen*	*Null*
Admitted	*Frozen*	*Enrolled*	*Assess*	*Admitted*	*Frozen*	*Assessed*
”	”	”	*Graduate*	*Graduated*	*Frozen*	*Enrolled*
Admitted	*Frozen*	*Assessed*	*Graduate*	*Graduated*	*Frozen*	*Assessed*
Graduated	*Null*	*Null*	*Offer*	*Graduated*	*Offered*	*Null*
Graduated	*Offered*	*Null*	*Freeze*	*Graduated*	*Frozen*	*Null*
Graduated	*Offered*	*Enrolled*	*Freeze*	*Graduated*	*Frozen*	*Enrolled*
Graduated	*Frozen*	*Enrolled*	*Assess*	*Graduated*	*Frozen*	*Assessed*

the first three columns to generate further transitions — except when the new state is final. The rows of Table 3.2 are ordered by the first triple, then by the event.

It would be possible to express Table 3.2 as a state-transition diagram, but most people would find a diagram with 15 states and 21 transitions too complex to be helpful. Instead, we may show the interactions between FSAs using the bi-partite graph of Figure 3.21.

In Figure 3.21, the circles represent states, and the thick bars represent events. States with edges leading to an event are the **input states** of the event, and states

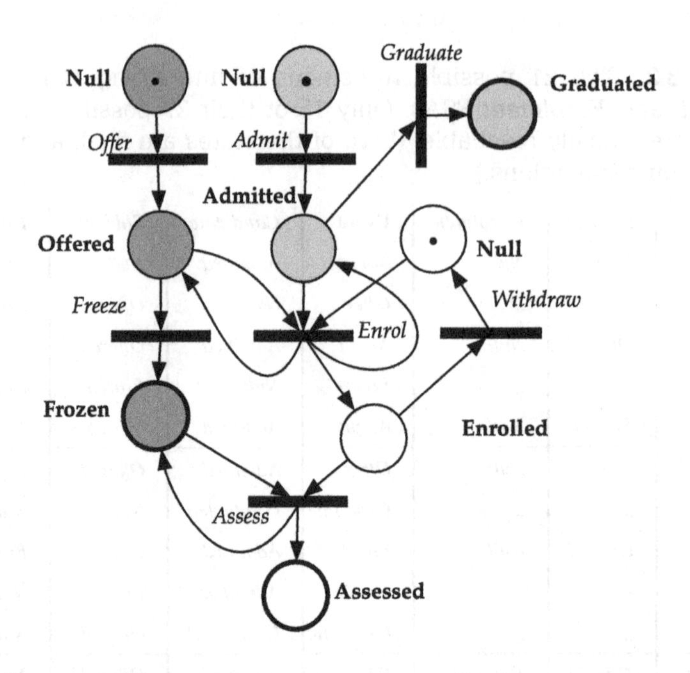

FIGURE 3.21

A bi-partite graph where circles represent states and bars represent events. The white circles represent the states of *Enrolment*, the dark grey circles represent the states of *Subject*, and the light grey circles represent the states of *Candidate*.

with edges from an event are its **output states**. Considering the dots as tokens, an event is **enabled** only when *all* its input states contain a token. If something in the environment occurs to **trigger** an *enabled* event, it **fires**, and a token flows from its input states along each in-edge and a token flows to its output states along each out-edge. It is possible for more than one event to be enabled; the sequence of real-world events determines the order in which transitions fire.

The diagram is a kind of **Petri net**. Because each transition has as many out-edges as it has in-edges, Figure 3.21 is **conservative**; tokens are neither created nor destroyed. In Petri net terminology, FSA states are called places, and events are called transitions. The **state of a Petri net** is determined by the numbers of tokens at each place.

It is reasonable to ask if anything can happen to prevent any of the original FSAs reaching their final states. This is indeed the case. The danger arises from the *Graduate* event. Figure 3.21 suggests that a candidate can graduate without ever enrolling. The reason for this is that the rules for graduating are complex and involve many assessments. As a result, they cannot be expressed as a simple interaction between FSAs. So it appears that the *Enrolment* FSA may never leave its *Null* state.

This is to be expected, because many possible enrolments never occur. The important issue is that an *Enrolment* FSA in the *Enrolled* state is always able to reach its final *Assessed* state.

Table 3.2 and Figure 3.21 show the interactions among only one subject, one candidate, and one enrolment. It is possible to visualise more atoms by imagining more tokens and colouring them according to which candidate or subject they represent. (*Enrolment* tokens could be striped with the two colours of their corresponding *Candidate* and *Subject*.)

We shall see in the next chapter that an enrolment system might need to model atoms other than those we have already discussed. For example, a *Candidate* might only be allowed to enrol in a *Subject* that was relevant to the candidate's study *Programme*. Even though both the *Programme* and the *Subject* might be available, that might not imply that the candidate could enrol, and a careful event-state analysis would reveal the need for an additional kind of atom — one that we shall call a *Component*.

Ultimately, it would be possible to construct a giant event-state matrix for an entire system. Computing the transitive closure of such a matrix would enable us to check that every desirable state was reachable. Interesting as that might be, the primary use of event-state analysis is to discover what atoms the analyst should model.

Incidentally, although the author isn't opposed to object-oriented programming, the nature of shared events suggests that it is important not to seek object-oriented solutions too soon. An *Enrol* event concerns a *Candidate*, a *Subject*, and an *Enrolment*. Of which should it be made a method? For example, should it be a method of *Candidate*, which invokes methods in *Subject* and *Enrolment*, or should it perhaps be a method of *Subject*, which invokes *Candidate* and *Enrolment*? This question is best settled by use-definition analysis, which we shall discuss in Chapter 7.

3.10 MODELLING ATOMS

From the foregoing discussion, we can see that an information system will need to store snapshots of atoms, which will have certain kinds of attributes, some of which we have already identified.

3.10.1 IDENTIFIERS

For a system to model atoms and events in the outside world, it must communicate with it using messages. Data must be collected from the system's environment, and results must be sent to it. The system needs to have a means of knowing which (real-world) atoms are associated with input data, and it needs to know where to send output data.

In a hard-wired system like an aircraft, its various sensors and controllers may be connected to the control system by a bus, and each will have a unique hardware address that identifies it. Alternatively, they may be connected to different hardware ports, which again have unique addresses.

In the case of systems that interface with humans, the sources of data are usually identified by an operator typing a code or identifier of some kind. On the other hand, the destinations of outputs are usually spelt out in full. For example, an input that records the decision by a candidate to enrol in a certain subject will normally make use of the candidate's admission number, a code of about half a dozen characters. In contrast, an output informing the candidate of his or her examination results is likely to include the candidate's full name and address. The main reason for this asymmetry is that most people can read a lot faster than they can type. It is also cheaper (at the time of writing), for example, to identify a candidate by swiping a card, than by scanning the candidate's name and address. Therefore, the arrangements for input and output are often asymmetrical. There are exceptions, of course: in electronic communications, both incoming and outgoing messages are identified by e-mail addresses.

Most systems need to be able to distinguish between different atoms of the same kind, e.g., different books in a library, or the different candidates of an institute of higher education. The snapshots of such atoms are usually stored within a database. To address the correct atom, typically, an operator will enter the short code that uniquely identifies it. In the following, we shall call an atom's *input* code its **identifier** and its *output* identifier its **description**. Each atom of a given kind must have a unique identifier. Identifiers are the means that events must use to address the atoms they update.

Apart from being unique, an identifier should also be both *short* and *stable*. Short, because identifiers usually need to be typed by people who use a system. Stable, because changing an identifier is confusing to humans and time-consuming for a computer.

Although they are unique, short, and stable, assigning arbitrary serial numbers to atoms isn't usually a good strategy;[26] it is better to choose identifiers that have some natural meaning. If no natural connection exists between the identifier and any property of the atom, it is easy to forget the identifier, use the wrong identifier, or mistakenly believe that an atom isn't already represented, and so assign it a new identifier.[27] It is important to reduce the chance that the same atom will be represented more than once. For example, numbers used to identify the goods in a sales catalogue are often derived from a description of the atom concerned.

Arbitrary serial numbers *can* sometimes be a valid choice, provided an external motive exists to avoid assigning the same atom more than one serial number. For example, an academic record system may safely assign serial numbers to candidates, because any candidate who is represented twice will eventually be asked to pay two fees. On the other hand, the subjects that are taught should have meaningful identifiers, to reduce the chance of mistakes. No candidate is likely to think that *Maths101* is a third-year psychology subject.

---

[26] Hard-wired systems are one possible exception.

[27] As a result, people have been able to defraud government agencies by registering to draw several pensions, each time acquiring a new identification number.

In interactive systems, short identifiers are less important. A candidate atom can be retrieved using a meaningful attribute, such as a person's name. Although two candidates might share the same name, the person in question can be established by displaying other attributes, such as the person's home address, telephone number, etc. Whatever collection of attributes is sufficient to uniquely identify the individual becomes the identifier.

Suppose we have chosen to identify candidates by name, address, and telephone number. Although such a choice is user-friendly, it has drawbacks: the identifier is neither stable nor short. A candidate who moves house, changes a name through marriage, or loses a mobile phone is likely to cause confusion. As we shall see later, other items in the database that refer to the candidate, such as enrolments made by the candidate, will also need to include the identifier of the candidate concerned. Since candidates typically make many enrolments, this means a massive duplication of name and address data and will make much unnecessary work when the identifier changes. In such a case, it can pay to use two identifiers: one for external messages and one used solely inside the computer system. Either identifier is capable of uniquely identifying the candidate. The internal identifier can be a serial number; since it isn't used externally, its lack of meaning isn't an issue.

3.10.2 REPRESENTING STATES

In the ordinary way, a state is best represented by a small integer or a single character, or if the programming language permits, by an enumerated variable.[28] But as we have seen, a state may be represented less directly by a variable that counts the number of objects in a certain state.

It is also possible, and sometimes useful, to represent atoms in different states by placing them in different sets. For example, we might choose to store information about *Graduated* candidates in a different data structure from *Admitted* candidates. The justification might be that *Graduated* candidates will eventually greatly outnumber *Admitted* candidates, but there are no events that can affect them. Therefore, access to *Admitted* candidates will be more efficient if they are stored separately. For technical reasons, the *Enrolments* for these two groups of candidates would also need to be stored separately. The drawback of such an arrangement is that to find the details of a candidate in an unknown state would require access to both sets, resulting in a more complex implementation.

What is often done, but should *never* be done, is to store state information implicitly. It might be argued that a candidate who has an enrolment in the current semester is *Admitted*, and a candidate who hasn't must therefore be *Graduated*. There are three objections to this practice: First, it leads to obscure computer programs. Checking for current enrolments is a very roundabout way of discovering a *Candidate's* state and is likely to mystify anyone attempting program maintenance. Second, it is prone to error. An *Admitted* candidate without enrolments might be on leave of absence; a *Graduated* candidate might have enrolled in a subject, but then

[28] In examples that follow, we use 10-character names to make the examples easier to understand.

discovered it wasn't needed in order to qualify. Third, even if the practice is valid at the time of implementation, subsequent changes to the specification may invalidate it. In short, ***don't do it!***

3.11 SUMMARY

Atoms are the most primitive objects that a system models. An atom has a behaviour: the sequences of events that can change its state. An event can only occur if *all* the atoms it inspects or affects are in suitable states. The relationship between events and states can be expressed in the form of an event-state diagram.

Atoms often, but not necessarily, become represented as data objects that have one or more identifiers and other attributes. Sometimes it is unnecessary to distinguish between similar atoms, in which case the number of atoms in a given state is represented by an integer, or in the case of stuff, by a real number. In some cases, e.g., 'coins' in an accounting system, atoms have so many possible states that each state is best represented by a different data object. Except at the boundaries of a system, atoms are conserved.

Sometimes the sequence of events can only be controlled correctly by introducing a new, so far undiscovered, type of atom. Discovering the events and states that affect each kind of atom is an interactive process called event-state analysis, which is an important fact-finding tool in systems analysis. The analyst must decide which events are important in the light of business requirements. As a rule, states are only important when they determine the permissible sequences of events.

3.12 FURTHER READING

Chapter 10 of 'Foundations of Computer Science: C Edition' by Alfred V. Aho and Jeffrey D. Ullman (1994, ISBN: 0716782847) covers the material on FSAs presented in this chapter and Appendix A. The book 'Introduction to Automata Theory, Languages, and Computation' by John E. Hopcroft, Rajeev Motwani, and Jeffrey D. Ullman (2001, ISBN 0-201-44124-1) will tell you more than you ever wanted to know about the theory of FSAs.

Petri nets, such as that in Figure 3.21, were first described by Carl Adam Petri in his 1962 doctoral thesis 'Kommunikation mit Automaten' (Communication with Automata). Petri nets have many applications, several of which are discussed in James L. Peterson's 1981 book, 'Petri Net Theory and the Modeling of Systems' (ISBN 0-13-661983-5).

3.13 EXERCISES

1. Consider Figure 3.9, which shows an FSA for books in a library. Sketch the figure and modify it to include an additional state, *On Hold*, and two additional events.

Books can be placed *On Hold* by a *Hold* event when they are in any of the states *On Loan*, *Shelved*, or *Archived*. A book in the *On Hold* state can become *On Loan* as the result of a *Lend* event.

2. Consider the DFA of Figure 3.7 on page 86, which detects if the last letter but one of a binary sequence is '1'. By analogy, how many states would you expect a DFA that detects if the last letter but *two* of a binary sequence is '1'? How many states would you expect the DFA to have in order to detect if the last letter but n of a binary sequence is '1'? What does this tell you about the tractability of converting an NFA to a DFA?
3. What regular expressions describe the FSAs of Figure 3.1 and Figure 3.2? Are they equivalent?

CHAPTER

Data-structure analysis

4

CHAPTER CONTENTS

INTRODUCTION

In Section 2.6 and Section 2.7, we discussed many of the ways the mathematical objects we have described — sets, functions, relations, and graphs — can be represented in a computer system. My aim now is to convince the reader that these abstractions are the only ones we need to think about.

Table 4.1 A table showing some enrolments

Semester	*Candidate*	*Subject*	*State*	*Mark*
20yyA	20135468	acctg101	Assessed	54
20yyA	20135468	javaprog	Assessed	54
20yyB	20135468	graphthy	Enrolled	
20yyA	21136648	qtheory	Assessed	80
20yyA	24711128	qtheory	Assessed	30

A complex data structure that seems an inherent part of a problem can at most be part of a potential solution. Sets, functions, and relations are all we need. Rather than proceed directly from the abstract level of sets and relations directly to the implementation level of files or linked lists, it proves better to proceed indirectly through an intermediate, *conceptual level.* At the conceptual level, we shall consider just one simple method of implementation: **tables**.

Table 4.1 shows a table describing some enrolments. We can imagine the table to be implemented as an array to which additional **rows** can be added as new enrolments are made. The table does not contain rows for enrolments in the *Null* state. In this respect it behaves similarly to almost every possible implementation: storage must be allocated for new enrolments, but may be recovered if enrolments are deleted.[1] Thus, data design is split into three stages: discovering what relations need to be represented, deciding how they can best be expressed as tables, then choosing efficient physical data structures such as linked lists, sparse matrices, or indexed-sequential files to implement the tables.

Notice that Table 4.1 does not contain the personal name of each candidate, nor the full name of each subject. Although such information would be handy on a report, we do not choose to include it in the *Enrolments* table; this information can be found by joining it with information from other tables. Rather than being tempted into storing a data structure that reflects a desired output, the reader is advised to design a system using tables, and elaborate it later only if it proves to be necessary. Joining data from well-designed tables enables any number of desired outputs to be produced.

A second motive for using tables is that it is how data are described in relational databases — although the tables might be physically stored as B-trees. Tables give us a convenient way of talking about data structures without getting bogged-down in the details of implementation.

In what follows, we shall use the *conceptual-level* terminology of relational databases. Nonetheless, the important content of this chapter is not about the relational model itself, but about how we analyse real-world concepts and synthesise computer data structures to represent them.

---

[1] The only exception would be if storage were pre-allocated for every possible row, each row having been initialised with the *Null* state.

4.1 CONCEPTUAL SCHEMAS

The collection of data objects a computer application needs is called its **database**. The predominant technology for storing and accessing a database is a relational **database management system**, or relational **DBMS**. It is therefore important to understand the capabilities of this technology and to know its terminology. Even if we don't plan to use a relational DBMS ourselves, we can still learn a great deal.

The **relational model** of databases assumes that anything we want to represent can be modelled as a collection of relations. The assumption is valid because relations can represent predicates, and predicate calculus can represent any finite set of facts.

A **relational database** comprises one or more tables. The number of **columns** in a table is fixed,[2] but the number of rows may vary. Table 4.1 has five columns: *Semester*, *Candidate*, *Subject*, *State*, and *Mark*.

Each table typically records the properties of a set of similar atoms. The columns of the table represent the properties of the atoms. Columns are also called attributes. Attributes must be simple unstructured data items; no arrays, linked lists, etc., are permitted.[3] Any such data structures that might seem necessary must be represented by additional tables and relationships between tables. It is possible for the values of attributes to be optional; when they are missing, as occurs in the case of *Mark*, they are **null** and have no value.

One or more of the attributes of a table (a subset of its columns) form its **primary key**. No two rows may have the same primary key value. *A primary key value uniquely identifies a single row within a given table*. In the case of Table 4.1 the primary key consists of the first three columns. Primary key attributes can never be **null**; every row must have a unique primary key. If a key consists of more than one column, it is said to be composite, otherwise it is **simple**.

In the degenerate case that the primary key comprises all the columns, there is always one implicit two-valued attribute: if the row exists, its state is not *Null*. Such a table represents a set if its key is simple, or a relation if its key is composite. The values of its implicit attribute define its characteristic function, sometimes alternatively referred to as a θ-function.

Each row represents a single fact, usually a snapshot of the atom identified by the value of its primary key. As events occur, rows are updated, inserted, or deleted.

The relational model does not specify the order in which rows are stored. In practice, it is usually in order of primary key value.

A description of a system's tables and their columns is a conceptual schema called the **database schema**. The set of facts that **populate** the tables forms the **database instance**.

A database schema is traditionally represented as a set of **table schemas** of the form

$$T(\underline{A},\underline{B},\ldots C,D,\ldots)$$

[2] Fixed, but not necessarily for all time.

[3] Some database systems allow columns to represent pictures, sounds, etc. The internal complexities of these structures are ignored. They are treated as binary strings.

where T is the name of the table, and $A,B,C,D,\ldots$ are its attributes. The underscored attributes $(\underline{A},\underline{B},\ldots)$ form its primary key, and the remaining attributes are non-key attributes. For example, Table 4.1 has the schema

$$\mathit{Enrolments}(\underline{\mathit{Semester}, \mathit{Candidate}, \mathit{Subject}}, \mathit{State}, \mathit{Mark}).$$

We may think of each table as a collection of time-dependent *functions* whose common domain is its set of primary key values and whose ranges are its non-key attributes. In database terminology, such a time-dependent function is called a **functional dependency (FD)**, its domain is called its **determinant**, and its range is called its **dependent attribute**. If a particular function is partial, its dependent attribute is allowed to be **null**.

Tables may be inter-related. A column or columns of one table may contain the primary key of a different table. Such a cross-reference is called a **foreign key**.[4] The row containing the foreign key is called a **child**, and the row containing the primary key is called its **parent**. For example, a child row in an *Enrolments* table that refers to both a parent row in a *Candidates* table and a parent row in a *Subjects* table would have *two* foreign keys. The primary key of this *Enrolments* table would be the composite (*Semester*, *Candidate*, *Subject*); its two foreign keys, *Candidate* and *Subject*, would be simultaneously part of its primary key.

4.1.1 CONSTRAINTS

Primary keys and foreign keys are examples of **constraints**. Database management systems can enforce other types of constraint:

Column constraints limit the values that columns may contain. They include **type constraints** (e.g., **integer**) and **range constraints** that control their sets of possible values. One useful column constraint is **not null**, which requires every row to contain a value in that column.

Row constraints enforce arbitrary conditions between the columns of a given row, for example, whether a certain attribute is allowed to be **null** might be determined by the value of another attribute.

Unique constraints forbid two rows to share the same value of a given attribute or set of attributes. A primary key is therefore a special kind of **unique** constraint. Conversely, any set of attributes that is unique is a potential primary key, or **candidate key**.

In a modern DBMS,[5] we expect to see the following support for primary and foreign key constraints:

---

[4] Thus, all links between tables are regarded as symbolic, rather than physical or relative addresses. Conceptually, this allows a table to be moved in storage, or copied, without any need to treat the links in any special way, greatly simplifying our later discussion.

[5] We are particularly thinking of DBMSs that support *SQL* (Structured Query Language).

- A row cannot be inserted into a table if it has the same primary key as an existing row in the same table.
- A child row cannot be added to a table unless its parent row already exists.
- A parent row cannot be deleted without one of three conditions being satisfied: it already has no existing child rows, all its existing child rows are deleted along with the parent, or the foreign keys in all its child rows are set to **null**.[6]
- In addition, updates that alter primary key values are treated with similar caution.

The result of these rules is that a foreign key in a child row must either refer to exactly one parent row or be **null**. It cannot refer to a non-existent parent. This property is called **referential integrity**. The advantage of a DBMS enforcing referential integrity is that programming errors that might put the database into a meaningless or inconsistent state will be detected at source.

Referential integrity means that a foreign key models either a total or a partial function from the child domain to the parent codomain.

In addition to constraints that the DBMS can enforce, there may be other **general constraints** (or **database invariants**) that can only be enforced by application programs. For example, a candidate's *Admission No* might need to be divisible by 13.

It is important to distinguish between the conceptual view of a relational database as a set of tables from the physical reality; the DBMS is free to use whatever efficient implementation it chooses.[7] The technology of the implementation is concealed from the user of the database. Typically, when a row is located using its primary key, a table is not searched from top to bottom; a file is searched using an index, perhaps a B-tree. A foreign key might actually contain the relative storage address of its parent row. In turn, the parent row might keep a linked list of all its children's addresses. (Among other things, this makes it possible to enforce foreign key constraints efficiently.) The analyst can temporarily ignore these issues; they are hidden within the DBMS. Whatever the details of the implementation, a database user never needs to worry about the physical location of records, only the values of keys. This means that a database may be copied or reorganised without affecting programs that use it. Any physical addresses that may need to be updated are solely the responsibility of the DBMS.

Because the relational model is independent of its implementation, and because it is so closely related to predicate calculus and the theory of relations, the reader is advised, whatever the final implementation of a database in a particular situation, to first analyse a problem using tables. In many cases, an existing DBMS can be employed. When that is impossible and special software has to be developed, it often

[6] If the database is implemented using home-grown software, the parent row might need to store a reference counter recording the current number of its children, or perhaps a linked-list of its children, as in the graph data structure of Figure 2.34.

[7] Usually a privileged user, called the **database administrator** (**DBA**), can guide the DBMS into choosing a good implementation. This might include creating secondary indices or deciding between different file structures, etc.

remains best to design an interface that presents the database as a set of tables. Indeed, a prototype can be made using a simple implementation of the tables and elaborated later only if it proves necessary. Consequently, in what follows, we shall use the terminology of relational databases.

4.1.2 TABLES, ATOMS, AND STATES

Atoms in the *Null* state are rarely stored in a database. Therefore, an event that moves an atom *from* the *Null* state is modelled by a new row being inserted into a table. Likewise, events that move the atom *to* the *Null* state cause its row to be deleted. Consequently, an atom in the *Null* state has no matching table row. When we say that an FD is total, we mean that every *row* in its domain has a dependent attribute value in its range. If we say that an FD is onto, we mean that every value in its range appears in at least one row.

We may therefore define the domains (determinants) of our FDs as follows,

$$Candidates = \{c \in Admission\ Nos \mid Candidate\ State(c) \neq Null\}$$
$$Enrolments = \{e \in Semesters \times Admission\ Nos \times Subject\ Codes \mid Enrolment\ State(e) \neq Null\}$$
$$Subjects = \{s \in Subject\ Codes \mid Subject\ State(s) \neq Null\}$$

In making this distinction between *Candidates* and *Admission Nos*, etc., we have anticipated some decisions that we shall justify later, namely, that *Admission Nos* will identify *Candidates*, *Subject Codes* will identify *Subjects*, etc. Strictly, we should have said that *Candidates* is a subset of 'whatever we later choose to identify candidates'.[8]

For example, a total FD such as

$$Personal\ Name : Candidates \xrightarrow{+\ \ 1} Personal\ Names$$

associates a *Personal Name* with every row of *Candidates*, and there are no other values of *Personal Name* stored in the database. In contrast, a partial FD such as

$$Mark : Enrolments \xrightarrow{+\ \ ?} 0 \ldots 100$$

means that even *Enrolments* in a *non-null* state don't always have a *Mark* — in fact, only those in the *Assessed* state do. Similarly, an into FD such as

$$Subject : Enrolments \xrightarrow{*\ \ 1} Subjects$$

means that every *non-null Enrolment* has a *Subject*, but some *Subjects* may have no *Enrolments*.

---

[8] If we had done so, the outcome would be the same, but the diagrams in this chapter would have been messier.

In the ordinary way, a foreign key attribute is almost certain to be associated with an FD that is *into*, because the parent row has to exist before the child row can reference it. Other dependent attributes are typically *onto*, because the only codomain values that exist are those expressed within the table itself. The majority of FDs prove to be *total* and *onto*. To highlight the exceptions, we shall omit the decorations '+' and '1'.[9] For example,

$$\mathit{Personal\ Name} : \mathit{Candidates} \rightarrow \mathit{Personal\ Names}$$

means the same as

$$\mathit{Personal\ Name} : \mathit{Candidates} \xrightarrow{+\ \ 1} \mathit{Personal\ Names}$$

and so on.

4.2 DERIVING FUNCTIONAL DEPENDENCIES

The example that follows is of an unrealistically simple Academic Records database for the (imaginary) National University of Technology and Science (NUTS). This example was chosen because most readers will have experienced a similar system. The system we describe is simple enough to illustrate basic principles, yet complicated enough to pose some problems — without overloading you with needless detail. The solution we present will be less than perfect. One can learn a lot from other people's mistakes.

As we have seen, a conceptual schema comprises a set of tables related by foreign keys. Foreign keys implement many-to-one functions. In Figure 2.23, we sketched a schema that described an Academic Records System. The diagram is reproduced below as Figure 4.1. Let us develop this sketch and discover the set of FDs needed to define a database. Part of the process is formal, and part depends upon the advice of experts.

The process begins with armchair analysis, in which the analyst uses experience and common sense to sketch the schema shown in Figure 4.1.

It seems obvious that a *Degree* may have many *Candidates* and will comprise at least one *Subject*. Further, each *Subject* must be part of at least one *Degree*. It is also reasonable that a *Candidate* would need to be *Enrolled In* several *Subjects* to obtain a *Degree*, but might perhaps be absent and not enrolled in any. It is also possible that a *Subject* might attract no *Candidates* or be cancelled. The analyst's next task is to check this perception against reality, by talking to a domain expert, a worker in the existing Enrolment Office.[10] The process is a two-way street: the analyst's schema generates the questions, the answers generate the schema.

[9] In practice, to reduce the risk of error, the reader may prefer to retain them.

[10] In real life, it might be necessary to speak to more than one worker. In any department of two or more people, nobody knows everything.

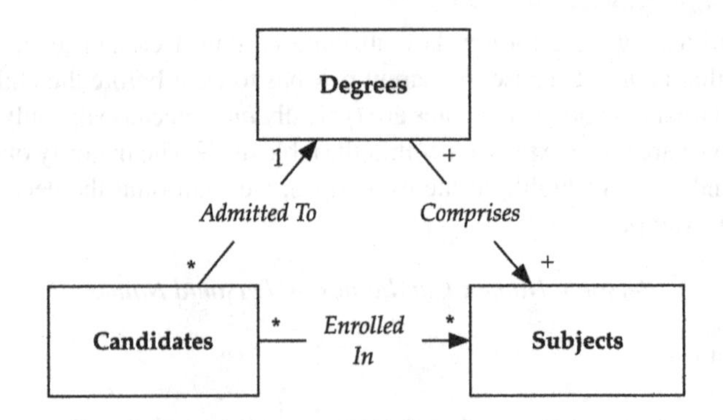

FIGURE 4.1

A schema describing the Academic Records database. *Candidates* are *Admitted To* exactly one *Degree*, and may be *Enrolled In* any number of *Subjects*. Each *Degree Comprises* at least one *Subject*, and every *Subject* must be part of at least one *Degree*.

One point the analyst wants to check is the nature of the *Comprises* relation. Perhaps *Degrees* are so highly structured that *Subjects* are confined to only one *Degree*. Perhaps they are so unstructured that it would be better to store the **complement** of *Comprises*, a relation that lists forbidden choices. We can imagine the following conversation:

> 'I want to ask you about the way subjects relate to degrees. Do subjects belong to just one degree, or to several?'
>
> 'Most belong to only one, although there are a few subjects, like *Maths 101*, that are used in several'.
>
> 'How many subjects would a typical degree offer?'
>
> 'About a hundred'.
>
> 'So candidates have quite a lot of choice?'
>
> 'Oh no! Because of the technical nature of our programmes, and the requirements of accreditation with professional bodies, a candidate has very few choices. Usually just a couple of options in the final year'.
>
> 'Surely they don't have to study a hundred subjects!'
>
> 'No, only about twenty. Look, here is the e-Commerce programme'. (See Figure 4.2.)[11]

[11] The NUTS logo is a pastiche of images from openclipart.org. The Latin motto translates as, "Knowledge Is Power."

National University of Technology and Science

Graduate Diploma in Information Technology

Electronic Commerce Programme (ecom)

Level 1 Subjects:

acctg101	Basic Accountancy	$500.00
fnclmgta	Financial Management A	$500.00
graphthy	Graph Theory	$500.00
javaprog	Java Programming	$700.00
maths101	Foundation Mathematics	$500.00
security	Internet Security	$600.00

Level 2 Subjects:

dataanal	Data Analysis	$550.00
eeconomy	The Electronic Economy	$600.00
hmnbhvra	Human Behaviour A	$500.00
marketec	Marketing for e-Commerce	$650.00
netmgt1	Network Management I	$700.00
projmgta	Project Management A	$650.00
qtheory	Queueing Theory & Simulation	$550.00
sqlprgrm	SQL Programming	$600.00

Level 3 Subjects:

dbadmin1	Database Administration I	$850.00
fnclmgtb	Financial Management B	$550.00
htmlprog	HTML & PHP Programming	$600.00
intrface	Interface Design	$750.00
mrktrsch	Market Research Methods	$800.00
projmgtb	Project Management B	$650.00
services	On-line Services & Cloud Comp.	$700.00
socmedia	Exploiting Social Media	$650.00

To qualify for the Graduate Diploma in Information Technology, a candidate must pass 18 subjects, including 6 subjects at Level 3.

FIGURE 4.2

The official programme for the Graduate Diploma in Information Technology (Electronic Commerce).

At this point, the analyst does some thinking:

> 'When you say "programme", is that the same as a "degree"?'
>
> 'No, there are lots of programmes that lead to the same degree. For example, a candidate for the *Diploma in Information Technology* can study the *e-Commerce* programme, or the *Network Management* programme, and so on'.
>
> 'Right! So what you are telling me is that a candidate can take the *e-Commerce* programme and get a *Diploma in Information Technology*, and in that *programme*, there are few choices. Does a given programme always lead to the same degree?'
>
> 'That's correct'.

The analyst now wonders if *Subjects* can be forced into a many-to-one relation with *Programmes*, which would simplify the database.

> 'I know that subjects like *Maths 101* are shared between several programmes, but would it be possible to regard it as several subjects, like *Maths for e-Commerce*, *Maths for Network Management*, and so on?'
>
> 'Well, they all share the same lectures, so they *are* the same really'.
>
> 'Yes, but can't people enrol in *Maths for e-Commerce*, then attend the same lecture as those enrolled in *Maths for Network Management?*'
>
> 'It would make things awkward for us. There's a limit to how many people we can fit into a lecture room, so there's a pretty tight quota in *Maths 101*. How would we allocate seats to *e-Commerce*, *Network Management*, and so on?'
>
> 'You're right, I hadn't thought of quotas. Thank you'.

This is typical of interactions with a domain expert.[12] The analyst learns that the *Comprises* relation really is many-to-many, that *Degrees* are properties of *Programmes*, and, also, that quotas exist. Time for the analyst to think again, resulting in the decision not only to store the *Quota* for each subject in the database, but also to store the number of places that have been *Filled*. This will allow enrolment events to check if places remain.[13] Strictly speaking, *Filled* is redundant, because its value can be discovered by counting the number of existing enrolments, but the analyst foresees that this might prove both awkward and time-consuming.[14]

---

[12] No real interview ever goes this smoothly. With some interviewees, the analyst needs to assert authority by mimicking the clothing and short haircut of a TV newsreader. With others, it pays to be more informal. Deciding to either visit, or summon, the interviewee also affects the dynamic. The analyst should study the dress codes of the interviewees and their managers.

[13] This is an example of a database invariant that a DBMS can't be expected to check.

[14] At this point, an experienced analyst would *also* foresee that the two paths from *Candidates* to *Subjects* in Figure 4.1 will need to agree, which may cause some problems. In practice, such parallel paths are rare, and most real-world schemas prove to be trees.

4.2.1 MANY-TO-ONE RELATIONS

Ultimately, whatever the analyst discovers, the various relations in the schema must be converted into functions, because a database can only express other kinds of relation indirectly. The necessary synthesis depends on the type of each relation.

Four types of many-to-one relations are possible. Because they are functions, they are already FDs:

$$R_1 : X \xrightarrow{+\ \ 1} Y \ (\textit{total} \text{ and } \textit{onto}),$$

$$R_2 : X \xrightarrow{*\ \ 1} Y \ (\textit{total} \text{ and } \textit{into}),$$

$$R_3 : X \xrightarrow{+\ \ ?} Y \ (\textit{partial} \text{ and } \textit{onto}), \text{ and}$$

$$R_4 : X \xrightarrow{*\ \ ?} Y \ (\textit{partial} \text{ and } \textit{into}).$$

All four types are easily modelled by a foreign key in the child table (containing X) referring to the primary key Y of its parent table. In the case of the total relations, the foreign key should have the **not null** constraint. For the partial relations, the foreign key should be allowed to be **null**. Consequently, if the child's parent row is deleted, it would usually make sense to choose the option that sets the foreign key of the child to **null**.

Therefore, for example, the many-to-one relation from *Candidates* to *Programmes* poses no problems and can be implemented by a foreign key in the *Candidates* table referencing the primary key of the *Programmes* table and being assigned the **not null** constraint.

4.2.2 ONE-TO-MANY RELATIONS

A one-to-many relation has a converse that is a function (i.e., an FD). Therefore, we may replace a one-to-many relation by its converse many-to-one relation, then model it as above:

$$R_1 : X \xrightarrow{1\ \ +} Y \text{ is replaced by } R_1^{-1} : Y \xrightarrow{+\ \ 1} X,$$

$$R_2 : X \xrightarrow{1\ \ *} Y \text{ is replaced by } R_2^{-1} : Y \xrightarrow{*\ \ 1} X,$$

$$R_3 : X \xrightarrow{?\ \ +} Y \text{ is replaced by } R_3^{-1} : Y \xrightarrow{+\ \ ?} X, \text{ and}$$

$$R_4 : X \xrightarrow{?\ \ *} Y \text{ is replaced by } R_4^{-1} : Y \xrightarrow{*\ \ ?} X.$$

In terms of the schema, these rules merely amount to changing the direction of the edge and replacing the name of the relation by a suitable name for its converse.

4.2.3 ONE-TO-ONE RELATIONS

One-to-one relations, or correspondences, are functions that have an inverse. To express this, we can make both the original relation and its converse into FDs:

$$R_1 : X \xrightarrow{1\ \ 1} Y \text{ is replaced by } R_1 : X \xrightarrow{+\ \ 1} Y \text{ and } R_1^{-1} : Y \xrightarrow{+\ \ 1} X,$$

$$R_2 : X \xrightarrow{1\ \ ?} Y \text{ is replaced by } R_2 : X \xrightarrow{+\ \ ?} Y \text{ and } R_2^{-1} : Y \xrightarrow{*\ \ 1} X,$$

$$R_3 : X \xrightarrow{?\ \ 1} Y \text{ is replaced by } R_3 : X \xrightarrow{*\ \ 1} Y \text{ and } R_3^{-1} : Y \xrightarrow{+\ \ ?} X, \text{ and}$$

$$R_4 : X \xrightarrow{?\ \ ?} Y \text{ is replaced by } R_4 : X \xrightarrow{*\ \ ?} Y \text{ and } R_4^{-1} : Y \xrightarrow{*\ \ ?} X.$$

We have to treat both FDs as many-to-one relations, because that is all that foreign keys support. Fortunately, their converses then constrain both FDs to be one-to-one. Conveniently, a one-to-one relation can often be implemented simply by making X a primary key and assigning Y the **unique** attribute. For example, we discover from our domain expert that *Subject Codes* and *Subject Names* are in one-to-one correspondence. In a *Subjects* table, *Subject Code* can be its primary key, and *Subject Name* can be declared as **unique**.

4.2.4 MANY-TO-MANY RELATIONS

The many-to-many relation in Figure 4.1 (page 118) from *Candidates* to *Subjects* cannot be expressed either as an FD from *Candidates* to *Subjects* (which would imply a *Candidate* could study only one *Subject*) or vice versa (which would imply that no class could contain more than one student).

Many-to-many relations are more problematic, as there is no direct support for them in the relational model. In particular, the model forbids making lists of edges; columns cannot contain arrays or lists. Consequently, the only way to express a many-to-many relation is as a list of pairs stored in a separate table. A multi-valued relation $R : X \to Y$ would therefore be expressed as a set of $(x \mapsto y)$ pairs.[15] In other words, R becomes a new table with primary key (X,Y), typically having a foreign key X referring to table X, and a foreign key Y referring to table Y.[16]

In the schema, we replace the edge from X to Y with a new set labelled R, and two edges from R: an edge labelled π_X from R to X, and an edge labelled π_Y from R to Y, where π_X and π_Y are projection functions, or, in database terminology, **trivial FDs**:

$$R_1 : X \xrightarrow{+\ \ +} Y \text{ is replaced by } \pi_X : R_1 \xrightarrow{+\ \ 1} X \text{ and } \pi_Y : R_1 \xrightarrow{+\ \ 1} Y,$$

$$R_2 : X \xrightarrow{+\ \ *} Y \text{ is replaced by } \pi_X : R_2 \xrightarrow{*\ \ 1} X \text{ and } \pi_Y : R_2 \xrightarrow{+\ \ 1} Y,$$

---

[15] As, indeed, would every other representation of a relation.

[16] There is no technical problem with part of a primary key also being a foreign key.

$R_3 : X \xrightarrow{* \quad +} Y$ is replaced by $\pi_X : R_3 \xrightarrow{+ \quad 1} X$ and $\pi_Y : R_3 \xrightarrow{* \quad 1} Y$, and

$R_4 : X \xrightarrow{* \quad *} Y$ is replaced by $\pi_X : R_4 \xrightarrow{* \quad 1} X$ and $\pi_Y : R_4 \xrightarrow{* \quad 1} Y$.

The resulting projection functions are always total; the only rows of R are the (x, y) pairs. In the case of R_2, notice that the decoration $(^+)$ associated with X becomes associated with π_Y, not π_X. (Every y value is associated with at least one x value, therefore π_Y must be *onto*.) A similar situation holds for R_3.

In this example, we must express the many-to-many relation from *Candidates* to *Subjects* by creating an *Enrolments* table.

4.2.5 EXPRESSING FDS IN THE RELATIONAL MODEL

After converting relations into FDs, they are usually expressed within the relational model as follows:

$F_1 : X \xrightarrow{+ \quad ?} Y$: X is the primary key of a table having Y as an ordinary attribute.

$F_2 : X \xrightarrow{+ \quad 1} Y$: X is the key of a table having Y as an attribute with the **not null** constraint.

$F_3 : X \xrightarrow{* \quad ?} Y$: X is the key of a table having Y as a foreign key.

$F_4 : X \xrightarrow{* \quad 1} Y$: X is the key of a table having Y as a foreign key with the **not null** constraint.

F_1 and F_2 are guaranteed to be *onto* functions, because a value in Y can only exist if it is in a row with a primary key in X.

The reason why F_3 and F_4 require Y to be a foreign key is that they are *into* functions, so there must be values in Y that are absent from the table having X as its primary key. Therefore, the set Y must exist as the primary key of a separate table.

In the case of F_1 and F_2, it may also be necessary for Y to be a foreign key. This would arise only when Y was the determinant (domain) of at least one other FD. In such a case, a relational DBMS cannot enforce the *onto* property.

The rule that a parent row must exist before any of its children means that every FD is an *into* relation, at least transiently. In order to maintain an *onto* relation correctly, care must be taken when deleting a child row to ensure that the parent retains at least one child. Typical DBMSs do not enforce such constraints. Instead, application programs can, for example, maintain a **reference count** in each parent row, recording the number of its children. If this falls to zero, the parent should be deleted.

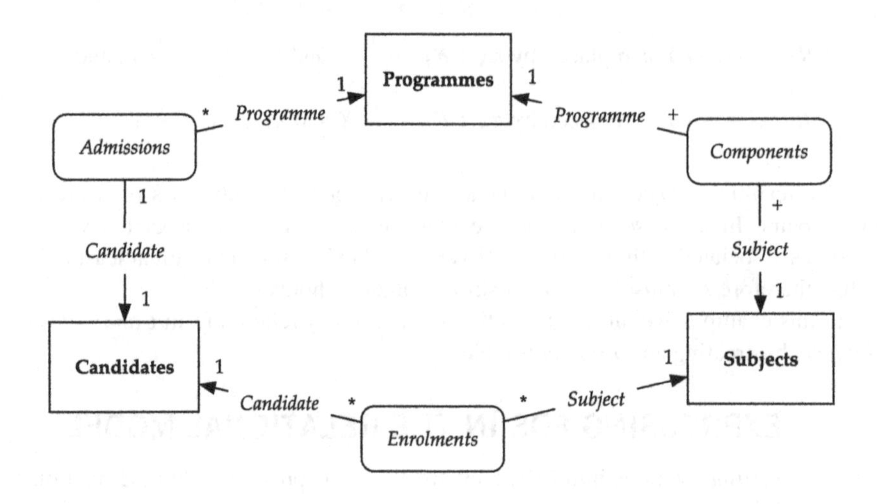

FIGURE 4.3

An E-R diagram describing the Academic Records database. The *Admitted To* relation is reified as *Admissions*, *Enrolled In* as *Enrolments*, and *Comprises* as *Components*.

4.3 ENTITY-RELATIONSHIP ANALYSIS

The process of making an object out of a relation, discussed in Section 4.2.4 to deal with many-to-many relations, is called **reification**. Reification is the basis of a design methodology called **entity-relationship analysis. E-R analysis** uses a bipartite graph called an **E-R diagram** (or entity-relationship diagram).

In the E-R diagram of Figure 4.3, the rectangles represent **entities**, and the rounded rectangles represent **relationships** between entities.[17]

An E-R diagram is essentially the same as what we, in Section 2.5.6, called an abstract schema, except that the edges of the schema are now redrawn as vertices. This leads to an important shift in our point of view: by regarding the relationships as entities in their own right, the analyst may then question whether a relationship has properties additional to its trivial projection functions. In the case of *Enrolments* this indeed proves to be the case. We saw in Section 3.9 (page 100) that *Enrolments* have states. Additionally, when a *Subject* is *Assessed*, the enrolment acquires a *Mark*.

In effect, E-R analysis can achieve an outcome similar to that of event-state analysis (Section 3.9), where we discovered that *Enrolment* atoms were needed to control the sequence of *Enrol* and *Withdraw* events. Clearly, the Academic Records System needs an *Enrolments* table.

On consulting the domain expert from the Enrolment Office, we learn that the primary key of *Enrolments* cannot simply be (*Candidate*, *Subject*) because candidates

[17] Often, ellipses are also drawn to represent *properties* of entities, but they are ignored in this discussion.

are allowed to reattempt subjects that they fail. Naturally, a candidate can only be enrolled in a subject once at one time, so the addition of *Semester* makes the primary key unique.

E-R analysis doesn't restrict itself to reifying many-to-many relations. We also see that the reification of the *Admitted To* relation as *Admissions* raises the question of whether a *Candidate* can be associated with more than one *Admission*, a question we now address.

How did the analyst come to the conclusion that a candidate could be admitted to only one programme? Are there no exceptions?

Our domain expert admits that there *are* cases where a candidate is admitted to a second degree programme, usually having already completed another. So we ask:

> 'Would it be better to speak of *Admissions*, identified by both *Candidate* and *Programme*?'
>
> 'Yes, but that would be cumbersome. What we do is to give the same candidate a new admission number'.

In other words, the system pretends that the same person is two different candidates so that the *Admitted To* relation becomes many-to-one. This means that it is only necessary to ask for one code to identify a candidate's admission to a particular programme. From the point of view of the database designer, this means that it is unnecessary to reify the *Admitted To* relation, thus saving the need for an additional table. But, strictly speaking, what we call *Candidates* are really *Admissions*.

This decision has its drawbacks. For example if a candidate moves to a different postal address, only the current row for the candidate is likely to be updated. The rows for any previous admissions will be in error. Luckily, our expert thinks this a trivial problem, because postal addresses are only needed to communicate with current candidates. The office makes no effort to track the postal addresses of graduates. Decisions like this clearly demand the advice of a domain expert and cannot be decided from the analyst's armchair.[18]

4.3.1 RELATIONSHIPS AS ENTITIES

Reification of relationships makes them into entities in their own right.[19] *In effect the bi-partite graph becomes a homogeneous graph whose edges are functions.*

As a result of treating relationships as entities, we may then consider relationships *between* relationships. Consider the E-R diagram shown in Figure 4.4.

The only change from Figure 4.3 is that Figure 4.4 makes *Enrolments* a relationship between *Admissions* and *Components* instead of between *Candidates* and *Subjects* (an option that only becomes obvious *after* we have reified the *Comprises*

---

[18] A possible compromise is to assign a candidate a permanent *Candidate ID* that is stored in each *Admissions* row associated with the candidate. This could then be used to discover all the admissions associated with the candidate.

[19] Don't get philosophical about the difference between entities and relations! What matters is the resulting set of tables.

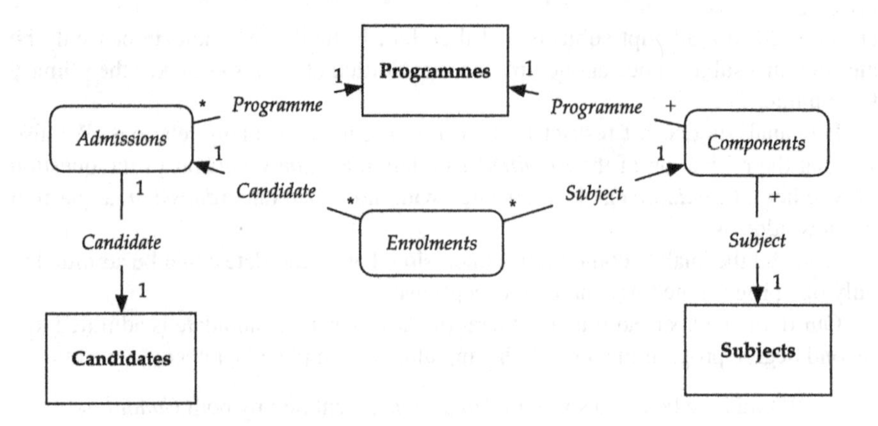

FIGURE 4.4

A modified E-R diagram describing the Academic Records database. *Enrolments* is now considered as a relation between *Admissions* and *Components*.

and *Admitted To* relations). This alternative E-R diagram would require *Enrolments* to contain the primary keys of *Components*, (*Programme*, *Subject*), and of *Admissions*, (*Programme*, *Candidate*), as foreign keys.

This has both advantages and disadvantages. An implication of giving *Admissions* a composite primary key is that the *Enrolments* table could then include a column for *Programme*, common to both its foreign keys. As a result, the DBMS could automatically check that a candidate enrols only in subjects valid for the candidate's programme. This would be a useful integrity check.

On the other hand, given that the expert we consulted considers *Candidates* and *Admissions* to correspond one-to-one, it makes little sense to store both tables. The question is whether the remaining table should have *Programme* as part of its primary key. The resulting problem is that, given a candidate's admission number, it would become necessary to also know the candidate's programme to locate the candidate's row in the *Admissions* table, an option that our domain expert already dismissed as being cumbersome.[20]

The choice between *Enrolments* referring to *Components* or *Subjects* also has an impact. Because the Academic Records System is intended to preserve the entire history of *Enrolments*, the principle of referential integrity requires the rows of whichever table is chosen as the foreign key of *Enrolments* to be preserved in perpetuity. Inevitably, this means storing information about subjects that are no longer taught. In consequence, if *Enrolments* were to reference *Components*, it would also be

[20] There is a sneaky compromise: we can make (*Programme*, *Candidate*) the primary key of *Candidates* and also give *Candidate* the **unique** attribute. This will create an index to the table giving rapid access to a candidate's record given only the *Admission No*. Despite this being a good idea, for simplicity, we are going to ignore this possibility.

necessary to store (*Programme*, *Subject*) combinations that are no longer current. On the other hand, if *Enrolments* references *Subjects*, obsolete rows of the *Components* table won't need to be stored.

We have a strong tutorial reason for choosing the second alternative. The resulting *Components* table has only two states, one of which is *Null* and the other of which is therefore implied. The *Components* table then becomes an interesting degenerate case in which the table contains a primary key, but no other attributes. The decision to remove obsolete components also implies that Figure 4.4 is no longer entirely correct. It shows the *Programme* and *Subject* FDs as onto. Although a *Subject* must be a *Component* of at least one *Programme* and a *Programme* must include at least one *Subject*, the system needs to preserve the *Programme* and *Subject* information for graduates long after their *Programmes* become obsolete. Unfortunately, the *Components* that relate them will no longer be stored. There are two alternatives: we may revise the *Programme* and *Subject* FDs to be into rather than onto, or we allow *Components* to enter an *Obsolete* state. Here, for the tutorial reason just explained, we choose the former.

In Figure 4.4, there are *two* directed paths from *Enrolments* to *Programmes*. Since each child table must contain its parent's primary key as a foreign key, *Enrolments* must inherit the same *Programme* as part of *two* foreign keys.[21] But in Figure 4.3, the paths are undirected, and the DBMS is unable to enforce the constraint. If it were intended for users to update the database directly by issuing *SQL* commands, we would have a good case for choosing Figure 4.4 over Figure 4.3. Nonetheless, in the context of this system, the *Enrol* event can ensure that subjects for the correct *Programme* are chosen, so having the constraint enforced by the DBMS is of little value. In what follows, despite its deficiencies, we shall consider the earlier and simpler E-R diagram of Figure 4.3.[22]

From the above discussion, the reader should realise that database design is always a tricky compromise between the complexity of the real world and simplicity of implementation.

4.4 SYNTHESISING A DATABASE SCHEMA BY COMPOSITION

Given a set of FDs, we may synthesise the schema of a relational database algorithmically. We can do this in either of two ways: by composition or by decomposition. We first discuss two compositional methods: a graphical method and an algebraic method.

[21] Not every DBMS can support this.

[22] As we said, this is for tutorial reasons: it simplifies the database, but also causes some interesting difficulties. Any E-R diagram containing parallel paths can cause a similar need for compromise.

4.4.1 FUNCTIONAL DEPENDENCY GRAPHS

A **functional dependency graph** (**FD graph**) is a special kind of abstract schema. The vertices of an FD graph are domains, either simple or composite, and its edges are always functions between domains, rather than more general relations. The FD graph is closer to a database schema than an E-R diagram, because a relational database can only model functional dependencies.

Because a relational database instance excludes rows for atoms in a *Null* state, we shall define the domains as follows:

$$Candidates = \{c \in Admission\ Nos \mid Candidate\ State(c) \neq Null\}$$
$$Enrolments = \{e \in Semesters \times Admission\ Nos \times Subject\ Codes \mid Enrolment\ State(e) \neq Null\}$$
$$Subjects = \{s \in Subject\ Codes \mid Subject\ State(s) \neq Null\}$$
$$Programmes = \{p \in Programme\ Codes \mid Programme\ State(p) \neq Null\}$$
$$Components = \{c \in Subject\ Codes \times Programme\ Codes \mid Component\ State(c) \neq Null\}$$

Figure 4.5 shows an FD graph for the NUTS Academic Records System derived from the E-R diagram of Figure 4.3 by converting all relations into FDs, and adding attributes we ignored in the E-R diagram. The projection functions from *Enrolments* and *Components* have been given meaningful names: *Subject*, *Semester*, *Candidate*, and *Programme*. Because *Subjects* is in correspondence with *Subject Names*, and *Programmes* is on correspondence with *Programme Names*, it is not clear to which *Components* should refer. We resolve this matter shortly.

Because almost all the edges of an FD graph are total, many-to-one, and *onto*, only the exceptions are highlighted. *Phone* and *Mark* are both *partial*, the projections from *Components* are *into*, *Subject Names* corresponds with *Subjects*, and *Programme Names* corresponds with *Programmes*.

A full alphabetical list of attributes follows.

Balance Owing: The total of tuition fees the candidate has incurred but not yet paid for.

Candidate: The *Admission No* of an actual candidate.

Candidate State: Either *Admitted* or *Graduated*.

Component State: If the row exists, implicitly *Offered*.

Date of Birth: A candidate's birth date, sometimes needed to disambiguate personal names.

Degree Name: The full name of the degree to which a study programme leads.

Enrolment State: Either *Enrolled* or *Assessed*.

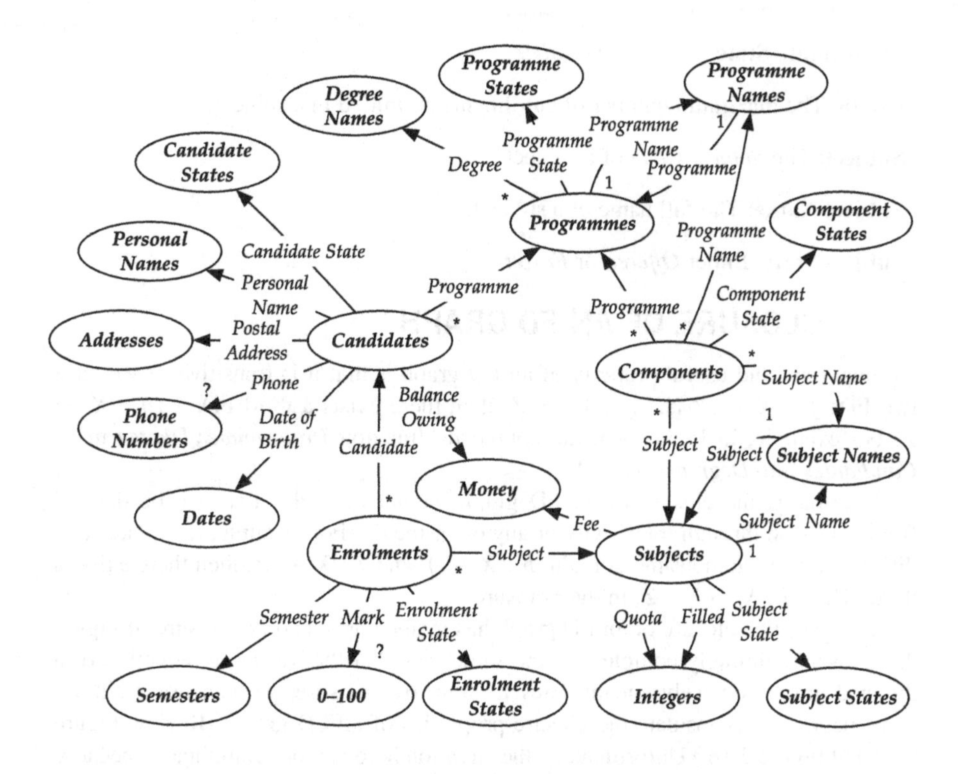

FIGURE 4.5

An FD graph describing an Academic Records database for the *National University of Technology and Science*. The vertices are domains, and the edges are functions. To highlight exceptions, undecorated arrows represent FDs that are *total* and *onto*.

Fee: The tuition fee for a subject.

Filled: The current number of enrolments in a subject.

Mark: Either a value in the range 0–100 or **null**.

Personal Name: The name of a candidate, as it should appear in correspondence.

Phone: The candidate's telephone number, needed for informal communications.

Postal Address: The address of a candidate, as it should appear on a postal envelope.

Programme: The *Programme Code* of a study programme.

Programme Name: The full name of a study programme.

Programme State: *Active* or *Obsolete*.

Quota: The maximum number of enrolments permitted in a subject.

Subject: The *Subject Code* of a subject.

Subject Name: The full name of a subject.

Subject State: Either *Offered* or *Frozen*.

4.4.2 CLOSURE OF AN FD GRAPH

An important and useful property of an FD graph is that it is transitive. If we have two FDs $f : X \to Y$ and $g : Y \to Z$, then there exists a third FD, $f\,;g : X \to Z$. For example, in Figure 4.5, the composite function $Programme\,;Degree$ maps *Candidates* into *Degree Names*.[23]

Essentially, the closure of an FD graph includes all FDs that can be derived from the graph through transitivity or any other means. For example, if we have two different FDs with the same domain, $h : X \to Y$ and $j : X \to Z$, then there exists a third FD, $h||j : X \to Y \times Z$, in their closure.

The transitive closure of an FD graph has many edges, and the closure of Figure 4.5 is overwhelmingly complex. In the same way that the transitive reduction of a directed graph concisely encapsulates its transitive closure, we can use a similar construction to encapsulate the closure properties of an FD graph. (Review Figure 2.14 and Figure 2.18.) Unfortunately, the situation here is more complicated because the edges of an FD graph are labelled. This means that if there are parallel edges from X to Y, say $f : X \to Y$ and $g : X \to Y$, where $f \neq g$, they are different edges, and both need to be retained in the reduction (as in the case of *Quota* and *Filled*).

We can highlight the parallel and composite edges of a labelled graph by finding a minimal transitive reduction ignoring the labels, then subtracting the reduction from the original graph. The resulting set of edges are those that parallel the edges of the reduction. Some may prove to be redundant, as in the case of the parallel edges from *Components* to *Subject Names*. The FD $Subject\ Name : Components \to Subject\ Names$ is the composition of $Subject : Components \to Subjects$ and $Subject\ Name : Subjects \to Subject\ Names$.

Before we discuss how to remove redundant edges from an FD graph, we formally define the **closure of a set of FDs**:

> The **closure F^+ of a set F of FDs** is the set of all FDs that can be derived from F by the following rules of inference (where $f, g, h \ldots$ represent individual FDs, and $X, Y, Z \ldots$ represent either simple or composite domains):

---

[23] We admit to some ambiguity in our use of undecorated arrows. In much of the following discussion, an undecorated arrow signifies *any* type of FD. Despite this, in Figure 4.5 and elsewhere, it specifically means an FD that is *total* and *onto*. Since database theory tends to assume that all FDs are *total* and *onto*, the reader is assured that the ambiguity will prove harmless.

$$f : X \times Z \to Y \Rightarrow g : Z \times X \to Y \tag{4.4.1}$$
$$f : X \to Y \times Z \Rightarrow g : X \to Z \times Y \tag{4.4.2}$$
$$f : X \to Y \land g : Y \to Z \Rightarrow f;g : X \to Z \tag{4.4.3}$$
$$Z = X \times Y \Rightarrow \pi_X : Z \to X \tag{4.4.4}$$
$$Z = X \times Y \Rightarrow \pi_Y : Z \to Y \tag{4.4.5}$$
$$f : X \to Y \Rightarrow g : X \times Z \to Y \tag{4.4.6}$$
$$f : X \to Y \land g : X \to Z \Rightarrow f \| g : X \to Y \times Z \to Y \tag{4.4.7}$$
$$f : X \to Y \times Z \Rightarrow g : X \to Y \land h : X \to Z \tag{4.4.8}$$
$$g : Y \times Z \to W \land f : X \to Y \Rightarrow h : X \times Z \to W. \tag{4.4.9}$$

These rules are explained as follows[24]:

Commutativity 4.4.1, 4.4.2: It is immaterial in which order we are given the parts of a determinant of an FD or in which order we learn its dependent attributes. This justifies treating X, Y, and Z as *sets of attributes* rather than ordered products. In the following, $X \times Y$ is not merely the Cartesian product of X and Y, but either $X \times Y$ or $Y \times X$.

Transitivity 4.4.3: If *Candidates* determines *Programmes* and *Programmes* determines *Programme Names*, then (in two steps) *Candidates* determines *Programme Names*.

Projection 4.4.4, 4.4.5: If Z is a composite domain then we know its components; e.g., given a value in *Candidates* × *Subjects*, we certainly know the value of *Subjects*. This is called a **trivial dependency**. Projection functions are often written using a special notation, for example, the projection *Candidates* × *Subjects* → *Subjects* may be written as $\pi_{Subjects}$.

Augmentation 4.4.6: This is the converse of projection. We can always add redundant information; if *Personal Names* depends on *Candidates*, then it also depends on *Subjects* × *Candidates*.

Additivity 4.4.7: If *Programmes* determines *Programme Names* and *Programmes* determines *Degree Names*, then *Programmes* determines the parallel composition *Programme Names* × *Degree Names*.

Simplification 4.4.8: This is the converse of additivity. If *Programmes* determines *Programme Names* × *Degree Names*, then *Programmes* determines *Programme Names* and *Programmes* determines *Degree Names*.

Pseudo-transitivity 4.4.9: This is a combination of transitivity and augmentation. If *Subjects* × *Candidates* × *Semesters* → *Marks*, and *Subject Names* → *Subjects*, then *Subject Names* × *Candidates* × *Semesters* → *Marks*.

[24] The reader should note the similarity between these rules and the rules of inference in Section 2.1.4. For example, additivity corresponds to **or** introduction.

The purpose of defining the closure F^+ of a set of FDs F is that if we have any other set G of FDs and $G^+ = F^+$, we know that G and F are, in some sense, equivalent. G is called a **cover** of F. In particular, if no cover H of G exists such that $H \subset G$, then G is called a **minimal cover** of F. Although a minimal cover is similar to a minimal transitive reduction of an unlabelled homogeneous graph, it isn't the same. In particular, if $f : X \to Y$, $g : Y \to Z$ and $h : X \to Z$, we cannot eliminate $h : X \to Z$ unless $h = f\,;g$.

We can also define the **closure of a set of attributes**: the set of all attributes that are functionally dependent on them.

> The closure X^+ of a set of attributes X under the set of dependencies F is the set of all attributes Y, such that the FD $f : X \to Y$ is in F^+ for some f. That is, the closure includes all the attributes that are directly or transitively dependent on the set X.

An important use of the closure of a set of attributes is to find a candidate key, which we now define formally:

> If Y is a set of attributes, a candidate key K of Y under the set of FDs F is any minimal set of attributes such that $f : K \to y$ is in F^+ for all y in Y. That is, every attribute in Y is either in K itself or is dependent on K.
>
> It is possible for Y to have more than one candidate key. An attribute that is a member of *any* candidate key is called a **prime attribute**.

We define an FD $f : X \to Y$ to be **simple** if Y, its codomain (dependent attribute), is simple (not composite).

Given any FD $f : X \to Y$, we can always derive another FD, $g : X \times Z \to Y$, by **augmenting** its domain to form a Cartesian product (f is then a projection of g). The domain (determinant) of f is **minimal** if f is *not* the augmentation of some other FD, say g. (This does not imply that the domain of f cannot be composite.)

4.4.3 MINIMAL FD GRAPHS

It is unnecessary to form the closure of an FD graph to find a minimal cover. This is fortunate, because, given any single FD, its closure contains every possible augmentation of its determinant. The determinant can be augmented to contain any subset of the set of all possible attributes, and there are 2^N subsets of N attributes; finding the closure of a set of FDs is therefore intractable.

To eliminate any redundant vertices and edges from an FD graph to find a minimal cover, which we call a **minimal FD graph**, we work backwards from simple attributes to their minimal determinants. Algorithm 4.1 is a tractable procedure for finding a minimal cover.

We can see that the cover Algorithm 4.1 derives is valid, as follows:

- Because the algorithm ensures that every simple attribute is in the cover, we have no need for the simplification rule, and the additivity rule ensures that we can then derive every possible product of attributes that is in the closure.

Algorithm 4.1 Deriving a minimal FD graph from a set of simple FDs.

1. Draw a vertex representing each simple attribute domain — including those needed to express whether the state of an atom is *Null*, e.g., *Component States*.
2. For each simple FD, construct an edge from its minimal determinant to its dependent attribute. If its minimal determinant is composite, construct a new vertex V representing the set of its components (unless V has already been created in a previous step). Further, for every existing vertex W that is a projection of V, create an edge $V \to W$ representing the trivial FD, $\pi_W : V \to W$. (To represent the FD *Component State*: *Programmes* × *Subjects* → *Component States*, we must construct the vertex *Programmes* × *Subjects* and its two out-edges, to *Programmes* and to *Subjects*.)
3. While there exists an edge $X \underset{f}{\to} Z$, and non-empty paths $X \underset{g}{\to^+} Y$ and $Y \underset{h}{\to^+} Z$ such that $f = g\,;h$, delete the edge $X \underset{f}{\to} Z$. (There are two paths in Figure 4.5 from *Components* to *Programme Names*: *Programme Name* itself, and *Programme* ; *Programme Name*. The rule suggests we should eliminate *Programme Name* : *Components* → *Programme Names*.)
4. While there exist two paths $X \underset{f}{\to^*} Y$ and $X \underset{g}{\to^+} Y$ such that $f \neq g$, consider removing an edge from one of the paths. (Step 3 is a special case of Step 4.)

- Provided we interpret the components of composite domains as unordered, that takes care of the commutativity rule.
- By using minimal determinants, we can derive all possible determinants using the augmentation rule,[25] and we have no need for the projection rule.
- By imagining (but not drawing!) the transitive closure of a minimal FD graph we implement the transitivity rule, which, in conjunction with the augmentation rule, implements the pseudo-transitivity rule.

[25] Although the FD graph emphasises that a composite key has projections onto its components, it doesn't show that knowing each component separately is equivalent to knowing the composite. For example, starting from a row of *Enrolments*, we can find the corresponding rows of *Programmes* and *Subjects*. These two rows taken together identify a row of *Components*. (This *happens* to be the row relevant to the enrolment, but only because the *Enrol* event forces *Programme* to be consistent; it isn't implied by the FD graph.)

Virtually every approach to data-structure analysis fails to show that composite domains can be formed from their components. It is rarely important, so we make little apology for its omission.

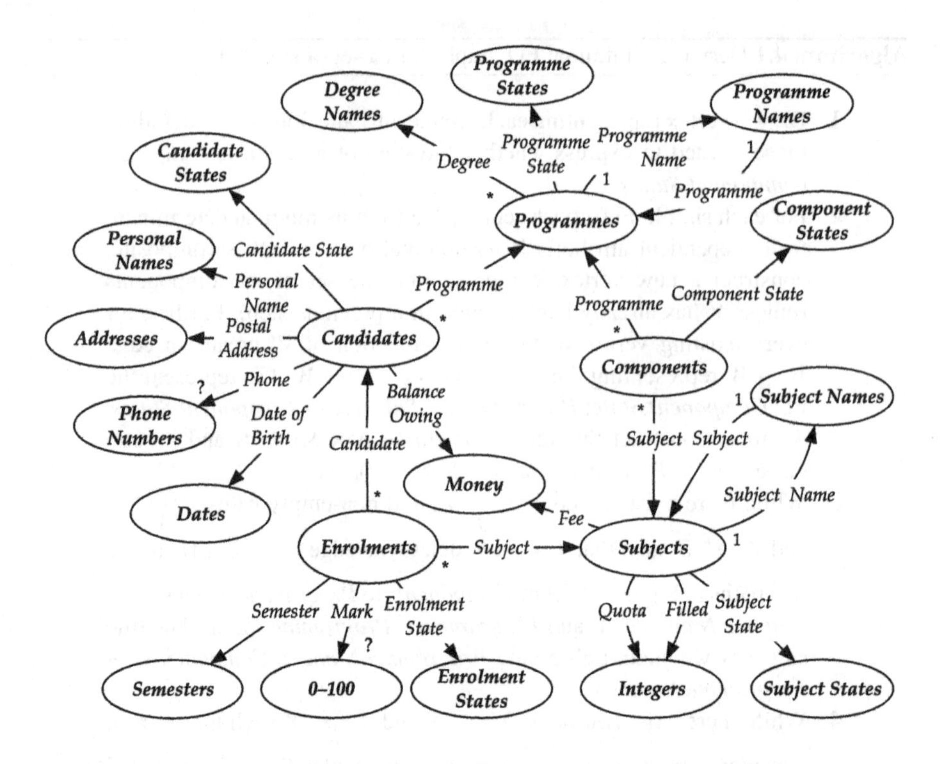

FIGURE 4.6

A minimal FD graph describing the NUTS Academic Records database. Compared with Figure 4.5, it contains no edges from *Components* to *Subject Names* and *Programme Names*.

A possible result of the algorithm is shown in Figure 4.6.[26]

The algorithm does not call for vertices to be drawn to represent tables. The correct choice of tables is deduced from the FD graph. A straightforward implementation is that every non-terminal vertex is the primary key of some table, and each of its outgoing edges is a dependent attribute.[27] Where the edge leads to a non-terminal vertex, it is a foreign key.

In defining a table schema, we shall use the convention that an attribute's name is the same as the corresponding FD. If it has no corresponding FD, it must be part of the key, and its name is the same as its domain, but in the singular. Thus, the primary key of the *Candidates* table is called *Candidate*.

---

[26] Although there are two undirected paths from *Enrolments* to *Programmes* in Figure 4.5 (page 129), there are no *directed* paths. Therefore Step 4 of the algorithm won't eliminate an edge from either path.

[27] There can sometimes be a case for splitting the set of edges, leaving a vertex between two or more tables. See the caption to Figure 7.12 on page 241.

From Figure 4.6, by assuming that each non-terminal vertex is the primary key of a table, we can deduce that there should be the following tables, where primary keys are underlined and foreign keys are in bold type:

Programmes (*Programme*, **Programme Name**, *Programme State*, *Degree*),
Subjects (*Subject*, **Subject Name**, *Subject State*, *Fee*, *Quota*, *Filled*),
Components (**Programme**, **Subject**, *Component State*),
Candidates (*Candidate*, *Personal Name*, *Postal Address*, *Phone*,
Date of Birth, *Balance Owing*, *Candidate State*,
Programme),
Enrolments (*Semester*, **Candidate**, **Subject**, *Enrolment State*, *Mark*),
Subject Index (*Subject Name*, **Subject**), and
Programme Index (*Programme Name*, **Programme**).

The last two tables are needed to enforce the uniqueness of *Subject Name* and *Programme Name*. Their dependent attributes will prove useful if the names are used in queries to discover the corresponding primary keys. In a DBMS that supports *SQLs* **unique** attribute, *Subject Name* and *Programme Name* can both be declared to be **unique** and the two index tables become redundant.[28]

4.4.4 THE CANONICAL FD GRAPH

The FD graph of Figure 4.6 treats *Programmes* (which is a subset of *Programme Codes*) and *Programme Names* asymmetrically. There is a hidden assumption that the *Components* table should use *Programme Codes* as part of its primary key, rather than *Programme Names*. The reason for this asymmetry was the arbitrary order in which the vertices and edges were drawn or, equally, the order in which FDs were considered. It happened that the FD *Programme* : *Components* → *Programmes* was considered before the FD *Programme Name* : *Components* → *Programme Names*, so the algorithm ignored the second FD because it was equivalent to the product (*Programme* ; *Programme Name*) : *Components* → *Programme Names*. If, on the other hand, the FD *Programme Name* : *Components* → *Programme Names* had been considered first, *Programme* : *Components* → *Programmes* would have been ignored instead. To correct this asymmetry, we consider the strongly connected components of the FD graph, deriving what we call the **canonical FD graph**, shown in Figure 4.7. The vertices of the canonical FD graph represent *sets* of domains, and its edges combine the edges that linked single domains in the minimal FD graph.

The canonical FD graph treats each member of a strong component equally, so we are left with a free choice of which attributes should be primary keys. Each possibility is a candidate key.

[28] Or rather, the DBMS will maintain them itself in order to enforce the **unique** constraint.

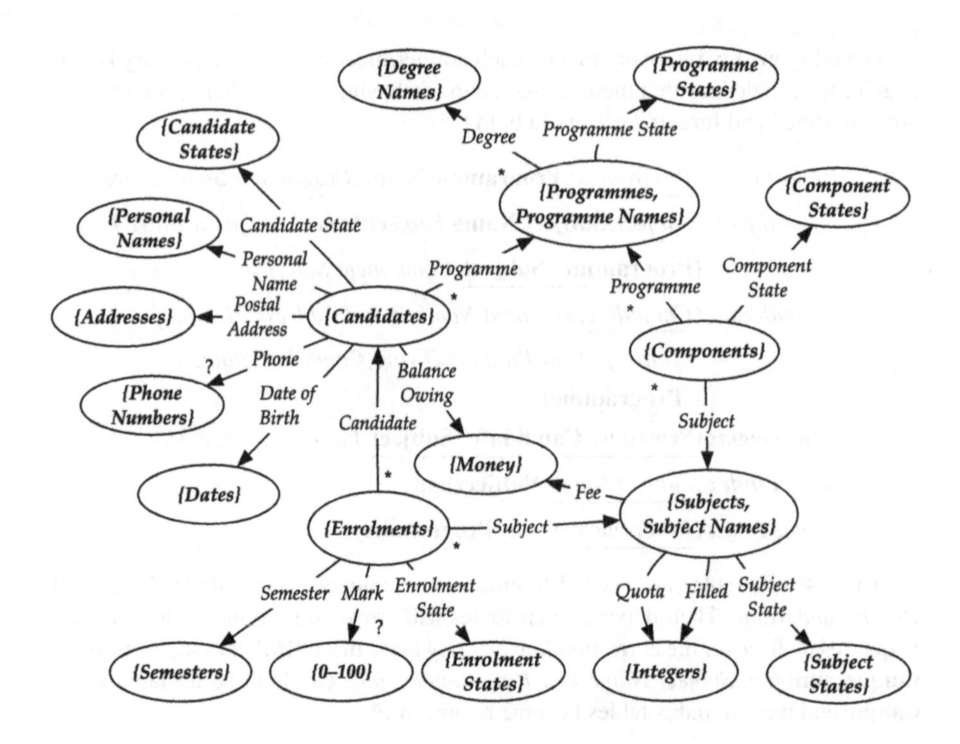

FIGURE 4.7

The canonical FD graph describing the Academic Records database. The vertices are *sets* of domains, and the edges are functions.

The *Subjects* table has a choice of *Subject* or *Subject Name*. Likewise, *Programmes* has a choice of *Programme* or *Programme Name*. These choices propagate, so that *Components* now has four candidate keys:

$$Programme \times Subject,$$
$$Programme \times Subject\ Name,$$
$$Programme\ Name \times Subject,$$
$$Programme\ Name \times Subject\ Name.$$

Enrolments has two candidate keys:

$$Candidate \times Subject \times Semester,$$
$$Candidate \times Subject\ Name \times Semester.$$

Because foreign keys must reference primary keys, the choices need to be consistent.

How do we decide which attributes should be chosen as primary keys?

There are two main criteria: shortness and stability. Because the primary key of the *Subjects* table will appear as a foreign key in the *Enrolments* and *Components* tables, it will be stored many times, which argues in favour of shortness. Further, every time its value is changed, all the references to it will need to be updated too, which argues in favour of stability.[29] When we learn from a domain expert that *Subject Names* can change (as when 'System Design' became 'System Design Project', while retaining the same *Subject Code*[30]), we opt in favour of using *Subject* as the primary key. Likewise, we favour *Programmes* over *Programme Names*. Thus, after making these choices, we actually derive the same minimal FD graph we derived in Figure 4.6 — but by design rather than accident.

4.4.5 CYCLES

Now consider the domains *within* the strongly connected components of Figure 4.7. Although *Subject Names* and *Subjects* are in correspondence, in general, the domains within a strongly connected component need not always correspond one-to-one. For example, it is common for final-year or higher-degree candidates to act as tutors to first-year candidates. Consider a situation in which some candidates are recorded as paid employees within a Payroll system. As such, they would be identified by *Payroll Nos*. Now consider a database that integrates the Academic Records System with the Payroll system. Such a database will need a partial one-to-one FD *Employee* : *Admission Nos* $\rightarrow$ *Payroll Nos* and another partial one-to-one FD *Candidate* : *Payroll Nos* $\rightarrow$ *Admission Nos*. *Employees* and *Candidates* are overlapping subsets of *Persons*. (See Figure 4.8.) How do we implement this?

One approach is to store separate tables for *Employees* and *Candidates*, making *Candidate* a nullable foreign key in the *Employees* table and *Employee* a nullable foreign key in the *Candidates* table. This has a drawback: the personal details of an employed candidate are stored in both tables. There is a danger that these details may get out of step, which can only be overcome by complicating the events that update the *Candidates* and *Employees* personal data.

Alternatively, there can be a *Persons* table, which contains personal details, cross-referenced by both the *Employees* and *Candidates* tables, which no longer contain personal details but only information specific to a person's role as a candidate or employee. But this alternative now complicates every query involving the personal details of either *Candidates* or *Employees*, by requiring access to the additional *Persons* table. On the other hand, in the case that a candidate has studied for more than one qualification, this solution would allow multiple admission numbers to be associated with the same row of *Persons*, thus ensuring that postal addresses remained consistent.

[29] Even so, a suitably cunning implementation of the DBMS might store a pointer to the parent record, rather than the actual value of its primary key, which negates both arguments.

[30] Not a wise decision! A candidate who studied 'Systems Design' in the past would now be recorded as having studied 'Systems Design Project'. It would have been better to have created a new *Subject* row with a new code.

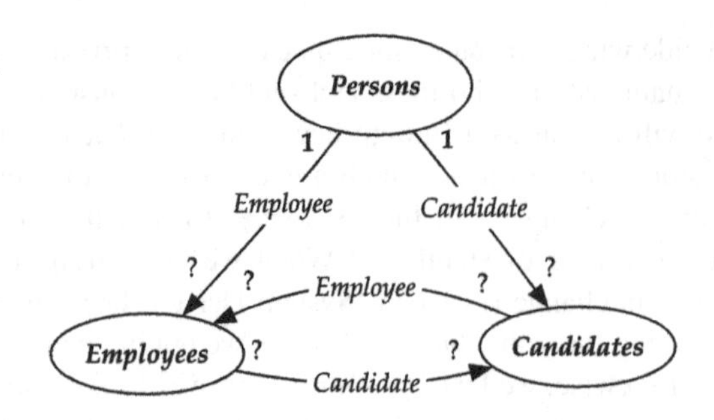

FIGURE 4.8

Subset relationships between *Persons*, *Candidates*, and *Employees*. Some *Persons* might be *Persons*, and some might be *Candidates*, and some might be both.

A third option is also possible, which is to store *only* the *Persons* table, making employee-specific and candidate-specific information nullable attributes of *Persons*. This has the drawback of needing a bigger table, with slightly longer access times, but perhaps more importantly, meaning that amendments to the Payroll system software might require matching amendments to the Academic Records System, and *vice versa*. Further, it fails to allow a person to ever study for more than one qualification.

The third option suggests a general strategy, which is to find (or even invent) a domain that is a superset of all the domains within the connected component. The problem then is to find a primary key for it. Should the *Persons* table have *Payroll No* or *Admission No* as its primary key? One solution would be to give everyone a uniform *Personal ID*.

At the other end of the spectrum would be to 'solve' the problem by ignoring it and keeping the Payroll database and Academic Records database entirely separate.

In short, no solution is entirely satisfactory. On the brighter side, the integration of *SQL* with object-oriented techniques, in the form of, e.g., *SQL-3*, overcomes some of these problems. Objects (tables) are allowed to inherit properties from more general objects. For example, the *Candidates* and *Employees* tables could inherit personal details from a *Persons* table. (A likely implementation is that *Candidates* and *Employees* would contain hidden foreign keys that reference *Persons*, as in our second alternative above. Consequently, the cross-references would now be maintained by the DBMS and no longer be the responsibility of the application programmer.)

4.4.6 ALGEBRAIC COMPOSITIONAL METHODS

The algebraic compositional approach is essentially the same as using an FD graph, so we do not need to describe it in detail here. The reader may consult Philip A. Bernstein's article referred to in *Further Reading* (Section 4.7). He begins with a minimal set of simple FDs, groups them according to their determinants, then constructs a table for each group. In the case of a domain that does not appear as the determinant of any FD (such as the primary key of *Components*), he creates an FD — which he calls a θ-function — to represent its characteristic function.

Bernstein's approach differs in three minor ways from the graphical method we have described:

- First, it does not distinguish FDs from their codomains, so that it cannot accurately describe the fact that *Quota* and *Filled* are both integers.
- Second, it does not give a tractable method for finding the minimal cover of a set of FDs, but since this normally only consists of deleting a few redundant FDs, it is rarely a challenge.
- Third, it assumes that any member of a strongly connected component is equally valid as a primary key, which we have seen is only the case if all the FDs involved are *total* and *onto*.

4.4.7 SUMMARY OF COMPOSITIONAL METHODS

We have presented a compositional or synthetic approach to database design. Armed with a preliminary schema or E-R diagram, and following several discussions with a domain expert, we obtain a set of FDs. We must be careful to recognise, as in the case of *Component State*, that even when an FD's state can be shown implicitly by the existence or non-existence of rows in a table, it must still be included in the FD graph. Prematurely omitting its domain (*Component States*) before constructing the graph would lead to a faulty design.

Once we have a set of FDs, we then use simple domains and minimal determinants to synthesise a canonical FD graph. This step of the process is formal. Deriving the canonical FD graph is tractable, because the graph can have no more edges than the number of FDs.

On the other hand, we have learned that when an FD graph contains non-trivial connected components, we must make some choices that can only be based on factors not expressed by the FD graph.[31]

The final step of converting the FD graph into a database schema is again straightforward.

The *hardest* part of the synthesis can sometimes be choosing what domains and FDs to model. I am reminded of an ecological database that recorded interactions between organisms. As the client specified it, there were to be tables for *Insects*, *Vertebrates*, *Plants*, and *Fungi*. The tables were to be linked by relationships such as *Eats*, *Pollinates*, *Parasitises*, *Symbiont*, and so on. Already, the database was looking surprisingly complex. *Vertebrates* can eat other *Vertebrates*, they can eat *Insects*, they can eat *Plants*, and they can eat *Fungi*. Unfortunately, because each of these foods would belong to a different table, *Eats* became four distinct relations with four

[31] The step of choosing a vertex from each strongly connected component to become its primary key is theoretically intractable. If we had C components each containing V vertices, there would be V^C possible choices. In practice, V^C is small (only 4 in the case of the Academic Records database), and the choices for each component need to be consistent, so the complexity of this step is actually $O(V \times C)$ in practice.

distinct foreign keys. Since *Insects* can also eat a wide variety of foods, they added another four distinct *Eats* relations. A similar story applies to the other relationships. As time went on, and as the client kept introducing more classes of organism and more interactions, the proposed database schema soon got out of hand.

The solution? Just four domains: *Organisms*, *Classes of Organism*, *Interactions*, and *Types of Interaction*. *Classes of Organism* included *Insects*, *Vertebrates*, *Plants*, and *Fungi* as values. *Types of Interaction* included *Eats*, *Pollinates*, *Parasitises*, and *Symbiont* as values. Only four tables were needed:

Organisms (Organism, Class of Organism, Name of Organism),

Interactions (Subject Organism, Object Organism, Interaction Type),

Classes of Organism (Class of Organism, Class Name), and

Types of Interaction (Interaction Type, Interaction Name).

Apart from simplicity, this solution had two other advantages: adding new types of organism or interaction became the user's responsibility rather than a programming problem, and the homogeneity of the database made it easier to trace transitive chains of interaction.

Choosing the right level of abstraction and the right level of detail is tricky. The world is infinitely complex: the more detail is modelled, the more a system will cost to develop. The less detail is modelled, the greater the danger that the database schema will prove short-sighted and inadequate. If we were to ask the proposed Academic Records System, 'Which candidates have been admitted to more than one programme?' it would be unable to answer.

Finally, to be honest, although the canonical FD graph of Figure 4.7 (page 136) is useful for tutorial purposes, a systems analyst would rarely bother to draw the leaves of the diagram or document all the possible primary keys. The analyst would more likely sketch something like Figure 4.9, which omits everything the analyst dismisses as 'obvious'.

4.5 DESIGNING A DATABASE SCHEMA BY DECOMPOSITION

The *decompositional* approach is to begin with a badly designed set of tables or, in the extreme case, a single table containing every attribute, called the **universal relation**. Such badly designed tables often result from an analysis technique called **view integration**.

View integration is a form of reverse engineering in which the analyst examines forms and reports that the existing system generates and infers what set of tables they derived from. For example, a candidate's academic transcript lists the names of the subjects the candidate has studied and the marks that the candidate has achieved. Unfortunately, this set of attributes doesn't actually constitute a well-designed table, a problem that decompositional methods address.

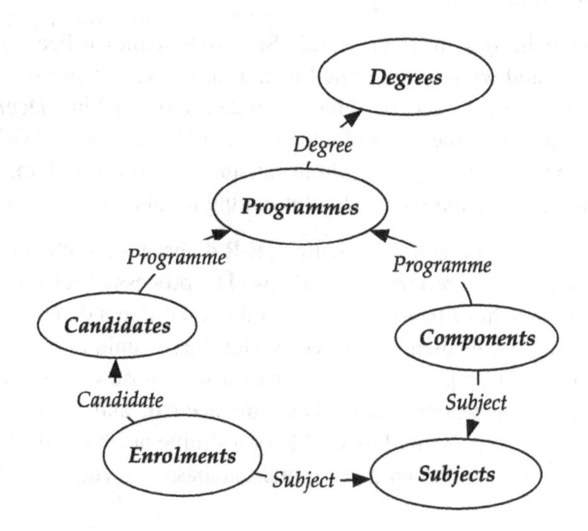

FIGURE 4.9

How a systems analyst might actually sketch the canonical FD graph of Figure 4.7.

Traditionally, database design has employed an analytic algebraic approach called **normalisation**. Normalisation is a process of designing table schemas informally, then *analysing* them to expose redundancies. Although analysing existing reports can be useful in discovering what attributes a system needs to record, the author does not believe that normalisation is an efficient way to synthesise a database schema.

Personally, I strongly advocate the sketching of FD graphs as a means of database design. But because a lot of database terminology stems from it, knowing about normalisation theory may prove important to you. If that is the case, read Appendix B. You don't need to understand normalisation to read the rest of this book.

4.6 SUMMARY

A database is where a system stores the states of the atoms it models. The relational model describes the database as a series of inter-related *tables*. A table is a *conceptual structure*, independent of its underlying implementation. As such, it allows a discussion of data structure that is independent of whether it will be implemented using arrays, linked lists, search trees, or whatever. Tables allow the design of a database to be split into two phases: a conceptual phase and an implementation phase. In many systems the physical structure will be a relational DBMS, and the analyst need never become too concerned about the actual representation of the data.

The primary aim of conceptual design is to eliminate redundancy, yielding a database that is easy to update consistently. In some cases, the analyst may consciously choose to include redundant information if the resulting speed up in data retrieval will make the system faster overall. Such redundancy is heavily exploited in data warehousing, and we have exploited it in a minor way in our use of the *Filled* attribute. We also chose to accept some redundancy by making *Degree Name* an attribute of *Programmes* rather than of an additional *Degrees* table. We assumed this would simplify the implementation without causing a serious problem.

There are two distinct approaches to eliminating redundancy:

- The compositional approach starts with an E-R diagram or a schema derived from event-state analysis and progresses, perhaps via a process of reification, to a point where all relations have been reduced to functional dependencies. A functional dependency graph can then be drawn, which has a unique canonical reduced graph. The reduced graph enables the analyst to choose the best identifiers for tables, to determine where foreign keys are needed, and to decide whether a property should be represented by a table or a simple attribute. In short, it reveals all that is necessary to design a relational database — without introducing any redundancies.
- The decompositional approach typically begins with a series of business documents that present views of the database. These views become potential table schemas. The potential schemas are examined for redundancies, then decomposed to remove them.

Contrasting the two approaches demonstrates the general point that a compositional, synthetic method is simpler and more effective than a decompositional, analytical method. Most of the reasoning in the synthetic approach takes place at the *abstract* level. Once an FD graph has been drawn, the rest is a formality. The normalisation approach prematurely rushes to the *conceptual* level of (probably incorrect) table schemas.

In practice, the two approaches can prove complementary: event-state analysis and E-R diagrams tend to help the analyst discover the determinants of FDs. Examining reports helps the analyst discover their dependent attributes.

The reader should be aware that the relational model is not the only possible conceptual model of databases. One alternative is the **network model**, which represents relations using sparse matrices (Figure 2.35). Unfortunately, this model too strongly suggests that procedures should follow chains of pointers. It is less abstract than the relational model and tends to restrict the implementor's thinking.

Another alternative, beloved of scientists and mathematicians, is to model data using vectors and matrices. Given that we explained how matrices can model relations in Section 2.5.5, this seems a sensible choice. As a result, a mathematician might too quickly decide to implement the *Enrolments* table as a three-dimensional matrix, with dimensions *Semesters*, *Candidates*, and *Subjects*. Assuming that there are 10,000 active undergraduates and 500 subjects, a direct implementation of an *Enrolments matrix* would contain 5,000,000 entries per semester. An *Enrolments table* would contain about 25,000 entries per semester. Neither alternative is a disaster in this situation, but it is easy for multi-dimensional matrix representations to replicate the same data many times and consume vast amounts of storage. A mathematician unaware of proper data-structure analysis techniques can seem to need the power of a supercomputer to solve a simple problem.[32]

4.7 FURTHER READING

The foundational paper on the relational model is by Edgar Frank Codd: 'A Relational Model of Data for Large Shared Data Banks', *Communications of the ACM* 13 (6): 377 (1970). Apart from showing that tables can always replace more complicated data structures, Codd's main contribution was to show that queries expressed in the relational algebra (essentially predicate calculus) could always be converted to operations in the relational calculus (i.e., **join**, **select**, and **project**, which are the basic operations used by a DBMS).

E-R diagrams were first described by Peter Chen in his 1976 paper, 'The Entity-Relationship Model — Toward a Unified View of Data', *ACM Transactions on Database Systems*, 1 (1), 9–36.

A commonsense approach to relational algebra is given in Chapter 8 of 'Foundations of Computer Science: C Edition' by Alfred V. Aho and Jeffrey D. Ullman (1994, ISBN: 0716782847). The book, 'Databases and Transaction Processing' by Philip M. Lewis, Arthur Bernstein, and Michael Kifer (ISBN: 0-201-070872-8), also discusses E-R analysis from a rather different perspective. (Conveniently, both books use a case study similar to ours.)

The material on functional dependency graphs is largely the author's own, although a related paper by Jonathan P. Bernick, 'A Graphical, Functional-Dependency Preserving Normalization Algorithm for Relational Databases', can be found on the Internet. This uses graphs, as we have done here, but the article doesn't distinguish domains from FDs, nor does it explain how to find a minimal cover. Bernick's graphs seem to be intended as a teaching aid rather than a contribution to database theory.

The graphical method we have used to derive a schema is almost identical to the syntactical method described in Philip A. Bernstein's 1976 article, 'Synthesising Third Normal Form Relations from Functional Dependencies', *ACM Transactions on Database Systems* 1 (4) 277–298 — including the importance of ensuring that every table derives from at least one FD, if necessary, by introducing θ-FDs to represent characteristic functions.

4.8 EXERCISES

1. Take the FD graph of Figure 4.6 and express it as a matrix whose columns are domains and whose rows are codomains.
2. In Figure 4.4 we considered *Enrolments* as a relation between *Admissions* and *Components*. What set of FDs would have resulted from this idea?

[32] Yes, I have met people who were proud to have 'reduced' a database structure to a set of massively redundant ten-dimensional matrices.

CHAPTER

Kernel specifications

5

CHAPTER CONTENTS

INTRODUCTION

The specification of a system kernel is a pivotal point in a system's creation. It marks the end of analysis, where the requirements have been discovered, and the beginning of synthesis, where design and construction begin. At this point we have discovered what atoms we need to model and learned about the events that change their states, but we have yet to consider what technologies will be used.

A *full* specification must address both the kernel and its interfaces, which call for different approaches. The kernel is best specified formally, using some kind of specification language; the interfaces are usually best described using pictures. We leave the question of interfaces for a later chapter and first discuss the kernel.

The reader may feel that I have not yet said enough about how the analyst should gather the information needed to specify the kernel, and that is true. It can be difficult to tell the difference between a genuine requirement of an existing system and a mere by-product of the way it is implemented. For tutorial reasons therefore, I have reversed the natural order of things:

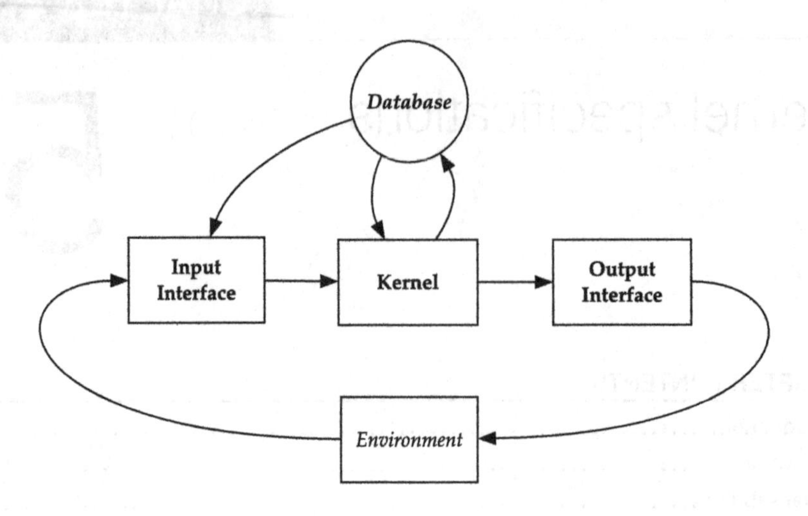

FIGURE 5.1

The pathways between the *Kernel*, the *Database*, and its *Interfaces*.

- First the reader is shown how to specify a system's requirements.
- Then the reader is shown how to synthesise a system from its specification.
- Third, the reader is shown how optimising a system design can obscure the original specification.
- Finally, it is hoped that the reader will thus become able to recognise the by-products of such optimisation when attempting to analyse an existing system.

5.1 THE KERNEL

Figure 5.1 shows how the *Kernel* interacts with other parts of the system. Events arrive from the system's *Environment* via the *Input Interface*. These events may cause the *Kernel* to update or interrogate the *Database* or to produce output. Output is sent back to the *Environment* via the *Output Interface*. The *Environment* typically responds by generating fresh events. In an interactive system, the *Input Interface* can usually interrogate the *Database* to allow an operator to select from a list of options, confirm a choice, and so on. Except for this, the system has no control of its inputs: the environment generates them as it pleases, and the system must deal with them as best it can.

The boundary between the *Kernel* and its interfaces is well defined. The boundary between the interfaces and the environment depends on one's point of view. For example, if a terminal operator is entering enrolment information on behalf of a candidate, is the operator part of the interface or part of the environment?

5.2 SERIALISABILITY

We make an important assumption here: that events can be **serialised**. In other words, we assume that events can be said to have happened in some particular order.

This might not seem like an assumption at all, but consider what happens when you make an international telephone call. Suppose you are calling a friend overseas, and at the exact same moment, your friend calls you. You might say that there is no such thing as 'the exact same moment', but the velocity of light is finite. Your local exchange might believe that you initiated the call before your friend did, but the overseas exchange might believe your friend initiated the call before you. As a result, you both might have to pay for the call.

There are two safe ways of dealing with such a situation: there can be an **arbiter** that judges the order of events, or both events can fail. For example, the telephone company may have established that some particular exchange on the path between you and your friend acts as an arbiter, and the order of arrival of the calls at that exchange is what decides the outcome. Alternatively, avoiding an arbiter, the telephone company might decide that you simply find that your friend's telephone is engaged, and your friend finds that your telephone is engaged. You will both have to try again later.[1]

You can see that this issue of simultaneity can arise whenever two events occur within less time than it takes a signal to cross the system. This issue isn't addressed here. We assume that an arbiter can always assign some order to events.

Sometimes the order of events matters, and sometimes it needn't matter. If I take all the money out of my bank account "at the same time that" my wife takes all the money out of her bank account, it doesn't matter which event is considered to have occurred first. But if, following some domestic argument, I take all the money out of our *joint* bank account "at the same time that" my wife takes all the money out of our joint bank account, then the order in which the events are considered to have occurred matters a great deal.

Along with the assumption that events are serialisable, we also assume that they are atomic. That is to say, there is no state where an event is half-finished. There is a state before the event, and there is a state after the event. Although the system's infrastructure may allow several events to be processed simultaneously, this is not a thing we shall need to worry about in the specification. It is the infrastructure's problem to make sure events don't interfere with one another.

5.3 INFRASTRUCTURE

What do we mean by infrastructure? **Infrastructure** is the technology within which rules defined by the kernel are embedded. A common infrastructure is a **client-server system** operating across the Internet. Typically, the server has access to a database and remote clients can view the database using a browser. The server

[1] Somewhat like the Ethernet protocol.

transmits database information to the client, whose browser renders the information according to *HTML* (Hypertext Mark-up Language) commands. A user of such a system interacts with the kernel by filling in *HTML* forms, which the client process transmits to the server. The server interprets the user's data and queries or updates the database as appropriate.

In an apparently similar situation, using a local-area network, the server merely allows access to a shared database, and the client computers contain the kernel logic. This offers the user closer interaction with the kernel logic.

In the early days of computing, a user might have interacted with the kernel by feeding a batch of punched cards into a reader, waiting for the job to be processed, and reading printed reports — often flagging errors in the input data. With this technology, user interaction was poor and very slow indeed.

Many systems are a mixture of technologies, with an interactive **front end** and a batch processing **back end**. For example, an automatic teller machine may record transactions in a database interactively as they occur, but a batch system may produce customers' monthly statements. In a retail context, customer orders may be recorded interactively, and at the end of the day, a batch system may prepare a report that enables store-hands to take goods in the most efficient manner and pack them for despatch next day. In preparing this book, I am using an interactive editor to create a text containing LaTeX commands, which is ultimately rendered into a printable format by a batch process.

An aircraft's infrastructure may contain several microprocessors, each dedicated to a specialised task. One might be dedicated to navigation, another to moving a control surface, a third to stabilising the aircraft's flight, and so on. The pilot interacts with the system by, among other things, moving a joystick. A car might contain an engine-management computer that controls the fuel-air mixture, ignition timing, and valve timing in response to engine speed, etc. The driver interacts with the system by controlling the throttle, the gear shift, and the brakes.

The reason why so many different infrastructures exist is that they aim to optimise different objectives: the efficient use of limited bandwidth, the efficient use of a shared resource, the efficient use of storage technology, the efficient use of human resources, the need for reliability, or the need to minimise product cost. In short, to optimise something at the expense of simplicity. The *cost of optimisation* is the additional software and hardware needed to allow different parts of a system to communicate. For example, in a batch processing system, transaction files are needed to transfer data between processing steps. These files — which can be rather complex — have to be defined, their records of various kinds have to be written and read, and the records have to be deciphered in order to call appropriate procedures. Similar communication software has to be written for almost every kind of infrastructure. Roughly speaking, about 90% of programming effort is devoted to the infrastructure and only 10% to the kernel logic.

This book doesn't have space to deal with every kind of infrastructure in depth. Indeed, it isn't the aim of this book to deal with infrastructure at all. But its examples and case study have to be set in some context, so it mainly focusses on a database server with multiple clients. Such an infrastructure shares most of the problems that many other infrastructures have. Even so, it is largely left to the reader to see the underlying unity of systems synthesis and adapt what is said to the reader's own speciality.

Ideally therefore, the *specification* of a system should be independent of the technology that it currently uses or of any technology that it might use in the future. This will free the systems analyst from being constrained by what technology currently exists, while allowing maximum freedom in implementation. System synthesis is therefore the act of *combining* the specification with a suitable technological infrastructure. Synthesis is 10% inspiration and 90% perspiration.

A specification will consist of two parts: a description of its state, or database, and a description of the events that can cause its state to change or can cause it to produce output. Ideally, the kernel specification should avoid all mention of infrastructure. The only time when the specifier must consider infrastructure is when designing the user interface.

5.4 AN ACADEMIC RECORDS SYSTEM

In this section we continue the development of the Academic Records System for the National University of Technology and Science (NUTS) whose database we analysed in the previous chapter, deriving the specific set of functional dependencies documented at the end of Section 4.4.1. Before we attempt to specify the system in a formal way, we first define the problem in plain language:

> The **Academic Records System** is required to keep track of what subjects each candidate enrols in and what examination marks the candidate obtains. Subject to certain entrance requirements (outside the scope of this system), a *Candidate* is *Admitted* to a specific study *Programme*. A programme leads to a *Degree*, e.g., 'Diploma in IT'.
>
> To *Graduate*, the candidate must gain sufficient marks in subjects specific to the chosen programme. The rules for determining when a candidate has graduated are beyond the scope of this specification. Currently, an examination board determines when the candidate has satisfied the programme's requirements.
>
> Each *Subject* is a *Component* of one or more programmes, and a candidate may *Enrol* only in subjects that belong to the programme to which he or she is admitted. The candidate may choose which subjects to study. It isn't the business of this system to check if the choices are sensible, except to check that they are part of the candidate's programme; it is assumed that he or she has already been approved by a **programme advisor**. Enrolments can only be made for the current semester.
>
> To help the student make good choices, the candidate is given the opportunity to enrol in a subject provisionally. A candidate who wishes

> to *Withdraw* from the subject incurs no penalty, and it is as if the enrolment had never taken place. This flexibility is allowed only for a limited time, and at a certain date, called the billing date, enrolments in all subjects are *Frozen*, and the candidate can no longer withdraw. At this time, each candidate is sent an account detailing enrolment fees and demanding payment. *Before* the billing date, the candidate may enrol and withdraw as often as desired.
>
> At a time following the billing date, formal examinations are held, and each candidate is given a mark in each enrolled subject. There is therefore a temporal sequence in which subjects become *Offered*, become *Frozen*, become *Assessed*, and become history.
>
> Since the function of the system is to keep records, it makes no provision for removing information from the system. Even when a student has graduated, his or her academic transcript remains as a public document.

Clearly, this system is simpler than any real student enrolment system. For tutorial reasons, our case study has to be modest. Although the above description is consistent with the event-state models we discussed in Chapter 3 and the FD analysis of Figure 4.7, it incorporates some complications that we haven't discussed earlier.

The need for most types of event is self-evident, but others were obtained in interviews with enrolment staff. They concerned possible, hitherto unknown interactions between atoms, such as, 'Are there any conditions that might prevent a candidate from enrolling or withdrawing from a subject?' 'Can a subject be added to a programme at any time?' and so on.

5.5 A SPECIFICATION LANGUAGE

It is useful to have a formal notation with which we can define a system at a conceptual level. This notation must concern both data structures and events. Its aim is to allow the analyst to document business requirements, not the way the requirements are currently implemented, nor the way it is expected that they will be implemented in the future. As the author of this book, I have to admit that I am not fussy about what notation you, the reader, may choose to specify requirements. I have chosen a mixture of mathematical and programming conventions that I have found convenient, but for various reasons you may prefer others.

There are several popular ways of specifying events. The main division is between *procedural* and *declarative* specifications. A **procedural specification** is essentially similar to a procedure in a programming language. A **declarative specification** defines a relation between states before and after an event. It defines the effect of an event, without specifying how the effect should be achieved.

Suppose we want to specify an event that exchanges the values of x and y. A possible declarative specification could be:

$$x' = y \land y' = x$$

where x and y denote the values of x and y *before* the event, and x' and y' denote the values of x and y *after* the event.

On the other hand, a procedural specification needs to be more explicit about how the exchange takes place. It could introduce a new temporary variable, t, in a sequence of three assignments:

$$t = x \,;\, x' = y \,;\, y' = t.$$

There are arguments for and against both styles. The argument against the procedural specification is that we could equally have written

$$t = y \,;\, y' = x \,;\, x' = t.$$

More obscurely, in the case that x and y are numbers, we might write[2]

$$x' = x - y \,;\, y' = x + y \,;\, x'' = y' - x'.$$

In other words, a procedural style can over-specify requirements, making it seem that a temporary variable or that arithmetic is part of the problem, rather than part of the solution. As a result, a programmer may not feel free to choose the best way of implementing the event, because it is hard to distinguish what is important from what is just a side effect.

The declarative approach avoids this difficulty, but introduces difficulties of its own. First, a simple specification doesn't guarantee a simple implementation, or indeed, any implementation at all. Many intractable problems can be specified quite simply, as can many paradoxes. For example, the following simple specification,

$$x'^2 + 3x' + 3 = y$$

has a non-trivial implementation. Whether this is a suitable specification then depends on the mathematical abilities of the programmer. In most situations it would be better for the specifier to suggest a procedure for deriving the value of x'.

Second, when an event must generate sequential behaviour — a dialogue with a user, for example — some sort of grammar is needed to describe its behaviour. Since grammars use constructs for choice, iteration, and sequential composition, the distinction between procedural and non-procedural specification becomes blurred.

I offer here a procedural style of specification for two reasons: most readers will be familiar with at least one programming language, *and a procedural style enforces honesty*. With a non-procedural approach, it is easy to specify things that cannot be done. Specification in the form of a tractable procedure guarantees the problem has a usable solution.[3] But the final system doesn't need to use the same algorithm as the specification if a better one can be found.

In most cases, event procedures will be so simple that there is no problem in writing an algorithmic specification. Even so, there will perhaps be some cases in which it is necessary to write a non-procedural one, because no tractable algorithm is yet known. *Any situation of this kind must alert the analyst to the need for research.*

---

[2] Here, we are using x, x', and x'' to denote the successive definitions of x.

[3] Even if no tractable algorithm is known, I believe that an intractable algorithm is better than none at all.

A system cannot be said to have been fully specified when there remains doubt as to whether it has a tractable solution. We shall encounter two examples of this kind in later chapters: an expert system to advise students about their study programmes, and a system to schedule examination timetables. In both cases, the analyst can only specify the desired relationship between input and output. For example, the timetabling problem can be stated (informally) as 'Find an allocation of subjects to examination times and rooms such that no candidate must attend two examinations at the same time, and no examination room overflows'. This requirement specifies a *relation* — there may be any number of solutions. An algorithm would generate only one. Although such a requirement can be stated more formally, formality doesn't solve the problem: if such procedures are essential to the success of a system, *finding a sufficiently fast algorithmic solution then becomes a matter of high priority.*

To define control structure, we use regular expressions (Section 3.3 on page 87)[4]:

- If action A is followed by action B, we write $A\,;B$.
- If action A and action B are alternatives, we write $A \cup B$.
- If action A may either occur or not, we write $A^?$.
- If action A is repeated zero or more times, we write A^*.
- If action A is repeated one or more times, we write A^+.

An action may **succeed** or **fail**, for example $x = y$ succeeds if x and y are equal, and *fails* if they are not. An *event* succeeds if it contains a sentence in which every *action* succeeds. Being atomic, if an event *fails*, the database remains unchanged.

We indicate the final, updated values of variables with a prime, e.g., x' and y', and their initial values, x and y, without a prime. We do this because we want to stress the changes in state that are caused by an event. Thus, the specification $y' = x\,;x' = y$ *is* allowed, but the implementor will have to ensure that the assignment to y' doesn't destroy the initial value of y, which will need to be preserved somewhere until the assignment to x'.

In the case of database variables, unprimed names (e.g., *Filled*(s)) indicate values before the event, and primed names (e.g., *Filled*(s)$'$) denote values following the event. In the case of temporary variable t, its initial value is written as t, and its final value is written as t'.

The idea that a variable has only two values, initial and final, allows the assignment $x' = x + 1$, but forbids $x = x + 1$ or $x' = x' + 1$. The benefit of this restriction is that it is possible to solve for the final values of variables by algebraic substitution. For example, in the specification

$$t = x\,;x' = y\,;y' = t$$

we simply substitute x for t to obtain $x' = y$ and $y' = x$.[5]

---

[4] Because events change the states of atoms, it may seem inconsistent to use behavioural descriptions of event procedures rather than state-based ones. On the other hand, regular expression notation is close to most programming languages, and therefore easier for programmers to understand. Converting a procedure to state-driven (**go to**) code is usually considered the task of the programming language compiler.

[5] Because of the single assignment rule, ';', '∧', and '∩' are often equivalent.

We have a second way of interpreting a specification written as a regular expression: as explained in Section 3.3, each transition may be considered as a state transformer. The assignment $x' = y$ transforms the state by changing the value of x. A regular expression is therefore an algebra of relations statement about how an event changes the system state. Nonetheless, remember that events are atomic, so the *observable* state of the system isn't changed during the procedure, only when it completes.

Unfortunately, the idea that a variable has only two values causes a difficulty when a procedure involves iteration, because working variables may be updated many times, once per iteration. Single-assignment languages such as *Haskell* or *Prolog* offer one solution to this difficulty: to express iteration as recursion. To many programmers this approach will seem unfamiliar and cumbersome, so we handle this problem here by adopting a simple convention: we let the value of variable t at the start of the iterated statements be denoted simply as t, and its value at the end of the first iteration be denoted by t'. In the second iteration, the initial value should be t', and its final value should be t''. Naturally, we cannot write every possible step, so we simply provide the first step as a sample, and, *notationally*, we treat the situation as if there will always be exactly one iteration.

Here is an example: suppose we have a set, S, and we want to find its cardinality, $Size'$. We specify this as follows:

$$Size = 0\,;(s = \textbf{any}\ S\,;S' = S \setminus \{s\}\,;Size' = Size + 1)^*\,;S' = \{\}.$$

The procedure consists of three parts: the **initialisation**, the **loop**, and the **exit rule**.

- The *initialisation* sets the initial value of $Size$ to zero. (S is assumed to be a datum.)
- The parenthesised expression constitutes the *loop*. It is repeated once for each element of S. It uses the **any** operator to select s, an arbitrary element of S. It then removes s from S to give S'. $Size$ is incremented to give $Size'$. The loop then repeats, implicitly updating S to the value of S' and $Size$ to $Size'$.
- The *exit rule* is $S' = \{\}$. This ensures that the loop will not quit until all elements of S have been counted.

The final values of S and $Size$ are always denoted S' and $Size'$ even in the case that there are zero iterations of the loop.[6] This is a *specification* language, not a programming language. We want it to be concise.

How do we know that $Size = 0$ is an assignment to $Size$, but that $S' = \{\}$ is a condition? In the first case, $Size$ has no existing value, so it is initialised, but in the second case S' already has a value and cannot be given another. As far as the algebra of relations expression linking initial values and final values is concerned, there is no difference.

[6] Any subsequent reference to S would yield its original value, and any subsequent reference to $Size$ would yield zero.

We use the **any** operator when we want to stress that the order in which values are selected is unimportant. If we wanted to process the elements in ascending order we would substitute $s = \textbf{min}\ S$; in descending order, $s = \textbf{max}\ S$.

The above form of construction is so common, that we provide a special shorthand:

$$Size = 0; (\textbf{all}\ x \in S; Size' = Size + 1)^*.$$

As before, after the iteration, S' is empty and $Size'$ contains the cardinality of S.[7] When the order of selection of elements is important, we use **asc** or **desc** in place of **all**.

Finally, how does the specification language distinguish important effects from side effects? Simple. Assignments to database variables are important; assignments to other variables are not.

5.6 SPECIFYING ATOMS

Apart from the events that the system will handle, we need to specify the database in which the states of its atoms are stored. The easiest way to specify a database is as a set of functional dependencies.

In the specification language, we shall use an array or functional style of notation to access to the components of functional dependencies. If $(x \mapsto y) \in (f : X \to Y)$, then $f(x) = y$.[8]

X and Y above are sets of values. A specification may use named sets, such as *integers*, *reals*, *Booleans*, and *strings*; ranges, such as 0–100 or \$9,999,999.99CR–\$9,999,999.99; explicit sets of values, such as $\{Open, Closed\}$; or dynamic sets of values that are modified by events.

As in our discussion of FD graphs, we need to be clear about domains and ranges, because we want to distinguish whether functions are partial or total, into or onto, many-to-one or one-to-one. Traditionally, the domain and range of a function are defined in one of two ways: either as the sets of *possible* values that pairs in the function can be constructed from, or as the sets of *actual* values that appear in the pairs of the function. Neither of these is really satisfactory for our purposes.

Consider the function that specifies the *Programme* to which a *Candidate* is admitted. There are many possible *Programme Codes*, most of which haven't been allocated to any programme yet. There are many possible candidates, many of whom aren't interested in studying anything. If we consider domains to include all *possible* elements, all the functions we consider will be partial and into. On the other hand, if we consider, for example, only those candidates and programmes where a candidate is actually admitted to the programme, the functions we consider will all be total and onto. We want to express, for example, the idea that a candidate *must* be admitted to a programme, but a programme might not have any candidates in it. Such a function

---

[7] Any future attempt to assign a value to S' would be treated as a test for equality.

[8] An alternative, which the reader may prefer, is to use the object-oriented notation, $x.f = y$. Unfortunately, this style doesn't look so good when composite domains are used, e.g., $(x, y).g = z$.

would be total and into. To do this, we define *Candidates* as a subset of *Admission Nos*, and *Programmes* as a subset of *Programme Codes*,

$$Candidates = \{c \in Admission\ Nos \mid Candidate\ State(c) \neq Null\}$$
$$Programmes = \{c \in Programme\ Codes \mid Programme\ State(c) \neq Null\}.$$

As a result, although *Programme* : *Admission Nos* → *Programme Codes* is a partial function, *Programme* : *Candidates* → *Programmes* is a total function.[9]

As in the previous chapter, the undecorated FD $f : X \to Y$ is assumed to be *total*, *many-to-one*, and *onto*, i.e., $f : X\ {}^{+}{\to}^{1}\ Y$. In all other cases, the arrow should be decorated. For example, g: $X \to^{?} Y$ denotes an FD that is *partial*, *many-to-one*, and *onto*, and $h : X\ {}^{1}{\to}\ Y$ denotes a function that is *total*, *one-to-one*, and *onto*. Thus, an FD that relates candidates to programmes would be specified as *Programme* : *Candidates* $^{*}{\to}$ *Programmes*.

One way to present the FDs would be as an FD graph. Although such a graph is good for tutorial purposes, the graph for a real system is likely to be too complex to understand, and in any case, drawing and maintaining graphics is hard work. An alternative would be to present the same information as a set of table schemas. However, even this might suggest too strongly how the system should be implemented. Here, we shall simply list the FDs.

We begin with the FDs associated with *Candidates*:[10]

$$Candidate\ State : Admission\ Nos \to \{Admitted, Graduated, Null\} = Null.$$
$$Candidates = \{c \in Admission\ Nos \mid Candidate\ State(c) \neq Null\}.$$
$$Candidate\ State : Candidates \to \{Admitted, Graduated\}.$$
$$Personal\ Name : Candidates \to Personal\ Names.$$
$$Postal\ Address : Candidates \to Addresses.$$
$$Phone : Candidates \to^{?} Phone\ Numbers.$$
$$Date\ of\ Birth : Candidates \to Dates.$$
$$Programme : Candidates\ {}^{*}{\to}\ Programmes.$$
$$Balance\ Owing : Candidates \to Money.$$

The undecorated FDs, *Personal Name*, *Postal Address*, *Date of Birth*, and *Balance Owing* are total, many-to-one, and onto. This means that every member of *Candidates* must have an associated *Personal Name*, *Postal Address*, *Date of Birth*, and *Balance Owing*. What is more, because the functions are onto, the database shouldn't store

[9] This idea could be extended to *Undergraduates* = {*c* ∈ *Candidates* | *Candidate State*(*c*) = *Admitted*} and *Graduates* = {*c* ∈ *Candidates* | *Candidate State*(*c*) = *Graduated*}, but we shan't pursue that possibility here.

[10] The reader should carefully note the difference between *Candidate State* : *Candidates* → {*Admitted*, *Graduated*} and *Candidate State* : *Admission Nos* → {*Admitted*, *Graduated*, *Null*}.

values for those properties that have no associated member of *Candidates*. In contrast, *Phone* is partial, many-to-one, and onto, meaning that a candidate doesn't need to have a phone number, and *Programme* is total, many-to-one, and into, meaning that some *Programmes* may exist that have no *Candidates*.

Notice that the first *Candidate State* FD is assigned the initial value *Null*. This implies that the initial value of *Candidates* is {}, the empty set. In a DBMS implementation, this means the *Candidates* table would initially be unpopulated. If a different data structure were used, the values of *Candidate State* might need to be explicitly initialised.

The remaining FDs aren't given initial values, so their values remain undefined.

From this example, you will see that I prefer to use self-explanatory names for FDs, such as *Date of Birth*, rather than cryptic abbreviations such as *DOB*. You will also see that domains are given plural names, but no precise definitions at this stage, as the numbers of lines and characters in *Addresses*, for example, are really a matter for the *interfaces*; they might be determined by the height and width of the window in a postal envelope. In some cases (*Phone Numbers* and *Dates*) the choice is externally constrained, but the format is still left vague here.

In similar manner, we can define the attributes of *Programmes*:[11]

$$\begin{aligned}
\textit{Programme State} &: \textit{Programme Codes} \rightarrow \{\textit{Active, Obsolete, Null}\} = \textit{Null}.\\
\textit{Programmes} &= \{p \in \textit{Programme Codes} \mid \textit{Programme State}() \neq \textit{Null}\}.\\
\textit{Programme State} &: \textit{Programmes} \rightarrow \{\textit{Active, Obsolete}\}.\\
\textit{Programme Name} &: \textit{Programmes} \overset{1}{\rightarrow} \textit{Programme Names}.\\
\textit{Degree} &: \textit{Programmes} \rightarrow \textit{Degree Names}.
\end{aligned}$$

From which we see that *Programme Name* is total, one-to-one, and onto. In other words, every programme has both a unique code and a unique name.

Before we can proceed to specify subjects and enrolments, we need to decide whether being *Frozen* is a property of an enrolment, of a subject, of a semester, or of the system as a whole. After all, once enrolments become billable, *all* enrolments in *all* subjects become frozen. For tutorial reasons, we have chosen to make *Frozen* a property of *Subjects* so as to agree with the FSA of Figure 3.19 (page 101).

A *Subject* is identified by its *Subject Code*, or *Subject Name*, and may have one of five states:

$$\begin{aligned}
\textit{Subject State} &: \textit{Subject Codes} \rightarrow \{\textit{Inactive, Offered, Frozen, Obsolete, Null}\}\\
&= \textit{Null}.\\
\textit{Subjects} &= \{s \in \textit{Subject Codes} \mid \textit{Subject State}(s) \neq \textit{Null}\}.\\
\textit{Subject State} &: \textit{Subjects} \rightarrow \{\textit{Inactive, Offered, Frozen, Obsolete}\}.
\end{aligned}$$

[11] Personally, I am not against overloading names of FDs that have different domains, or even different ranges, providing that the particular FD referred to is clear from its argument. So both *Candidate State* and *Programme State* might more simply have been called *State*, but I have used different names here to reduce confusion in the mind of the reader.

$$\begin{aligned}
\textit{Subject Name} &: \textit{Subjects} \xrightarrow{1} \textit{Subject Names}.\\
\textit{Quota} &: \textit{Subjects} \rightarrow \textit{Integers}.\\
\textit{Filled} &: \textit{Subjects} \rightarrow \textit{Integers}.\\
\textit{Fee} &: \textit{Subjects} \rightarrow \textit{Money}.
\end{aligned}$$

Quota is the number of students allowed to enrol in the subject, a number that is often limited by the size of a laboratory or lecture theatre. *Filled* is the current number of enrolments — which must never exceed *Quota*. As remarked earlier, *Filled* is redundant, being equal to the number of enrolments in the subject. For virtually *any* infrastructure that is finally chosen, the analyst considers it will prove easier and faster to keep *Filled* up to date than to count enrolments. *Fee* is the amount that a candidate will have to pay for studying the subject.

A *Component* is a *Subject* in the context of a particular *Programme* and is identified by the combination of *Subject* and *Programme*:

$$\begin{aligned}
\textit{Component State} &: \textit{Programmes} \times \textit{Subjects} \rightarrow \{\textit{Null, Offered}\} = \textit{Null}.\\
\textit{Components} &= \{c \in \textit{Programmes} \times \textit{Subjects} \mid \textit{Component State}(c) \neq \textit{Null}\}.\\
\textit{Component State} &: \textit{Components} \rightarrow \{\textit{Offered}\}.
\end{aligned}$$

A member of *Components* has only one possible state: *Offered*. In a DBMS, if a *Component* row existed, it would necessarily be in the *Offered* state. Consequently, the *Component State* column would be redundant and not needed at all.

It is useful to give names to *Components'* projection functions,

$$\begin{aligned}
\textit{Programme} &: \textit{Components} \xrightarrow{*} \textit{Programmes}.\\
\textit{Subject} &: \textit{Components} \xrightarrow{*} \textit{Subjects}
\end{aligned}$$

meaning that some programmes may exist that currently have no component subjects, and some subjects might not currently belong to any programme.

Finally, we need to consider enrolments. A candidate can only be enrolled in a given subject once per semester, therefore an enrolment can be uniquely identified by the triple *Semesters* × *Candidates* × *Subjects*:

$$\begin{aligned}
\textit{Enrolment State} &: \textit{Semesters} \times \textit{Candidates} \times \textit{Subjects} \rightarrow\\
&\quad \{\textit{Enrolled, Assessed, Null}\} = \textit{Null}.\\
\textit{Enrolments} &= \{e \in \textit{Semesters} \times \textit{Candidates} \times \textit{Subjects}\\
&\quad \mid \textit{Enrolment State}(e) \neq \textit{Null}\}.\\
\textit{Enrolment State} &: \textit{Enrolments} \rightarrow \{\textit{Enrolled, Assessed}\}.\\
\textit{Semester} &: \textit{Enrolments} \xrightarrow{*} \textit{Semesters}.\\
\textit{Candidate} &: \textit{Enrolments} \xrightarrow{*} \textit{Candidates}.\\
\textit{Subject} &: \textit{Enrolments} \xrightarrow{*} \textit{Subjects}.\\
\textit{Mark} &: \textit{Enrolments} \xrightarrow{?} 0\text{–}100.
\end{aligned}$$

The decorations on the arrows make it clear that every member of *Enrolments* must have an *Enrolment State*, but not every enrolment needs to have a *Mark*.[12] The projection functions make it clear that, for example, a *Subject* may have zero enrolments.

5.7 SPECIFYING EVENTS

An event is modelled by a procedure, usually having one or more parameters. In this section, we first focus on the **kernel events** that change the state of the database.

An event is always atomic, but it should also be *minimal*. An event procedure should specify the least action that leaves the system in a consistent state. For example, an event procedure might change a candidate's *Postal Address*, but not both the candidate's *Personal Name* and *Postal Address*, because this can safely be split into two more primitive actions. Conversely, an event shouldn't be too small: updating only the first line of an address might leave the system in an inconsistent state, in which the address as a whole is invalid. If, owing to a previous typographical error, only part of an address needs to be changed, it isn't hard to design an interface that allows an existing address to be edited.

The *Admit* procedure below has the parameters c, the *Admission No* of a new candidate, and p, the *Programme Code* of the programme to which the candidate is to be admitted. How the procedure is invoked is a matter for the interface: in times past, c might have been a number punched on a card or typed on a command line; more recently, c might have been selected using a graphical interface.

The main objective of *Admit* is to add the admission number c to the set of *Candidates*. Additionally, since all the FDs with domain *Candidates* (except for *Phone*) are specified as total, they need to be initialised. The procedure must check that the candidate is not already admitted and that the candidate's proposed study programme is currently *Active*:

$$
\begin{aligned}
Admit(c,p) \Leftarrow\ & Candidate\ State(c) = Null\,; \\
& Programme\ State(p) = Active\,; \\
& Candidate\ State(c)' = Admitted\,; \\
& Programme(c)' = p\,; \\
& Personal\ Name(c)' = \text{'Unknown'}\,; \\
& Postal\ Address(c)' = \text{'Unknown'}\,; \\
& Phone(c)' = \mathbf{null}\,; \\
& Date\ of\ Birth(c)' = 0\,; \\
& Balance\ Owing(c)' = 0.
\end{aligned}
$$

[12] If we wanted to be fussy about this, we could define $Assessments = \{e \in Enrolments \mid Enrolment\ State(e) = Assessed\}$, then make *Mark* a total function of *Assessments*.

The above is an algebraic style of specification and the one that the author prefers. Even so, the reader is free to adopt a different style, provided it is unambiguous. For example, the following style might better suit readers familiar with *SQL* — and it strongly suggests the need to add a row to the *Candidates* table:

$$\begin{aligned}
Admit(c,p) \Leftarrow\ & c \notin Candidates\,; \\
& p \in Programmes\,; \\
& Candidates' = Candidates \cup \{c\}\,; \\
& c.Candidate\ State' = Admitted\,; \\
& c.Programme' = p\,; \\
& c.Personal\ Name' = \text{'Unknown'}\,; \\
& c.Postal\ Address' = \text{'Unknown'}\,; \\
& c.Phone' = \mathbf{null}\,; \\
& c.Date\ of\ Birth' = 0\,; \\
& c.Balance\ Owing' = 0.
\end{aligned}$$

This version contains a superfluous term. Setting *Candidate State*$(c)'$ to *Admitted* already implies that c should be added to *Candidates*.

Mathematically, the two styles are the same: the condition *Candidate State*(c) = *Null* is equivalent to $c \notin$ *Candidates* by definition, and the relational notation *c.Phone* is equivalent to the functional notation *Phone*(c) — provided, of course, that *Phone* is a function.

An important aspect of the *kernel* specification is what it *fails* to specify. It does not mention what happens if an event's pre-conditions (e.g., $c \notin$ *Candidates*, $p \in$ *Programmes*) fail. It is assumed that the *interface* deals with failure according to the current technology. In the days of punched cards, a message might have been printed, but using a more modern graphical interface, the candidate and programme might have already been selected from lists in such a way that no error could arise. Error messages are discussed in a later chapter, but foreshadowing what is to come, let us agree that a message that says what *is* so is superior to one that says what is *not* so: 'John Doe has already graduated from Diploma in IT' is superior to 'John Doe is not admitted'.

The attentive reader will have noticed that the types of *c, p,* etc., aren't defined in the parameter list of a procedure. That is because their types are self-evident. In an assignment such as $x' = y$, x and y are assumed to have the same type, so if either type is known, the other is implied. On the other hand, where there is ambiguity, the types of working variables may be defined *globally* for all events:

$$c \in Admission\ Nos.$$

$$p \in Programme\ Codes.$$

$$n \in Personal\ Names.$$

$$a \in Postal\ Addresses.$$

$$ph \in Phone\ Numbers.$$
$$dob \in Dates.$$
$$m \in Money.$$

This convention imposes a burden on the specification writer always to use the same name to denote the same kind of thing — in short, not to be deliberately confusing.

The *Admit* procedure needs to be supplemented by a set of procedures to set the additional attributes of a candidate, such as

$$Set\ Name(c,n) \Leftarrow (Candidate\ State(c) \neq Null\,;Personal\ Name(c)' = n)$$

and so on.

Of course, it would have been possible for the *Admit* event itself to accept additional parameters specifying the candidate's name, address, etc. But after discussion, this more modular approach was chosen. One reason is that *Admit*, *Set Name*, etc., each keep the database in a consistent state, so why combine them? Further, there is no problem with an input form invoking a succession of kernel events to record all a candidate's details in what appears to be a single operation.[13]

After discussion with enrolment staff, we learn that a candidate cannot *Graduate* if any fees are outstanding:

$$\begin{aligned} Graduate(c) \Leftarrow\ & Candidate\ State(c) = Admitted\,; \\ & Balance\ Owing(c) = 0\,; \\ & Candidate\ State(c)' = Graduated. \end{aligned}$$

(*Admitted* is really a family of candidate states, depending on the value of *Balance Owing*. As a result, the *Graduate* event may fail. This kind of detail is hard to show in an event-state graph such as that of Figure 3.18.)

Consequently, we need an event to allow a candidate to *Pay*, otherwise graduation is impossible:

$$\begin{aligned} Pay(c,m) \Leftarrow\ & Candidate\ State(c) = Admitted\,; \\ & Balance\ Owing(c)' = Balance\ Owing(c) - m. \end{aligned}$$

The specifications of events that change the state of a *Subject* are straightforward:

$$\begin{aligned} Create\ Subject(s,sname) \Leftarrow\ & Subject\ State(s) = Null\,; \\ & sname \notin Subject\ Names\,; \\ & Subject\ State(s)' = Inactive\,; \\ & Subject\ Name(s)' = sname\,; \end{aligned}$$

[13] Later, we shall learn that in order to populate the database sufficiently for examination timetables to be prepared, these parameters are unnecessary and unwanted.

$$\begin{aligned}&\mathit{Quota}(s)' = 0\,;\\&\mathit{Filled}(s)' = 0\,;\\&\mathit{Fee}(s)' = 0.\end{aligned}$$

$$\mathit{Set\ Fee}(s,m) \Leftarrow \mathit{Subject\ State}(s) = \mathit{Inactive}\,;\mathit{Fee}(s)' = m.$$

$$\begin{aligned}\mathit{Set\ Quota}(s,\mathit{size}) &\Leftarrow \mathit{Subject\ State}(s) = \mathit{Inactive}\,;\mathit{Quota}(s)' = \mathit{size}.\\\mathit{Set\ Quota}(s,\mathit{size}) &\Leftarrow \mathit{Subject\ State}(s) = \mathit{Offered}\,;\\&\quad \mathit{size} \geq \mathit{Filled}(s)\,;\mathit{Quota}(s)' = \mathit{size}.\end{aligned}$$

$$\mathit{Offer}(s) \Leftarrow \mathit{Subject\ State}(s) = \mathit{Inactive}\,;\mathit{Subject\ State}(s)' = \mathit{Offered}.$$

$$\begin{aligned}\mathit{Freeze} \Leftarrow\ &\mathit{Offered\ Subjects} = \{s \in \mathit{Subjects} \mid \mathit{Subject\ State}(s) = \mathit{Offered}\}\,;\\&(\mathbf{all}\ s1 \in\ \mathit{Offered\ Subjects}\,;\mathit{Subject\ State}(s1)' = \mathit{Frozen})^{*}.\end{aligned}$$

The *Set Quota* procedure illustrates why we use the $\Leftarrow$ operator to define events. An event succeeds *if* any right-hand side expression succeeds — not *if and only if*. This allows for rules to be added piecemeal, allowing for error-processing rules to be added later on. In this case, there are already two rules: the first allows the *Quota* to be set to any value if the subject is currently *Inactive*, and the second prevents the *Quota* being reduced below the number of places already filled if the subject is currently *Offered*.

The *Freeze* event procedure is an example of how the specification language expresses iteration. First, a subset of *Subjects*, *Offered Subjects*, is formed, then the bracketed expression is repeated zero or more times, until it fails. On each iteration, the **all** operator chooses an arbitrary subject $s1$ from *Offered Subjects*.[14] The *Subject State* of $s1$ is then *Frozen*. Once all the members of *Offered Subjects* have been selected, the iteration stops.

Since *Enrol* events concern only the *Current Semester*, it is convenient to record it globally:

$$\mathit{Current\ Semester} \in \mathit{Semesters}.$$

Beginning a new semester is a matter of setting the new value of *Current Semester*. For each subject, we also reset *Filled* to zero. Naturally, this shouldn't be done while any subject is still *Offered*.

$$\begin{aligned}\mathit{Begin\ Semester}(\mathit{Sem}) \Leftarrow\ &\{s \in \mathit{Subjects} \mid \mathit{Subject\ State}(s) = \mathit{Offered}\} = \{\}\,;\\&\mathit{Current\ Semester}' = \mathit{Sem}\,;\\&(\mathbf{all}\ s1 \in \mathit{Subjects};\mathit{Filled}(s1)' = 0)^{*}.\end{aligned}$$

[14] We could have used s here instead of $s1$. The use of s in the definition of *Offered Subjects* is local.

Or, equivalently,

$$\begin{aligned}
&\mathit{Begin\ Semester(Sem)} \Leftarrow \nexists s\ (\mathit{Subject\ State}(s) = \mathit{Offered});\\
&\quad \mathit{Current\ Semester}' = \mathit{Sem};\\
&\quad \mathit{Frozen\ Subjects} = \{s1 \in \mathit{Subjects} \mid \mathit{Subject\ State}(s1) = \mathit{Frozen}\};\\
&\quad (\mathbf{all}\ s1 \in \mathit{Frozen\ Subjects};\ \mathit{Filled}(s1)' = 0)^*.
\end{aligned}$$

Enrolments can be made only after satisfying several conditions. These conditions necessarily include those in the *Enrol* row of Table 3.2 and the *Enrol* transition in Figure 3.21:

$$\begin{aligned}
\mathit{Enrol}(c,s) \Leftarrow\ & \mathit{Candidate\ State}(c) = \mathit{Admitted};\\
& \mathit{Subject\ State}(s) = \mathit{Offered};\\
& \mathit{Filled}(s) < \mathit{Quota}(s);\\
& p = \mathit{Programme}(c);\\
& \mathit{Component\ State}(p,s) = \mathit{Offered};\\
& e = (\mathit{Current\ Semester}, c, s);\\
& \mathit{Enrolment\ State}(e) = \mathit{Null};\\
& \mathit{Enrolment\ State}(e)' = \mathit{Enrolled};\\
& \mathit{Mark}(e)' = \mathbf{null};\\
& \mathit{Filled}(s)' = \mathit{Filled}(s) + 1;\\
& \mathit{Balance\ Owing}(c)' = \mathit{Balance\ Owing}(c) + \mathit{Fee}(s).
\end{aligned}$$

The procedure specifies that if c is *Admitted* to *Programme*(c), the state of s is *Offered*, s is part of the programme to which c is admitted, c isn't currently enrolled in s, and at least one place remains in the subject,[15] then the new state of the enrolment is assigned the value *Enrolled*, a place in the subject is filled, and the candidate's debt is incremented by the fee. In short, much of the logic of the *Enrol* event is implicit in Table 3.2 and Figure 3.21 and is a direct result of event-state analysis.

Note the difference between **null**, which means that *Mark* has no value, and *Null*, which means that *Enrolment State* has the value *Null*.

The *Withdraw* event undoes the work of *Enrol*, provided that the subject hasn't yet been frozen:

$$\begin{aligned}
\mathit{Withdraw}(c,s) \Leftarrow\ & e = (\mathit{Current\ Semester}, c, s);\\
& \mathit{Enrolment\ State}(e) = \mathit{Enrolled};\\
& \mathit{Subject\ State}(s) = \mathit{Offered};\\
& \mathit{Enrolment\ State}(e)' = \mathit{Null};
\end{aligned}$$

[15] Here, the author has been guilty of anticipating part of the implementation: rather than writing *Filled*(s), we should be writing $|\{e \in \mathit{Enrolments} \mid \mathit{Subject}(e) = s \land \mathit{Semester}(e) = \mathit{Current\ Semester}\}|$. None of us is perfect.

$$Filled(s)' = Filled(s) - 1\,;$$
$$Balance\ Owing(c)' = Balance\ Owing(c) - Fee(s).$$

In this case, it might be expected from Table 3.2 that the procedure should check that *Candidate State(c)* ≠ *Null*, but this is unnecessary, because only a candidate in the *Admitted* state can enrol, and there is no path in the *Candidate* DFA from *Admitted* to *Null*. Even so, the inclusion of the pre-condition *Candidate State(e)* ≠ *Null* would certainly be harmless, and might prove valuable as a defence against possible programming errors.

Candidates are assessed and assigned a mark only *after* a subject has been frozen:

$$\begin{aligned} Assess(c,s,score) \Leftarrow\ & Subject\ State(s) = Frozen\,;\\ & e = (Current\ Semester, c, s)\,;\\ & Enrolment\ State(e) = Enrolled\,;\\ & Enrolment\ State(e)' = Assessed\,;\\ & Mark(e)' = score. \end{aligned}$$

Once all the candidates in a subject have been assessed, the subject can become *Inactive* again. This involves checking that no enrolment is still in the *Enrolled* state:

$$\begin{aligned} Deactivate(s) \Leftarrow\ & Subject\ State(s) = Frozen\,;\\ & \nexists c\ (Enrolment\ State(Current\ Semester, c, s) = Enrolled)\,;\\ & Subject\ State' = Inactive. \end{aligned}$$

We also need the means to add subjects to programmes, and to remove them:

$$\begin{aligned} Create\ Programme(p, pname, deg) \Leftarrow\ & Programme\ State(p) = Null\,;\\ & pname \notin Programme\ Names\,;\\ & Programme\ State(p)' = Active\,;\\ & Programme\ Name(p)' = pname\,;\\ & Degree\ Name(p)' = deg. \end{aligned}$$

$$\begin{aligned} Permit(p,s) \Leftarrow\ & Programme\ State(p) \neq Null\,;\\ & Subject\ State(s) \neq Null\,;\\ & Component\ State(p,s) = Null\,;\\ & Component\ State(p,s)' = Offered. \end{aligned}$$

$$\begin{aligned} Forbid(p,s) \Leftarrow\ & Component\ State(p,s) = Offered\,;\\ & Component\ State(p,s)' = Null. \end{aligned}$$

Most of the printed reports that any system generates are used internally and are really part of the solution, not the problem. On the other hand, the NUTS Academic

Records System is legally required to generate two reports on demand: an academic transcript, which lists all a candidate's results, and a *mark sheet*, which lists each candidate's marks in a given subject:

Print Transcript$(c) \Leftarrow c \in$ *Candidates* ;
  Transcript.Heading$(c,$ *Personal Name*(c), *Postal Address*(c), *Programme*$(c))$;
  When $= \{$*Semester*$(e) \mid e \in$ *Enrolments* $\wedge$ *Candidate*$(e) = c\}$;
  (**asc** *Sem* $\in$ *When* ;
    What $= \{e \in$ *Enrolments* $\mid$ *Semester*$(e) =$ *Sem* $\wedge$ *Candidate*$(e) = c\}$;
    (**asc** $e1 \in$ *What* ;
      $s =$ *Subject*(e) ;
      (*Enrolment State*$(e1) =$ *Assessed* ;
       Transcript.Result(*Sem*, s, *Subject Name*(s), *Mark*$(e1)$)
)
    $\cup$ (*Enrolment State*$(e1) =$ *Enrolled* ;
       Transcript.Current(*Sem*, s, *Subject Name*(s))
)
    $)^*$
  $)^*$;
  Transcript.All Done.

The first thing to explain is how the *Print Transcript* event communicates with its output interface. The output subsystem that formats the academic transcript is a separate program object called *Transcript*. It provides four *methods*: *Heading*, *Result*, *Current*, and *All Done*. Each method is responsible for the correct rendering of the data it is given onto a printed form, but it is *not* responsible for retrieving it. Output methods have no direct interface with the database.

The *Print Transcript* procedure contains two, nested, loops. The outer loop steps through the semesters during which the candidate made at least one enrolment. Its function is purely to impose temporal order on the contents of the transcript. The inner loop steps through the enrolments within the semester selected by the outer loop. It distinguishes between those subjects that have been assessed and those that have not, invoking *Result* or *Current* accordingly. After the completion of the outer loop it invokes *All Done* so that *Transcript* can terminate the output promptly and correctly.

The *Print Marks* procedure is similar:

Print Marks$(s) \Leftarrow$ *Subject State*$(s) =$ *Frozen* ;
  Mark Sheet.Heading(s, *Subject Name*(s), *Current Semester*, *Filled*(s)) ;
  Marks List $= \{e \in$ *Enrolments* $\mid$

$$Subject(e) = s \wedge Semester(e) = Current\ Semester\}\ ;$$

(**asc** $e1 \in Marks\ List$;
  $c = Candidate(e1)$;
  $p = Programme(c)$;
  $Mark\ Sheet.Result(c,\ Personal\ Name(c),\ Mark(e1), p,\ Programme\ Name(p))$
)* ;
Mark Sheet.All Done.

Notice that *Marks List* fixes the values of *Subject* and *Semester* — but not *Admission No*, so the individual enrolments and marks for the current semester will be listed in order of *Admission No* rather than *Personal Name*. *Mark Sheet* is an output object that provides methods that correctly format and print the results. (For examples of what mark sheets might look like, consult Figure 8.5 and Figure 8.6 on pages 264 and 265.)

There are several other procedures needed to complete the kernel specification, e.g., to set the value of a candidate's phone number, but these have no additional tutorial value.

5.8 SUMMARY

The documentation of business requirements is an important part of system development. It is important not to over-specify requirements. For example, the texts of error messages and the designs of interfaces are not business requirements, but are influenced by the technology of implementation. It is important that a specification doesn't anticipate what infrastructure will be used for implementation, as the choice of infrastructure should be a separate consideration. Also, the specification should only consider atomic events, not complex business procedures. For example, enrolling a new candidate consists of admitting the candidate to a degree programme, recording personal information, and enrolling the candidate in several subjects. Each of these steps can be a separate event. If it is later decided that these steps will be accomplished by the candidate completing a single form, that is an implementation decision, not a business requirement. An event is a *smallest* useful step that preserves the consistency of the database. Even recording personal information can be broken down into several smaller events.

We have seen that event-state modelling is an important step in developing the specifications of update events. An analysis such as that of Table 3.2 or Figure 3.21 spells out the values state variables should have before and after each event. Event procedures must also preserve database constraints, such as referential integrity and key constraints. In addition, there are more subtle constraints, such as the need to make sure that *Filled*(s) always reflects the actual number of enrolments in subject s. It is a good idea to document all such database constraints and then prove mathematically that the event procedures preserve them, but we shall not pursue that idea here.

With so many constraints, specifying event procedures becomes almost mechanical. As a result, writing a specification can be an early phase of analysing a system and can document important requirements and decisions. How the specification should be re-packaged into business or computer processes is a matter for later deliberation.

Inspection events, which merely report the state of a database, will be discussed more fully in a later chapter.

System specifications benefit from being expressed in a formal language, making them concise and unambiguous. There are many possible choices for such a language. The author has used a style that has both a mathematical and a procedural interpretation, and which uses a mixture of both mathematical and programming conventions. Feel free to adapt it or invent your own!

5.9 FURTHER READING

There are several well-known specification languages. The two most widely known are the Z specification language and VDM (Vienna Development Method). I haven't used these notations here, first, because I didn't want to devote time and space to explaining them and, second, because I wanted to use a consistent and minimal set of notations throughout the book — even if they aren't the ones normally used in particular fields.

Those interested in Z might wish to read Jonathan Bowen's book 'Formal Specification and Documentation Using Z: A Case Study Approach' (1996, ISBN: 1-85032-230-9), Ben Potter, Jane Sinclair, and David Till's 'An Introduction to Formal Specification and Z' (1991, ISBN: 0-13-478702-1), or Jonathan Jacky's 'The Way of Z: Practical Programming with Formal Methods' (1997, ISBN: 0-521-55976-6).

Those interested in VDM might prefer Cliff Jones's 'Software Development: A Rigorous Approach' (1980, ISBN: 0-13-821884-6).

My personal interest in using the algebra of relations in programming and specification was stimulated by my old tutor, John G. Sanderson, whose ideas are recorded in 'A Relational Theory of Computing', *Lecture Notes in Computer Science, 82* (1980, ISBN: 3-540-09987-5).

A restricted form of *HTML* was described by Tim Berners-Lee on the Internet in 1991. *HTML* is now supported by *ANSI* and *ISO* standards.

5.10 EXERCISES

1. It is usual to provide a separate update event for every attribute in a database, with the possible exception of those that can change a primary or foreign key. In the case of the Academic Records System, which attributes do you think should be allowed to be updated, and which should not?
2. Specify an event that will set the value of *Phone*, and a second event that will allow it to be restored to **null**.

3. It is proposed to allow candidates to protect their data using passwords. Passwords are usually stored in encrypted form, in this case, as an integer.
   Specify an FD that allows each candidate to store an encrypted password.
   Assuming that you already have an *encrypt* function available, specify an event to check the correctness of a password, and a second event to change a password. The change should only be allowed if the existing password is known and the new password isn't the same as the old one.
   Which existing procedure will need to be modified?

CHAPTER

Database technology

6

CHAPTER CONTENTS

INTRODUCTION

In this chapter, our motives are first to show how the tables used to represent a system's state can be manipulated to answer complex queries or to produce reports and, second, to show how multiple users can update the system state concurrently without mutual interference and confusion.

We shall discuss these topics using the notation of *SQL*. *SQL* is sufficiently close to predicate calculus to be abstract, but it has the advantage that many systems are indeed built around an *SQL* database management system (DBMS).[1]

This chapter is not intended to be an *SQL* tutorial, but it does contain enough of an introduction to *SQL* for the reader to follow the remaining discussion. We are less interested in the language than in how a DBMS works. We have at least three reasons for this:

- Not every system is built using a commercial DBMS, and its database must be built from scratch.
- Features common to *all* shared data, whether an *SQL* database or not, profoundly affect system design.
- Not every *SQL* DBMS is of the highest quality, and the programmer may have to guide it carefully.

6.1 THE *SQL* DATA DEFINITION LANGUAGE

For the most part, relational databases are accessed and updated using *SQL*, **Structured Query Language**. As its name suggests, *SQL* has powerful support for answering queries. Its support for updates is rather limited. *SQL* isn't a programming language; it is a *sub-language* that must be embedded in some other procedural language. (A popular choice is *PHP*.) *SQL* itself is **non-procedural**. When the user executes a query, the DBMS creates a procedure to answer it efficiently. One motive for describing *SQL* here is that the language closely mirrors what a DBMS can do and leads us naturally to the question of how a database schema can be optimised.

SQL includes a **data-definition language**. We illustrate its use by showing how the NUTS Academic Records System database would be declared.

SQL prefers the definitions of parent tables to precede those of their children, so we begin with the *Programmes* table, which has no parent:[2]

```
create table Programmes (
    Programme              char(4),
    Programme_State        varchar(10) not null,
    Programme_Name         varchar(32) unique,
    Degree                 varchar(32) not null,
    check (Programme_State = 'Active'
        or Programme_State = 'Obsolete'),
    primary key (Programme)
    );
```

[1] I am *not* suggesting that an implementor who uses some other technology should develop a general-purpose query language equivalent to *SQL*, but should merely provide special-purpose interfaces having a similar level of abstraction.

[2] *SQL* keywords are shown in **bold**.

To create the definition, we have had to commit to some decisions that weren't included in the kernel specification. First, we had to decide that a *Programme* would be uniquely identified using a 4-character code, which we have made the primary key of the table. This decision was made because 4-letter abbreviations are already in use within the existing academic records system. Second, *SQL* doesn't support variables with enumerated values such as {*Null*, *Active*, *Obsolete*} directly. Instead, we declare *Programme State* as a ten-character string and **check** it is assigned a correct state.[3] (We use **varchar**(10) rather than **char**(10) so that we can write 'Active' rather than 'Active   '.) We don't need to include *Null* in an *SQL* context; *Null* is merely a way of saying that the corresponding value of *Programme* isn't represented by an actual row in the table. Third, if a row does exist, *Programme State* and *Degree* must be specified, and cannot be **null**. Fourth, *Programme Name* is given the **unique** attribute to enforce its correspondence with *Programme*. We also have decided *Degree* is to be at most 32 characters in length. Finally, although *Programme* cannot be **null**, *SQL* doesn't require this to be specified; it is already implied by it being the primary key of the table.

The *Subjects* table also has no parent:

```
create table Subjects (
        Subject                 char(8),
        Subject_State           varchar(10) not null,
        Subject_Name            varchar(32) not null unique,
        Quota                   integer not null,
        Filled                  integer not null,
        Fee                     decimal(7,2) not null,
        check (Subject_State = 'Inactive' or Subject_State = 'Offered'
                or Subject_State = 'Frozen' or Subject_State = 'Obsolete'),
        constraint Fee_Range check (Fee between 100.00 and 1000.00),
        primary key(Subject)
        );
```

Specifying that *Subject Name* is **unique** means that no two subjects can have exactly the same name. This is intended to reduce confusion in the system's environment and eliminate a potential source of error. The example also defines *Fee* as having seven digits with two decimal places, so that it may have a value as high as $99,999.99.[4] This makes room for a certain degree of inflation, but if the cost should ever exceed this maximum, *SQL* provides means by which the format can be changed. In addition, the constraint *Fee Range* limits the values of *Fee* still further. Because the constraint is named, a programmer can easily change the range with a single *SQL* command.

---

[3] In reality, states would almost certainly be represented by cryptic one-byte codes, but we are trying to keep the examples easy to understand.

[4] Yes, *Fee* is almost certain to be a whole number of dollars, but we are illustrating a point of *SQL* syntax here!

The *Components* table specifies which *Subjects* are part of which *Programmes*:

```
create table Components (
        Programme        char(4),
        Subject          char(8),
        primary key (Programme, Subject),
        foreign key (Subject) references Subjects(Subject),
        foreign key (Programme) references Programmes(Programme));
```

This example shows how foreign keys are declared, preventing a *Component* from referring to an imaginary *Subject* or an imaginary *Programme*.

Notice that *Components* doesn't need a *Component State* attribute. *Components* have only two states: *Active* and *Null*. If the row exists, *Component State* must be *Active*.

The *Candidates* table has *Programmes* as its parent:

```
create table Candidates (
        Candidate          char(8),
        Candidate_State    varchar(10) not null,
        Personal_Name      varchar(32) not null,
        Street             varchar(32) not null,
        Suburb             varchar(32),
        State              varchar(24),
        Post_Code          varchar(8),
        Phone              varchar(12),
        Date_of_Birth      date not null,
        Programme          char(4) not null,
        Balance_Owing      decimal(7,2) not null,
        check (Candidate_State = 'Admitted'
              or Candidate_State = 'Graduated'),
        check (Balance_Owing <= 0
              or Candidate_State <> 'Graduated'),
        primary key (Candidate),
        foreign key (Programme) references Programmes(Programme)
        );
```

In this table, *Postal Address* has become four columns: *Street*, *Suburb*, *State*, and *Post Code*. This is because *SQL* supports only simple attributes, and it was decided that it will prove more convenient to split the address into separate lines, as discussed in the next section. *Date of Birth* is stored as an *SQL* **date** (a format in which the integer part represents whole days since a certain starting date, and the fractional part represents the time of day). Two **checks** are declared: one ensures that *Candidate State* has a sensible value, and the second ensures a candidate cannot graduate while money is still owed.

In practice, not every DBMS supports **check** clauses, and provided the event procedures have been specified and implemented correctly, the clauses should never detect an error, so they could safely be omitted.

Now that its parents have been declared, we may declare the *Enrolments* table:[5]

```
create table Enrolments (
    Candidate               char(8),
    Subject                 char(8),
    Semester                char(5),
    Enrolment_State         varchar(10) not null,
    Mark                    smallint,
    check (Enrolment_State = 'Enrolled'
        or Enrolment_State = 'Assessed'),
    check (Enrolment_State = 'Enrolled' or Mark is not null),
    check (Mark is null or Mark between 0 and 100),
    primary key (Candidate, Subject, Semester),
    foreign key (Candidate) references Candidates(Candidate),
    foreign key(Subject) references Subjects(Subject)
    );
```

Mark cannot have the **not null** constraint because *Mark* only has a value after the enrolment has been assessed.

We shall assume that a semester is represented by a five-character string containing the four-digit year and a single letter 'A' or 'B', e.g., '20yyA'. We need the value of *Current Semester* to be stored because enrolments can only be made for the current semester. Unfortunately, many DBMSs cannot store isolated variables, so it is sometimes necessary to create a table containing one row. We shall call the table *Globals*. It isn't unusual to store several such system-wide items, for example, *Current Date*, used to fool the system into believing the current date is different from the actual date — which is especially useful during testing, debugging, or error recovery:

```
create table Globals (
    The_Key                 smallint,
    Current_Semester        char(5) not null,
    Current_Date            date not null,
    check The_Key = 1,
    primary key The_Key
    );
```

Since the *Enrolments* table has *Semester* as part of its primary key, it is tempting to make *Current Semester* a foreign key in the *Enrolments* table. On the other hand,

---

[5] The primary key of *Enrolments* has three components. The order of the components of a composite key usually determines how rows will be clustered within the database. For this reason, the choice of the order is important. As written, we can expect that all the rows for a given *Candidate* will lie within the same page or the same few pages.

remember that the *Enrolments* table stores rows for previous semesters, which would then have no parent. This, in turn, might tempt us into storing previous values of *Current Semester* in the *Globals* table, but this is not only pointless, it makes it harder to know which row represents the current semester.[6]

6.2 FROM SCHEMA TO *SQL*

Non-prime attributes are those that don't appear as determinants in any FD, are not candidate keys of tables, and are terminal vertices of an FD graph. In most cases they are simply numbers or text strings, but the NUTS Academic Records System illustrates some exceptions.

First, we treated *Postal Address* as a simple attribute earlier, but addresses are usually displayed or printed on several lines. The line endings could either be indicated by special characters embedded in a single text string or by dividing the string into lines. Thus, in the *SQL* example, *Postal Address* appeared as *Street*, *Suburb*, *State*, and *Post Code*, it having been decided that this should make the programming task simpler.[7]

Second, *Degree Name* has a small set of possible values, and other values should be forbidden. One option would be to use an *SQL* constraint to prevent wrong values being chosen. A second option would be to use a look-up table. A **look-up table** is an (usually small) *SQL* table whose only purpose is to define a set of possible values.[8] Among other things, such a table can be used to create a drop-down list from which an operator can choose a value on a form. The choice between these two options depends on several factors, but the main factor is stability. If the set of values is unlikely to change an *SQL* constraint is often the better choice. For example, the names of the days of the week are unlikely to change, and could be defined as an *SQL* constraint. On the other hand, new values of *Degree Name* are likely to arise in the future, and by storing the set in a look-up table, it can easily be updated using *SQL*'s **insert** command.[9] Since either of these alternatives adds to the burden of implementation, a third alternative should be considered, which is to do nothing. In this case, it is a good choice: the *Programmes* table is small and rarely updated, and misspelling a degree name would have little impact.

Third, since *Component State* has only one non-null value, it needn't be stored. It is sufficient for a row of *Components* to exist to know that *Component State* isn't *Null*.

Fourth, and most problematic, is *Semester*. It is tightly constrained, because enrolments can only be made in the current semester. It is better for the value of

[6] It would be possible to make *Current Semester* a constraint in the *Enrolments* table, but this isn't a good idea. A constraint can tell us when a value is wrong, but a constraint cannot be interrogated to discover what a value should be.

[7] A similar treatment could be given to the candidate's name, but we are trying to keep the example short.

[8] The *Components* table is little more than a look-up table with a composite key.

[9] A variation on the look-up table uses a short code to access the value of a long text string. This is useful when it saves space in the database, which will happen if each string value appears many times.

Current Semester to be supplied by the system, rather than be supplied by hand. The correct value could be specified by a look-up table containing exactly one value, or derived from the current date. We choose *not* to use the date, because it is sometimes necessary to lie about dates, as when external paperwork isn't completed on time, and so on. Instead, we choose to use a look-up table. Our motive is to illustrate a common practice: a system often needs to store global variables and constants; often the only convenient way to store them is in an *SQL* table with a single row, called *Globals*, for example. (If the NUTS Academic Records System were planned to be used in more than one institution, the name of the institution could be also stored in *Globals*, along with other site-specific information.)

Finally, I must confess that the *Frozen* state of *Subjects* is really a state of *Globals*: all subjects are frozen at the same time in each semester. The *Frozen* property was assigned to *Subjects* for a tutorial reason: the discussion of shared events in Section 3.9 would have been even more confusing if a fourth interacting atom had been involved.

6.3 *SQL* QUERIES

We shall look at *SQL*'s **query language** to see how it relates to event specifications. But our ultimate interest isn't in *SQL* itself, but in how the computer answers the queries.

An *SQL* query can combine information from several tables in almost any way imaginable. This means that the designer of the database schema is freed from considering the reports and forms that will display what is in the database and can focus on the logical relationships between the atoms it represents.

SQL was originally intended for non-programmers to be able to ask for information from complex databases. For that reason, it has a sophisticated syntax for expressing queries. The basic form of an *SQL* query is

select *column1*, *column2*, ...
    from *table1*, *table2*, ...
    where *selection condition*
    order by *columnA*, *columnB* ...;

where {*columnA*, *columnB*, ... } is a subset of {*column1*, *column2*, ... }.

As an example, consider a query to find the names and marks of all candidates in all subjects that have been assessed, in ascending order by subject and candidate name:

```
select s.Subject_Name, c.Candidate_Name, e.Mark
    from Subjects s, Candidates c, Enrolments e
    where e.Subject = s.Subject
    and e.Candidate = c.Candidate
    and e.Enrolment_State = 'Assessed'
    order by asc s.Subject_Name, asc c.Candidate_Name;
```

It may seem curious that we need to specify that the *Candidate* and *Subject* in the *Enrolments* rows match the primary keys of the *Candidates* and *Subjects* rows, given that the three tables are linked by foreign keys. But *SQL* forces us to specify the obvious because, just occasionally, we might want to do something different.

A **where** clause supports the usual arithmetic and comparison operators and also has a **like** operator that compares character strings with patterns containing wild cards. This allows searches for strings containing a given word, starting with a given letter, and so on.

In our specification language, the above **select** statement would be equivalent to

$$\begin{aligned}\{(&\mathit{Subject\ Name}(s), \mathit{Personal\ Name}(c), \mathit{Mark}(e))\\ &\mid s \in \mathit{Subjects} \wedge c \in \mathit{Candidates} \wedge e \in \mathit{Enrolments}\\ &\wedge \mathit{Subject}(e) = \mathit{Subject\ Code}(s)\\ &\wedge \mathit{Candidate}(e) = \mathit{Admission\ No}(c)\\ &\wedge \mathit{Enrolment\ State}(e) = \mathit{Assessed}\}\end{aligned}$$

from which it can be seen that *SQL* queries are equivalent to predicate calculus expressions. However, the **order by** clause is an additional feature, because elements of a set are unordered. In our specification language, this particular **order by** clause would be implemented by using the **asc** operator to select elements starting with the smallest.

Notionally, a **select** statement forms the Cartesian product of the tables in its **from** clause, selects the rows in the product that satisfy the condition in its **where** clause, projects the rows onto the columns in the opening **select** clause, then sorts them according to its **order by** clause. *Notionally*. If there were 100,000 candidates (mostly graduates), 500 subjects, and 2,000,000 enrolments (mostly in earlier semesters), the Cartesian product would contain 100,000,000,000,000 rows. In practice, a **query optimiser** within the DBMS finds an efficient way to answer the query. In this case the optimiser has merely to take each *Enrolments* row and find its matching *Candidates* and *Subjects* rows, using the foreign keys in the *Enrolments* table. We discuss query optimisation shortly.

SQL queries can also *summarise* information, using **group by** and **having** clauses. These can use three statistical functions: **sum**, **avg**, and **count** (the number of rows in the group).

The following query will highlight all subjects in which the average performance of candidates has been poor:

```
select s.Subject_Name, Average = avg(e.Mark)
    from Subjects s, Enrolments e
    where e.Subject = s.Subject
    and e.Semester = '20yyA'
    and e.Enrolment_State = 'Assessed'
    group by s.Subject_Name
```

```
having Average < 50
order by asc Average;
```

6.3.1 SUB-QUERIES

The reason for the word 'structured' in 'Structured Query Language' is that the result of one query can be used in the **where** or **having** clause of an enclosing query. For this to be possible, the enclosed **sub-query** must return a single value or a set of values from a single column. We can adapt the previous query so that it always refers to the current semester:

```
select s.Subject_Name, Average = avg(e.Mark)
    from Subjects s, Enrolments e
    where e.Subject = s.Subject
    and e.Enrolment_State = 'Assessed'
    and e.Semester =
        (select Current_Semester
            from Globals
            where The_Key = 1)
    group by s.Subject_Name
    having Average < 50
    order by asc Average;
```

To find all candidates who scored better than average in *Maths101*, we could write

```
select c.Personal_Name, e1.Mark
    from Candidates c, Enrolments e1
    where c.Candidate = e1.Candidate and e1.Subject = 'Maths101'
    and e1.Mark >
        (select avg(e2.Mark)
            from Enrolments e2
            where e2.Subject = 'Maths101')
    order by asc c.Personal_Name;
```

In this example, the sub-query must be evaluated first, yielding the average mark in *Maths101*. Only then can the outer query be answered.[10]

To answer the same question for *all* subjects requires only a small change:

---

[10] In contrast, the following query fails:

```
select c.Personal_Name, e.Mark
    from Candidates c, Enrolments e
    where c.Candidate = e.Candidate and e1.Subject = 'Maths101'
    and e.Mark > avg(e.Mark)
    order by asc c.Personal_Name;
```

The reason is that e.Mark > **avg**(e.Mark) cannot be evaluated in a single inspection of the table rows. This example shows why, despite the best intentions, computer literacy remains a prerequisite to using *SQL*.

```
select s.Subject_Name, c.Personal_Name, e.Mark
    from Enrolments e1, Subjects s, Candidates c
    where e1.Candidate = c.Candidate and e1.Subject = s.Subject
    and e1.Mark >
        (select avg(e2.Mark)
            from Enrolments e2
            where e2.Subject = e1.Subject)
    order by asc s.Subject_Name, c.Personal_Name;
```

Unfortunately, in this case the inner sub-query can no longer be evaluated independently, because the subject to be averaged depends on the outer query. A situation where the inner query must be evaluated many times is called a **correlated sub-query**. The alternative case is called an **uncorrelated sub-query**. The two cases can easily be distinguished; a correlated sub-query always refers to variables defined in the outer query. An uncorrelated query never does.

The following query finds the *Subject Codes* of all *Subjects* in which the average mark was below 50:

```
select s.Subject
    from Subjects s, Enrolments e
    where e.Subject = s.Subject
    group by s.Subject
    having avg(e.Mark) < 50;
```

The result of the following inner query is a *set* of values that can be referred to in an outer query using the **in** (∈) operator. It lists all candidates who scored a mark of at least 50 in those subjects where the average mark was below 50.

```
select s1.Subject_Name, c.Personal_Name, e1.Mark
    from Enrolments e1, Subjects s1, Candidates c
    where e1.Candidate = c.Candidate and e1.Subject = s1.Subject
    and e1.Mark >= 50 and s1.Subject in
        (select s2.Subject
        from Subjects s2, Enrolments e2
        where e2.Subject = s2.Subject
        group by s2.Subject
        having avg(e2.Mark) < 50
        )
    order by asc s1.Subject_Name, asc c.Personal_Name;
```

6.3.2 VIEWS

By default, the results of *SQL* queries are sent to the user's monitor, or they can be directed to a printer. Additionally, it is possible to use the result of a query to create a new table, in this example, one that contains the subjects where the average mark was below 50:

```
create table Weak_Subjects_Snapshot as
    select s.* from Subjects s, Enrolments e
        where e.Subject = s.Subject
        group by s.Subject
        having avg(e.Mark) < 50;
```

where s.∗ means *all* columns of the *Subjects* table. The result could then be used in a more complex query.

An alternative would be to create a **view**:

```
create view Weak_Subjects_View as
    select s.* from Subjects s, Enrolments e
        where e.Subject = s.Subject
        group by s.Subject
        having avg(e.Mark) < 50;
```

A view can be used in a query wherever a table can be used. The difference between a view and a table is that a view is a *procedure*, whereas a new table is a snapshot. When the *Enrolments* table is updated, *Weak Subjects Snapshot* will continue to reflect the marks as they were when the snapshot was created. In contrast, *Weak Subjects View* will be freshly evaluated.

A common use of views is to bring together columns from different tables to provide missing detail:

```
create view Mark_Sheets_View as
    select s.Subject_Name, c.Personal_Name, e.Mark
        from Subjects s, Candidates c, Enrolments e
        where e.Subject = s.Subject and e.Candidate = c.Candidate;
```

Views are a key feature of *SQL*, which an implementor would do well to imitate even if special home-made data structures (rather than a DBMS) are used. The ability to use views means that tables do *not* need to be designed with reports or forms in mind. For example, the names of candidates don't need to be recorded in each *Enrolments* row just because they appear on every line of a mark sheet. They can be recorded once only, in the *Candidates* table.

6.3.3 EMBEDDED *SQL*

SQL is a sub-language, designed to be **embedded** in a **host language**. The host language passes *SQL* statements to the DBMS. Usually, a DBMS also has a built-in facility to allow users to interact with the database through a command-line or graphic interface. When it does so, the results of queries are often displayed on a monitor or printed. Unfortunately, such an interface allows too much freedom for secure use, and special application programs are usually written to ensure that the database is correctly managed.[11] *SQL* typically interfaces with the application program by means of **return codes** and **cursors**.

---

[11] A database administrator (DBA) can control who is allowed to issue various *SQL* commands.

An *SQL* return code typically indicates whether a query was successful and, if not, why it failed.

A cursor is a pointer to what is effectively a row in a table returned by a query. A cursor may be advanced one row at a time, so that each row may be processed in whatever way is desired.

Recall the specification of the *Freeze* event for the Academic Records System:

$$Freeze \Leftarrow Offered\ Subjects = \{s \in Subjects \mid Subject\ State(s) = Offered\}\ ;$$
$$(\textbf{all}\ s1 \in Offered\ Subjects\ ; Subject\ State(s1)' = Frozen)^*.$$

This might be translated into embedded *SQL*, as follows:[12]

```
declare
        Buffer Subjects row type;
        cursor Offered_Subjects as
                select * from Subjects
                        for update of Subject_State
                        where Subject_State = 'Offered';
begin
        open Offered_Subjects;
        fetch Offered_Subjects into
                Buffer.Subject, Buffer.Subject_State, Buffer.Subject_Name,
                Buffer.Quota, Buffer.Filled, Buffer.Fee;
        loop
                exit when Offered_Subjects not found
                update Subjects set Subject_State = 'Frozen'
                        where current of Offered_Subjects;
                fetch Offered_Subjects into
                        Buffer.Subject, Buffer.Subject_State,
                        Buffer.Subject_Name, Buffer.Quota,
                        Buffer.Filled, Buffer.Fee;
        end loop;
        close Offered_Subjects;
        commit;
end;
```

The example begins by declaring *Buffer* to be a data structure with the same format as a row of the *Subjects* table. The second declaration is of a **cursor**, *Offered Subjects*, which can be thought of as declaring a table to contain the result of the **select** statement, which will retrieve all the rows of *Subjects* whose state is *Offered*. The **for update of** clause signals the intention to update *Subject State*, which, as

[12] The language used in this example is broadly similar to *PL/SQL*. Because the syntactic rules of *SQL* sometimes conflict with those of the host language, an example in *PHP*, say, would be full of syntactic clutter to distinguish *SQL* statements from host language statements.

we shall see in a later section, will prevent interaction with other users who may be simultaneously trying to use the same column.

The procedure itself begins by **opening** the cursor, which has the effect of executing the **select** statement to actually load the table. The **fetch** statement then transfers the first row of the table into *Buffer*. There then follows a **loop**, which will **exit** when an *SQL* return code indicates that there are no more rows to fetch. Within the loop, the **update** statement sets the *Subject State* of the row of the *Subjects* table that matches the current row of *Offered Subjects*. This is followed by a **fetch** statement to refresh the buffer, and the loop repeats. After the loop exits, the cursor is **closed**, freeing the space occupied by the results of the query. Finally, the **commit** statement is executed, and the procedure ends.

The purpose of the **commit** statement is to make the procedure atomic. Changes to the *Subjects* table are buffered until the **commit**, then written out as a whole. If the procedure fails before the **commit**, the database remains unaffected. The **commit** statement also relinquishes control of *Subject State*, making it free for other users to access.

It should be clear from this example that the style of our kernel specification language is well suited to conversion to embedded *SQL*. The assignment of a set expression to *Offered Subjects* in the *Freeze* event procedure corresponds pretty closely to the **select** clause, and the use of the **all** operator corresponds to **fetch**.[13]

It is worth noting that in addition to the **update** command, *SQL* provides **insert** and **delete** commands to create new rows and to remove existing rows.

6.4 QUERY OPTIMISATION

The power of the *SQL* query sub-language is that there is always a simple procedure that can find the result:

1. Form the Cartesian product of a set of tables.
2. *Select* a subset of the resulting rows.
3. *Project* the result onto a subset of their attributes.
4. *Sort* the result into order.

We have four tools at our disposal: **Joining** is a process of combining information from two tables. **Selection** is the action of choosing rows that satisfy a certain condition. **Projection** is the act of ignoring all columns except those of interest. Finally, sorting is the means by which the **order by** clause is satisfied.

Query optimisation is the art of combining these operations most effectively. There are four ways this can be achieved:

1. By careful programming of the query,
2. By the database administrator creating indices and choosing file structures that speed up queries,

[13] The aim of this example is to show how embedded *SQL* is used. In fact, the *Freeze* event can be implemented by a single *SQL* command: **update** Subjects s **set** s.State = 'Frozen' **where** s.State = 'Offered';

3. By designing the most suitable database schema,
4. By the DBMS's query optimiser selecting the most efficient method.

Consider the following example, designed to list marks for the current semester, which is 20yyA:

```
select p.Programme_Name, s.Subject_Name, c.Personal Name, e.Mark
    from Subjects s, Candidates c, Enrolments e, Programmes p
    where e.Subject = s.Subject
    and e.Candidate = c.Candidate
    and c.Programme = p.Programme
    and e.Enrolment_State = 'Assessed'
    and e.Semester = '20yyA'
    order by asc p.Programme_Name,
        asc s.Subject_Name,
        asc c.Personal_Name;
```

Suppose that the database instance contains 500 *Subjects* rows, 100,000 *Candidates* rows (all but 10,000 of them for past graduates), 2,000,000 *Enrolments* rows (all but 20,000 for previous semesters), and 100 *Programmes* rows. Then the Cartesian product will contain 10^{14} rows, of which at most 2,000,000 can possibly satisfy the query, and only 20,000 actually will.

An example of the first strategy would be for the programmer to recognise that the query can only concern current candidates, so adding the condition *c.Candidate_State* = 'Admitted' would help to narrow the query and perhaps give a quicker result. In fact, the programmer can do much more, as we shall discuss below.

An example of the second strategy might be for the database administrator to provide an index to the *Enrolments* table by *Semester*, so that the current enrolments may be found more quickly — or, what would be almost equivalent, to make *Semester* the first part of the *Enrolments* table's primary key. This will cause all the *Enrolments* rows for the *Current Semester* to be **clustered** together and reduce the number of pages that need to be examined.

An example of the third strategy might be to store enrolments for *current* candidates in a separate table from enrolments for graduates, reducing the number of rows more than tenfold.

Finally, we consider what the query optimiser of a DBMS can do. One good strategy would be for the DBMS to immediately discard all *Enrolments* rows except those with *e.Enrolment_State* = 'Assessed' and *e.Semester* = '20yyA'. Let us assume that this leaves 20,000 *Enrolments* rows. Because *Enrolments* and *Subjects* are in a foreign key-primary key relationship, their join should create an intermediate result with 20,000 rows. From this result, all columns except *e.Candidate*, *s.Subject_Name*, and *e.Mark* can be discarded. The intermediate result is then joined with *Candidates*, retaining the columns *c.Name*, *c.Programme*, *s.Subject_Name*, and *e.Mark*. Finally, the join with *Programmes* yields the required result. At no point does an intermediate result exceed 20,000 rows.

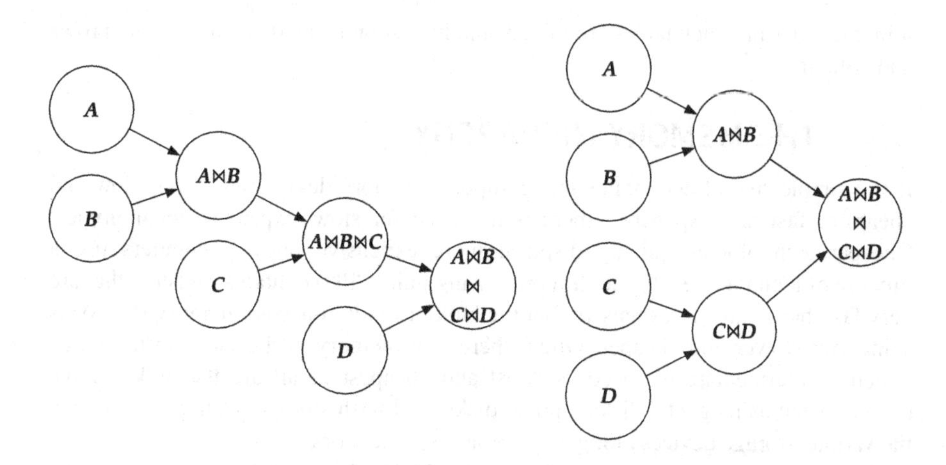

FIGURE 6.1

The two distinct join trees with four leaves. The notation *A* ⋈ *B* denotes the join of *A* and *B*. *A, B, C,* and *D* may be assigned to *Subjects*, *Candidates*, *Enrolments*, and *Programmes* in 24 possible ways.

In contrast, to begin by joining *Enrolments* and *Programmes* would be a bad move, because they have no foreign key-primary key relationship, and it would lead (even after eliminating unwanted *Enrolments* rows) to an intermediate result with 2,000,000 rows. Likewise, joining *Candidates* and *Subjects* is a bad move, giving an intermediate result with 50,000,000 rows. (Fortunately, after such a bad start, the *final* join can be made efficiently by matching both *Subject* and *Candidate*.)

These examples show the importance of choosing a good sequence of joins. Regrettably, the number of possible **join trees** grows intractably with the number of tables involved.

Four tables, as in this example, have two distinct join trees, shown in Figure 6.1. In the left-hand tree, tables are added stepwise to an intermediate result. In the right-hand tree, two separate intermediate results are formed and joined. There are 4! = 24 ways *Subjects*, *Candidates*, *Enrolments*, and *Programmes* can be assigned to *A*, *B*, *C*, and *D*. In the left-hand tree the leaves *A* and *B* are interchangeable, so only 12 labellings are distinct, and in the right-hand tree, once we decide which table should be joined with *Subjects*, the rest is settled. There are therefore 15 join trees in all. Clearly, a query optimiser must attempt to minimise the sizes of intermediate results. Potentially, these results will use a lot of storage. Even if the results can be **streamed**[14] rather than stored, they still take time to be generated and processed. A query optimiser therefore considers three factors: the best algorithm to make each

[14] That is, consumed as quickly as they are generated, avoiding the need to store more than one current row.

join, the order in which tables are joined, and how soon it can discard unwanted rows and columns.

6.4.1 THE MEMORY HIERARCHY

It is a simple rule of economics that computer memory devices are either slow and cheap, or fast and expensive; there is no place for slow, expensive technologies. Cheaper technologies quickly displace more expensive ones. Computers use a mixture of technologies: the central processing unit (CPU) contains registers that are very fast, but relatively expensive, but the bulk of random-access memory (RAM) is somewhat slower and cheaper. Often, there is a memory cache close to the CPU, which is intermediate in speed. Slowest and cheapest of all are the bulk storage devices, such as magnetic disks, optical disks, and flash storage. Data flow between the various storage devices along a bus or across a network.

Modern computers often have enough RAM for the whole of a database to be stored in main memory. Unfortunately, RAM is **volatile**; when power is interrupted, data are lost. Secondary media are **persistent**; data are retained almost indefinitely or until they are overwritten by new data. Therefore, even a **main-memory database** needs to be shadowed on a persistent medium.[15]

Data are retrieved from secondary storage in *blocks* or *pages*, as we outlined in Section 2.6.4, when discussing *B-trees*. Blocks contain **check sums**, redundant bytes that verify that data bytes aren't corrupted when blocks are read or written. Check sums depend on *all* the data in a block, not individual bytes. Therefore it is impossible to read less than a whole block without risk of error.

The time it takes to read or write data consists of two components: **latency** and **transfer time**. Latency is the time it takes to locate the block concerned, and transfer time is how long it takes to move the data to RAM. During the latency period, no data are moved. Once the latency has expired, there is little further cost in moving a whole page (one or more blocks), rather than just a few bytes. For example, a hard drive might have an average latency of 10 milliseconds (during which the read-write head is moved to the correct track, and the disk rotates to the correct position), a transfer rate of 100 MB/s, and a block size of 1 KB. Thus it would take 10.01 milliseconds to read 1 KB, 11 milliseconds to read 100 KB, 12 milliseconds to read 200 KB, and so on.

A complicating issue is the use of **virtual memory**. In a virtual memory system, RAM is subdivided into **pages** (1 KB, for example). A page may either be stored in RAM or on a secondary medium. The idea is that when the RAM becomes full, idle pages are **swapped out** of RAM to secondary storage, and some RAM is freed for a new page. If a page that has been swapped out contains data that the CPU needs, it will be **swapped in** again, and a different page will need to be swapped out to make room for it. Virtual memory systems work well when the number of page swaps is

---

[15] This may be done by simply saving it to disk from time to time, but more sophisticated means are also used.

small.[16] Usually, the delays caused by swapping are acceptable if they occur when a user switches from one task to another, but not when swapping occurs within a single task.

The performance of virtual memory often proves better than one might expect because of the **80:20 rule**, which states that 80% of the accesses will concern 20% of the data. The rule applies recursively, so that 80% of 80% of the accesses concern 20% of 20% of the data. Conversely 64% of the data is only accessed 4% of the time. The 'rule' is an experimental one and has no theoretical foundation — but it works.

Unfortunately, because the allocation of data pages isn't under the control of the programmer, virtual memory isn't usually the most effective way to use secondary storage when accessing a large database. It is better to choose a data structure that respects block boundaries. As a result, data structures and algorithms that are efficient in RAM can be poor choices for block structured media.[17]

6.4.2 JOIN ALGORITHMS

There are three basic join algorithms: **Nested Loops**, **Table Look-Up**, and **Sort-Merge**. Of these, *Nested Loops* is the most general, being capable of **cross joins** (many-to-many joins, including full Cartesian products), but it is the least efficient. *Table Look-Up* and *Sort-Merge* are more efficient, but can only make many-to-one joins, typically between a foreign key and a primary key. There are other hybrid algorithms that can be derived from them.

6.4.2.1 *Nested Loops*

Consider tables $T_1(\underline{A},B,C)$ and $T_2(\underline{X},Y,Z)$ with join condition $T_1.C\ \theta\ T_2.Z$ yielding the table $J(\underline{A,X},B,C,Y,Z)$. ($\theta$ is an arbitrary relationship.) The *Nested Loops* algorithm (Algorithm 6.1) considers every row of table T_2 once for each row of table T_1, thus finding every pair of the Cartesian product. Nonetheless, only those pairs that satisfy the join condition are added to the result.

The join condition can be any relationship between any set of attributes. Unfortunately, the number of iterations of the inner loop is $|T_1|\times|T_2|$, which makes it inferior to the two algorithms that follow.

6.4.2.2 *Table Look-Up*

Consider child table $T_1(\underline{A},B,C)$ and parent table $T_2(\underline{X},Y,Z)$ with join condition $T_1.C = T_2.X$, yielding the table $J(\underline{A},B,C,Y,Z)$. C is a foreign key in T_1 that references

---

[16] Suppose that it takes 1 nanosecond for the CPU to access a word in RAM, but 10 milliseconds to swap in a word from secondary storage. This ratio is 10,000,000:1. If only one access in 10,000 causes a swap, the average access time will be 1,000 times worse than it would be with sufficient RAM.

[17] Even so, many applications (for example, an editor such as the one I am using now) are much more easily programmed by assuming that all data are in RAM. As I make additions and corrections, almost all activity occurs within a small area of text. I don't expect swapping to occur except perhaps when I typeset or search the whole text.

Algorithm 6.1 The Nested Loops join.

1. Initialise the join to be empty.
2. For each row R_1 of T_1,
   - For each row R_2 of T_2,
     - If R_1 and R_2 satisfy the join condition ($T_1.C\ \theta\ T_2.Z$),
       Add a new row to the join containing the attributes of R_1 and R_2.

primary key X of T_2. The *Table Look-Up* algorithm (Algorithm 6.2) considers every child row in table T_1 and matches it with its *unique* parent row in table T_2.

Algorithm 6.2 The Table Look-Up join.

1. Initialise the join to be empty.
2. For each row R_1 of child table T_1,
   - Use the foreign key (C) in R_1 to find the parent row R_2 in T_2 with the matching primary key (X).
     - Add a new row to the join containing the attributes of R_1 and R_2.

The algorithm makes two assumptions: First, that every row of T_1 is associated with exactly one row of T_2, which will be true if the DBMS enforces child-parent referential integrity. Second, that selected rows of T_2 can be accessed efficiently. Assuming that T_2 is indexed using a B-tree, or similar structure, this should take time $O(\log |T_2|)$. The complexity of *Table Look-Up* is then $O(|T_1| \times \log |T_2|)$.

6.4.2.3 Sort-Merge

Again consider child table $T_1(\underline{A}, B, C)$ and parent $T_2(\underline{X}, Y, Z)$ with join condition $T_1.C = T_2.X$, yielding the table $J(\underline{A}, B, C, Y, Z)$. The *Sort-Merge* algorithm begins by sorting table T_1 into ascending order by foreign key C. It is assumed here that since X is its primary key, T_2 is already in ascending order by X and doesn't need to be sorted.

Algorithm 6.3 takes each row of the parent table (T_2) and joins it with every row of the child table (T_1) that has the matching foreign key. Because the child table has been sorted on the foreign key, this amounts to a single sequential pass through the sorted table.

Although *Sort-Merge* has two nested loops, and the total number of iterations of the outer loop is $|T_2|$, the total number of iterations of the inner loop is only $|T_1|$. Each

Algorithm 6.3 The Sort-Merge join.

1. Sort the child table T_1 into sequence by its foreign key (C) referencing the primary key (X) of T_2, creating the sorted table S_1.
2. Initialise the join to be empty.
   - For each parent row R_2 of T_2
     - For each child row R_1 of S_1 whose foreign key matches the primary key of R_2,
       Add a new row to the join containing the attributes of R_1 and R_2.

child row is examined exactly once. Additionally, sorting T_1 to produce S_1 takes time $O(|T_1| \times \log |T_1|)$. The complexity of *Sort-Merge* is therefore $O(|T_1| \times \log |T_1|)$.

6.4.2.4 Comparison of Sort-Merge with Table Look-Up

Whether *Sort-Merge* or *Table Look-Up* is faster depends on the sizes of T_1 and T_2. Nonetheless, it would be simplistic to compare the terms $|T_1| \times \log |T_1|$ and $|T_1| \times \log |T_2|$ because the constant factors for the algorithms are wildly different.

In most situations, retrieving pages from secondary storage dominates the time a database query takes to complete. If a page of a table has **blocking factor** B, that means the page contains B rows. When we examine all the rows of a table in *sequential* order of its primary key, we can expect to fetch each page of the table exactly once. Each time we fetch a page, we fetch B rows, so the total number of page fetches for N rows is $N \div B$. On the other hand, when we fetch a row *at random*, we are forced to fetch the whole page that contains it. Therefore, to fetch N rows we need N page fetches. (For simplicity, we dismiss the possibility that two consecutive fetches happen to retrieve the same page.)

Let us assume that T_1 and T_2 have blocking factors B_1 and B_2 respectively: both *Sort-Merge* and *Table Look-Up* fetch $|T_1|$ rows sequentially, either from table T_1 or from table S_1, implying that both make $|T_1| \div B_1$ page fetches from T_1. In contrast, *Table Look-Up* fetches $|T_1|$ rows from table T_2 at random, implying $|T_1|$ page fetches, but *Sort-Merge* fetches all $|T_2|$ rows from T_2 sequentially, implying $|T_2| \div B_2$ page fetches. Ignoring the time taken to sort T_1 to give S_1, the question is, which is greater: $|T_1|$, or $|T_2| \div B_2$?

Since $|T_2| \div B_2$ is the number of pages in table T_2, this gives the following rule of thumb:

> If the number of *rows* in the child table exceeds the number of *pages* of the parent table, *Sort-Merge* is faster than *Table Look-Up*, but if the number of *rows* in the child table is fewer than the number of *pages* of the parent table, *Table Look-Up* is faster than *Sort-Merge*.

In the case of a many-to-one foreign key-primary key join, it is almost certainly the case that $|T_1| > |T_2|$, so *Sort-Merge* is faster even for $B_2 = 1$. On the other hand, if, owing to an earlier *selection* process, only a small fraction of the child rows participate in the join, *Table Look-Up* can be faster.

What about the terms we neglected in making the comparison? Since the output of both algorithms is the same, the time to create the join rows is the same. This leaves just one factor: the time taken to sort T_1 to create S_1. Since *Table Look-Up* is faster than *Sort-Merge* only when T_1 is small, the rule of thumb remains accurate because the sorting time is then negligible. (Even so, if T_1 were *extremely* large, the time to sort it would eventually dominate all other factors, but this never[18] happens in practice.)

6.4.2.5 Hybrid Algorithms

A useful idea is to combine the sorting step of *Sort-Merge* with the direct access by primary key used by *Table Look-Up*. This means that rows are read from the parent file in an orderly fashion rather than at random. When $|T_1| < |T_2| \div B_2$, this **Skip-Sequential algorithm** behaves like *Table Look-Up*. If $|T_1| > |T_2| \div B_2$, it behaves like *Sort-Merge*.

Sorting can also be used to improve the performance of *Nested Loops*. If two tables are joined on a common attribute K (that isn't a primary key), both tables should first be sorted on K. If K has k different values, this partitions each table into k clusters. In the best case, where each value of K is equally likely, *Nested Loops* performs k cross joins that are each $1/k^2$ times smaller, reducing the number of iterations of the inner loop by a factor of k. The speed-up will be less when the distribution is non-uniform.

In some DBMSs, foreign key integrity is implemented by linking the parent rows to their child rows, perhaps using a linked list. This permits a fourth distinct algorithm: the parent table is scanned sequentially row by row, and then, for each parent row, each associated child row is found by consulting the list. Since this implies random access to child rows, which almost certainly outnumber parent rows, this algorithm is essentially *Table Look-Up* using the wrong table and cannot be recommended. On the other hand, if the primary key of the child row is composite, and includes the parent key as its first component, then the child rows will be clustered by the parent key, and the algorithm behaves like *Sort-Merge*. It is most useful when the selection criteria limit the query to just a few parent rows.

The use of a cache further complicates the choice of a join algorithm. For example, if the parent table in *Table Look-Up* can be stored entirely in RAM, then slow random access to secondary storage is replaced by fast random access to RAM, and the only transfer cost is the time taken to copy the parent table sequentially into RAM. Indeed, in many applications the entire database will easily fit into RAM, and the choice of a join algorithm hardly matters.

---

[18] Well, hardly ever. Only for truly vast databases.

6.4.3 JOIN TREES

As we saw above, there are 15 possible join trees for a query involving four tables. Even if we consider only those trees that join tables to the current result one at a time, given N tables, there are still $N! \div 2$ orders in which they can be added, so the number of possible trees grows intractably. Fortunately, in our above analysis of join algorithms, we have discovered a useful heuristic: *consider natural joins first*, i.e., parent-child relations enforced by a foreign key and primary key, which therefore avoid cross joins and the use of *Nested Loops*.

If we again consider the query

```
select p.Programme_Name, s.Subject_Name, c.Personal_Name, e.Mark
    from Subjects s, Candidates c, Enrolments e, Programmes p
    where e.Subject = s.Subject
    and e.Candidate = c.Candidate
    and c.Programme = p.Programme
    and e.Enrolment_State = 'Assessed'
    and e.Semester = '20yyA'
    order by asc p.Programme_Name,
        asc s.Subject_Name,
        asc c.Personal_Name;
```

we see that there are three natural joins: *Enrolments* ⊗ *Subjects*, *Enrolments* ⊗ *Candidates*, and *Candidates* ⊗ *Programmes*.[19] Fortunately, three joins are enough to construct a join tree for four tables, so we don't have to consider silly alternatives, such as finding the Cartesian product of *Subjects* and *Candidates*, and can focus on the six possible sequences of natural joins.

Different DBMSs differ in their knowledge of the database instance. Some may keep track of exactly how many *Enrolments* rows have *e.Enrolment State* = 'Assessed' and how many have *e.Semester* = '20yyA'; others may not. But almost every DBMS will know how many rows each table contains.[20]

Suppose, as we did earlier, that the database instance contains 500 *Subjects* rows, 100,000 *Candidates* rows (all but 10,000 of them for past graduates), 2,000,000 *Enrolments* rows (all but 20,000 for previous semesters), and 100 *Programmes* rows.

Because the joins are many-to-one, we can be sure that the join *Enrolments* ⊗ *Subjects* will yield no more than 2,000,000 rows, the join *Enrolments* ⊗ *Candidates* will also give at most 2,000,000 rows, and the join *Candidates* ⊗ *Programmes* will give exactly 100,000 rows. The kernel specification ensures that current enrolments cannot belong to past graduates, but no DBMS can deduce this.

Let us suppose that the DBMS believes that only 20,000 *Enrolments* rows have *e.Semester* = '20yyA'. Then, after selecting those particular rows from *Enrolments*, the join *Enrolments* ⊗ *Candidates* will contain at most 20,000 rows. Joining this

[19] We are using the notation *Child* ⊗ *Parent* to denote the join of *Child* and *Parent*.

[20] An exception might occur when an interface tool is used to make a collection of *Cobol* (say) files appear to be a relational database.

result with *Programmes* can then yield at most 20,000 rows. Since this is better than the 100,000 rows that would result from the uninformed join *Candidates* ⊗ *Programmes*, clearly, the tree (*Enrolments* ⊗ *Candidates*) ⊗ *Programmes* is better than *Enrolments* ⊗ (*Candidates* ⊗ *Programmes*).

With this constraint, there are only three possible trees:

((*Enrolments* ⊗ *Candidates*) ⊗ *Programmes*) ⊗ *Subjects*

((*Enrolments* ⊗ *Subjects*) ⊗ *Candidates*) ⊗ *Programmes*

((*Enrolments* ⊗ *Candidates*) ⊗ *Subjects*) ⊗ *Programmes*

There is little to choose between these alternatives. Because the only *selection* step occurs at the start, they all carry forward the same number of rows in their intermediate results. The final choice depends on the numbers of bytes in each row, in other words, it depends on *projecting* only those columns that are needed for the result. Since *Subject Name*, *Programme Name*, and *Personal Name* are all 32 characters long, it doesn't matter in which order they are added to the result. The third alternative has a marginal disadvantage, in that it carries *Programme Code* through two stages rather than one. Of the remaining two alternatives, the first is slightly better, because an *Admission No* is 4 bytes longer than a *Subject Code*. Figure 6.2 shows the optimum join tree.

The placement of *selection* and *projection* steps in a join tree is straightforward. They are placed as early in the tree as possible, remembering that *projection* needs to preserve not only the columns that appear in the final result, but also those used in selection or join conditions.

We may contrast the procedure outlined in Figure 6.2 with a situation in which a DBMS doesn't keep statistics about the values of attributes. This less-informed procedure would probably join the *Candidates* and *Programmes* tables first, generating an intermediate result with 100,000 rows. This isn't a disaster, but it is less than optimal.

As a result of the weaknesses of some DBMSs, it is sometimes necessary for the programmer to specify the join sequence. *SQL* permits the syntax

```
select p.Programme_Name, s.Subject_Name, c.Personal_Name, e.Mark
    from Enrolments e
        join Candidates c on e.Candidate = c.Candidate
            join Programmes p on c.Programme = p.Programme
                join Subjects s on e.Subject = s.Subject
    where e.Enrolment_State = 'Assessed'
    and e.Semester = '20yyA'
    order by asc p.Programme_Name, s.Subject_Name,
            c.Personal_Name;
```

spelling out the join sequence. Unfortunately, the **join** syntax still does not make it clear how soon the relevant *Enrolments* rows are selected.

The programmer can also control selection, projection, and the join sequence by creating a succession of *views*. The join tree could be expressed as follows,

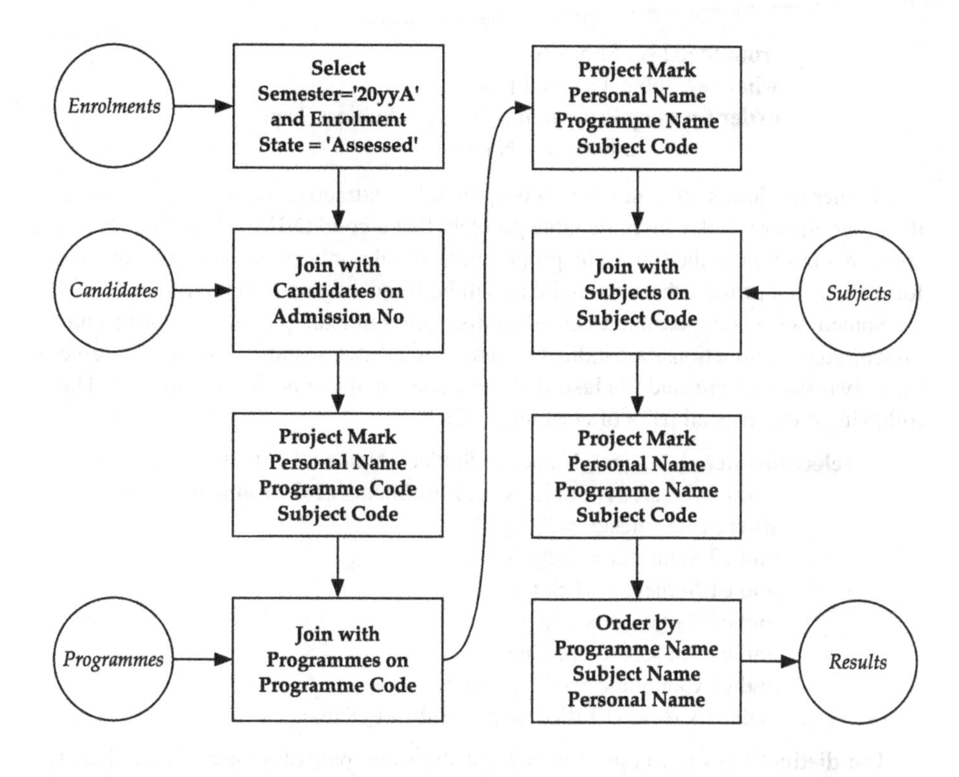

FIGURE 6.2

An optimum join tree. Circles represent tables; rectangles represent *join, select, project,* and *sort* processes. Intermediate results flow along the arrows. The joins are best made using the *Sort-Merge* algorithm, which involves additional *Sort* operations, but they aren't shown.

```
create view Step1 as
    select c.Programme, e.Subject, c.Personal_Name, e.Mark
        from Enrolments e, Candidates c
        where e.Candidate = c.Candidate
        and e.Enrolment_State = 'Assessed'
        and e.Semester = '20yyA';

create view Step2 as
    select p.Programme_Name, s1.Subject,
        s1.Personal_Name, s1.Mark
        from Step1 s1, Programmes p
        where s1.Programme = p.Programme;

select s2.Programme_Name, s.Subject_Name,
    s2.Personal_Name, s2.Mark
```

```
        from Step2 s2, Subjects s
        where s2.Subject = s.Subject
        order by asc p.Programme_Name, s.Subject_Name,
                c.Personal_Name;
```

Neither the **join** syntax nor the **view** approach is attractive, because they require the programmer to solve an intractable problem that a good DBMS should be able to solve in a fraction of the time. The programmer should only resort to such strategies for queries that prove to be slow and that will be frequently used in the future.

Sometimes it is impossible to avoid a *cross join*. For example, in scheduling end-of-semester examinations, a candidate cannot attend two examinations at the same time. Two subjects are said to **clash** if at least one candidate is enrolled in both. The following query lists all pairs of clashing subjects:[21]

```
select distinct s1.Subject_Name, s2.Subject_Name, s1.Filled
        from Subjects s1, Subjects s2, Enrolments e1, Enrolments e2
        where e1.Semester = '20yyA'
        and e2.Semester = '20yyA'
        and e1.Subject = s1.Subject
        and e2.Subject = s2.Subject
        and e1.Subject <> e2.Subject
        and e1.Candidate = e2.Candidate
        order by desc s1.Filled, asc s1.Subject_Name;
```

The **distinct** keyword is used to prevent the same pair of subjects being listed many times — once for every candidate who is enrolled in both of them.

The join of the *Enrolments* table with itself (*e1* and *e2*) is a cross join. It can be made efficiently by first selecting only the enrolment rows for the correct semester, then sorting both *Enrolments e1* and *Enrolments e2* in order by *Candidate*. Since there are only a handful of subjects in which any one candidate will be enrolled, the set of *e1.Subjects* can be cached and joined with *e2.Subjects*. Subsequently, the subject pairs can be sorted by (*e1.Subject, e2.Subject*) to eliminate duplicates and be merged with their parent *s1.Subject Name*. Then, the pairs can be sorted again by (*e2.Subject, e1.Subject*) to eliminate duplicates and be merged with their parent *s2.Subject Name*. Finally, the pairs can be sorted into order by *s1.Filled*.

Surprisingly, there is a still better solution. The trick is to observe that the join of *Enrolments e1* with *Subjects s1* gives exactly the same result as the join of *Enrolments e2* with *Subjects e2*. Therefore there is no point in producing both joins, as one is merely a copy of the other. Further, in the cross join, by sorting the join of *Enrolments e1* and *Subjects s1* by *Candidate*, the subjects for each candidate become clustered,

[21] The problem of assigning examinations to the fewest sessions is intractable. Producing the timetable itself is beyond the scope of *SQL*. A useful heuristic is to consider the largest classes first, as these tend to clash with more subjects than others — and take longer to mark. Therefore the results are sorted in descending order by *Filled*. A special program can then assign the largest class to the first session, then each following subject to the earliest session with which it doesn't clash and for which there is space available. This yields a good but not always optimal timetable.

and can be cached and joined with *themselves*, because the set of subjects in the *Enrolments e2* and *Subjects s2* join would be identical to what is in the cache. As a final step, the cross join can be sorted by *s1.Filled*, *s1.Subject Name*, and *s2.Subject Name*, and the duplicates eliminated.

In a situation like this, it would be a clever query optimiser that could spot the symmetry between *Enrolments e1* and *Enrolments e2*, and it might take the programmer a lot of work to make the query efficient. Fortunately, this query is only likely to be made once per semester, and speed may not matter much. My personal advice to the implementor is not to worry about optimising *anything* until it has been proved to be unacceptable. There are always more important things to do.

6.5 TRANSACTIONS

In the *embedded SQL* example of Section 6.3.3, we saw the use of **commit** to mark the completion of an event procedure. It was stated that the purpose of **commit** was to make the procedure atomic. By atomic, we mean that the database has a defined state *before* the event and a defined state *after* the event, but its state shouldn't be inspected during the event. In the context of a DBMS, the change from the *before* state to the *after* state is called a **transaction**. In most cases, 'transaction' and 'event' are synonymous, but it is possible to imagine events that comprise more than one transaction. Indeed, it wouldn't matter greatly if the *Freeze* event of Section 5.7 updated the state of each subject as a separate transaction.

6.5.1 THE NEED FOR LOCKING

The purpose of making transactions atomic is to prevent different transactions from interfering with one another. This can only happen if more than one user process is trying to access the database at the same time, as when a server allows several clients to use it concurrently. The simplest way to enforce atomicity is for the DBMS to refuse to start any transaction until the previous one has committed. Unfortunately, this can be too restrictive, especially if the transaction needs to interact with a user. While one user is dithering, several other users could be served. For example, while a candidate enrols in one subject, other candidates enrolling in other subjects needn't be held up. In contrast, let us consider what happens if two candidates simultaneously attempt to enrol in the *same* subject.

Candidate *A* queries subject *S* to check that *Filled* < *Quota*. *Filled* has the value 10 and *Quota* has the value 100. Nonetheless, candidate *A* dithers for a while, and during that time candidate *B* enrols in *S*, creating a new *Enrolments* row and increasing *Filled* to 11. Candidate *A*, possessing a now out-of-date copy of *Filled*, creates a new *Enrolments* row and updates *Filled* to 11 — again. There are now 12 *Enrolments* rows for subject *S*, but *Filled* shows only 11. This is an example of the **lost update problem**. Ultimately, this can result in a situation where more candidates enrol than the lecture room can hold. Therefore it is necessary to *warn* the DBMS that candidate *A* has the intention of updating *Filled* (which is signalled using the '**for update of**' phrase, as illustrated in Section 6.3.3). When candidate *B* signals the same intention, candidate *B* must wait until candidate *A* **commits**. Candidates enrolling in other subjects aren't held up.

Sometimes, waiting can turn into **deadlock**, as when candidate *B* is waiting for candidate *A*, and candidate *A* is waiting for candidate *B*. No amount of waiting can break a deadlock, and the only recourse is for one transaction to be abandoned. This is done by the transaction issuing a **rollback** command. Like **commit**, this ends the transaction — but the database is left unchanged.

The existence of the **rollback** command creates another potential problem. Suppose, as before, that candidate *A* queries subject *S* to check that *Filled* < *Quota*. *Filled* has the value 10. Candidate *A* updates *Filled* to 11, then, candidate *B* enrols in *S*, correctly updating *Filled* to 12. Now assume for some reason candidate *A* has to roll back before creating an *Enrolments* row. There are now only 11 *Enrolments* in subject *S*, but *Filled* shows 12. (If *A*'s transaction carefully restored *Filled* to the value it had beforehand, it would have the value 10, which is equally wrong.) This is called the **uncommitted update problem.**

Finally, consider an attempt to produce a reconciliation report to verify that *Filled* accurately represents the number of *Enrolments* rows in subject *S*. This requires a query to count the number of *Enrolments* rows in subject *S* and a second query to find the value of *Filled*. Imagine, too, that while this is happening, candidate *A* enrols in *S*. There are several possible sequences in which things could happen: perhaps the reconciliation transaction completes counting *Enrolments* rows, then candidate *A* enrols and **commits**, then the reconciliation checks the value of *Filled*. It thus determines that *Filled* is more than it should be. (The problem isn't solved by checking the value of *Filled* first, because its value might then prove *less* than it should be.) This is called the **inconsistent analysis problem.**

The examples show that an update transaction should **lock** all the data it uses to prevent other transactions accessing them. During the update, the database may be in an *inconsistent state*. The **commit** and **rollback** commands mark points in time when the data the transaction has updated should again be consistent, either in the *after* state or restored to the *before* state.

6.5.2 THE TWO-PHASE PROTOCOL

The *lost update* and *uncommitted update* problems show how two update transactions can interact, but the *inconsistent analysis* problem shows that an update can also interact with a query that doesn't modify the database. Although the update locks the data it uses, and the reconciliation transaction always sees the database in a consistent state, unfortunately, it sees it partly in the *before* state and partly in the *after* state. Therefore the reconciliation transaction itself ought to lock the database to force the *update* to wait.

Clearly, any number of queries can access the database at the same time providing they don't update it. This suggests using two kinds of lock: a read-only, or **shared lock**, and an update, or **exclusive lock**. If a transaction tries to put *any* lock on data that already have an *exclusive lock*, it must wait. If a transaction tries to put an *exclusive lock* on data that already have a *shared lock*, it must also wait. But, if a transaction tries to put a *shared lock* on data that already have a *shared lock*, it can proceed.

As we have said, allowing each transaction to lock the entire database certainly guarantees that the database remains consistent. In practice this can prove over-restrictive, and there is no reason why two or more transactions that inspect or update unrelated data shouldn't be allowed to proceed at the same time. Suppose transaction *A* begins before transaction *B*, but finishes after transaction *B*. Although we cannot say whether *A* occurred before *B* or after *B*, provided they don't interact, it doesn't matter: the outcome is the same in both cases. The events are serialisable; the sequence [*A*, *B*] and the sequence [*B*, *A*] are both valid interpretations of what occurred. In fact, events are *partially ordered* (Section 2.5.1); only when two events interact does their order matter. Any topological sort of the partial ordering would lead to the same final state.

If each transaction locked only the data it needed to use, this would preserve consistency. If each transaction locked all its data at its outset, no deadlock could possibly occur. Unfortunately, in many cases it is only possible for a transaction to know what data it *might* use, rather than what it *needs* to use, especially if it offers choices to a user interactively. Consequently, such interactive transactions might need to lock entire tables, seriously limiting concurrency. It is more practical for a transaction to lock data resources as it actually needs them. All its locks are then released simultaneously when the transaction either rolls back or commits. This is called the **two-phase locking protocol**.[22]

6.5.3 DEADLOCK

As we hinted in Section 6.5.1, the two-phase locking protocol can lead to deadlock, which no amount of waiting can resolve. It is therefore important that a DBMS can determine when deadlock occurs, so that it can tell a transaction that it must roll back.

The simplest strategy is to assume that if a transaction waits longer than a certain time, then it is deadlocked. But this isn't reliable: a transaction that updates an entire table (such as *Freeze*) might take a long time to complete, and during that time any transaction that tries to query the table must wait — but it is not necessarily deadlocked.

A better method is for the DBMS to maintain a data structure representing a bipartite graph whose two types of vertex represent transactions and data resources. Figure 6.3 shows that data *R* and *S* are exclusively locked by transaction *A*, datum *T* is exclusively locked by transaction *B*, and datum *U* is shared by transactions *C* and *D*. Transaction *A* is waiting for *B* to release its lock on *T*, *C* is waiting for *A* to release its lock on *S*, and transaction *D* is waiting for *A* to release its lock on *R*. As things

[22] **Optimistic locking** is an alternative to the two-phase protocol. When a transaction commits, the time when the transaction first inspected a datum is compared with the time when it was most recently updated by another transaction. If the datum hasn't been updated since it was inspected, all is well, otherwise the transaction must be rolled back.

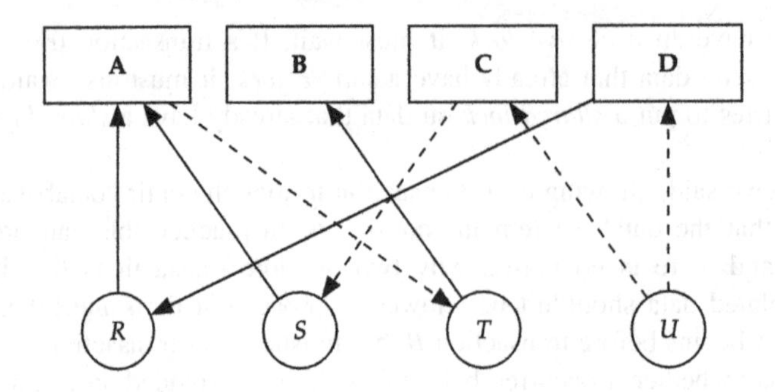

FIGURE 6.3

Detecting deadlock. A square represents a transaction process and a circle represents a datum. An edge from a datum to a transaction indicates that the datum is locked by the transaction. An edge from a transaction to a datum indicates that the transaction is waiting to lock the datum. A solid edge represents an exclusive lock. A dashed edge represents a shared lock.

stand, if B commits, it will release its lock on T allowing A to continue, eventually unlocking R and S, allowing C and D to continue.

In contrast, suppose B's next action is not to commit, but to attempt to put an exclusive lock on datum U. This is shown in Figure 6.4. B cannot continue until C and D release their shared locks on U. But C cannot continue until A releases S, and D cannot continue until A releases R. A, however, cannot continue until B releases T, which it certainly won't do, because B is waiting for U. In short, there are cycles $B \to U \to C \to S \to A \to T \to B$ and $B \to U \to D \to R \to A \to T \to B$. *Any* cycle is enough to cause deadlock.

A DBMS can maintain a central shared data structure to represent such a graph. Using *depth-first search*, the DBMS can detect a cycle in time $\mathbf{O}(|V| + |E|)$, where $|V|$ and $|E|$ are the numbers of vertices and edges in the deadlock graph. For example, when B tries to lock U, a DFS of vertex U will lead back to B, either via the path $U \to C \to S \to A \to T \to B$ or via $U \to D \to R \to A \to T \to B$.

To break the deadlock, one transaction involved in the cycle must be nominated as the **victim** and be *made to fail*.[23] The victim will usually be selected in one of two ways: either it will be the transaction whose request completes the cycle, or, more fairly, it will be the transaction that began most recently. The first strategy tends to cause transactions that retrieve many rows to fail; the second tends to help them complete. Which of these should be favoured depends on the application, but the choice is rarely under the programmer's control.

We saw, in the *inconsistent analysis problem* of Section 6.5.1, that it is not enough to lock data as they are retrieved and unlock them when a transaction commits. In

[23] This is the reason the **rollback** command is necessary.

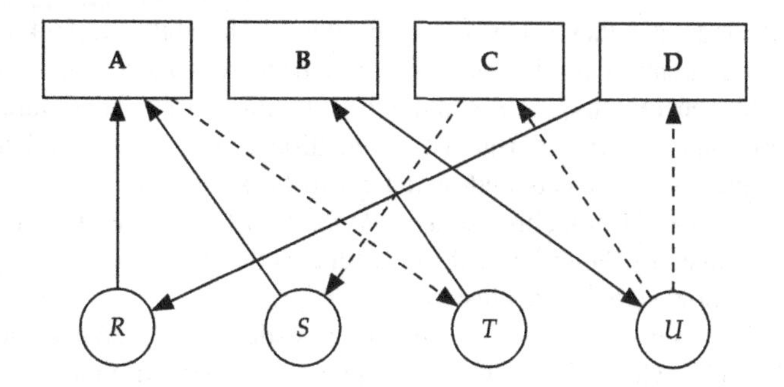

FIGURE 6.4

A deadlock. Transaction B is now waiting for datum U, completing the cycles $B \rightarrow U \rightarrow C \rightarrow S \rightarrow A \rightarrow T \rightarrow B$ and $B \rightarrow U \rightarrow D \rightarrow R \rightarrow A \rightarrow T \rightarrow B$.

that example, candidate A enrolled in subject S while a reconciliation transaction was counting the *Enrolments* rows for subject S. Since the row that A creates doesn't exist at the start of the reconciliation report, it is not enough for the reconciliation to lock the *Enrolments* rows that *currently* exist; it is necessary to lock the rows that *potentially* exist. This could be done by locking the entire *Enrolments* table, or, better, subject S within the *Enrolments* table.

There are two distinct ways this can be done. The first is to store the **where** clause associated with the reconciliation report, and check that no other transaction has an overlapping **where** clause. This approach is called **predicate locking**. Checking whether two **where** clauses overlap cannot be done without reference to the database instance, so the DBMS must constantly check if, for example, transaction A tries to lock a row that satisfies the **where** clause of transaction B. This method has the drawback that the time taken to check each locking attempt grows in proportion to the number of concurrent transactions, so it is little used.

A second approach, the **hierarchical locking** method, is more practical, and *is* widely used. This divides a table into a hierarchy of regions. In the case of *Enrolments*, a possible hierarchy might be *Semester/Subject/Candidate*. If the DBMS can determine that a transaction concerns only the rows for *Subject S* in the current semester, then it can lock only those rows for S in the current semester. If it can determine that a transaction concerns only the current enrolment for *Subject S* and *Candidate C*, it can lock just one row. On the other hand, if a transaction involves candidate C's enrolments in all subjects in the current semester, because the scheme is hierarchical, the DBMS is forced to lock the enrolments for the current semester in *all* subjects. Similarly, if a transaction involves enrolments in different semesters, the DBMS must lock the entire table.

The choice of the hierarchy is important and should depend on the way the database will be used. For example, if it is intended that enrolments will be made by the candidates themselves over the Internet, the ability to lock candidates will cause no contention for *Enrolments* rows, because no two concurrent transactions should concern the same candidate.[24] In this case, the hierarchy might be *Semester/Candidate/Subject*. But if the intention is for faculty staff to register enrolments, since both subjects and staff tend to be specific to programmes, then the better choice would probably be *Semester/Subject/Candidate*. The hierarchy is typically determined by the order of the components of a table's primary key, which also determines how its rows are clustered.

Another important contribution to the risk of deadlock is the size of the resource that a DBMS locks, which, depending on its quality, may vary from a single attribute to a whole row, or even to the page of the database containing the row. The smaller the resource, the less often deadlock will occur.

6.5.4 CORRECTNESS OF THE TWO-PHASE PROTOCOL

To see that the combination of the two-phase protocol and either predicate or hierarchical locking preserves consistency, assume that the database is always in a consistent state at the start of any transaction and that transactions *A* (which reads *X* and *Y*) and *B* (which modifies *Y*) each preserve consistency when they are processed one at a time. Can *B* modify *Y* so as to make the values of *X* and *Y* seen by *A* inconsistent?

The two-phase protocol implies that once transaction *A* has read a given data resource *X*, *X* cannot be updated by a concurrent transaction until *A* commits. But what if *B* modifies *Y* after *A* has read datum *X* but before *A* has read datum *Y*? We know that *B* cannot *modify* *X* until after *A* has read *Y* and has committed. There are two possibilities:

- *B* does *not* read *X* before it modifies *Y*, in which case either the value of *Y* must be independent of *X* and the question of consistency cannot arise, or the change to *Y* is one that is guaranteed to preserve consistency irrespective of the value of *X*.
- If *B does* read *X*, provided *B* maintains consistency, it can only update the value of *Y* so that it remains consistent with *X*. Because *B* puts an exclusive lock on *Y*, *A* cannot read *Y* until *B* has committed. After *B* commits, it cannot have updated *X*, and *Y* must still be consistent with *X*.

The effect is that, although *A* began before *B*, *B* is deemed to have preceded *A*. In general, the order in which transactions begin is unimportant, it is the order in which transactions commit that matters.

As a concrete example of this situation, consider *A* to be an *Enrol* event and *B* to be a modified *Set Quota* event — different in detail from the event specification given earlier. *A* must read *Quota* before updating *Filled* to ensure that *Filled* ≤ *Quota*. Provided *B increases* the value of *Quota* then *Filled* ≤ *Quota* must remain true and

---

[24] In isolation. But because *Enrol* events update *Filled*, the *Subjects* row will be locked anyway.

B need not inspect *Filled*. On the other hand, if *B* wants to *decrease Quota* then it can only do it after inspecting (and thus locking) *Filled*.

6.6 BACK-UP AND RECOVERY

In order to preserve the atomicity of transactions, it is necessary for the DBMS to ensure that the database instance is updated atomically. This isn't simple. If several rows of the database are updated, then several blocks of secondary storage must be physically updated one after another.[25] This takes considerable time, during which a power failure or hardware failure may occur. The DBMS therefore maintains a **log** or **journal**,[26] a *sequential* file that lists the updates made by each commit, followed by a special marker called a **checkpoint record**. Once the checkpoint record is written to the journal, the **commit** is deemed to have succeeded.

The entries in the journal permit error recovery. They can consist (in some form) of a snapshot of a datum before the update and a snapshot of the datum after the update, called its **before-image** and **after-image**. In the event of error recovery from a temporary power failure, the *after-images* can be used to bring the existing database to the most recent consistent state, as indicated by the last checkpoint record in the journal. (Any images *following* the last checkpoint are ignored.) In the case of a more serious failure, the *after-images* can bring an earlier **back-up** copy of the database up to date. In contrast, if the database has been corrupted by faulty application software, the *before-images* in the journal may be used to **roll back** the database to an earlier, error-free, state.

Since the journal has more authority than the database itself, it is important that the database is not updated until the corresponding journal entries have been safely written to a persistent (non-volatile) storage device. Otherwise, the database might contain changes that cannot be accounted for and cannot be undone.

It might seem that the journal imposes an overhead that might slow the DBMS, but the opposite is usually the case. The journal is written sequentially, one block at a time, which is the most efficient way to deal with secondary storage media. Updates to the database itself become less urgent, because they can always be recovered from the journal. They may be buffered and dealt with lazily, provided the buffer is never allowed to overflow. (It follows that the buffer must be consulted every time a transaction queries the database, in case a change has been written to the journal but not yet written to the database.)

Buffering allows the *peak* rate of updating to exceed the average rate for a short time, without causing excessive delay. To make this efficient, changes that haven't yet been written to secondary storage are buffered in RAM. In the extreme case, a *main-memory database system* maintains an image of the entire database in primary

[25] But not necessarily in the order specified by the DBMS: in the case of a hard disk drive, the updates may be transferred to a cache within the drive, then written to disk using the **elevator algorithm**, which minimises motion of the read-write head by sorting requests by ascending address and descending address alternately, similar to the way an elevator (lift) services floors in a building.

[26] As do many modern file systems.

storage and writes the journal to secondary storage, but the database is *never* updated on secondary storage, except to create an occasional back-up copy. In the event of a system failure, the journal provides the sole means to recover the current state of the image from the most recent back-up.

6.7 SUMMARY

We have used the terminology of *SQL* to describe several important aspects of databases at a conceptual level. We have done so because it introduces us to many useful concepts, such as view, join, transaction, deadlock, etc., at a reasonably abstract level. These concepts are useful to any database — however it is implemented.

Of particular interest are the join algorithms used in query optimisation, and the methods used to ensure the atomicity of transactions. In later chapters we shall see how these basic ideas can be adapted to different technologies and infrastructures.

6.8 FURTHER READING

Chapter 8 of 'Foundations of Computer Science: C Edition' by Alfred V. Aho and Jeffrey D. Ullman (1994, ISBN: 0-7167-8284-7) gives an excellent, *SQL*-free, introduction to query optimisation. The 2002 textbook 'Databases and Transaction Processing: An Application-Oriented Approach' by Philip M. Lewis, Arthur Bernstein, and Michael Kifer (ISBN: 0-201-070872-8) is a good source of material about database technology and includes plenty of material about the *SQL* language and its implementation.

SQL (originally called *SEQUEL*) was initially developed by IBM researchers in the 1970s. The language is now the subject of *ANSI* and *ISO* standards.

PHP, which we have cited as a widely used host language for *SQL*, was created by Rasmus Lerdorf in 1995. *PHP* was originally an acronym for 'Personal Home Page'. It is free software.

PL/SQL, used in the example of embedded *SQL*, is a product of the Oracle Corporation.

6.9 EXERCISES

1. On page 192 we discussed a query that could produce a list of timetable clashes. It was suggested that only a very clever optimiser would spot that a cross join was being made between two identical sets of data.

   Suppose that the DBMS's query optimiser proves inadequate and the resulting query takes too long. It is decided to write an embedded *SQL* procedure to create the cross join from a single instance of the data. This would involve fetching a list of enrolments for one candidate at a time, storing them in an array — ten locations would be more than enough — and inserting all the required pairs into a new table. Write a query that will produce the required lists of enrolments.

2. Candidate C is enrolling in subjects S and T and has locked the corresponding rows of the *Subjects* table for updating. Concurrently, candidate D is updating subject U. Candidates E and F are currently reading subject V, but have no intention of updating it. Candidate E wants to read subject T, but must wait for candidate C to commit. Additionally, candidate F wants to update subject S but must also wait for C to commit. At the same time, candidate C is waiting to read subject U.
   Is the database deadlocked? If not, in what order can the candidates' transactions commit?
   Suppose that the DBMS detects that a candidate's transaction is about to create a cycle in the graph. Can the potential deadlock be avoided without killing one of the transactions?
3. Consider the query on page 177, to find all candidates who scored better than average in *Maths101*.

```
select c.Personal_Name, e1.Mark
    from Candidates c, Enrolments e1
    where c.Candidate = e1.Candidate
    and e1.Subject = 'Maths101'
    and e1.Mark >
        (select avg(e2.Mark)
            from Enrolments e2
            where e2.Subject = 'Maths101')
    order by asc c.Personal_Name;
```

   What sequence of *select*, *project*, *join*, and *sort* operations would you expect a query optimiser to choose to answer the query?
   Which join algorithms should be used?
   Can you suggest an improvement?

CHAPTER

Processes

7

CHAPTER CONTENTS

INTRODUCTION

Use-definition analysis means discovering the way in which one variable depends on another. For example, if a procedure contains the assignment $y' = f(x)$, then the value of x must be known before the value of y' can be found. We say that data *flow* from x to y', or equivalently, that y' *Depends On* x, or that x *Determines* y', where *Determines* = *Depends On*$^{-1}$. Such data flows often determine the structure of information systems on both large and small scales, as we shall learn. They are particularly important in the design of batch processing systems, and batch processing systems are important because that is what so many business systems have evolved to become.

The *Determines* relation is transitive. In the sequence $(y' = x ; z' = y')$, x *Determines* y', y' *Determines* z', and transitively, x *Determines* z'. Strictly speaking, the *Determines* relation concerns not variables, but *definitions* of variables. For example, in the sequence $(y' = x ; z' = y)$, x *Determines* y' and y *Determines* z', but since y' doesn't affect y, x does *not* affect z'.

A variable receives a new **definition** when it is the object of an assignment statement. The assignment $y' = f(x)$ *uses* x to create the *definition* y'. In our specification language, a system variable is allowed at most two definitions, an initial definition retrieved from the database, and a final definition that becomes stored in the database. Internal variables are treated similarly, and although loops may cause a variable to assume many definitions dynamically, the specification language distinguishes only the value before an iteration (e.g., x) from the value after the iteration (x'). In reality, there can be a long series of definitions (x, x', x'', and so on), and a proper analysis of a loop needs to take two iterations into account.

7.1 USE-DEFINITION ANALYSIS

7.1.1 THE *DETERMINES* RELATION

We begin by listing the cases when one datum *Determines* another:

assignment: If the value of x is used to define a new value of y, as in $y' = f(x, \ldots)$, then x *Determines* y'.

control: If the value of x determines whether y is given a particular definition, as in $(x > 1 ; y' = 0)$, then x *Determines* y'.

selection: If the value of x is used to determine which value of y is selected to define the local value of $y(x)$, then both x *Determines* $y(x)$ and y *Determines* $y(x)$.

update: If the value of x determines *which* element of y' is defined, as in $y(x)' = \ldots$, then both x *Determines* y' and $y(x)'$ *Determines* y'.

quasi-update: In most cases where the value of x determines which element of y is selected to define $y(x)$, x *Determines* y.

iteration: If x is the initial value of x at the start of an iteration, and x' is defined within the loop, then at the end of the iteration, x' *Determines* x. Likewise, at the end of an event, x' becomes the new value of x for the next event.

In the **assignment rule** and **control rule** it is obvious that the value of y depends on the value of x. In the **selection rule**, we treat $y(x)$ as a local variable derived from

y and x, as in $y(x) = f(x, y)$. In the **update rule**, we treat $y(x)'$ as a local variable used to update element x of y', as in $y' = f(y, x, y(x)')$.

The **quasi-update rule** is more problematic: strictly speaking, this is not a case where x *Determines* y, because y itself is unchanged. We shall see shortly that the main use of the *Determines* relation is to determine in what order variables are inspected or updated. Although such an inspection doesn't modify y itself, it is usually convenient to treat it as if it does. That is to say, we usually need to know the value of x before selecting the element of y to inspect. We can, of course, snapshot every element of y before we know x, then select the desired element, but this would usually cause unnecessary work. Bearing in mind that other concurrent events may be trying to update other elements of y, an alternative to a snapshot is to put a shared lock on the *whole* of y, but this might cause concurrent update events to wait. Nonetheless, it is worth remembering that the quasi-update rule can be broken if locking or snapshotting isn't an issue, for example, if y, like *Subject Name*, is very rarely updated.

The **iteration rule** is needed because of the way we have chosen to specify loops, in which x' at the end of the loop implicitly becomes the new value of x at the start of the loop. An important consideration is whether one iteration *Determines* the next. For example, in the *Freeze* event, in which each iteration deals with a single subject, it is clear that there is no case in which the attributes of one subject affect the attributes of any other subject. Each subject is treated independently, and the state of one subject cannot affect another.

The kernel specification of the NUTS Academic Records System offers no example where one iteration of a loop does affect the next, so we here consider a hypothetical enhancement of the system in which candidates are forbidden to enrol in subjects for which they lack background knowledge. This could be expressed by a many-to-many *Prerequisite* relation between subjects, such that *S Prerequisite T* if a candidate cannot enrol in subject *T* without having previously passed subject *S*.

If we now consider a case where a candidate wants to enrol in *T* but hasn't passed *S*, they should be advised to first study *S*. Consequently, in this case, the state of one enrolment (in subject *S*) affects a *different* enrolment (in subject *T*). In addition, it may be that *R Prerequisite S*, so that the candidate might first need to study *R*, so the enrolment procedure would need to contain a loop. Clearly, the subjects involved in this loop must be inspected in the proper order, whereas the *Freeze* event can inspect subjects in *any* order or even concurrently.

7.1.2 THE DETERMINES GRAPH

Every procedure generates a *Determines* relation between its variables, which may be drawn as a homogeneous graph.[1] Strictly, the *Determines* relation depends on

---

[1] Programming-language compilers construct similar graphs to aid code optimisation. They are also exploited by spreadsheet programs. If Cell A2 of a worksheet contains the formula '=A1+1', an edge A1→A2 is added to the graph. If A1 is updated, a depth-first search of the graph will allow all its dependent cells to be updated too. Depth-first search will also detect circular references.

the procedure, not the problem that the procedure solves. In practice however, the *Determines* relation is robust, in the sense that any similar procedure for solving the problem generates the same relation. For example, the procedure $(a' = b\,; c' = d)$ generates the relation $\{b \mapsto a', d \mapsto c'\}$, and so does the procedure $(c' = d\,; a' = b)$. In fact, as we shall see shortly, we are less interested in the *Determines* relation between variables *within* a procedure than on the relation it transitively induces on system variables.

As a useful application of use-definition analysis, consider again the specification of the *Enrol* event given in Section 5.7:

$$
\begin{aligned}
Enrol(c,s) \Leftarrow\ & Candidate\ State(c) = Admitted\,;\\
& Subject\ State(s) = Offered\,; Filled(s) < Quota(s)\,;\\
& p = Programme(c)\,; Component\ State(p,s) = Offered\,;\\
& e = (Current\ Semester, c, s)\,; Enrolment\ State(e) = Null\,;\\
& Enrolment\ State(e)' = Enrolled\,;\\
& Mark(e)' = \mathbf{null}\,;\\
& Filled(s)' = Filled(s) + 1\,;\\
& Balance\ Owing(c)' = Balance\ Owing(c) + Fee(s).
\end{aligned}
$$

In essence, the *Enrol* procedure consists of two steps: five pre-conditions are tested, then, if they all are satisfied, four database variables are *defined*. It is intuitively obvious that there are many orders in which the conditions could be tested and that the assignments could also be made in any order. In other words, the procedure over-specifies the requirement. A transitive reduction of the *Determines* graph for the *Enrol* event is shown in Figure 7.1. To show its derivation, Figure 7.1 contains dummy vertices for the five pre-conditions that must be satisfied, and another dummy vertex, labelled '∧', to represent their intersection. Strictly, to comply with the rules above, these vertices should have been omitted, and every variable that determines a pre-condition ought to have been directly connected to each definition of an affected variable, but the resulting edges would have created too much visual clutter. Likewise, *Filled′* should have had both *Filled(s)′* and *s* itself as antecedents, and so on. We have been able to simplify the drawing because the *Determines* relation is transitive.

The graph forms a partial ordering — as any loop-free specification must — in which pre-conditions precede assignments. Likewise, the *Enrol* event imposes no ordering between its assignments, nor between its pre-conditions, except that *Programme* must be inspected before *Component State*, and *Current Semester* must be inspected before *Enrolment State*.[2] The *Enrol* specification above therefore displays only one possible topological sort of the graph. There are many other topological sorts of Figure 7.1, any one of which could be the framework for an implementation of the *Enrol* procedure. This flexibility can prove useful.

[2] Because of the quasi-update rule.

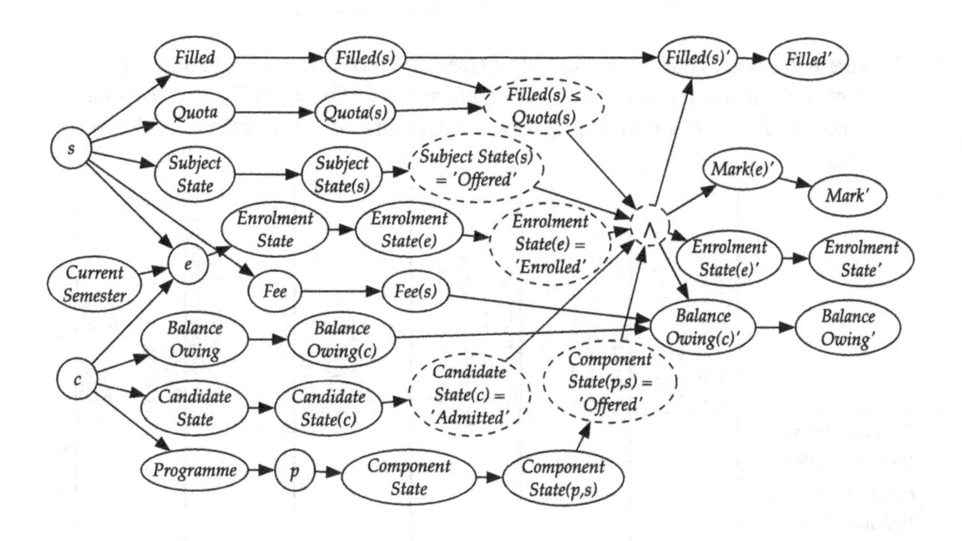

FIGURE 7.1

The *Determines* graph of the *Enrol* event procedure. Some additional vertices with dotted outlines have been added for tutorial purposes.

Table 7.1 presents the same information as Figure 7.1. The reader is advised to study the entries in Table 7.1 to see how the *Determines* relation is constructed from the rules.

For most purposes, both Figure 7.1 and Table 7.1 show too much detail. They have been presented here for tutorial reasons. Now that the reader understands how to derive the *Determines* relation, we can proceed to the next section.

7.1.3 THE STATE DEPENDENCY GRAPH

Usually, the analyst is interested only in the *Determines* relation between system variables. Figure 7.2 shows a graph derived from Figure 7.1. Formally, Figure 7.2 can be derived by finding the transitive closure of Figure 7.1, deleting all the local variables (and their edges), and then finding a transitive reduction of the resulting subgraph. Informally, and in practice, an analyst would construct Figure 7.2 directly, by inspection of the *Enrol* specification.

Figure 7.2 can be simplified further by merging the initial states of attributes (such as *Filled*) with their final states (such as *Filled'*). The justification for this is that *Filled'* silently replaces *Filled* when the event commits ready for a following event. In other words, *Filled' Determines Filled* — applying the iteration rule across a sequence of events.

The corresponding *Determines* relation between state variables, called a **state dependency graph**, is shown in Figure 7.3 on page 209.

Table 7.1 The adjacency matrix of the graph of Figure 7.1. An entry x in Row r, Column c indicates that r *Determines* c through Rule x, where a = assignment, c = control, q = quasi-update, s = selection, and u = update

	Balance Owing	Balance Owing(c)	Balance Owing(c)′	Balance Owing′	c	Candidate State	Candidate State(c)	Component State	Component State(p, s)	e	Enrolment State	Enrolment State(e)	Enrolment State(e)′	Enrolment State′	Fee	Fee(s)	Filled	Filled(s)	Filled(s)′	Filled′	Mark(e)′	Mark′	p	Programme State	Programme State(p)	Quota	Quota(s)	s	Subject State	Subject State(s)	Current Semester
Balance Owing	-	-	s	-	-	-	-	-	-	-	-	-	-	-	-	-	-	-	-	-	-	-	-	-	-	-	-	-	-	-	-
Balance Owing(c)	-	-	a	-	-	-	-	-	-	-	-	-	-	-	-	-	-	-	-	-	-	-	-	-	-	-	-	-	-	-	-
Balance Owing(c)′	-	-	-	u	-	-	-	-	-	-	-	-	-	-	-	-	-	-	-	-	-	-	-	-	-	-	-	-	-	-	-
Balance Owing′	-	-	-	-	-	-	-	-	-	-	-	-	-	-	-	-	-	-	-	-	-	-	-	-	-	-	-	-	-	-	-
c	q	s	-	u	-	q	s	-	-	a	-	-	-	-	-	-	-	-	-	-	-	-	-	q	s	-	-	-	-	-	-
Candidate State	-	-	-	-	-	-	s	-	-	-	-	-	-	-	-	-	-	-	-	-	-	-	-	-	-	-	-	-	-	-	-
Candidate State(c)	-	-	c	-	-	-	-	-	-	-	-	-	c	-	-	-	-	-	c	-	c	-	-	-	-	-	-	-	-	-	-
Component State	-	-	-	-	-	-	-	-	s	-	-	-	-	-	-	-	-	-	-	-	-	-	-	-	-	-	-	-	-	-	-
Component State(p, s)	-	-	c	-	-	-	-	-	-	-	-	-	c	-	-	-	-	-	c	-	c	-	-	-	-	-	-	-	-	-	-
e	-	-	-	-	-	-	-	-	-	-	q	s	-	u	-	-	-	-	-	-	-	u	-	-	-	-	-	-	-	-	-
Enrolment State	-	-	-	-	-	-	-	-	-	-	-	s	-	-	-	-	-	-	-	-	-	-	-	-	-	-	-	-	-	-	-
Enrolment State(e)	-	-	c	-	-	-	-	-	-	-	-	-	c	-	-	-	-	-	c	-	c	-	-	-	-	-	-	-	-	-	-
Enrolment State(e)′	-	-	-	-	-	-	-	-	-	-	-	-	-	u	-	-	-	-	-	-	-	-	-	-	-	-	-	-	-	-	-
Enrolment State′	-	-	-	-	-	-	-	-	-	-	-	-	-	-	-	-	-	-	-	-	-	-	-	-	-	-	-	-	-	-	-
Fee	-	-	-	-	-	-	-	-	-	-	-	-	-	-	-	s	-	-	-	-	-	-	-	-	-	-	-	-	-	-	-
Fee(s)	-	-	a	-	-	-	-	-	-	-	-	-	-	-	-	-	-	-	-	-	-	-	-	-	-	-	-	-	-	-	-
Filled	-	-	-	-	-	-	-	-	-	-	-	-	-	-	-	-	-	-	s	-	-	-	-	-	-	-	-	-	-	-	-
Filled(s)	-	-	c	-	-	-	-	-	-	-	-	-	c	-	-	-	-	-	a,c	-	c	-	-	-	-	-	-	-	-	-	-
Filled(s)′	-	-	-	-	-	-	-	-	-	-	-	-	-	-	-	-	-	-	-	u	-	-	-	-	-	-	-	-	-	-	-
Filled′	-	-	-	-	-	-	-	-	-	-	-	-	-	-	-	-	-	-	-	-	-	-	-	-	-	-	-	-	-	-	-
Mark(e)′	-	-	-	-	-	-	-	-	-	-	-	-	-	-	-	-	-	-	-	-	-	u	-	-	-	-	-	-	-	-	-
Mark′	-	-	-	-	-	-	-	-	-	-	-	-	-	-	-	-	-	-	-	-	-	-	-	-	-	-	-	-	-	-	-
p	-	-	-	-	-	-	-	q	s	-	-	-	-	-	-	-	-	-	-	-	-	-	-	-	-	-	-	-	-	-	-
Programme State	-	-	-	-	-	-	-	-	-	-	-	-	-	-	-	-	-	-	-	-	-	-	a	-	-	-	-	-	-	-	-
Programme State(c)	-	-	-	-	-	-	-	-	s	-	-	-	-	-	-	-	-	-	-	-	-	-	a	-	-	-	-	-	-	-	-
Quota	-	-	-	-	-	-	-	-	-	-	-	-	-	-	-	-	-	-	-	-	-	-	-	-	-	-	s	-	-	-	-
Quota(s)	-	-	c	-	-	-	-	-	-	-	-	-	c	-	-	-	-	-	c	-	c	-	-	-	-	-	-	-	-	-	-
s	-	-	-	-	-	-	-	-	-	a	-	-	-	-	q	s	q	s	-	u	-	-	-	-	-	q	-	s	q	s	-
Subject State	-	-	-	-	-	-	-	-	-	-	-	-	-	-	-	-	-	-	-	-	-	-	-	-	-	-	-	-	-	s	-
Subject State(s)	-	-	c	-	-	-	-	-	-	-	-	-	c	-	-	-	-	-	c	-	c	-	-	-	-	-	-	-	-	-	-
Current Semester	-	-	-	-	-	-	-	-	-	a	-	-	-	-	-	-	-	-	-	-	-	-	-	-	-	-	-	-	-	-	

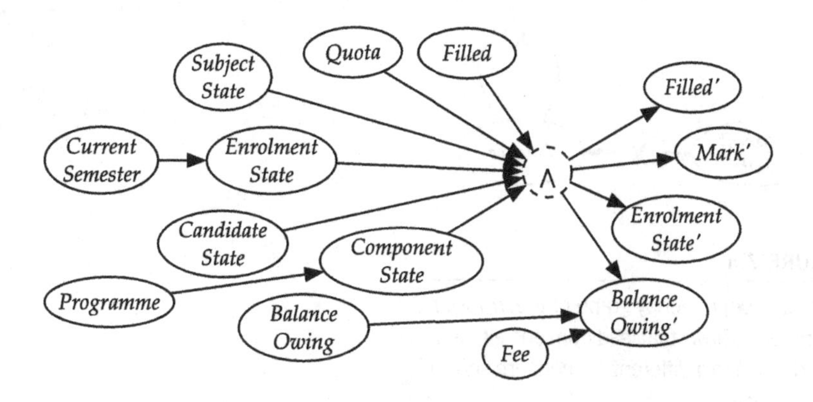

FIGURE 7.2

The *Determines* graph between system variables of the *Enrol* event procedure.

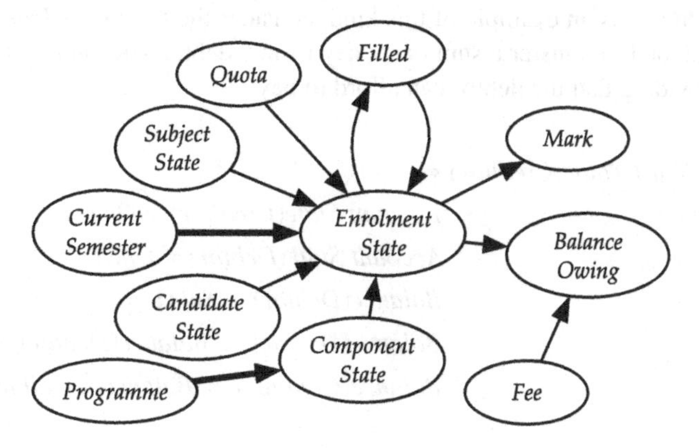

FIGURE 7.3

The state dependency graph of the *Enrol* event. The edges required by the quasi-update rule are drawn with thicker lines.

Figure 7.3 contains one non-trivial strongly connected component:

$$\{Filled,\ Enrolment\ State\}.$$

This captures the fact that *Enrolment State*(e) cannot be updated unless *Filled*(s) < *Quota*(s), but *Filled*(s) may only be incremented if *Enrolment State*(e) = *Null*.

Although *Filled*(s)′ directly depends on *Filled*(s), Figure 7.3 does *not* show a loop on *Filled* (likewise, neither on *Balance Owing* nor *Enrolment State*). This is because we are interested in the partial ordering imposed by the *Determines* relation, and loops don't influence it. Instead, we reserve the use of loops to highlight **cross flows**,

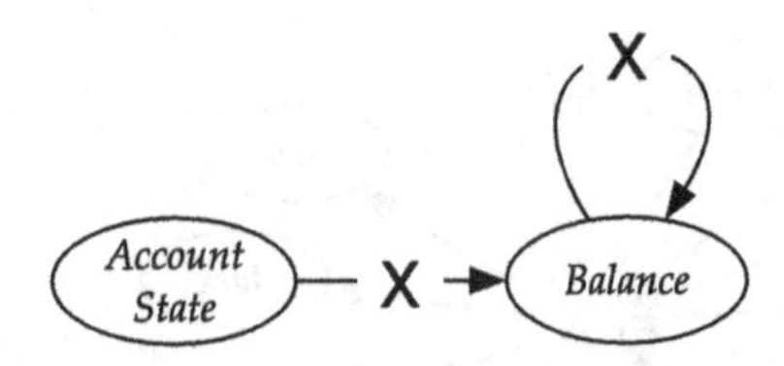

FIGURE 7.4

The state dependency graph of the *Transfer* event. Cross flows that lead from one element of a domain to a different element are shown with a cross.

those cases when one element depends on a *different* element, as in $A(x)' = A(y)$ or $A(x)' = B(y)$. As an example of this kind, consider the following *Transfer* event, which is intended to transfer a sum of money from a debtor's account to a creditor's account, providing that the debtor can afford to pay:

$$
\begin{aligned}
&Transfer(Sum, Debtor, Creditor) \Leftarrow \\
&\qquad (Account\ State(Creditor) = Open\,; \\
&\qquad Account\ State(Debtor) = Open\,; \\
&\qquad Balance(Debtor) \geq Sum\,; \\
&\qquad Balance(Debtor)' = Balance(Debtor) - Sum\,; \\
&\qquad Balance(Creditor)' = Balance(Creditor) + Sum \\
&\qquad).
\end{aligned}
$$

Account State(*Creditor*) *Determines* *Balance*(*Debtor*) (and *vice versa*) through the control rule, so the state dependency graph of the *Transfer* event (Figure 7.4) shows a cross on the edge from *Account State* to *Balance*.[3] It also shows a loop on *Balance* because *Balance*(*Debtor*) *Determines* *Balance*(*Creditor*)′.[4]

7.2 PREVENTING DEADLOCK

In Section 6.5.3, we found that deadlock can arise when concurrent transactions are competing for database resources. Deadlock is an embarrassment. Candidates

[3] Strictly speaking, every edge that links attributes with different domains is a cross flow too. I haven't marked them because I don't think it is necessary to point out the obvious. Feel free to think differently!

[4] The reader may wonder why we don't draw separate vertices for *Balance*(*Debtor*) and *Balance*(*Creditor*). Part of the reason is that subscripts like *Creditor* and *Debtor* are local variables, and a multiplicity of vertices might result simply because different names are used in different events. The other part of the reason is that the resulting graph has less clutter, and emphasises the cross-dependence.

may spend time enrolling online, making enrolment choices, only to find that their interaction with another candidate has caused a deadlock, forcing their transaction to fail. Deadlock is less embarrassing when an event merely produces a report, because the transaction can be restarted from the top without the user even knowing — provided it doesn't fail repeatedly and cause significant delay.

7.2.1 RESOURCE PRIORITIES

Is it possible to avoid deadlock completely?

Yes it is — provided one is willing to accept the penalty. Depending on the system, the penalty may be unimportant, or it may be worse than the occasional deadlock. The trick is to assign every data resource a **priority**. Then, provided that resources are locked in priority order, no deadlock can occur.

Suppose we have two subjects, *javaprog* and *maths101*. Candidate *C* chooses to enrol in *javaprog*; candidate *D* chooses to enrol in *maths101*. Candidate *C* now also chooses subject *maths101*, but must wait for *D's* transaction to complete. Meanwhile *D* has chosen *javaprog*, and must wait for *C*.

One solution to the deadlock would be to make *C* lock the entire subject table, forcing *D* to wait, and preventing *D* from locking *maths101*. But we can see that this isn't a good option. If we assume that it takes a candidate an average of 30 seconds to make choices, only 120 candidates would be able to enrol per hour, and it would take nearly 84 hours for all 10,000 candidates to enrol — seven 12-hour days.

Now assume that *javaprog* has priority over *maths101*. *C* proceeds as before, but *D* must lock *javaprog before* locking *maths101*, even though *maths101* was *D's* first choice. This can be done if we buffer each candidate's choices, sort them into order by *Subject Code*, then activate the transaction only when the candidate submits the form. In this example, rows of the subject table would be locked for only a fraction of a second, and the system throughput would be more than adequate. The *penalty* is that, because no subject is locked until the form is submitted, a candidate might believe that the last place in a subject was still available, only to discover it had vanished while the form was being completed.[5]

Actually, there is a perfect solution to this problem. Enrolling in a *set* of subjects is not a kernel requirement. Each subject can be enrolled in independently. So it is wisest to make each enrolment a separate transaction.

Now reconsider the *Transfer* event of Figure 7.4. If transaction *T* is transferring money from row *A* to row *B*, while transaction *U* is transferring money from *B* to *A*, deadlock can still result. On the other hand, once we prioritise rows, the matter is resolved. Assuming that the first access to an attribute causes it to be locked, we may implement the *Transfer* procedure as follows:

$$\begin{aligned} \mathit{Transfer}(\mathit{Sum}, \mathit{Debtor}, \mathit{Creditor}) \Leftarrow ((&\mathit{Creditor} < \mathit{Debtor}\,; \\ &\mathit{Creditor\ State} = \mathit{Account\ State}(\mathit{Creditor})\,; \end{aligned}$$

[5] Have you ever tried to book theatre or airline seats from a plan, only to find that once you have chosen the seats you want, they are no longer available?

$$
\begin{aligned}
&\quad\quad \textit{Debtor State} = \textit{Account State}(\textit{Debtor})\,;\\
&\quad\quad \textit{Creditor Balance} = \textit{Balance}(\textit{Creditor})\,;\\
&\quad\quad \textit{Debtor Balance} = \textit{Balance}(\textit{Debtor}))\\
&\quad \cup\,(\textit{Debtor} < \textit{Creditor}\,;\\
&\quad\quad \textit{Debtor State} = \textit{Account State}(\textit{Debtor})\,;\\
&\quad\quad \textit{Creditor State} = \textit{Account State}(\textit{Creditor})\,;\\
&\quad\quad \textit{Debtor Balance} = \textit{Balance}(\textit{Debtor})\,;\\
&\quad\quad \textit{Creditor Balance} = \textit{Balance}(\textit{Creditor}))\,;\\
&\textit{Debtor Balance} \geq \textit{Sum}\,;\\
&\textit{Balance}(\textit{Debtor})' = \textit{Debtor Balance} - \textit{Sum}\,;\\
&\textit{Balance}(\textit{Creditor})' = \textit{Creditor Balance} + \textit{Sum}),
\end{aligned}
$$

thus always taking the lower account number first. Here, the only penalty is a more complex implementation.

Only the edges caused by the quasi-update rule are important when considering deadlock. For example, although it is natural here to inspect *Account State* before *Balance*, the order can be reversed. Naturally, if *Account State* = *Null*, *Balance* isn't a meaningful datum, but provided steps are taken to ensure the procedure doesn't fail, the reverse order can be made to work.

Deadlock can also arise because of the order in which *tables* are accessed. Suppose that the *Enrol* event, which confirms both that the *Quota* is not filled and that the enrolment doesn't already exist, first locks the *Subjects* table, and then locks the *Enrolments* table. On the other hand, the *Withdraw* event doesn't need to check *Quota*, so it locks and updates the *Enrolments* table *before* updating *Filled* in the *Subjects* table. We now have the conditions for deadlock: candidate *C* enrols, locking the *Subjects* table, and candidate *D* concurrently withdraws, locking the *Enrolments* table. *C* cannot access *Enrolments*, and *D* cannot access *Subjects*.

The trick is to ensure that each type of event always locks tables in the same order. Although the logic of the *Withdraw* specification suggests accessing *Enrolments* before *Subjects*, it isn't a requirement, as use-definition analysis reveals.

If we apply use-definition analysis to the *Enrol* specification, we see that any topological sort of Figure 7.3 will yield a suitable list of priorities. Of particular importance are the edges required by the quasi-update rule: from *Programme* to *Component State*, and from *Current Semester* to *Enrolment State*. If *Programme* were to have lower priority than *Component State*, it would be necessary to lock all *Component State* rows for subject *s*, rather than the single row for the candidate's programme. Similarly, if *Current Semester* were not to precede *Enrolment State*, it would be necessary to lock the relevant rows of *Enrolment State* for all semesters. With the hierarchical locking scheme of *Semesters/Candidates/Subjects*, this would lock the entire table, meaning that *Enrol* events could never be dealt with concurrently, severely limiting the system's performance.

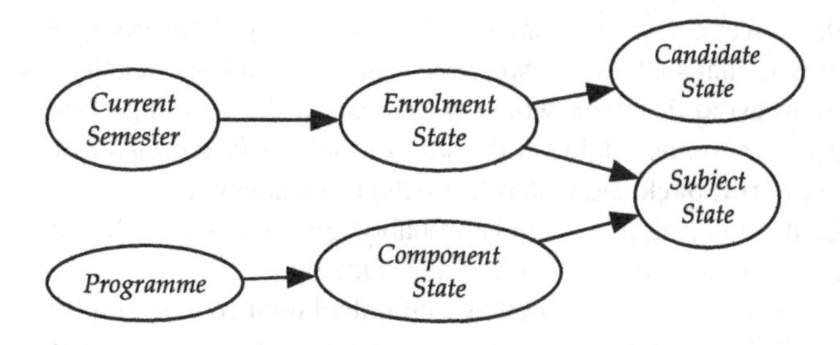

FIGURE 7.5

All edges caused by the quasi-update rule for the NUTS Academic Records System. (The edge from *Component State* to *Subject State* is not yet required by any event in the kernel specification.)

Of course, for the whole system to be deadlock-free, *every* event procedure must follow the same set of priorities. For the most part, this isn't a problem: the ordering imposed by the quasi-update rule arises from child-parent relationships and is therefore closely related to the database schema.[6] Figure 7.5 shows all the edges caused by the quasi-update rule in the specification of the NUTS Academic Records System, including those created by foreign keys that form part of a primary key. The edges *Programme* → *Component State* and *Current Semester* → *Enrolment State* are required by the *Enrol* event, the edge *Enrolment State* → *Subject State* is required by *Print Transcript*, and the edge *Enrolment State* → *Candidate State* is required by *Print Marks*. In addition, although the specification doesn't call for it, the edge *Component State* → *Subject State* is included. Some (as yet unplanned) report might later be needed to display all *Subjects* that are *Components* within a particular *Programme*.[7]

It is clear from Figure 7.5 that the attributes define a partial ordering.

If, as is quite likely, the DBMS cannot lock individual columns but only whole rows,[8] then, since *Candidate State* and *Programme* occupy the same row, the best possible ordering would be [*Current Semester*, *Enrolment State*, *Candidate State*, *Programme*, *Component State*, *Subject State*].

If it later proved necessary to produce a *List Programmes* report listing all programmes in which a given subject is offered, unfortunately, there would then need

---

[6] Although the vertices of a state dependency graph have the same names as FDs, they aren't FDs, they are database columns, or attributes.

[7] Figure 7.5 may seem counter-intuitive. Since it is necessary to know a value of *Admission No* and *Subject Code* to uniquely identify an *Enrolment* row, the reader might think *Enrolment State* should depend on *Subject State* and *Candidate State*. As it turns out, in those event procedures that examine *Enrolment State* (e.g., *Enrol*) both *Admission No* and *Subject Code* are given parameters, so there is no need to first know *Subject State* or *Candidate State*.

[8] If this should prove to be a serious problem, a single table can always be split into two or more tables with the same primary key.

to be a further edge, *Component State* → *Programme*, and the graph would become cyclic. Assuming that such events were rare, and the resulting penalty not too great, the best way to avoid deadlock would be to respect the ordering above, and either lock the entire *Programmes* table at the start of the *List Programmes* event, or allow the procedure to **roll back** and restart if it detects a deadlock.[9]

Although the ordering just given is probably optimum, *any* ordering will prevent deadlock, provided that all events follow the same one.

If a system cannot deadlock, locking and unlocking become simpler. This hardly matters if an *SQL* DBMS is used, but if the database has to be implemented some other way, it saves the need to maintain a graph like that of Figure 6.3 (page 196). Cycles cannot occur, so it is sufficient merely to count the number of shared locks and exclusive locks on each datum.

7.2.2 REAL-TIME SYSTEMS

Real-time systems are those that handle events as quickly as they occur, with minimal delay. Obviously, they need to have sufficient processing power to deal with the incoming stream of events. Typically, they consist of several interleaved processes, often running on different hardware processors. The processes communicate by passing messages.

Consider a situation in which processes *A* and *B*, which we can think of as sensors, send messages to process *C*. The processes run asynchronously, meaning that *C* has no control over when *A* or *B* will send a message, this being determined by how long it takes them to make measurements.

There are two possible kinds of communication:

- In *synchronous communication*, *C* regularly polls *A* and *B*, requesting an input measurement. In effect, *C* sends a message to *A* asking for the next input, and must then wait until *A* has replied, before requesting input from *B*. Roughly speaking, *C* can do no useful work while waiting for *A* or *B*, and only one processor tends to be active at a time.
- In *asynchronous communication*, *A* sends a message to *C* whenever it is ready, typically *interrupting* any processing done by *C*. *A* may either send messages autonomously or because of a previous request from *C*. Either way, when the message from *A* arrives, *C* is probably occupied on urgent business of its own and won't be ready to deal with it. Typically, the message from *A* will cause an **interrupt**, invoking a procedure within *C* that temporarily seizes control.

Interrupt procedures are usually short. A typical action is to post a notice that the message is available, so that the main process in *C* can deal with it at a suitable time. For example, *C* might execute a loop. At the start of the loop *C* looks for an incoming message and deals with it according to its source and content. *C* then returns to the start of the loop. When it has more than one source, *C* may find that more than one

[9] Unfortunately, many database systems don't let the application programmer choose which transaction to roll back.

message is waiting, in which case *C* will deal with the messages according to their priorities.

For the purpose of this example, suppose that *C* maintains a first-in-first-out queue in which each message is stored as it arrives. Such a queue is often implemented using a **circular buffer**, an array of fixed length. As messages arrive, each is stored in a higher location, but when the last location has been used, the first location is reused. A similar method of indexing is used to retrieve entries from the buffer. Writing is controlled by a *write pointer* and reading by a *read pointer*. Provided that the processor has sufficient speed to deal with messages faster than they arrive *on average*, the buffer will smooth out any short-term peaks, and the write pointer will never lap the read pointer.

As a result, assuming the buffer occupies locations 1 – *N*, a typical interrupt routine might be the following:

```
Write_Pointer = (Write_Pointer mod N) + 1;
Buffer(Write_pointer) = Incoming_Message;
```

This seems harmless enough, but there is a danger that an interrupt can itself be interrupted. If we imagine that a message from *B* interrupts a message from *A*, we might see the following sequence,

```
Write_Pointer = (Write_Pointer mod N) + 1;
Write_Pointer = (Write_Pointer mod N) + 1;
Buffer(Write_pointer) = Incoming_Message_B;
Buffer(Write_pointer) = Incoming_Message_A;
```

in which the message from *A* overwrites the message from *B*, and *A*'s intended location typically contains an old message. This is an example of a lost update, in which the message from *B* is lost. It also illustrates a more general problem:

Wherever data flows converge, they cause timing problems.

Clearly, this problem needs to be resolved. Interrupts are often given *priorities*, so that an interrupt can only override one of lower priority. In such a case, we use a separate buffer for each level of interrupt. Even so, we must make sure that interrupts with equal priority don't conflict.

The general solution to this problem is to use a flag, called a **semaphore**. At the start of the interrupt, the procedure tests to see if the flag has been raised. If so, it must wait for it to be lowered. Then, before it proceeds, it raises the flag and doesn't lower it again until it has completed. The stretch of code between raising and lowering the flag is called a **critical region**. Only one process should be in its critical region at one time. A semaphore is typically implemented as a bit that is 1 when the flag is raised and 0 when it is lowered. Despite this, there remains a danger that a process will see that the flag is lowered and raise it, not realising that a second process is doing exactly the same thing at the same time. For this reason, real-time hardware must provide a single, atomic instruction, **test and set**, that indivisibly both tests the flag and raises it. The result of executing *test and set* is that it returns the state of the semaphore before the instruction, and after the instruction, the semaphore is always raised. (Lowering the flag causes no such difficulty. Only one process can be in a position to lower it.) A *semaphore* is therefore the ultimate arbiter of the order of events.

Fortunately for most programmers, these difficulties are usually already solved within the operating system kernel or even the hardware. Such problems usually only concern people who write operating systems or who deal directly with microprocessors.[10]

7.3 PROCESS GRAPHS

A system often contains several **processes**, which communicate with one another by passing **messages**. For example, the NUTS have both an Enrolment Office and a Bursary. The function of the Enrolment Office is to keep track of enrolments and assessments. The function of the Bursary is to deal with financial matters, including the collection of fees from students. When a candidate enrols in a subject or withdraws from a subject, the Enrolment Office sends a message to the Bursary so that the Bursary can invoice the candidate for the fee. It doesn't matter if the activities in the Bursary lag behind those in the Enrolment Office — and indeed they do: the Bursary doesn't act on the information received from the Enrolment Office until subjects are *Frozen* and enrolments and withdrawals cease. The Enrolment Office and the Bursary use two different (and potentially concurrent) **subsystems**, both of which are parts of the Academic Records System as a whole.

For a network of processes to be *correct*, it should behave as if it were a single process that dealt with one event at a time. We say that it must be **real-time equivalent**. Alternatively, we can say that it processes events in **first-come, first-served** (**FCFS**) order. We can guarantee this to be true if the history of every state variable is the same as in the equivalent real-time system. This can be ensured if each variable is *owned* by some process, and no other process may inspect or update the variable. In addition, this process must receive messages in first-come, first-served order. If messages aren't acted on immediately, they should be kept in a first-in-first-out store, or queue, so that they are eventually dealt with in the correct order.[11] A *process* (as we use the term here) therefore has exclusive access to one or more system variables. The processes therefore *partition* the variables.

A **process graph** shows the assignment of system variables to processes, and the message queues that link them. Figure 7.6 shows a possible process graph for the *Enrol* event. Vertices represent processes, and edges represent queues or data flows.

[10] A DBMS has to observe a similar discipline to prevent two concurrent updates locking the same datum.
[11] Compare a fast-service restaurant that hangs order chits on a carousel, with one that piles them onto a spike!

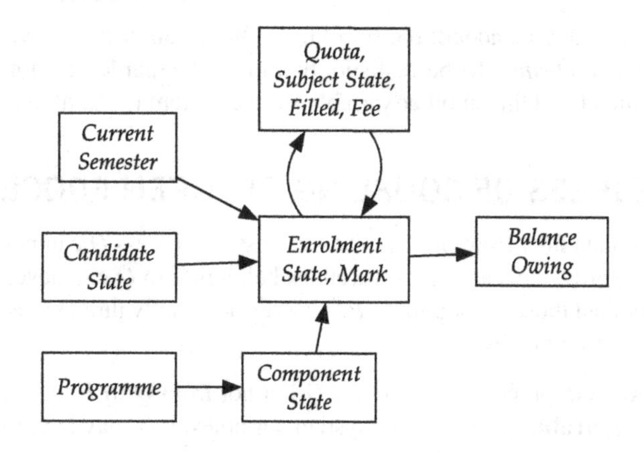

FIGURE 7.6

A possible process graph for the *Enrol* event. The graph contains a cycle.

As a rule, different events have different process graphs. Fortunately, the graphs for different events are often similar. For example, the process graph for the *Withdraw* event is a subset of the graph for the *Enrol* event.

> **The process graph for a *system* is valid if the graph for each individual event is a sub-graph of the system process graph.**

In Figure 7.6, with the exception of *Balance Owing*, the processes have been chosen to correspond to database tables. The cycle between $\{Mark, Enrolment\ State\}$ and $\{Quota, Subject\ State, Filled, Fee\}$ means that these two processes are closely coupled and will need careful synchronisation. On the other hand, because *Balance Owing* and *Candidate State* are in separate processes, the states of *Balance Owing* may lag those of *Candidate State*, even though they belong to the same table.

A snapshot of the system as a whole may reveal it to be in an inconsistent state. For example, the value of *Balance Owing* recorded by the Bursary won't always reflect the state of the candidate's enrolments recorded by the Enrolment Office, but may lag behind it.

There are three solutions to this apparent problem:

1. Only inspect the system when it is **quiet**, that is, when all its message queues are empty.
2. Send the message requesting the value of *Balance Owing* along the same path as the messages that concern enrolments and withdrawals, thus subjecting it to the same delays.
3. Don't expect a global snapshot of a system to reveal a consistent state!

A fourth solution may be added: the Enrolment Office can keep its own (unofficial) record of *Balance Owing*. To be realistic, this may be regarded as not solving the problem, but merely adding another variable for the system to maintain.

7.3.1 DEGREES OF COUPLING BETWEEN PROCESSES

Observe that if data flow from process *P* to process *Q*, process *Q* can never be ahead of *P*. In other words, the state of the variables belonging to *Q* can never correspond to a later event than those belonging to *P*. As a result, exactly three kinds of coupling are possible between processes:

- If data flow from process *P* to process *Q* but not from *Q* to *P*, then *P* and *Q* are said to be **separable**. The states of system variables in *Q* may lag those of *P*, but those of *P* can *never* lag those of *Q*.
- If data flow in both directions between *P* and *Q*, neither can lag the other, and they are **tightly coupled** or **synchronous**.
- If no data flow between *P* and *Q* in either direction, either directly or transitively, they are **independent**. *P* may lag *Q*, or *Q* may lag *P*, or both may be concurrent.

In Figure 7.6, processes {*Programme*} and {*Component State*}, for example, are *separable*; processes {*Mark, Enrolment State*} and {*Quota, Subject State, Filled, Fee*} are *tightly coupled*; and processes {*Current Semester*} and {*Candidate State*} are *independent*.

When two processes are tightly coupled, the data flows between them permit no lag, so in most cases it is pointless to keep them separate and better to combine them into a single process. Unavoidable exceptions occur when the two processes must be physically separated or must use incompatible technologies. For example, database management systems often run a central server that supports many (possibly remote) client processes concurrently. Generally speaking, when a client process issues a query, it can do little else except wait for the DBMS to answer it. The client and server are tightly coupled.

A similar situation exists in a system that controls the position of a control surface in an aircraft. Typically, a sensor informs a microprocessor of the actual position of the control surface, and the microprocessor informs a motor to adjust its position to comply with the demand made by the pilot or autopilot. The data flow is linear and simple: sensor to microprocessor and microprocessor to motor. Even so, once we take the behaviour of the control surface into account, we see that there is a cycle: motor to control surface, control surface to sensor, sensor to microprocessor, and microprocessor to motor. Although the control surface subsystem is hardware rather than software, it is still a *process* that is tightly coupled to the microprocessor. We shall have more to say about the dynamics of such control systems in Section 10.2.

On the other hand, when processes are separable or independent, data flows merely impose a temporal ordering on them. Processes downstream can lag those upstream. In Figure 7.6, *Balance Owing* can lag all the other processes, thus making it possible for enrolment information to flow from the Enrolment Office to the

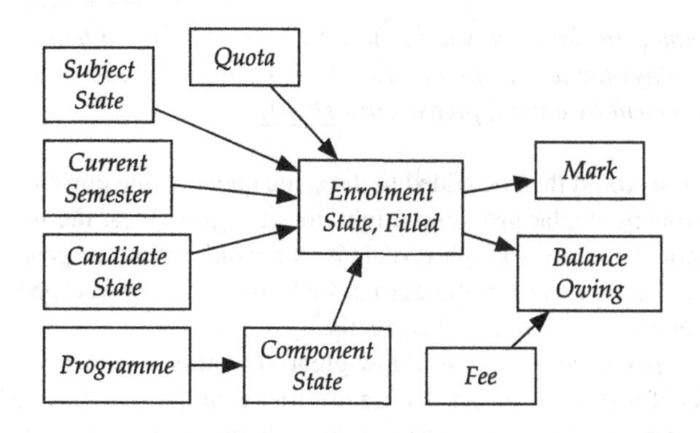

FIGURE 7.7

The canonical process graph of the *Enrol* event.

Bursary. If *Balance Owing* had been tightly coupled to the other processes, sharing the enrolment data between the two departments would have been much harder to organise.

7.3.2 THE CANONICAL PROCESS GRAPH

The minimal reduction of a cyclic graph isn't unique, but the transitive root of its strongly connected components is. (Review Sections 2.5.1 and 2.5.2, pages 44–48.) We call the resulting reduced graph a **canonical process graph**. The transitive root of the strongly connected components of the state dependency graph of Figure 7.3 is shown in Figure 7.7. Figure 7.7 is therefore the *canonical process graph* of the *Enrol* event.

A canonical process graph contains the maximum possible number of separable processes. Vertices can be merged to form **aggregate processes** (including the case when all the vertices are merged into a single process), but they cannot be split without creating cycles. For this reason, the vertices are referred to as **minimal processes**.

Like any reduced graph, a canonical process graph is acyclic and therefore imposes a partial ordering on its minimal processes. It is only useful to cluster vertices in ways that leave the reduced graph acyclic.

As an example, consider grouping all the vertices together, except the vertices for *Fee* and *Balance Owing*. When a successful enrolment is made, a record containing the *Admission No* of the candidate and the *Subject Code* of the subject concerned could be placed on a queue for later processing. At some later time, *Balance Owing* could be updated (by reference to *Fee*). During this time, the database as a whole would be inconsistent; new enrolments would be recorded, but not yet billed for. A candidate querying the database might see a warning, such as

> *The balance displayed may not reflect recent enrolments. Please allow 5 working days for fees to be processed. If you wish to receive an up-to-date statement by e-mail, please click HERE.*

Any such request would then be added to the same queue as the enrolment messages.

Although this particular grouping might not be a good idea, the important thing is that the canonical process graph reveals it as a valid option. In general, provided events arrive at each process in the correct first-come, first-served order, the whole database will eventually be updated correctly.

Although *independent processes* allow great flexibility, in practice, implementation is simpler if a process graph is organised as a simple pipeline; wherever two flows meet, timing problems can arise. Where there are parallel paths in the data flow, it is possible for a later event to overtake an earlier one, known as a **race condition**. A process may therefore stall, unable to act on *any* event because it cannot know whether some earlier event may yet arrive by a slower pathway. This can never happen if processes form a pipeline. On the other hand, it does mean that processes may receive messages for events that don't concern them, but still have to pass those messages transparently to processes downstream — not a big problem as a rule.

Another way to deal with parallel paths is to use dummy messages. This can either be done by each event sending dummy messages along otherwise unused paths or by periodically sending dummy messages along all pathways to flush them.

Even when the independent processes are final vertices of the graph, a problem still remains if their variables can be inspected and compared by an external observer, who may see them in inconsistent states.[12] Formally, we can deal with this by regarding the observer as a terminal vertex added to the graph.

7.3.3 PROCESS OPTIMISATION

An **optimal process graph** can be *synthesised* from a canonical process graph using the following two heuristics:

- Consider merging two processes only if there is *no* compound directed path between them. This implies that merging the two processes cannot create a cycle.
- Consider merging two processes only if the set of rows accessed by the first process is a subset of the set of rows accessed by the second process (including the case that the two sets are the same).[13] This implies that the first process doesn't add to the access cost already incurred by the second process.[14]

The heuristics may be applied using Algorithm 7.1.

---

[12] As when my son arrives home before his holiday postcards.

[13] The reader might suspect that merging processes two at a time is a restriction, but it can be proved that any process graph that can be obtained by merging three or more processes at a time can also be obtained by merging processes in pairs, although in more steps.

[14] Which, in turn, implies that we should exclude cross flows.

Algorithm 7.1 Optimising a process graph.

> **while** there remain two processes P and Q *not* connected
> by a compound directed path,
> **and** the set of rows accessed by P is a subset of
> the set of rows accessed by Q,
> *merge* P and Q.

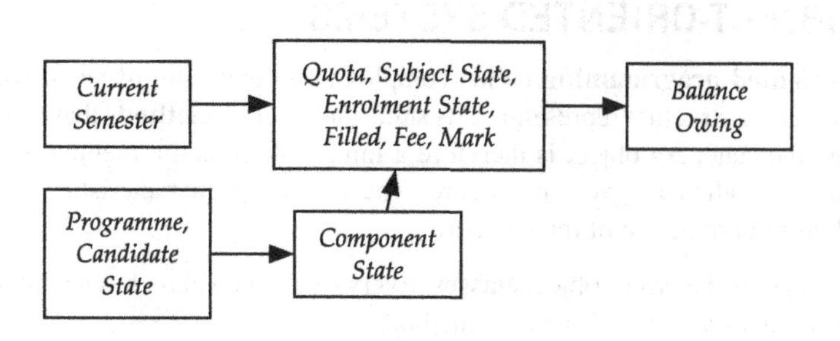

FIGURE 7.8

The optimal process graph obtained by applying Algorithm 7.1 to the canonical process graph of the *Enrol* event in Figure 7.7.

Algorithm 7.1 can be non-deterministic: there might be several possible choices of P and Q. Let us apply it to the canonical process graph of the *Enrol* event, shown in Figure 7.7:

1. Merge *Quota* with *Subject State*. Both access row s.
2. Similarly, merge {*Quota, Subject State*} with *Fee*.
3. Merge *Mark* with {*Filled, Enrolment State*}, which already needs to access row e.
4. Merge {*Subject State, Quota, Fee*} with {*Filled, Enrolment State, Mark*}. *Filled* already needs to access row s.
5. Merge *Programme* with *Candidate State*. Both access row c.

Although *Balance Owing* accesses the same row as {*Programme, Candidate State*}, merging them would create a cycle and would not usually be advisable.

Irrespective of the order in which it chooses to merge attributes, Algorithm 7.1 leads inexorably to the optimal graph of Figure 7.8.

Usually, as in this case, the order in which processes are merged doesn't affect the outcome. Even so, in some (rare) situations there may be a choice between two or more optimisation steps that ultimately lead to different outcomes, and the problem of finding the optimal graph can be mapped onto the Shortest Super-sequence problem.

Finding the optimum is therefore intractable in theory, but is almost always trivial in practice.

This example clearly illustrates the value of drawing a process graph that concerns *attributes* rather than *tables*. *Balance Owing* shares the same domain as *Candidate State* and *Programme*, but if all three attributes had been assigned to the same *Candidate* process, only *Current Semester* would have been separable from the remaining aggregate {*Programme*, *Component*, *Candidate*, *Subject*, *Enrolment*} process.

7.4 OBJECT-ORIENTED SYSTEMS

Object-oriented programming builds complex programs from **objects**. An object comprises a data structure representing its state, and a set of **methods** that can inspect or change the state. An object is therefore a mini-system, and a method is close to what we have called an event procedure. The following characteristics of object-oriented programming are of interest here:

- Objects are instances of object **classes**. Every object in a class has a similar set of state variables and the same set of methods.
- Objects communicate by **method calls**, which are similar to procedure calls. In what follows, *Candidate.Enrol(sem, c, s)* denotes a method call directed to the *Enrol* method of object *Candidate*. The call carries the parameters (*sem, c, s*).
- Object classes can **inherit** properties from more basic classes. For example, a *Candidate* class, *Subject* class, etc., might inherit properties from some more general class, perhaps called *Table*. They will, of course, need to have additional methods of their own, such as *Set Name*, *Set Quota*, etc. We might even take this idea further, perhaps creating sub-classes of *Candidate* according to *Candidate State*, such as *Undergraduate* and *Graduate*. *Undergraduate* objects would have methods, such as *Enrol*, that *Graduate* objects would not share. We do not explore such possibilities in what follows.

7.4.1 SYNTHESIS OF AN OBJECT-ORIENTED SYSTEM

We here consider an object-oriented approach to the design of a system for the Enrolment Office. Since the entire *NUTS* database will occupy less than 25 MB, it is feasible to store it in RAM. The intention is for the database to be restored from sequential files at the start of the day's business and for the updated state of the database to be written to new copies of the files at the end of the day. During the working day, each successful transaction will be logged to a sequential file. This file provides a means of recovery in the event of a power failure; the database can be restored from copies of the files used at the start of the day, then updated from the log file.

We assume that the implementors of the system have chosen a representation of the *Table* class that allows rapid access to object instances by primary key. In what follows, we shall not concern ourselves with these details; we shall be solely interested in the ways in which objects will communicate with each other.

The enrolment system will need to have methods such as *Enrol*, *Withdraw*, etc. This raises the interesting question of whether these particular methods should belong to the *Candidate* class, the *Subject* class, or the *Enrolment* class. More specifically, assuming that external events initially invoke methods in a single *Academic Records* object, should this object directly invoke a *Subject* object, a *Candidate* object, or what?

Before answering this question, we remind the reader that if the system allows events to be processed concurrently, locking resources is just as necessary in an object-oriented approach as in any other. In such a case, the *Table* class must support inheritable *Lock* and *Unlock* methods. If we want to avoid deadlock, we need to respect the canonical process graph of Figure 7.7.

Another important question — even when deadlock isn't an issue — is how we factor event procedures between objects. It would be possible to make the *Academic Records* object responsible for all the logic in the kernel specifications, leaving the *Subject* object, *Candidate* object, etc., with nothing to do but provide *Get* and *Set* methods for their attributes. Nonetheless, if the system is being developed by a team, it might be better to spread the implementation work across several objects, so that part of the *Enrol* event is implemented by a method in a *Candidate* object, and part by a method in a *Subject* object. If this is the case, we not only want to make sure the interfaces between objects are as simple as possible, but also to ensure that there is a sensible order in which objects can be developed and tested.

Suppose we were to choose a design in which *Enrol* events first invoked the *Candidate.Enrol* method. After verifying that *Candidate State* was *Admitted*, *Candidate.Enrol* could then invoke *Subject.Enrol* to check that *Filled* < *Quota*. *Subject.Enrol* could then invoke *Component.Enrol* to check if the subject was a valid component of the candidate's programme, and *Enrolment.Enrol* to check that *Enrolment State* was *Null* and then to update it to *Enrolled*. Finally, *Enrolment.Enrol* could invoke *Candidate.Charge* to update *Balance Owing*.

We call this design a **development sandwich**: we cannot test the *Subject* object until we have implemented the *Candidate.Charge* method, and we cannot test the *Candidate* object until we have implemented the *Subject.Enrol* method. Because of the cyclic relationship between them, neither the *Candidate* object nor the *Subject* object can be tested independently.

Is there a way to improve the design?

One solution, inspired by the optimised process graph of Figure 7.8, is to create a new *Ledger* class, whose sole purpose is to store and update *Balance Owing*. It then becomes possible to test *Ledger.Charge* independently, and thus *Subject.Enrol*, *Candidate.Enrol*, etc., in turn. A less radical solution is to develop the *Candidate* class in two stages, first implementing and testing its *Charge* method, then (after developing the *Subject* class) its *Enrol* method.

Given that the optimal process graph contains the strongly connected component {*Quota*, *Subject State*, *Enrolment State*, *Filled*, *Fee*, *Mark*}, how is it to be dealt with? It implies that the *Enrolment* and *Subject* objects are tightly coupled, requiring close synchronisation and two-way communication.

Fortunately, there is a straightforward solution: we make one object *subordinate* to the other. We might either make *Enrolment* subordinate to *Subject* or *vice versa.* In the first case, all events involving the *Enrolment* class would first have to invoke a method in the *Subject* object, which would then invoke a corresponding method in the *Enrolment* object. Alternatively, the *Enrolment* object could invoke the *Subject* object. Only the first choice is feasible. *Subject* and *Enrolment* are related as parent and child. A parent row may have no children, but referential integrity requires that every child row has a parent. If we make *Subject* subordinate to *Enrolment*, a *Subject* row with no children has nothing to be subordinate to. On the other hand, making *Enrolment* subordinate to *Subject* causes no problems.

> *In general, we can conclude that if two tightly coupled objects have a parent-child relationship, the child should be subordinate to the parent.*

Given such a parent-child arrangement, there is a surefire way to implement any method that involves the child class. When the parent method is invoked, it should call the child method, passing to it any parent attributes that the method requires. The child should then execute the whole text of the method, returning to the parent the values of any parent attributes that the method is capable of updating. For example, when an *Enrol* event occurs, the *Subject.Enrol* method would be invoked. This would call the *Enrolment.Enrol* method, passing it, among other things, the current values of *Quota* and *Filled*. The *Enrolment* object can then compare these values and also check the value of *Enrolment State*. If the *Enrol* event then succeeds, the *Enrolment* object should then update *Enrolment State* and return the updated value of *Filled* to its parent. The parent *Subject* object would then update *Filled*. In the case that the event fails, the *Enrolment* object should return the unmodified value of *Filled*. The *Subject* object can then blindly update *Filled* as before.

7.4.2 EVENT SEGMENTATION

When we split the processing of an event procedure across several objects, it usually means that each object implements only part of the procedure. We call this fragmenting of a procedure **event segmentation**.

We shall now show how to factor kernel specifications into class methods. If expanding the method calls recovers the original kernel specification, we can be sure that the methods correctly implement it — provided the infrastructure preserves the atomicity of events.

Asynchronous calls need to be distinguished from synchronous calls; asynchronous calls need not delay the caller but synchronous calls must. To make the distinction clear, the specification language uses **tell** to indicate an asynchronous call, and **ask** to indicate a synchronous call. When an object's event procedure calls one of its own internal methods, neither keyword is used, as the call is necessarily synchronous.

The following example shows a possible objected-oriented design. The implementor has decided to keep the *Ledger* object separate from the *Candidate* object. Initially, we consider a situation where each object class corresponds to a complete table and has only one instance.

In this example, as in the kernel specification, pre-conditions are tested, but, with one exception, no error processing is described. This keeps the example short and emphasises the important points. The reader will have to imagine additional rules to deal with other errors.

The chosen locking hierarchy is [*Global*, *Candidate*, *Component*, *Subject*, *Enrolment*, *Ledger*]. Each object provides *Read Lock*, *Update Lock*, and *Unlock* methods, which control access to the whole object. (A refinement would be to lock each row and each attribute independently.) If we assume that all events use the same locking hierarchy, then detecting deadlock isn't an issue, and a graph structure like that in Figure 6.3 (page 196) won't be needed; it is merely necessary to count the numbers of shared locks and exclusive locks. Although the locking protocol that follows has no global commit operation, it is still safe here — except in the event of a power failure or a system failure.

The implementor decided that only the *Candidate*, *Ledger*, *Subject*, and *Enrolment* classes were worthy to have *Enrol* methods; *Global* and *Component* merely provide *Get*... and *Set*... methods for each attribute. Only the *Enrol* events are described here. The treatment of the *Withdraw* and other events would follow the same pattern.

The outermost *Academic Records* level retrieves the value of *Current Semester* and then hands over the event to the *Candidate* object. Because the two processes are separable, *Academic Records.Enrol* can use an asynchronous (**tell**) procedure call to invoke the *Enrol* event procedure in the *Candidate* object:

$$\begin{aligned} \textit{Academic Records.Enrol}(c,s) \Leftarrow\ & \textbf{ask}\ \textit{Global.Read Lock}\,;\\ & \textbf{ask}\ \textit{Global.Get Semester}(sem')\,;\\ & \textbf{tell}\ \textit{Candidate.Enrol}(sem',c,s)\,;\\ & \textbf{ask}\ \textit{Global.Unlock}. \end{aligned}$$

The *Candidate* object checks *Candidate State* and *Component State* before passing control to the *Subject* object.[15]

$$\begin{aligned} \textit{Candidate.Enrol}(sem,c,s) \Leftarrow\ & \textit{Candidate.Read Lock}\,;\\ & \textit{Candidate.Candidate State}(c) = \textit{Admitted}\,;\\ & p = \textit{Candidate.Candidate Programme}(c)\,;\\ & \quad \textbf{ask}\ \textit{Component.Read Lock}\,;\\ & \quad \textbf{ask}\ \textit{Component.Get State}(p,s) = \textit{Offered}\,;\\ & \quad \textbf{ask}\ \textit{Component.Unlock}\,;\\ & \quad \textbf{tell}\ \textit{Subject.Enrol}(sem,c,s)\,;\\ & \textit{Candidate.Unlock}. \end{aligned}$$

The method *Subject.Enrol* first checks that a place remains in the subject and then invokes the *Enrolment* object to update the value of *Filled*(s) to *Filled*$'(s)$. To show

[15] In some object-oriented languages, the variable *Candidate State* would be referred to within the object *Candidate* as **this**.*Candidate State*, **self**.*Candidate State*, or simply as *Candidate State*. Here, we refer to the state as *Candidate.Candidate State*. This is meant to make the example easier to understand.

how error processing would be handled, we illustrate the case that if *Enrolment.Enrol* fails, it must then copy the value of *Filled(s)* into *Filled′(s)* unchanged. This means that *Subject.Enrol* doesn't need to check whether *Enrolment.Enrol* succeeds, but can update *Filled(s)* blindly, simplifying their interface.

Where two processes are separable, only input parameters are needed, but if they are tightly coupled (as in the case of *Subjects* and *Enrolments*), a parameter has to be *updated*, and needs to be passed both as an input and as an output.[16] Therefore *Subject.Enrol* must use a synchronous (**ask**) call to invoke *Enrolment.Enrol*:

$$
\begin{aligned}
\mathit{Subject.Enrol(sem,c,s)} \Leftarrow\ & \mathit{Subject.Update\ Lock}\,; \\
& \mathit{Subject.Filled(s)} < \mathit{Subject.Quota(s)}\,; \\
& \textbf{ask}\ \mathit{Enrolment.Enrol(sem,c,s,Subject.Fee(s),} \\
& \qquad\qquad \mathit{Subject.Filled(s), Subject.Filled'(s))}\,; \\
& \mathit{Subject.Unlock}.
\end{aligned}
$$

The *Enrolment.Enrol* method checks that the *Enrolment* object doesn't already exist, sets the value of *Enrolment State*, then passes the value of *fee* to the *Ledger* object:[17]

$$
\begin{aligned}
& \mathit{Enrolment.Enrol(sem,c,s,fee,filled,filled')} \Leftarrow \\
& \qquad \mathit{Enrolment.Update\ Lock}\,; \\
& \qquad \mathit{Enrolment.Enrolment\ State(sem,c,s)} = \mathit{Null}\,; \\
& \qquad \mathit{Enrolment.Enrolment\ State'(sem,c,s)} = \mathit{Enrolled}\,; \\
& \qquad \mathit{filled'} = \mathit{filled} + 1\,; \\
& \qquad \textbf{tell}\ \mathit{Ledger.Enrol(c,fee)} \\
& \qquad \mathit{Enrolment.Unlock}.
\end{aligned}
$$

On the other hand, if the enrolment *does* already exist, then *Filled′* must be given a suitable value so that *Subject.Enrol* will update *Filled′* correctly, and a message must be passed to an *Error* object to cause a suitable error message to be displayed:

$$
\begin{aligned}
& \mathit{Enrolment.Enrol(sem,c,s,fee,filled,filled')} \Leftarrow \\
& \qquad \mathit{Enrolment.Update\ Lock}\,; \\
& \qquad \mathit{Enrolment\ State(sem,c,s)} \neq \mathit{Null}\,; \\
& \qquad \mathit{filled'} = \mathit{filled}\,; \\
& \qquad \textbf{tell}\ \mathit{Error.Already\ Enrolled(sem,c,s)}\,; \\
& \qquad \mathit{Enrolment.Unlock}.
\end{aligned}
$$

---

[16] We use two separate variables here: *Subject.Filled* and *Subject.Filled′*. In many programming languages a single updated variable would be used instead.

[17] Where two objects are tightly coupled, locking one effectively locks the other. In this situation, locking *Subject* makes it unnecessary to lock *Enrolment*. But we must do this consistently: in the case of the *Assess* event, *Subject* must be locked for *update*, even though it is not *Subject*, but *Enrolment*, that is updated.

Finally, the *Ledger* object updates the value of *Balance Owing*.

$$\begin{aligned} Ledger.Enrol(c,fee) \Leftarrow\ & Ledger.Update\ Lock\,; \\ & Ledger.Balance\ Owing' = Ledger.Balance\ Owing + fee\,; \\ & Ledger.Unlock. \end{aligned}$$

The *Ledger* class doesn't have to be distinct from the *Candidate* class, provided *Ledger.Enrol* is distinct from *Candidate.Enrol*. If the two objects were one, *Ledger.Enrol* would have to be renamed, to *Candidate.Add Fee* perhaps.

Algorithm 7.2 Segmenting specifications in an object-oriented design.

- Each method of an object should use its input parameters and the variables the object owns to evaluate any conditions it can.
- The method should make any assignments to the object's variables that it can.
- If there remain more assignments to make or more conditions to evaluate, it must call a suitable method in a downstream or subordinate object. If the invoked method needs the values of variables known to the current method, it should pass them to it as *input* parameters.
- If the current method has insufficient data to make an assignment to a variable that the object owns, it must pass that variable as an *output* parameter to a subordinate object.

Algorithm 7.2 offers a general strategy for correctly segmenting specifications between objects. This strategy was used in the above example, with two exceptions:

- The *Global* object was invoked directly from *Academic Records*.
- The *Component* object was implemented within the *Candidate* object. The indentation in the text shows where the boundary would have fallen if the objects had been separate.

With the exception of the call from *Subject* to *Enrolment*, the method calls are asynchronous (**tell**). This means that the objects form a pipeline:

[*Academic Records*, *Candidate*, *Subject/Enrolment*, *Ledger*].

Because of locking, each object can be safely executed as a concurrent task, and up to four events could be chasing each other along the pipeline at any one time. In general, if two processes are separable, it is worth checking whether it is acceptable for them to be pipelined. (In contrast, the situation at the end of the procedure for *Subject.Enrol* is *not* so flexible, because *Subject* cannot safely be unlocked until after $Filled'(s)$ has been updated.)

It is interesting to contrast the above situation with one where there are multiple instances of *Candidate*, *Subject*, and other objects. As a result, several *Candidate*,

Subject, and other objects could be active concurrently, increasing the possible throughput of the system. This would be significant if candidates were allowed to enrol online. With only one *Candidate* object, only one candidate could interact with the system at a time. If each candidate spent time dithering for 30 seconds, the system would only be able to process 120 enrolments per hour — not enough.

In this situation, when a candidate enrols in a subject, instead of a database row being created, a new *Enrolment* object will be instantiated, and if the candidate withdraws, it will be destroyed. Where previously we spoke of database rows, we now speak of object instances. As in the case of database rows, the specification language expresses instantiation and destruction of an object as a change of state: from *Null*, or to *Null*.

Technically, the important change is that there would now be many possible data pathways instead of the simple pipeline. Asynchronous calls made by two *Candidate* instances could converge onto the same *Subject* instance, creating a race condition. The race condition could be avoided by replacing all the asynchronous (**tell**) calls by synchronous (**ask**) calls. In practice, a race condition might not prove dysfunctional. If we make the calls within the *Candidate* objects *asynchronous*, and two candidates try to enrol in the last place in a subject at the same time, it will be the first call to reach the *Subject* object that matters.[18] The *Subject* object would then be the arbiter of the order in which enrolments were made: in effect, the order in which candidates completed their choices rather than the order in which candidates began to make them.

A mixed approach is also possible: there could be multiple instances of *Candidate* and single instances of the other objects. This would be a good choice if candidates were allowed to make their enrolment choices online, perhaps over the Internet. Once an enrolment choice was made, it would be passed asynchronously to a single *Subject* object, at the head of a pipeline.

The above examples demonstrate that, in the search for efficiency, synthesis methods often lead to implementations that obscure their origins. The unfortunate thing about event segmentation is that the kernel specification of the *Enrol* event has become distributed across several objects. This can be avoided. The technique the *Subject* object used above to call the *Enrolment* object can be exploited ruthlessly to defer all processing to the most subordinate object in the chain — which we shall call the *Kernel* object. Algorithm 7.3 describes a strategy that works for event procedures that don't contain iteration.

There is a technical difficulty, but it is easily overcome. For example, if *Candidate State* ≠ *Admitted*, the *Candidate* object has no sensible value of *Programme* to pass to the *Subject* object (as it would now be obliged to do). As it happens, any value will do, because when the value finally reaches the *Kernel* object, it will ignore it.

Unfortunately, placing all kernel processing in the final object foregoes the opportunity to exploit pipelining, and all its predecessors are reduced to merely providing *Get* and *Set* methods.

---

[18] In our specification language, an asynchronous call to method *Enrol* in instance *s* of *Subject* would be written as **tell** *Subject*(*s*).*Enrol*(*sem*, *c*).

Algorithm 7.3 Preserving the kernel specification in an object-oriented design.

- Organise the objects as a topological sort of the process graph. Each object has access to the variables it owns, plus those passed to it as parameters.
- Each object (except the last) should call the next object in the sequence, passing the values of the variables it can access as *input* parameters.
- If the object might need to make an assignment to a variable that it can access, it must pass that variable to its successor as an *output* parameter.
- If the object is the last in the sequence, it should implement the kernel specification, returning the updated values of any variables belonging to upstream objects.

7.5 BATCH PROCESSING

The design of batch processing systems offers a strong demonstration of the power of use-definition analysis. It also illustrates how optimisation, in the sense of being able to process transactions at a higher rate, can make a system hard to understand and lead to compromises and work-arounds.

> ***Given that many systems have evolved as batch systems, understanding their problems is an essential tool in systems analysis.***

A **batch processing system** is one that does not act on events as they occur, but records them to process later in **batches**. One motive for processing events in batches is that it can often be done more efficiently than processing them as soon as they occur. Another is that, if we want to explain the *history* of a system, we first need to *reify* events.

Although the NUTS Academic Records System database records the balance that each candidate owes, it does not retain enough information to resolve a dispute. If your bank tells you that your account is overdrawn, you are likely to ask the bank for a statement that details your recent transactions. Likewise, a candidate should be entitled to ask to see a record of his or her recent enrolments, withdrawals, and payments.

As we know, collecting fees is not a function of the Enrolment Office but a function of the Bursary. Following the enrolment period, once enrolments are *Frozen*, the Bursary sends each candidate a *Statement*, detailing the amount the candidate already owes from earlier semesters, the fees for the candidate's current enrolments, the payments the candidate has recently made, and the balance that the candidate now owes. (See Figure 7.9.)

National University of Technology and Science

Hazel Filbert
123 Walnut Grove
Mongongo
Macadamia 54321

Admission No: 20135467

Tuition Fees Payable

Date	Code	Details	Fees	Balance
1 Jan 20xx		Carried forward from previous statement	$3,300.00	$3,300.00
3 Feb 20xx		Payment received - thank you	$3,300.00CR	$0.00
25 Mar 20xx	acctg101	Basic Accountancy	$500.00	$500.00
25 Mar 20xx	dataanal	Data Analysis	$550.00	$1,050.00
25 Mar 20xx	hmnbhvra	Human Behaviour A	$500.00	$1,550.00
25 Mar 20xx	prjmgta	Project Management A	$650.00	$2,200.00
25 Mar 20xx	qtheory	Queueing Theory & Simulation	$550.00	$2,750.00
28 Mar 20xx	qtheory	*Withdrawal*	$550.00CR	$2,200.00
30 Apr 20xx		**Payment Due**		**$2,200.00**

FIGURE 7.9

A sample statement sent by the Bursary to a candidate's postal address after enrolments are frozen. The *Heading* includes everything down to (and including) the line dated '1 Jan'. The line dated '3 Feb' is a *Payment*. The lines dated '25 Mar' are *Enrolments*. The line dated '28 Mar' is a *Withdrawal*. The line dated '30 Apr' is a *Footing*.

Unfortunately, the Academic Records database we have described doesn't allow us to work backwards from the candidate's current *Balance Owed* and recent enrolments to the amount previously owed because it doesn't record individual *Pay* events, only their total effect. Instead, the Enrolment Office sends the Bursary a list of *all* the events affecting *Balance Owing* that occurred during the enrolment period, up until the time that enrolments were *Frozen*. The Bursary, then (in a sense) replicates the work of the Enrolment Office, but can supply the candidate with the necessary details, including situations where a candidate has enrolled in subjects and withdrawn again in some confusing way, a matter that is often the cause of dispute. Once the candidate has paid all or part of the fees owing, the Bursary issues a receipt and forwards a copy to the Enrolment Office to allow the payment to be recorded in their database. (Therefore, the payment won't be acknowledged until the candidate's *next* statement.)

For now, we assume that the NUTS Academic Records System writes a record of each *successful* event to a sequential file or to an additional database table. (Presumably, there would be no point in including unsuccessful events.) Each record would comprise an indication of the type of event, its parameters, and the time when it occurred.[19] In a database, events must be stored in a table indexed by time — their only possible primary key[20] — and have attributes representing the event type and all possible event parameters. Typically, for any given event, many of the possible parameters would be **null**, only those relevant to the particular event being present. If we include the *Pay* events transcribed from the receipts that the Bursary forwards to the Enrolment Office, we can describe the relevant event records by the following table schema

$$Events(\underline{When},\ Event\ Type,\ Candidate,\ Subject,\ Amount)$$

where *Event Type* is 'Enrol', 'Withdraw', or 'Pay'. (*Subject* = **null** if *Event Type* = 'Pay', and *Amount* = **null** otherwise.)

With this schema, we can specify the *Print Statements* event as follows[21]

Print Statements ⇐
- *Undergraduates* = {*c* ∈ *Candidates* | *Candidate State*(*c*) = *Admitted*} ;
- **tell** *Statements.Begin*(*Current Semester*) ;
- (**asc** *c* ∈ *Undergraduates* ;
  - *bal* = *Balance Owing*(*c*) ;
  - **tell** *Statements.Heading*(*c*, *Personal Name*(*c*), *Postal Address*(*c*), *bal*) ;
  - *Transactions* = {*t* ∈ **dom**(*Events*) | *Candidate*(*t*) = *c*} ;
  - (**asc** *t* ∈ *Transactions* ; *event* = *Event Type*(*t*) ;
    - (*event* = 'Enrol' ; *s* = *Subject*(*t*) ; *bal′* = *bal* + *Fee*(*s*) ;
      - **tell** *Statements.Enrol*(*t*, *s*, *Subject Name*(*s*), *Fee*(*s*), *bal′*)
    - ∪ *event* = 'Withdraw' ; *s* = *Subject*(*t*) ; *bal′* = *bal* – *Fee*(*s*) ;
      - **tell** *Statements.Withdraw*(*t*, *s*, *Subject Name*(*s*), *Fee*(*s*), *bal′*)
    - ∪ *event* = 'Pay' ; *bal′* = *bal* – *Amount*(*t*) ;
      - **tell** *Statements.Payment*(*t*, *s*, *Subject Name*(*s*), *Amount*(*t*), *bal′*)
    -)
  -)* ;

---

[19] This is essentially the same information the DBMS records in its log, but a DBMS log is unlikely to be accessible for use by an application program.

[20] Assuming that an arbiter can assign an order to events, the time of each event must be unique.

[21] Strictly speaking, printing a statement is a *meta-event*, an event about events.

```
        tell Statements.Footing(bal′)
    )* ;
    tell Statements.All Done.
```

where *Statements* is a process that formats the report to match Figure 7.9.

- The *Statements.Begin* procedure stores the value of *Current Semester* and calculates the correct values of the dates shown in Figure 7.9 as '1 Jan' and '30 Apr', which are the same for every candidate. It also does anything else that might be necessary to initiate the printing process.
- The *Statements.Heading* procedure produces the details at the top of Figure 7.9, down as far as 'Carried forward from previous statement'.
- The *Statements.Enrol*, *Statements.Withdraw*, and *Statements.Payment* procedures produce the detail entries that update the balances.
- The *Statements.Footing* procedure produces the final balance at the end of a statement.
- Finally, the *Statements.All Done* procedure does whatever is needed to wrap up the *Statements* process.

(It is assumed that the calls to *Statements* are queued and served in first-in first-out order.)

How should the Bursary use the information that the Enrolment Office provides? One solution would be to set up its own database, similar to that used by the Enrolment Office, complete with an *Events* table. The information necessary to print the statements could be satisfied by the following *SQL* query:

```
select t.When, t.Event_Type, t.Candidate, t.Subject, t.Amount,
    s.Subject_Name, s.Fee,
    c.Personal_Name, c.Postal_Address, c.Balance_Owing
    from Events t, Subjects s, Candidates c
    where t.Subject = s.Subject and t.Candidate = c.Candidate
    and c.Candidate State = Admitted
    order by t.Candidate, t.When;
```

In itself, this query cannot generate the *Statements* report, because each row of the result would be associated with the same value of the candidate's *Balance Owing*. Consequently, the query should be *embedded* in a procedure that will update the balances correctly and post the final value of *Balance Owing* to the *Candidates* table. (See Section 6.3.3 on page 179.)

Although batch processing systems *can* use a DBMS, they frequently use *sequential* (SAM) or *indexed-sequential* (ISAM) files, as suggested in Figure 7.10.[22]

In preparation for the batch process shown in Figure 7.10, the Enrolment Office would send the Bursary a file recording *Enrol* and *Withdraw* events and up-to-date

---

[22] 'SAM' is a frequently used abbreviation for 'sequential access method, and 'ISAM' for 'indexed-sequential access method'.

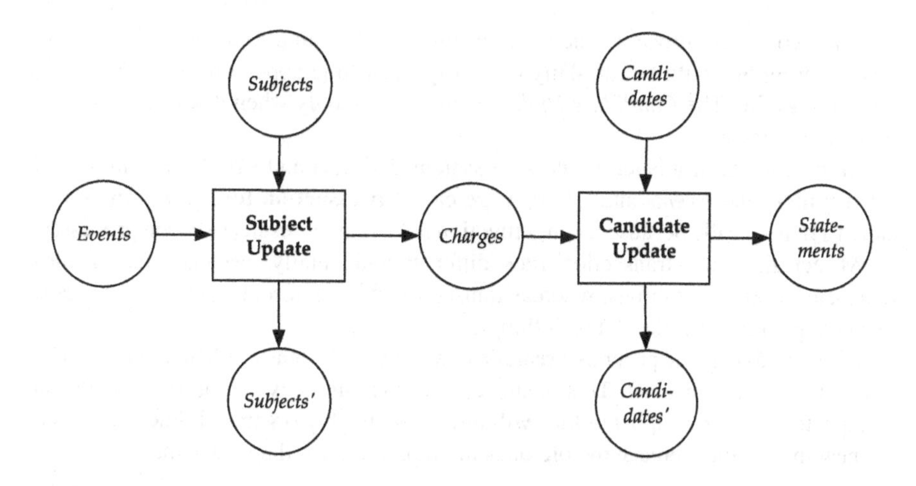

FIGURE 7.10

A bi-partite graph, showing how the *Subject Update* and *Candidate Update* programs are connected. *Event Records* are passed to the *Subject Update* program, which by reference to the *Subjects* file is able to calculate the correct *Charges*. The *Candidate Update* program then uses *Charges* to update *Balance Owing*, and print *Statements*. During the process, new copies of the *Subjects* and *Candidates* files are produced, containing updated information.

snapshots of the *Candidates* and *Subjects* tables. These snapshots allow the Bursary to ensure that its own *Candidates* file contains the latest *Personal Name* and *Postal Address* for each undergraduate and that its *Subjects* file has the current values of *Fee*. On the other hand, the Bursary retains its own (previous) value of *Balance Owing* and ignores the updated value from the Enrolment Office.[23] Finally, the Bursary prepares a *Payments* file that records the payments received from candidates, which it merges with the *Enrol* and *Withdraw* events from the Enrolment Office to form an *Events* file.

Then, as Figure 7.10 suggests, the *Events* file is used to inspect the *Subjects* file. In this process, events are **extended** (joined) with data from the *Subjects* file; for example *Enrol* and *Withdraw* events are *extended* with the *Fee* and *Subject Name* of the subject concerned. The resulting *Charges* file is then used to both update the *Candidates* file and print the statements in a single **pass** of the file. Both the steps can use the sequential update algorithm, which we shall describe shortly. Fortunately, the loop structure of the *Print Statements* event procedure nicely matches that of the sequential update algorithm in *Candidate Update*.[24]

[23] A useful check of the correctness of the system would be to compare the *final* balances that the Bursary calculates with those the Enrolment Office has calculated.

[24] This isn't a coincidence. The purpose of a statement is to trace the updating process, as it affects each individual candidate.

The two update processes needn't be run with the same frequency: the *Subject Update* program could be run daily or hourly, each time appending new records to the *Charges* file. The *Candidate Update* would be run only when it was necessary to prepare statements.

In the context of a batch processing system, *Subjects* and *Candidates* are called **master files**, and *Events* and *Charges* are called **transaction files**. *Charges* is also called a **transfer file** because its only function is to transfer results between processes.

Master files and transaction files differ fundamentally, because master files represent the *states* of things, whereas transaction files represent sequences of *events* that change or inspect the states of things.

A sequential update program creates a new copy of the master file it updates, without destroying the original. This means that if anything goes wrong, the update can be repeated. The new, updated files will always be freshly organised. Each semester, the new master files replace the old ones in preparation for the next semester.

7.5.1 UPDATE ALGORITHMS

There are two basic update algorithms, random-access update, based on the *Table Look-Up* algorithm (page 186), and sequential update, based on the *Sort-Merge* algorithm (page 187). It is usual for these algorithms to use files rather than database tables. In both cases, the transaction file containing the event records takes the role of the child table, and the master file (*Subjects* or *Candidates*) takes the role of the parent table.

- In a **random-access update**, events are processed in the order in which they originally occurred. Each event procedure reads and updates master records as it requires. In case anything goes wrong, it is usual to make a back-up copy of the master files first.
- In a **sequential update**, event records are sorted into time order within the primary key. They are then merged with the master file records. As with the Sort-Merge join, this means that each master record is accessed only once. All the updates to one master record are dealt with as a group — first-come, first-served. The sequential update algorithm always produces a new copy of the master file, and the existing master file can be retained as a back-up.

Operations on records take place in a buffer. The buffer is initialised with the existing master record if it exists, otherwise it is initialised with a dummy record whose state is *Null*. Event records invoke procedures that update the buffer. If, *after* the buffer has been updated, the state of the record has become *Null*, in the case of the random-access update, any existing master record is deleted, or, in the case of a sequential update, no record is written to the new master file.

As with the join algorithms, there are hybrid forms of the random-access update, such as skip-sequential, or clustered random access.

It is possible to use the random-access algorithm to update any number of master files, but it isn't usually possible to combine a sequential update with random access.

For example, if events were sorted into order by *Candidate*, updates to each *Subject* would occur in order by *Admission No*, rather than *When*. In an enrolment system, if there were contention for the last place in a subject, this would give priority to the candidate with the lower *Admission No* rather than the one who tried to enrol first.

It is instructive to compare how long a sequential update procedure would take compared with the time it took the Enrolment Office to update the same events when they actually occurred. Suppose, as we did earlier, that the database records 500 *Subjects*, 10,000 active *Candidates*, 20,000 *Enrolments*, 10,000 *Payments*, and 1,000 *Withdrawals*. The *Enrol* kernel event requires five records to be read and three records to be updated, The *Withdraw* event requires three records to be read and three records to be updated, and the *Pay* event requires one record to be read and updated — altogether, 186,000 inspections and updates. If each operation accessed a new block, taking 10 mS, the total time taken would be just over 30 minutes.

Assuming the Bursary is only interested in undergraduates, the *Candidates* file would occupy about 1,700 KB, the *Subjects* file would occupy about 33 KB, and the *Events* file would occupy about 750 KB. Using the *sequential update* algorithm, the *Events* and *Subjects* files would be merged by sorting the 750-KB *Events* file into order by [*Subject, When*] (in RAM) and reading the 33-KB *Subjects* file sequentially. The extended event records in the 800-KB *Charges* file would then be sorted into [*Candidate, When*] order and merged with the *Candidates* file sequentially, requiring 1,700 KB to be read. Including the writing and reading of the *Charges* transfer file, and writing the new copies of the master files, less than 5,000 KB need to be read or written. With 1-KB blocks and a 10-mS access time, this would take about 50 seconds. On the other hand, using the random-access algorithm, it being in effect a replay of the events recorded by the Enrolment Office, would take 30 minutes.

Even with 1-KB blocks, the sequential process is already over 30 times faster than the random-access equivalent. It can be made even faster by increasing the block size. Each doubling of the block size would halve the access time.

Increasing the block size wouldn't speed up the random-access update much until the block size became comparable with the size of the master file. For example, if the *Subjects* file occupied only two blocks, one-half of the time the desired master record would already be in RAM, so the number of accesses to the *Subjects* file would be halved. In practice, the automatic caching of blocks by a disk drive or by the operating system might achieve a similar effect.

In this application, both the random-access and the sequential updates are easily fast enough, but in very high-volume transaction processing, a sequential update may be the only option.

7.5.2 VALIDITY OF BATCH PROCESSING

It is implicit in the system flowchart of Figure 7.10 that *Subject Update* is executed before *Candidate Update*, among other things, to ensure that the current values of *Fee* are used to update each candidate's *Balance Owing*. We observe that data can flow *via* the *Charges* file *from* the *Subjects* file *to* the *Candidates* file, but they cannot flow in the reverse direction. Nor does the *Sort-Merge* algorithm allow data to flow arbitrarily between different *Subjects* records or different *Candidates* records. We therefore need to be sure that the process graph doesn't demand any reverse flows or cross flows.

The Bursary subsystem has simpler event specifications than the equivalent kernel events, because if the events have already succeeded, they must concern an *Admitted* candidate and an *Offered* subject. (Indeed, the Bursary doesn't need to keep track of *Subject State* at all.) On the other hand, since the Bursary itself creates the records for *Pay* events, they haven't already been validated. The specification therefore needs to check that the *Admission No* that the Bursary supplies concerns a candidate who has been *Admitted*, but not yet *Graduated*.

Let's consider the Bursary's specifications of the *Enrol*, *Withdraw*, and *Pay* procedures:

$$\begin{aligned} \mathit{Bursary.Enrol}(t, c, s) \Leftarrow\ & e = (\mathit{Current\ Semester}, c, s)\,; \\ & \mathit{Enrolment\ State}(e)' = \mathit{Enrolled}\,; \\ & \mathit{Mark}(e)' = \mathbf{null}\,; \\ & \mathit{Filled}(s)' = \mathit{Filled}(s) + 1\,; \\ & \mathit{Balance\ Owing}(c)' = \mathit{Balance\ Owing}(c) + \mathit{Fee}(s). \end{aligned}$$

$$\begin{aligned} \mathit{Bursary.Withdraw}(t, c, s) \Leftarrow\ & e = (\mathit{Current\ Semester}, c, s)\,; \\ & \mathit{Enrolment\ State}(e)' = \mathit{Null}\,; \\ & \mathit{Filled}(s)' = \mathit{Filled}(s) - 1\,; \\ & \mathit{Balance\ Owing}(c)' = \mathit{Balance\ Owing}(c) - \mathit{Fee}(s). \end{aligned}$$

$$\begin{aligned} \mathit{Bursary.Pay}(t, c, \mathit{Amount}) \Leftarrow\ & \mathit{Candidate\ State}(c) = \mathit{Admitted}\,; \\ & \mathit{Balance\ Owing}(c)' = \mathit{Balance\ Owing}(c) - \mathit{Amount}. \end{aligned}$$

Figure 7.11 displays the canonical process graph for these events. It shows that the attributes of the *Subjects* and *Candidates* tables are separable. This means that the flowchart of Figure 7.10 on page 233 is valid. The absence of cross flows means that the records within the *Subjects* file are independent, and the records within the *Candidates* file are independent. In turn, this means that the sequential update algorithm is a valid (and efficient) way to update both files.

The flowchart of Figure 7.10 and the sequential update algorithm form the *infrastructure* that supports the efficient processing of events. They shouldn't be specified as part of the system kernel — they are a solution, not the problem. Although the systems analyst might expect to use such a batch solution at an early stage of analysis, it would be a serious mistake to anticipate this decision by including it in the kernel specification. To do so would make the specification harder to write, harder to understand, and block the paths to alternative implementations.

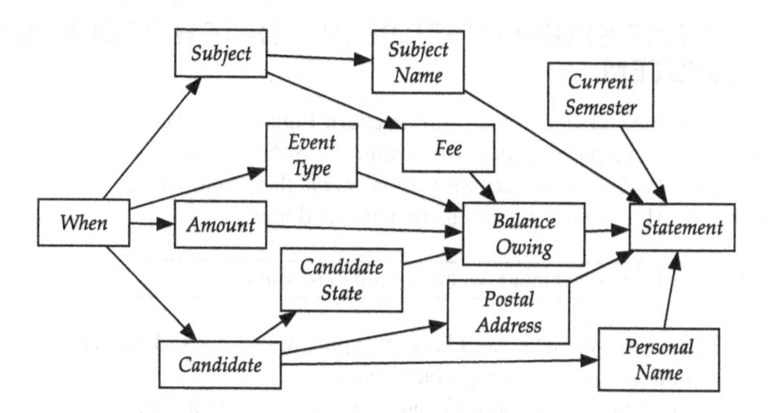

FIGURE 7.11

The canonical process graph for the Bursary subsystem. It is convenient to include the *Statement* process in the graph; it may be regarded either as an external observer or as a temporary table that is being updated.

It is worth noting that independent processes can usefully be executed in *parallel* using a multi-processor architecture. Such an infrastructure could partition the database evenly between the processors. For example, *Subjects* might be partitioned between processors on the basis of *Subject Code*, and *Candidates* might be partitioned on the basis of *Admission No*. In the best possible case, if events were spread evenly across k processors, the system would run k times faster. In practice, events would be spread randomly, and the speed-up would be rather less.

Unfortunately, one problem would remain: data flow from several *Subject Update* processors could converge at a single *Candidate Update* process. As we pointed out in Section 7.3, this can create a race condition, allowing later events to overtake earlier events. If the temporal order of events is to be respected, no *Candidates* row should be updated until all *Subjects* rows have been processed; otherwise a later event might overtake an earlier one for the same candidate, simply because its *Subject Update* process was less busy.[25] In general, the consequence is that data pathways need to be periodically flushed, so, in effect, events are processed in batches. Even so, in the context of the high-throughput systems where a multi-processor approach is needed, an efficiently sized batch can be collected in less than 1 second.

The same restriction also applies to the batch system of Figure 7.10. We cannot substitute a first-in-first-out queue for the *Charges* file. If we did, updates to each *Candidates* row would occur in order by *Subject Code* rather than *When*. We must collect and sort the updates so that they can be processed in the right order.

[25] In the particular case of the *Academic Records System* the order doesn't matter much.

7.5.3 EVENT SEGMENTATION IN A BATCH PROCESSING SYSTEM

Consider again the Bursary's batch subsystem of Figure 7.10: because the *Enrol* and *Withdraw* events are partly processed within the *Subject Update* and partly within the *Candidate Update*, it is necessary to *segment* them into two component event specifications. Algorithm 7.4 explains how this is done.

Algorithm 7.4 How to segment events for a batch system.

- Each event segment should use its input parameters and the variables it controls to evaluate any conditions it can.
- Each event segment should make any assignments it can.
- If there are more assignments or conditions still to evaluate, the current event segment should call a downstream or subordinate process. If that process needs the values of variables that are known to the current process, the current process should pass them to it as *input* parameters.

The result is as follows: the Bursary system calls the *Subject Update* process:

$$Bursary.Enrol(t,\ c,\ s) \Leftarrow \textbf{tell}\ Subject\ Update(s).Enrol(t,\ c).$$
$$Bursary.Withdraw(t,\ c,\ s) \Leftarrow \textbf{tell}\ Subject\ Update(s).Withdraw(t,\ c).$$
$$Bursary.Pay(t,\ c,\ Amount) \Leftarrow \textbf{tell}\ Subject\ Update(\textbf{null}).Pay(t,\ c,\ Amount).$$

Within the *Subject Update*, we have

$$Subject\ Update(s).Enrol(t,\ c) \Leftarrow s.Filled' = s.Filled + 1\ ;$$
$$\textbf{tell}\ Candidate\ Update(c).Enrol(t,\ s,\ Fee(s)).$$
$$Subject\ Update(s).Withdraw(t,\ c) \Leftarrow s.Filled' = s.Filled - 1\ ;$$
$$\textbf{tell}\ Candidate\ Update(c).Withdraw(t,\ s,\ s.Fee).$$
$$Subject\ Update(s).Pay(t,\ c,\ Amount) \Leftarrow$$
$$\textbf{tell}\ Candidate\ Update(c).Pay(t,\ Amount).$$

where we use the notation $s.Filled$ rather than $Filled(s)$ to emphasize that *Filled* is an attribute of the record with key s of the *Subjects* file.

Within the *Candidate Update*, we have

$$Candidate\ Update(c).Enrol(t,\ s,\ fee) \Leftarrow e = (Current\ Semester,\ c,\ s)\ ;$$
$$e.Enrolment\ State' = Enrolled\ ;$$
$$e.Mark' = \textbf{null}\ ;$$
$$c.Balance\ Owing' = c.Balance\ Owing + fee.$$

$$Candidate\ Update(c).Withdraw(t,\ s,\ fee) \Leftarrow e = (Current\ Semester,\ c,\ s);$$
$$e.Enrolment\ State' = Null;$$
$$c.Balance\ Owing' = c.Balance\ Owing - fee.$$
$$Candidate\ Update(c).Pay(t,\ Amount) \Leftarrow c.Candidate\ State = Admitted;$$
$$c.Balance\ Owing' = c.Balance\ Owing - Amount.$$

We can see that by replacing the procedure call in *Subject Update(s).Enrol* with the text of *Candidate Update(c).Enrol*, after appropriate substitution of parameters, we recover the full text of *Bursary.Enrol* given earlier in Section 7.5.2. Unfortunately, event segmentation has caused the texts of *Subject Update* and *Candidate Update* to become more obscure. Each process implements *parts* of all three events. No single segment of code corresponds to the whole of the *Enrol* or *Withdraw* event specification.

If we were being careful to keep the processing of events separate from the printing of statements — as in a case where printing them was an option — we would need to implement these component events as we have described. But since the printing of statements replicates all the work of the events that it reports, it is a simple optimisation to update *Balance Owing* from the final value calculated by *Print Statements*. So, in practice, these component events would be integrated into the production of the statements.

In the above decomposition, we have indexed the process names, *Subject Update(s)*, *Candidate Update(c)*, etc., to emphasise that there can be an independent process for each *Subject* and for each *Candidate*. Indeed, the correctness of the sequential update algorithm process requires that this is possible. On the other hand, the *Pay* event doesn't concern any particular subject, so *Bursary.Pay* specifies the subject as **null**.[26]

Furthermore, it is important that the procedures that *Subject Update* invokes in *Candidate Update* need no return parameters. There is no communication path from *Candidate Update* to *Subject Update*. Use-definition analysis proves that none is necessary.

7.5.4 SYNTHESISING A BATCH PROCESSING SYSTEM

Suppose that the NUTS Enrolment Office has decided to make it possible for candidates to enrol online. But rightly fearing what technically savvy youngsters might do to their database, the Enrolment Office decides to allow students to merely record their selections, and to process them under controlled conditions later using an efficient batch process. Such a system is said to have an interactive front end and a batch back end. The Enrolment Office's situation is now similar to that of the Bursary, but it also has to check whether, for example, a candidate only chooses subjects that are components of the candidate's programme.

---

[26] In a sequential update it doesn't matter what value of *Subject Code* is used; in a parallel-processing environment, it would be better to assign differing dummy values to *Subject Code* to spread the load evenly across the processors.

In this scenario, the Enrolment Office's canonical process graph would be that of Figure 7.7 (page 219). Compared with the Bursary system, the optimised form of this graph, which was shown in Figure 7.8, is less well suited to a batch processing infrastructure:

- First, the process {*Quota, Subject State, Enrolment State, Filled, Fee, Mark*} in Figure 7.8 contains attributes from two different files: *Quota, Subject State, Filled,* and *Fee* from the *Subjects* master file and *Enrolment State* and *Mark* from the *Enrolments* master file. (This does not result from optimisation; *Filled* and *Enrolment State* are tightly coupled.) Normally this would mean that *both* files would need to be accessed randomly in temporal order. Despite this, in this particular situation, because of their hierarchical relationship, a reasonably efficient implementation remains possible. If the *Enrolments* file is ordered by [*Semester, Subject Code, Admission No*], all the relevant enrolment rows for each subject will be clustered together, so that accessing them will involve only one or two blocks.[27]
- Second, although it is tempting to merge all the *Candidate* attributes, it isn't a good idea. Merging *Balance Owing* with {*Programme, Candidate State*} would make the process graph cyclic. The cycle would encompass attributes from the *Candidates, Subjects, Components,* and *Enrolments* tables. The cycle implies that all four tables would fall within the same strongly connected component, making all their processes tightly coupled. We mentioned earlier that tightly coupled processes can be a problem. In this case, the problem is that the resulting composite process could not benefit from using sequential access.
  On the other hand, it is relatively harmless to access the same file twice. Figure 7.12 shows how this can be done. It is a topological sort of the optimised process graph of Figure 7.8. (It assumes that all events involve the current semester, otherwise an additional process would be needed to update *Current Semester* and correctly extend each event.)

In the case of *Enrol* events, the resulting system works as follows:

1. After sorting events into order by [*Admission No, When*], *Candidate Update* copies the *Candidates* file sequentially, checking *Candidate State* and extending each event with the value of *Programme*, finally writing the result to *Transfer 1*.
2. *Component Update* sorts the extended events in *Transfer 1* into order by [*Programme Code, Subject Code, When*], then copies the *Components* file sequentially to check *Component State*.
3. The event records in *Transfer 2* are then sorted into order by [*Subject Code, When, Admission No*] (*not* [*Subject Code, Admission No, When*]). *Subject Update*

[27] Because of the difficulty caused by the tight coupling between *Filled* and *Enrolment State*, the reader may be questioning the wisdom of storing the current number of enrolments in *Filled*. The alternative is worse. It would require an access to every enrolment in the subject each time a new enrolment was made, rather than one sequential access to the subject itself.

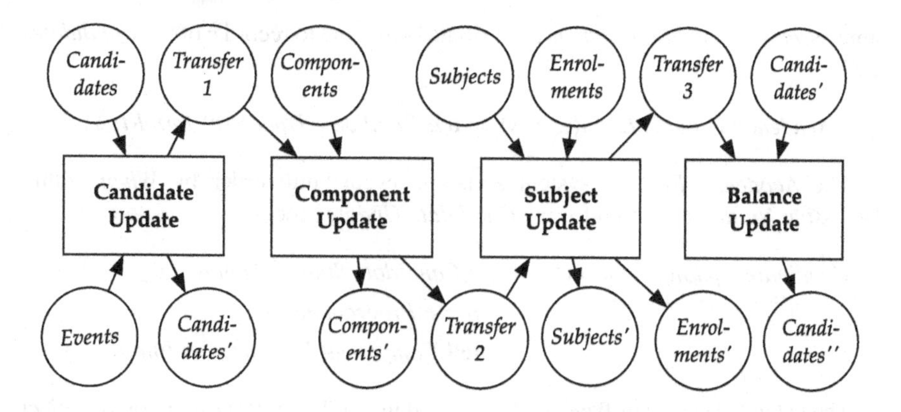

FIGURE 7.12

A batch system that could be used to process enrolments. Note that the *Candidates* file is processed twice. An alternative would be to have two different files, splitting or even duplicating the *Candidate* attributes. For example, the left-hand file might contain *Candidate State*, *Programme*, and *Phone*, whereas the right-hand file might contain *Balance Owing* and *Postal Address*. *Personal Name* could appear in both files.

uses sequential access to the *Subjects* file and clustered random access to the *Enrolments* file to check *Subject State* and *Quota* and update *Filled* and *Enrolment State*.

If we assume that the process that records the events never allows a subject to exceed its quota — and that all events concern the current semester, it would then become harmless to sort *Transfer 2* by [*Subject Code*, *Admission No*, *When*], allowing the *Enrolments* file to be accessed sequentially too. On the other hand, if the recording process allows there to be contention for the last place in a subject, and we want to allocate places on a first-come, first-served basis, this optimisation is not possible.

4. Finally, *Balance Update* uses sequential access to the updated *Candidates'* file to update *Balance Owing*.

As it happens, with the exception of *Begin Semester* and *Graduate*, this same batch infrastructure can be used to process *all* the events in the kernel specification, including *Print Statements*.

The *Candidates'* file that the first step generates will be inconsistent, containing updated personal details but the old values of *Balance Owing*. That won't matter. The *Candidates''* file will always be consistent; once it has been produced, *Candidates'* can be discarded.

To illustrate process segmentation in this system, we consider only the *Enrol* events. (We use the notation *c.Candidate State* rather than *Candidate State*(*c*) to emphasize that *Candidate State* is an attribute belonging to record *c* of the *Candidates* file.)

$$Academic\ Records.Enrol(t,\ c,\ s) \Leftarrow \textbf{tell}\ Candidate\ Update(c).Enrol(t,\ s).$$

The *Academic Records* system sorts the events into order by *When* within *Admission No* in preparation for the *Candidate Update* process:

$$\begin{aligned} Candidate\ Update(c).Enrol(t,\ s) \Leftarrow\ & c.Candidate\ State = Admitted\,; \\ & p = c.Programme\,; \\ & \textbf{tell}\ Component\ Update(p,\ s).Enrol(t,\ c). \end{aligned}$$

The extended events in *Transfer 1* are sorted into order by *Programme* and *Subject* in preparation for the *Component Update*. Within the *Component Update*, we have

$$\begin{aligned} Component\ Update(p,\ s).Enrol(t,\ c) \Leftarrow\ & (p, s).Component\ State = Offered\,; \\ & \textbf{tell}\ Subject\ Update(s).Enrol(t,\ c). \end{aligned}$$

The events in *Transfer 2* are sorted into order by *When* within *Subject*, permitting sequential access to the *Subjects* file and clustered access to the *Enrolments* file. Within *Subject Update*, we have

$$\begin{aligned} Subject\ Update(s).Enrol(t,\ c) \Leftarrow\ & sem = Current\ Semester\,; \\ & s.Filled < s.Quota\,; \\ & (sem,\ s,\ c).Enrolment\ State = Null\,; \\ & s.Filled' = s.Filled + 1\,; \\ & (sem,\ s,\ c).Enrolment\ State' = Enrolled\,; \\ & \textbf{tell}\ Balance\ Update(c).Enrol(t,\ s,\ s.Fee). \end{aligned}$$

Finally, after sorting events into order by *Admission No*, *Balance Update* adjusts *Balance Owing*.

$$Balance\ Update(c).Enrol(t,\ s,\ Fee) \Leftarrow c.Balance\ Owing' = c.Balance\ Owing + Fee.$$

Not surprisingly, given that it solves the same problem and shares the same optimised process graph, at the conceptual level of the specification language, this implementation is similar to that of the object-oriented system of Section 7.4.2.

Unfortunately, both examples paint too rosy a picture of event segmentation because all the segments for a single event have been presented in close proximity to one another in the text. In practice, these segments would appear in separate program modules. Any one module would contain segments of the *Enrol* event, *Withdraw* event, and so on. In a batch processing infrastructure each event procedure must share its input-output logic with the others, rendering its derivation from the specification even more obscure.

When we consider how an event procedure can become split into almost meaningless logical snippets, especially in a batch system, *and realise that many office systems are indeed batch systems*, it isn't surprising that it can be hard for a systems analyst to discover what office systems do.

We have already seen an example of this in the interface between the Enrolment Office and the Bursary. The trick is for the analyst to concentrate, not on the workings of a particular department, but on the processing of a particular kind of event, following its progress from one department to the next. Then the analyst should ignore *where* the work is done, and consider only *what* is done, as was suggested in Section 1.3. Usually, it pays to start with the most frequent kind of event (e.g., *Enrol*), because that is what the existing system was designed to handle efficiently. The design may not handle other kinds of event well, as we shall see in the following section.

7.6 MODES

Suppose that, in order to detect and deal correctly with erroneous events, the Bursary also uses a batch processing system like that just described in Figure 7.12 (page 241). In particular, the system checks the value of *Candidate State* before updating *Balance Owing*. The batch system proposed in Figure 7.12 accesses these two *Candidate* attributes in two separate update processes.

Now suppose that the Bursary becomes responsible for deciding whether candidates can graduate. This is reasonable, because it is a kernel requirement that a candidate cannot graduate until all fees have been paid — a financial matter, making it the responsibility of the Bursary. This requires *Balance Owing* to be inspected *before* updating *Candidate State* — the reverse of the order needed to process enrolments. The resulting state dependency graph is shown in Figure 7.13.

Compared with the state dependency graph of the *Enrol* event given in Figure 7.3, Figure 7.13 contains one new edge, from *Balance Owing* to *Candidate State*. Unfortunately, this creates a strongly connected component that encompasses attributes of *Candidates*, *Subjects*, and *Enrolments*.

Clearly, the batch system of Figure 7.12 is no longer viable. Despite this, rather than adopt a relatively slow random-access implementation, which might also need a lot of software to be rewritten, there is an alternative: to allow the use of two or more processing modes. A **mode** is a subsystem that implements a subset of all possible events. For example, the requirements of the *Pay* and *Graduate* events taken alone require access only to attributes of the *Candidates* table, which allows them to be implemented by a subsystem consisting of a single *Candidates* update.

The Bursary's strategy is now clear. Have two modes: *Statement mode* and *Graduation mode*. Having once printed statements in *Statement mode* and mailed them to candidates, the Bursary can switch to a new *Graduation mode*, in which new payments are received, receipts issued, and *Graduate* events are verified. Clearly, when in *Graduation mode*, the Bursary cannot deal with enrolments or withdrawals.

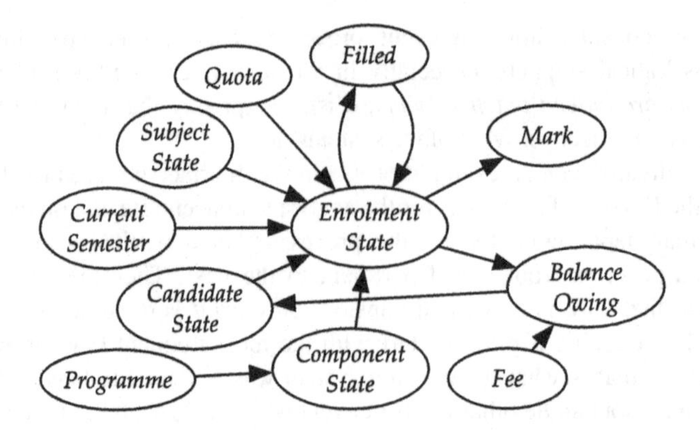

FIGURE 7.13

The state dependency graph for a Bursary system that includes *Enrol*, *Withdraw*, *Pay*, and *Graduate* events.

The use of modes is common in batch systems, but it is also a strategy that can be used to avoid deadlock or other inefficiencies. Even hardware systems have modes. There are usually at least two: *Test* mode and *Run* mode.

In most systems, parent tables are created and updated before their children. In the Academic Records System, *Candidates* and *Subjects* have to be in suitable states before *Enrolments* can be created. Typically, data tend to flow from parent to child. Difficulties arise when this flow has to be reversed.

Consider what would happen if, through unforeseen circumstances, the teaching of a subject had to be cancelled. It would be necessary to force the withdrawal of all candidates enrolled in the subject. The kernel specification doesn't provide for this contingency. The Enrolment Office would need to print a class list, then withdraw each candidate. Only when all the enrolments had been withdrawn could the subject be made *Inactive*. Ideally, this action would be atomic, otherwise candidates might mistakenly re-enrol in the subject while the Enrolment Office did its work. This is an example of a case where the ability to change the state of a parent depends on the states of its children — a reversal of the normal flow. Depending on a system's infrastructure, this can be a minor problem or a major problem. It would certainly be a major problem for a batch processing system and would require a special mode in which *Enrol* events could not be supported.

The kernel specification of the Academic Records System carefully avoided this sort of issue. For example, the *Set Quota* event is specified in a way that ensures *Quota* can never fall below *Filled*, therefore no existing enrolment can be affected, and the states of enrolments never control the state of the parent subject.[28]

[28] This is another advantage of redundantly storing the value of *Filled* rather than counting the number of enrolments.

Because of difficulties associated with maintaining parent-child relations, a database might become littered with parent records that have no children. The *Filled* attribute illustrates one way to deal with this. By keeping count of the number of children incrementally, it is easy to recognise when the state of a parent row can safely be modified. The alternative is to physically count the number of children through a search. This can be time-consuming, so it is best done when the system is quiet, in what is often referred to as a **weeding mode**.

Because weeding typically takes place infrequently, it is sometimes wise to put a parent row into a kind of limbo, a special state where some events are allowed but others are not. For example, a *Cancel Subject* event could give *Subject State* the special value *Cancelled*.[29] In the *Cancelled* state *Enrol* events would be forbidden but *Withdraw* events would be allowed. This would preserve the atomicity of cancelling a subject, while allowing the Enrolment Office plenty of time to weed out the subject's unwanted enrolments.

In reality, the Academic Records System already has modes: There is a time when the curriculum is being organised, which can overlap the time when candidates are being offered admission. There is a time when the curriculum has been frozen and admitted candidates may make enrolments and withdrawals. There is a time when enrolments are frozen and statements are posted by the Bursary. There is a time when candidates are expected to settle their debts, a time when subjects are assessed and results circulated, and a time when candidates may graduate. Then the cycle repeats. All these phases allow certain kinds of event but not others. These phases could be made explicit by a global variable, *Mode*, which we have crudely approximated by causing *Subjects* to be *Frozen*, and other obscure means.[30]

Finding the optimum set of modes is theoretically an intractable problem. In practice, it is a matter of considering the most common events first. For example, the most common event in the Academic Records System is *Enrol*. This event determines a certain state dependency graph. The next most common event is *Withdraw*. If this proves to have a graph that is compatible with that for *Enrol*, they can share the same mode. We continue to add events in this way. If any event has data dependencies that would make the processing of *Enrol* events less efficient, the difficulty needs to be resolved. This may happen in one of three ways:

- By accepting the inefficiency, perhaps by allowing the event to lock more resources than it needs, or, in a batch system, by replacing an efficient access method by a slower more general method.
- By allocating the event to a different mode — although this may not be possible if the event must be used interactively in the same session as *Enrol*.
- By harmlessly changing the specification, introducing a new state that puts an atom into a kind of limbo, as in the case of the *Cancelled* value of *Subject*

---

[29] It would be possible to set *Subject State* to *Inactive*, the *Cancelled* state being distinguished by the value of *Filled* being non-zero. Personally, I am against this sort of thing and prefer to see the state indicated explicitly.

[30] We could have introduced modes as part of the kernel specification, but at that point in the text the idea would have lacked motivation and increased the reader's confusion even further.

State. This would contaminate the system specification, making it depend on implementation details. It would be a price paid for efficiency.

7.7 FRAGILITY

Because the design of an efficient system — especially a batch system — depends so intimately on data flow, a small change to the data flow can invalidate a system design. Consider the batch back-end system the Enrolment Office uses, shown in Figure 7.12 on page 241. Now imagine a situation where the Bursary insists that no candidate may enrol in subjects that would bring *Balance Owing* to more than $5,000. This apparently small change in the specification of the *Enrol* event introduces a new edge into the state dependency graph of Figure 7.8, from *Balance Owing* to *Enrolment State*, because the ability to enrol in a subject now involves both the *Fee* for the subject and the existing value of *Balance Owing*. The new edge creates a cycle between *Balance Owing* and *Enrolment State*, making them tightly coupled. In this case introducing a new mode won't remove the cycle, because both the edges are caused by the same event.

The system of Figure 7.12 will no longer work, because it relied on *Balance Owing* and *Enrolment State* being separable. The *Subject Update* and *Balance Update* processes will have to be combined. The resulting composite process will need to access the *Subjects*, *Enrolments*, and the *Candidates* files. It could no longer continue to access the *Subjects* file sequentially or the *Enrolments* file using clustered random access; sorting the transactions by subject would now mean that updates to a given *Candidates* row would occur, not in correct temporal order, but in order by *Subject Code*.[31] The update would therefore need to access all three files randomly, in correct temporal order. This additional random access would severely impact on the efficiency of the back-end system, and, worse, it would require a major modification of the text of the *Subject Update* program. What seemed a simple change to the client becomes a major project for the IT department. We say that the system of Figure 7.12 is **fragile**.

7.7.1 DEFENCES AGAINST FRAGILITY

Clearly, a change to the process graph may need the *infrastructure* to be revised, but, ideally, the event procedures should not need to be changed. Given that some businesses have requirements that are much more complex than those of the NUTS Academic Records System, it is important to rewrite as little program code as possible.

There are at least three defences against changes to the infrastructure:

1. Make the original design **robust**: For example, avoid using the sequential update algorithm, or pipelining separable processes. Unfortunately, using direct access will usually be at least one order of magnitude slower than sequential access, so

[31] In this particular application the different order would be harmless — *but not in general.*

it will be feasible only if the less-efficient robust system can support its expected load.

2. Use **activation records**:
   Although the *optimised* process graph may change, the *canonical* process graph of an event cannot. By assuming that the infrastructure will interface with every *minimal* process, we may partition event procedures into their smallest possible segments.[32]
   Since the canonical process graph only imposes a partial ordering, *any* topological sort could become the basis of a valid pipeline. Revising the infrastructure might therefore change the ordering of the minimal processes, so it becomes impossible for one segment to use a method call to communicate with the next; the 'next' segment depends on the infrastructure. The alternative is for it to use an *activation record* to interface with the *infrastructure*. The infrastructure passes the activation record to whatever segment it deems to be next.
   Activation records contain the working variables that an event must retain from segment to segment. In this respect they are like the transfer records passed between update processes in a batch system — which in some cases is exactly what they might prove to be. The role of the infrastructure is to retrieve values from the database and pass them to the event procedures. In turn, the event procedures provide the infrastructure with output values to be stored in the database, and with the keys of attributes that future segments will need.
   A serious problem with this approach is that it complicates the texts of the event procedures. Although they will survive the infrastructure being revised, the revision may never happen, and the extra programming investment may never be repaid.
3. Persuade the client (the Bursary in this example) to accept an alternative specification that achieves, or at least partly achieves, the same goal, while retaining the original process pipeline. For example, because one semester's enrolments typically cost about $2,000, letters could be sent to candidates who already owe over $3,000, warning them that they will be suspended until their fees are paid. This would need a new *Suspended* value of *Candidate State* that allowed payments to be made, but forbade new enrolments.[33]
   This final strategy needs an additional event and state to be introduced that were not part of the kernel specification. Such work-arounds often challenge a systems analyst who is trying to understand an *existing* system — whether it is a manual system or a computer system. It can be difficult to disentangle work-arounds from genuine business requirements.

[32] In the case of the *Enrol* procedure, there would be segments for *Programme*, *Candidate State*, *Quota*, *Current Semester*, *Subject State*, and *Component State* in some order, followed by *Filled/Enrolment State*, *Mark*, and *Balance Owing*. *Fee* could be anywhere in the sequence.

[33] This is a similar strategy to using the *Cancelled* state to control which updates could be made to *Subjects*. We are guilty of changing the facts to fit the theory.

7.8 USER DIALOGUE

Use-definition analysis can also be valuable in designing how a server interacts with human clients. Typically, there are two delays that can be significant: transmission across a network (especially the Internet) and the time it takes a human to respond, particularly if difficult choices have to be made.

Consider a variant of the Academic Records System in which enrolments are made over a client-server network. If we assume that all 10,000 active candidates must be enrolled within a 40-hour period, then, on average, 250 must enrol every hour. It is therefore important that the average enrolment process doesn't take longer than about 10 seconds. It seems unlikely that a single operator could achieve such a performance, therefore the work must be distributed across several operators.

One solution is to make the candidates themselves responsible for recording their enrolments over the Internet. This could be done using *forms*. Typically, a **form** is a web page created by a server and sent to a client's web browser. The client enters data or selects alternatives then *submits* the form for processing by the server. If each candidate took about 1 minute to enrol, then an average of only about four dialogues would need to be current at any one time.

We can imagine that a candidate first accesses the NUTS web site and is directed to a form on which an enrolment can be recorded. The form will require the candidate to specify two data items: the candidate's *Admission No* and the *Subject Code* of the subject in which the candidate wishes to enrol.[34] Submitting this form would result in a new form being displayed, confirming whether the enrolment was successful, and inviting the candidate to make further enrolments — or not. Presumably, the server wouldn't ask for the candidate's *Admission No* a second time, but would remember it.

Can we be sure this approach can be made to work correctly? The answer is a simple 'Yes!' because we know from use-definition analysis that all the rows of the *Candidates* table are independent and can be assigned to independent processes. Since each *Enrol* event requires five rows to be read and three rows to be updated, the server can process an enrolment in under $\frac{1}{10}$ second, so it will be busy for less than half a minute in every hour, on average. Indeed, the server could handle a peak load of about 36,000 enrolments per hour. Clearly, the speed of the server won't be a problem.

In addition, we know that different *Subjects* rows and different *Enrolments* rows are independent. Can we exploit this in some way? We can, as we have seen, have several *Subject* processes operating concurrently, but this hardly seems necessary.

More important is the behaviour of a candidate: because *Subjects* and *Enrolments* rows are independent, several can be processed concurrently. Therefore all a candidate's choices could be specified in parallel. A single form could be used

[34] To prevent identity theft, the server should verify the identity of the candidate. With the present database schema, the server could require the candidate to submit his or her *Date of Birth*. Even so, it would be more secure to add a new attribute to the *Candidates* table, containing an encrypted version of the candidate's personally selected password.

to enter the candidate's *Admission No* and a *list* of *Subject Codes*. When the form was submitted, all the enrolments could be recorded, and the candidate informed about all the outcomes. This would both reduce Internet traffic and save the candidate time.

There are many possible refinements of this basic idea, which we shall discuss in the next chapter.

To deal *formally* with human interaction in use-definition analysis, we may regard the human as one or more functional dependencies: that is, as time-dependent functions that map questions (attribute names) onto answers (attribute values). Indeed, this viewpoint is quite obvious if we consider an interface with a mechanical or electronic subsystem. If the interaction forms part of a strongly-connected component of the data-flow graph, then we are *forced* to make the interface interactive. If, on the other hand, the resulting process graph reveals that the interface is separable, this gives us the freedom to make the interaction non-interactive, that is, a human user can be presented with a report, given time to make decisions then use a data-entry form to record the results — even making many decisions efficiently as a batch.

7.9 SUMMARY

Use-definition analysis has many applications in information systems. Although its most familiar uses are in the design of algorithms and the optimisation of computer programs, we are interested here in its impact on system structure. For this we are interested in the flows between database variables. We are concerned with local variables used within procedures only because of the dependencies they induce between state variables. We have described several rules used to construct a *Determines* relation. Of these, the quasi-update rule seems the most artificial, but often turns out to be the most important.

We summarise the use-definition analysis of a procedure in a state dependency graph. By finding the strongly connected components of a state dependency graph, we can derive a canonical process graph. The variables accessed within a strongly connected component determine whether sequential or parallel updating is possible. The outcome depends entirely on the event specification.

Some simple heuristics for merging minimal processes enable us to derive an optimal process graph.

One important use of process graphs is to design systems — including object-oriented designs — that cannot deadlock. A second use is to design efficient batch processing systems. Only by learning how to synthesise a batch system can an analyst hope to discover what specification an existing system embeds.

We learnt that, once a process pipeline has been derived, segmenting event specifications becomes a purely mechanical process. Given a specific technology — say object-oriented systems programmed in *Python*, or batch systems programmed using *Cobol* — a software tool could automatically generate program code. Automatic code generation is yet another way to defend against fragility.

We found that, because different event procedures generate different process graphs, it may be wise to use different *modes* in order to preserve efficiency.

Finally, we have seen that use-definition analysis influences the design of user dialogue.

7.10 FURTHER READING

Use-definition analysis is exploited in programming language compilers to allocate registers efficiently, among other things. This material is discussed in several textbooks, for example the classic 1986 *Dragon Book*, 'Compilers: Principles, Techniques, and Tools' by Alfred V. Aho, Ravi Sethi, and Jeffrey D. Ullman (ISBN: 0-201-10088-6). Even so, compilers aren't usually concerned with detecting the presence or absence of cross flows that determine whether the iterations of a loop are independent. For a discussion of how traditional use-definition analysis can be adapted to detect cross flows, see the author's doctoral thesis, 'The Automatic Design of Batch Processing Systems', *University of Adelaide* (1999), which also discusses the idea of modes. For more details of the sequential update algorithm, the reader should consult the same thesis, or the author's 1981 paper, 'One More Time — How to Update a Master File', *Communications of the ACM*, 24 (1), 3–8.

An influential paper distinguishing synchronous (two-way) and asynchronous (one-way) message passing was written in 1981 by Gregory R. Andrews, 'Synchronizing Resources', *ACM Transactions on Programming Languages and Systems*, 3 (4), 405–430.

The author's 1998 article, 'Separability Analysis Can Expose Parallelism', *Proceedings of PART '98: The 5th Australasian Conference on Parallel and Real-Time Systems*, 365–373 (ISBN: 9-814-02122-9), reports an experiment that exploits separability to sustain very high throughput in a distributed database system.

A more detailed discussion of object-oriented design can be found, for example, in the textbook by Grady Booch, 'Object-oriented Design with Applications' (3rd ed., 2007, ISBN: 0-201-89551-X).

The defence against fragility discussed in Section 7.7.1, using activation records, derives from ideas in Michael A. Jackson's 1982 book, 'System Development' (ISBN 0-13-880328-5).

7.11 EXERCISES

1. Consider the *Withdraw* event, which undoes the work of *Enrol*, provided that the subject has not yet been frozen:

$$
\begin{aligned}
\mathit{Withdraw}(c,\ s) \Leftarrow\ & e = (\mathit{Current\ Semester},\ c,\ s)\,;\\
& \mathit{Enrolment\ State}(e) = \mathit{Enrolled}\,;\\
& \mathit{Subject\ State}(s) = \mathit{Offered}\,;
\end{aligned}
$$

$$Enrolment\ State(e)' = Null\ ;$$
$$Filled(s)' = Filled(s) - 1\ ;$$
$$Balance\ Owing(c)' = Balance\ Owing(c) - Fee(s).$$

Construct a dependency matrix for the *Withdraw* event in the style of Table 7.1, marking the cells with the relevant rule or rules. (Use the same abbreviations as Table 7.1.)

2. Draw the state dependency graph for the *Withdraw* event in the style of Figure 7.3.
   Is the graph you obtained canonical?
3. Use Algorithm 7.1 to find an optimised process graph for the *Withdraw* event.
   Ignoring all other types of event, is it possible to use a batch system to process *Withdraw* events?
4. On page 225, it was stated that although the object-oriented hierarchy of processes has no global **commit**, it is still safe, except in the event of a power or system failure. Why?
5. In Section 7.4.2, we discussed an example of object-oriented design that used a particular series of classes: *Academic Records*, *Candidate*, *Subject*, *Enrolment*, *Ledger*. By analogy with the *Enrol* event, synthesise the segmented specifications for the *Withdraw* event.
6. At the end of Section 7.8, we suggested that a candidate could submit a form that made several enrolments in one hit. To prevent deadlock, is it necessary to prioritise subject updates, e.g., in order by *Subject Code*?

CHAPTER

Interfaces

8

CHAPTER CONTENTS

INTRODUCTION

System interfaces are multi-layered — at the risk of a cliché, like an onion. A candidate reads examination results that a postal worker delivered, that university staff posted, that a laser printer printed, that the NUTS Academic Records System software generated. We may therefore study a system at many levels: the interface between the software and the printer, between the printer and the staff, or between the letter and the candidate. In the following, we shall focus on the interfaces between humans and reports or forms.

Systems interact with their environments using **sensors** and **actuators**. Sensors sample the environment to provide inputs, and actuators convert outputs into effective actions. In an aircraft's flight control system, sensors include gyroscopes, altimeters, and airspeed indicators. Actuators include servo-motors and hydraulic valves. In an academic records system, sensors include the candidates or staff who provide the inputs that record events. Actuators include the postal workers who deliver letters to candidates bearing news of examination results, and the staff who post the letters. We may thus divide sensors and actuators into two categories: mechanical and human. Their primary difference is that mechanical transducers are less intelligent than the system controlling them, and human transducers are more intelligent than the system they believe they control.

Mechanical sensors and actuators are characterised by reliability and response times measured in milliseconds. Their behaviour is usually simple enough to be modelled by an FSA or a set of differential equations, and the host software system will usually include models of their behaviour so that, among other things, it can detect when they behave unexpectedly. Humans are characterised by unreliability and have response times measured in seconds. Their behaviour can only be modelled statistically. We shall have little to say about mechanical devices in this chapter, and move on to the study of humans. Using current technology, for the most part this will concern the eye, the ear, the hand, and the voice.

There was a time when one could draw a sharp boundary between hardware systems and software systems. The pilot of an aircraft would operate a mechanical control column, move mechanical levers, and read mechanical instruments, but someone playing with a flight simulator would use a computer joystick and look at pictures of instruments. But as time has gone on, the controls and instrument panels of an aircraft have become more and more like the simulator. Consequently, our main focus will be **graphical user interfaces**, or **GUIs**.

Even so, before we discuss the exciting topic of humans, we should first look at the interface from the inside: what the *system* can do.

8.1 REPORTS

Reports typically describe the database *instance*.[1] Their structure is often hierarchical, that is, it describes parent-child relationships. A report may have a *heading* for each parent, followed by a *detail* for each of its children. Parent-child relationships result from the database *schema*. For the reader's convenience, we reproduce the canonical FD graph of the *NUTS* database here as Figure 8.1.

[1] Except when they describe the history of events, as in the case of the Bursary's *Statement* report.

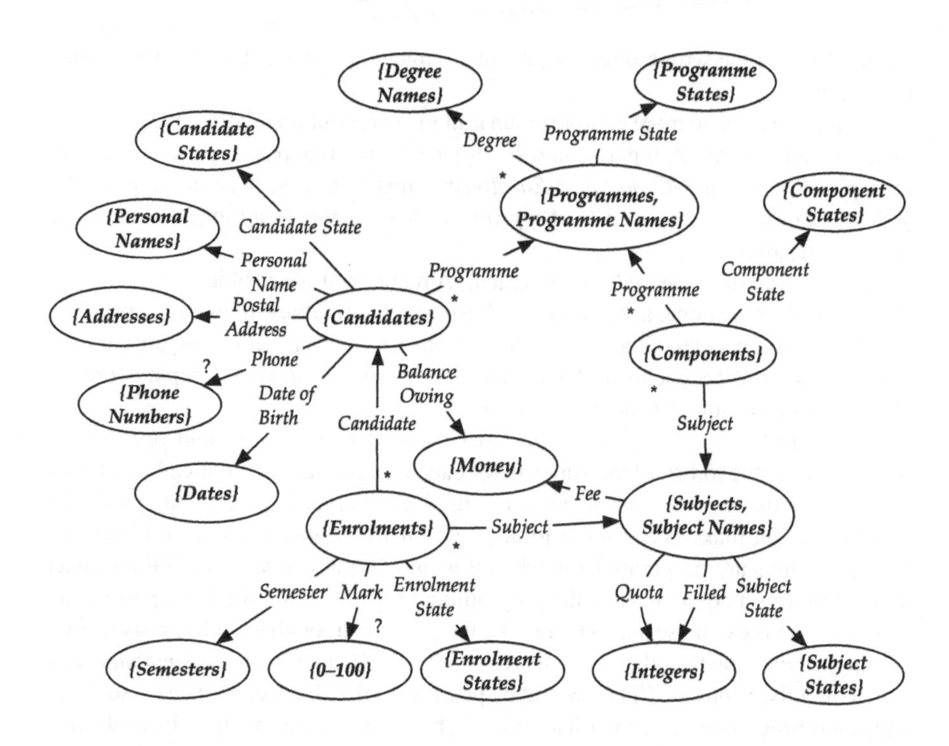

FIGURE 8.1

The canonical FD graph describing the Academic Records database, which determines what reports can be derived.

Consider a report that is focussed on *Enrolments*. *Enrolments* has two parents: *Subjects* and *Candidates*. A report about *Enrolments* can therefore be hierarchically organised by *Subject* or by *Candidate*. We have already discussed two such reports in Section 5.7. The first we called a *Mark Sheet*, the second we called a *Transcript*.

In the case of a *Mark Sheet* we made *Subjects* the parent, so we made *Candidates* subordinate; we treated attributes of *Candidates* as properties of each *Enrolment*. This was legitimate because the *Candidate* FD is many-to-one and *total*. Thus, for each *Enrolment* we may not only list a *Mark*, but also the candidate's *Admission No*, *Personal Name*, and any other attribute we choose.

We can take this idea a step further: we can follow the edges of Figure 8.1 transitively and report the candidate's *Program Name* and *Degree Name* too, if we wish.

On the other hand, if, as in the case of a *Transcript*, we make *Candidates* the parent, then we must make *Subjects* subordinate. This allows a *Transcript* to list the *Subject Name*, *Fee*, or any other subject attribute we wish. In this case, we are unable to follow the FD *Subject* : *Components* $\rightarrow$ *Subjects* transitively because the edge in

Figure 8.1 is in the wrong direction; a subject may be associated with several study programmes.

A hierarchy can be more complex than a single parent-child relation; it can involve three or more levels. A report could be organised as grandparent-parent-child, for example, *Programmes-Candidates-Enrolments*, that is to say, it could consist of a series of *Transcripts* grouped by *Programme*. Such a report might be useful to a programme advisor.

A report doesn't have to be hierarchical; sometimes it is sensible to display data as a matrix. For example, a programme advisor might find it useful to have a progress report that displays marks for a given *Programme* as a matrix, with a column for each *Subject* and a row for each *Candidate*. This would give the advisor a rapid overview of how various candidates are progressing.

Like *Enrolments*, *Components* has two parents: *Programmes* and *Subjects*. It would be possible to list all the subjects for each programme hierarchically, or to list all the programmes where each subject is used hierarchically, but in this particular system it might make better sense to display a matrix whose rows are subjects and whose columns are programmes. Each cell would be marked to show if the subject in that row was a component of the programme in that column. Such a report would certainly be more concise than a hierarchical report — although less obviously useful.

Programmes, on the other hand, has two *children*: *Candidates* and *Components* — and, indirectly, *Subjects*. Thus it would be possible to list all the candidates associated with each programme and then list all the subjects associated with it. In itself, this would serve no specific purpose and might better be dealt with as two separate reports: one listing candidates, the other listing subjects. Despite this, in this particular system, a *hierarchical* report, listing, for a given programme, each candidate followed by a list of *Offered* subjects makes sense if it is used to distinguish the subjects the candidate has already studied from those that the candidate might still need to study. Such a report could be useful to the candidate or to a programme advisor.

Reports are more useful if their contents are well organised. We have seen in Section 6.3 how **order by**, **select**, and **group by** clauses can be used to present results in a suitable order, select details of interest, or summarise simple statistics.

Ordering a list of candidates by *Balance Owing* can help the Bursary identify bad debtors. A *Mark Sheet* presenting *Enrolment* details ordered by *Mark* might help an examiner decide what mark best corresponds to a pass in an examination.[2] A *Transcript* ordered by *Semester* will give a better picture of a candidate's history than one ordered by *Subject*. In short, it is possible to order components of a report by any series of transitively dependent attributes. For example, *Enrolments* might be ordered by *every* domain in Figure 8.1, either singly or hierarchically.[3]

Selection could be used to identify those subjects whose *Quota* is full, and which might therefore need more resources in future. It could also be used to

[2] Examination results usually need to be rescaled to compensate for what may prove to have been an unintentionally difficult or unintentionally simple test.

[3] *Component* and *Component State* are a special case here. Although they aren't transitively dependent on *Enrolments*, both *Programme* and *Subject* are, and together they uniquely determine a *Component*.

identify candidates who have scored very poor marks and who might be in need of counselling.

Grouping and summarising could be used to count the numbers of students with each value of *Post Code* and gain some idea of their geographical distribution. It could be used to order subjects according to their average *Marks* and decide whether they contain too much or too little study material.

In conclusion, inspection of the database schema can suggest to the analyst the whole range of possible reports that can be generated. Focussing on some table, one can trace many-to-one or one-to-many relations, and decide what parent-child relationships a report might express. One must then verify that they serve a useful function or, more importantly, decide what report might best serve that function. Existing reports aren't a reliable guide to what is best; they are often constrained by the existing difficulties and costs in creating them. In most cases they could be improved by suitable ordering and selection.

We leave a discussion of the detailed layout of reports until we have described the human side of the interface.

8.2 FORMS

A **form** is the means by which an operator can update the database. If a form is used over the Internet, the client requests a server to send a blank form, the operator enters data onto the form, and then the operator *submits* it to the server. An application program in the server subsequently decides whether to update the database, to ask the operator to correct the form, or to reject the operator's request. This usually involves the server sending a new form to the client.

It is often possible to validate fields as they are entered. One commonly used method is to offer the user valid choices, for example, using selection lists or radio buttons. In such cases, the implementor must balance the amount of information that has to be transmitted from the server to the client against the advantage that the information offers: specifically, the time it takes the operator to successfully complete the form. When the server and client are on the same local network, transmission costs become less important.

A form is like an incomplete report. After it is completed, it displays how the database instance will look once the form is successfully submitted. Therefore forms are governed by the database schema in the same ways that reports are. In addition, they are governed by the *Determines* relation.

To illustrate, consider a form that allows a candidate to enrol in one or more subjects. We have seen in Section 7.8 that, because *Enrolments* are *independent*, a candidate could enrol in several subjects in parallel from the same form, submitting it only once. We also know that the subjects a candidate chooses must be components of the candidate's programme. Therefore we must access the correct row of the *Candidates* table to discover the value of *Programme* before validation can occur. With luck, this sequence may suggest to us the idea of restricting the candidate's

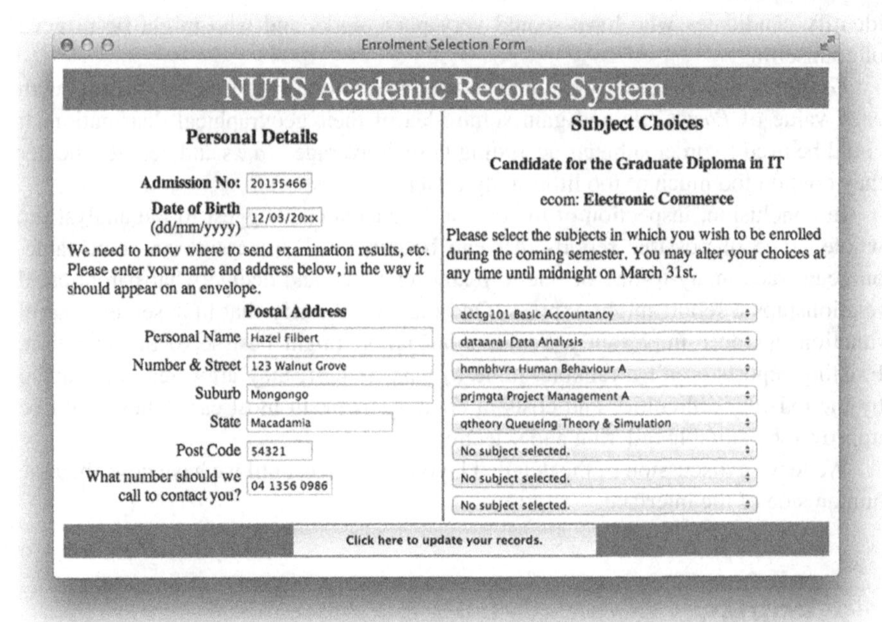

FIGURE 8.2

A form that might be used to allow candidates to choose their own enrolments. After entering *Admission No* and *Date of Birth* into the text boxes at the top left and submitting the form to log in, the candidate's personal details are displayed, and the candidate is presented with a series of selection lists. The lists only include about ten subjects: those that are part of the candidate's programme and offered in the current semester. The candidate can also update personal details, but cannot change the programme. (*Date of Birth* is a crude substitute for a password.)

choices on the form itself. Therefore our system would need to display a form asking for an *Admission No* that, when submitted, would lead to a second form that displayed only the subjects relevant to the candidate's *Programme*. The candidate would then be asked to select subjects from the list of valid choices.

This could be done either by displaying several *selection lists* offering the subjects the candidate might choose (Figure 8.2) or by listing all the component subjects and asking the candidate to mark *check boxes* to show which are chosen (Figure 8.3). This idea might be extended to offer radio buttons that allow the candidate to enrol, withdraw, or repeat a subject.

Offering the candidate a list of valid choices is an example of ignoring, or rather, *bending*, the quasi-update rule. We argued in Section 7.1.1 that it was sometimes possible to snapshot a table *before* knowing which element to select. In this case, we

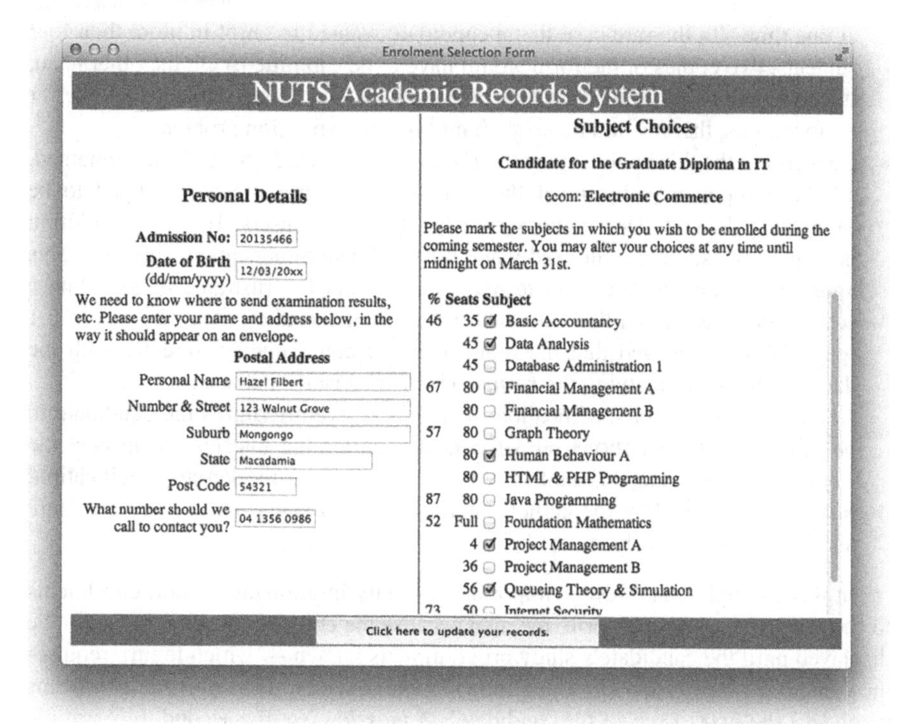

FIGURE 8.3

An alternative form that might be used to allow candidates to choose their own enrolments. This version displays all the subjects available in the current semester and all subjects the candidate has studied in previous semesters. In addition, it displays the number of places unfilled and any previous marks to help the candidate make good choices. The candidate is making a second attempt at *Basic Accountancy* and testing the waters of four Level 2 subjects.

plan not to snapshot a whole table, but a relevant subset of it. We need to be confident of three things: that the list of options won't overwhelm the candidate, that it won't take the server too long to transmit the data, and that locking all the options won't delay other users.

- As far as the first point is concerned, the candidate is merely asked to choose from a dozen subjects at most, so it shouldn't be a problem.
- As far as the second point is concerned, if selection lists are used, it will be necessary to transmit the same list several times, once for each enrolment the candidate might wish to make. In practice, the number of possible new enrolments could be limited, say to four, which is more than a normal candidate would study

at one time. (In the rare case that a candidate wanted to enrol in more than four subjects, two copies of the form would have to be submitted.) On the other hand, if a complete list of available subjects were displayed, along with check boxes or radio buttons, there would be no such multiple transmission problem.

- Concerning the third point, since the *Enrol* event causes *Filled* to be incremented, all the component subjects of the candidate's programme would need to be exclusively locked. This could seriously delay other users. If so, one solution is to leave the subjects unlocked while the candidate dithers about choices.[4] This approach means that a candidate may occasionally be frustrated by selecting a subject that becomes filled, but it is better than the alternative. In Figure 5.1 (page 146) we showed that the input interface could interact directly with the database without involving the *kernel*. This is precisely such a case.

  Nothing in the kernel specification of the *Enrol* event requires the candidate to be offered a list of options on a form, or requires the interface to inspect the *Components* and *Subjects* tables. Nor is it *implied* by the kernel specification; rather, it is *allowed* by the kernel specification, and suggested by data-structure and use-definition analysis.

In this example, data flow is important not only in allowing several enrolments to be made using the same form, but also because the component subjects cannot be displayed until the candidate's study programme is known — which in turn requires the candidate's *Admission No* to be known. Consequently, there must be *two* forms involved:[5] the first to accept the candidate's *Admission No*, the second to return the candidate's personal details and display the list of choices. In practice, the first form would normally require the candidate to enter a password, but our database schema hasn't made provision for one.

The need for dialogues that use several forms poses the question of what transitions between forms are possible. A sensible way to specify the transitions is as a finite-state machine. The corresponding state transition diagram can be drawn with vertices that resemble and define the forms themselves. Such a diagram is called a **story board**. In designing a story board, we need to be mindful of how much context a user has to remember when completing each form. We shall discuss this potential problem shortly, in Section 8.3.2.2.

8.3 HUMAN FACTORS

Human Factors or **ergonomics** may be divided into three broad categories:

Physiological: factors associated with the way the human body is made.
Cognitive: factors associated with the fixed wiring of the brain.
Behavioural: factors associated with the brain's ability to learn and adapt.

It isn't always easy to decide to which category a given factor should belong.

---

[4] If the DBMS always locks data automatically, it can be unlocked by an immediate **commit**.

[5] The two forms may look alike, simulating a single form that is updated.

8.3.1 PHYSIOLOGICAL FACTORS

If a person is to spend much time using a computer, it is important to provide a comfortable workstation.

- The seat should be high enough for the thighs to rest without cutting off circulation to the legs — usually about 50 cm above the floor.
- The keyboard should be at a height where the wrists are slightly below the elbows — usually about 70 cm above the floor.
- The centre of the screen should be slightly below eye-level — usually about 30 cm higher than the keyboard.
- The screen should be placed where bright lights or reflections won't distract the operator.

As a result, we can say that — for many westerners — most chairs are too low and most desks are too high. There should only be about 20 cm difference between seat height and keyboard height, so the keyboard should be virtually resting on the lap. Laptop computers satisfy this goal, but their screens are then too low, which can cause neck pain and even problems with blood supply to the brain.

If data are being copied from handwritten or printed forms, the forms are best positioned alongside the display to reduce eye movement, placed at the same distance to reduce refocussing, and given the same brightness to save repeated adjustment of the pupil.[6] Since focussing at a reading distance requires constant muscular effort within the eye,[7] it is also wise to provide rest periods when the eyes can focus at greater distances.

8.3.1.1 *Vision*

We tend to assume that the human eye works like a camera: as if its lens projected a photographic image onto the retina, which then became available to the brain to interpret. The truth is almost the reverse; the brain analyses the image on the retina into its component features, then synthesises a photographic picture for our scrutiny.[8] We have evolved to create such a photographic synthesis because we need an accurate spatial model of the real world.

We know the synthesis occurs for several reasons:

- First, the lens of the eye is no better than the plastic lens of a cheap telescope; the image on the retina is surrounded by coloured fringes, but we don't see them.
- Second, only a tiny area of the retina, the *fovea centralis*, is capable of seeing much detail. We are not aware of this because the eye is constantly moving in leaps, called 'saccades'. In contrast, the corner of the eye cannot see detail, but draws the attention to movement.

---

[6] Even better would be to use a webcam to show an image of the form on the display itself.

[7] Unless the viewer is short-sighted or wearing reading glasses.

[8] The synthesis takes time to mature; witness children's drawings, where arms and legs spring directly out of heads.

FIGURE 8.4

Discovering your blind spot. Close or cover your right eye and hold the chart about 35 cm in front of your left eye. Focus your attention on each number in turn. At certain distances the star will disappear and be replaced by the white background. There are no light sensors where the optic nerve leaves the back of the eye. You can see *nothing* in that area. The brain extrapolates what the eye doesn't see.

- Third, The left eye and right eye see different two-dimensional images, but the brain integrates both views to reveal a three-dimensional space. We only become aware of a double image if we deliberately cross our eyes.
- Fourth, we know that where the optic nerve passes through the retina, the eye is completely blind, but we aren't aware of this even with one eye closed, because the brain guesses what is missing and completes the image. (See Figure 8.4.)

What does this mean for the forms designer? Many things:

- The *fovea centralis* can resolve lines separated by about 1 minute of arc. This is usually taken to mean that the ratio of pixel size to viewing distance should therefore be about 1:3,500, so that a display screen viewed at 35 cm should have a dot pitch of better than $\frac{1}{10}$ mm (about 250 dpi), e.g., 1000 pixels should occupy no more than 10 cm. But for lines to be 'separate' requires at least one black and one white pixel per line. If a *sharp* eye is to be unable to detect individual pixels, we need about 500 dpi.
  Display resolution affects the quality of proofreading. The fraction of undetected errors decreases with improved resolution until the limit of the eye's own resolution is reached. What the eye doesn't see, the brain will guess — perhaps wrongly. (That is why, once a document is printed, one often spots errors that have escaped repeated proofreading during editing.)
- The *colour* resolution of the eye is poorer than its grey-scale resolution. The eye's colour sensors are spaced further apart than the grey-scale sensors. The resolution of the eye therefore depends both on colour and on colour combinations. The best combinations are those used by road signs: black on white or yellow, and white on green, red, brown, or blue are all good. Perhaps surprisingly, a good colour combination cannot usually be reversed. For example, white on black isn't equivalent to black on white, and is relatively poor.
  When choosing colours, remember that some people are colour-blind. There are several kinds of colour-blindness, but the most common is a failure to distinguish between red and green. Depending on where you live, between one in ten and one in fifteen males is red-green blind. (Because the genes for colour vision are on the

X chromosome — and females inherit two — the condition is rare in females.) It is therefore wise to design visual output so that it can be interpreted successfully without the use of colour and to use colour redundantly, for example, to highlight errors or mandatory inputs. Remember too that we have cultural conventions for the use of colour, e.g., red for 'danger'.

- Because only a small area can be seen sharply, don't expect people to notice what goes on in one part of a computer screen while they are focussed on another. The corner of the eye detects *movement*, not detail. Examine Figure 8.4 again, and notice that when the star disappears, the corner of the eye detects it immediately! Don't assume that because something is coloured red that the eye will automatically be drawn to it!
  Similarly, in a table, the saccade between items in adjacent columns shouldn't be too great, or the eye might drift to the wrong row. (See Figure 8.5.)
- Text is more readable in lower-case than in upper-case lettering because ascenders and descenders give words distinctive shapes that speed their recognition. Similarly, the serifs that decorate letters speed recognition by creating edges with which the brain can more easily detect the ends of strokes. Paradoxically, when display resolution is low, *sans serif* typefaces can actually prove to be more readable.[9]
- Legibility is also improved by including more white space (up to 30%) around blocks of text.

Related information should be arranged in blocks clearly separated from less related information. To gain most attention from Western readers, important text should be placed at the top left or bottom right of the display.

Based on these considerations, contrast the two reports shown in Figure 8.5 and Figure 8.6.

8.3.1.2 Voice

Many systems offer the option of conducting business over a telephone using **voice control**. At the time of writing, computer algorithms for recognising speech work best in a limited context.

Speech recognition algorithms are based on conditional probabilities. Speech sounds are built up of phonemes, which roughly correspond to single vowel or consonant sounds. For example, a speech recognition algorithm might decide that a phoneme has a similar probability of being an occurrence of either 'k' or 'g'. Such probabilities would then be weighted depending on what has gone before. Suppose the previous phoneme is strongly believed to be 'p'. The sequence of sounds 'p'–'k' is rare in English speech, although it does occur in 'cupcake'. The sequence 'p'–'g' is even rarer — but occurs in 'stopgap'. If the combined likelihood of both these sequences is low enough, the algorithm may consider that the 'p' itself was unlikely, and should be replaced by a similar sound, such as 'b'.

---

[9] Presumably, a clumsily drawn serif is worse than none at all. Print and display fonts might need to be different.

Examination Results
Semester 20xxA: 11 Candidates

Queueing Theory		**qtheory**
Admission No	**Personal Name**	**Mark**
21136648	Ada Lovelace	80%
22565335	Niklaus Wirth	5%
23265216	William Gates	42%
23352212	John Backus	63%
23560485	Donald E. Knuth	56%
23954892	Stephen Jobs	67%
24379797	John von Neumann	53%
24711128	Charles Babbage	30%
24717927	Alan Turing	94%
24860485	Timothy Berners-Lee	76%
25551214	Grace Hopper	48%

FIGURE 8.5

A badly designed report. The long saccade from *Personal Name* to *Mark* makes it easy to read a mark from the wrong row. Leaving space every five lines, as in Figure 8.6, would improve accuracy.

A second statistical model attempts to connect the most probable sequences of phonemes together to make words. The word sounds themselves can prove ambiguous, so a third statistical model considers the probability that one word might follow another. As a final flourish, the algorithm might try to decide what part of speech a word represents (verb, noun, adjective, etc.) and consider the various probabilities of one part of speech following another. From such statistical data, the algorithm selects the most likely utterance.

To be fair, speech recognition algorithms are more efficient than the above description suggests, but what they fail to do, in any deep sense, is to understand English. Although dictation programs can make reasonably accurate — although occasionally ludicrous — transcriptions of speech, they understand less than a parrot. Fortunately, in the context of an online enrolment system, almost any sentence that contains the words 'Basic Accountancy' is likely to be a request concerning *acctg101*. If the candidate hasn't already enrolled in *acctg101*, it is probably a request to enrol. If the candidate *has* already enrolled in *acctg101*, it is probably a request to withdraw. Thus a speech driven interface could reasonably ask, 'Do you wish to enrol in Basic Accountancy?' Deciding whether the answer is 'Yes' or 'No' should be straightforward.

Because speech recognition works best with a small vocabulary, it is normal to let the computer take control of the conversation, so that the user has only limited options. Thus, a conversation might begin,

Examination Results
Semester 20xxA: 11 Candidates

Queueing Theory **qtheory**

Mark	Admission No	Personal Name
94%	24717927	Alan Turing
80%	21136648	Ada Lovelace
76%	24860485	Timothy Berners-Lee
67%	23954892	Stephen Jobs
63%	23352212	John Backus
56%	23560485	Donald E. Knuth
53%	24379797	John von Neumann
48%	25551214	Grace Hopper
42%	23265216	William Gates
30%	24711128	Charles Babbage
5%	22565335	Niklaus Wirth

FIGURE 8.6

A well-designed report. There are several differences between this layout and the previous one. Placing *Mark* on the left avoids the long saccade. Sorting the rows by *Mark* serves two possible purposes: to help the lecturer to choose a pass mark, or to 'name and shame' poor candidates. You may form your own opinion about the difference between serif and sans serif fonts.

> 'Welcome to the *National University of Technology and Science* help line. If you want to speak, say "Speak!" If you want to use your telephone keypad, press the hash key'.
>
> *'Speak'.*
>
> 'Please choose one of the following options: "Change my enrolment details", "Change my personal details", "Enquire about my examination results", "Request a copy of my academic transcript", or say, "Something else"'.
>
> *'Change my enrolment details'.*
>
> 'Please say your admission number, digit by digit'.
>
> *'Two, two, five, six, double-three, five'.*
>
> 'Please say your password, letter by letter'.
>
> *'P, A, S, C, A, L, W, I, Z'.*
>
> 'Please say the name or code of a subject, or say "finished"'.

'Maths 101'.

'Do you wish to enrol in *maths101*, Foundation Mathematics? Yes or No'.

'Yes'.

and so on.

The main way that a conversation differs from a form is that the eye can quickly scan a form for what is relevant, whereas a telephone caller has to endure the entire conversation. It is good to keep the conversation short. In the above conversation, the computer didn't ask for an admission number and password until it was established that the candidate's response was not 'Something else', which would have directed the call to a human operator.

Although a conversation is designed roughly like a form, it pays to offer the user a fairly wide range of choices at each stage. Thus, the conversation we have described is like Figure 8.2 (page 258), rather than Figure 8.3 (page 259). This second form would have generated a boring conversation along the lines of,

'Do you wish to enrol in Basic Accountancy?'

'Yes'.

'Do you wish to enrol in Data Analysis?'

'Yes'.

'Do you wish to enrol in Database Administration?'

'No'.

'Do you wish to enrol in Financial Management A?'

'No'.

'Do you wish to enrol in Financial Management B?'

'No'.

and so on and so on.

So although checkboxes are a good option on a visual form, a spoken dialog works better when the choices are wider than a simple 'Yes' or 'No'.

If the user chooses to use the telephone keypad, the design of the conversation should consider the number of keystrokes required. If the typical candidate enrols in two subjects, each of which requires the typing of an eight-letter code, that should take about 3 or 4 seconds.[10] In contrast, answering ten yes/no questions would take ten keystrokes — only 2 or 3 seconds. Unfortunately, the additional time taken by the computer reading out ten subject names would easily outweigh the difference.

---

[10] If a telephone keypad is used, keystrokes could be matched against digitised versions of the subject codes. For example '22284101' matches *acctg101*.

8.3.1.3 Touch

Tablet computers and smartphones present special challenges. Neither has a keyboard that is suitable for touch-typing, therefore the operator's eyes must focus on the keyboard. This means their users should not be asked to transcribe information. Many people find it hard to use the tiny keyboard on a mobile telephone. On the other hand, touch-sensitive devices make it easy to select items by checking boxes, by just tapping the screen. Thus a form like the right-hand side of Figure 8.3 is better suited for a smartphone or a tablet than one like Figure 8.2. (Given that the left-hand sides of these two forms rarely need to be updated, text entry there is quite acceptable, especially as many smartphones allow text to be dictated.)

Smartphones have a small viewable area: a rectangular window through which only a part of a typical web page can be viewed. This suggests that related information should be grouped together, a principle that we have already observed in Figure 8.3, the only additional requirement being that the grouping should have roughly the same aspect ratio as the smartphone's screen.

It is common practice for special applications to be written to support mobile devices. One advantage of such applications is that they can store information *within* the device, e.g., a candidate's *Admission No*. Instead of a candidate needing to locate the NUTS enrolment form and type an admission number, an application could immediately display the relevant check boxes.

8.3.2 COGNITIVE FACTORS

8.3.2.1 Miller's Law

Our senses differentiate about 7 ± 2 levels of signal, i.e., with about 3-bit accuracy. For example, we talk of the seven colours of the rainbow. There are seven tones in a diatonic musical scale (not counting the octave). We are able to make similarly accurate judgements of loudness, brightness, salinity, and position. This phenomenon is known as **Miller's Law**, or the **magical number seven, plus or minus two.**

We can see Miller's Law at work in Figure 8.6. By printing the rows in blocks of five, each row is easy to trace. It is the line 'at the top', 'below the top', 'in the middle', 'above the bottom', or 'at the bottom'. We could have extended this idea to seven rows: it is equally easy to trace 'the line above the middle' and 'the line below the middle'. Some people would be happy with nine: 'the line halfway between the top and the middle' and 'the line halfway between the bottom and the middle'. Using blocks of five rows makes it easy to count them.

There is one notable exception to the magical number seven: perhaps because of our familiarity with (analogue) clocks, we can easily distinguish '1 o'clock', '2 o'clock', etc. Those familiar with a navigational compass can distinguish N, NE, E, etc., and the points in between, such as NNE, etc. — 16 values in all.

A second instance of Miller's Law is that we can store only 7 ± 2 'chunks' of data in immediate memory.[11] Surprisingly, the chunks can be of any complexity. We can

---

[11] Incidentally, the variation of ±2 depends on intelligence — or perhaps it is the other way around.

store seven binary digits, a 7-digit telephone number, or a list of seven words. The amount of *information* depends on the number of bits per chunk. We can store more information by recoding it into larger chunks. For example, the binary number '1010 0001 0101 1101' will overflow short-term memory, but the equivalent hexadecimal number 'A15D' is easily memorised.[12] A six-*digit* code number contains just under twenty bits of information; a six-*character* alphanumeric code contains over thirty.

As a result of the magical number seven, eight-digit admission numbers will overflow most people's short-term memories; they will have to be copied in two sections. In addition, accuracy will be poor. One consequence is that they should contain redundant information, or **check digits**.[13]

As a simple example of this idea, suppose we make the rule that every *Admission No* must be divisible by 13. Since one-thirteenth of all numbers divide by 13, one might suppose that an erroneous *Admission No* has exactly one chance in 13 of mistakenly being considered valid. In fact, the chances are much lower, because human errors aren't made at random.

There are two main categories of error:

- **Transcription errors** arise through misreading digits, especially if they are handwritten. For example, the digit '1' and the digit '7' can be confused, especially by people from different cultures — or when handwriting is recognised using optical character recognition (OCR).
- **Transposition errors** arise when one reverses the order of digits, as in confusing '12435' with '12345'.

Many check digit systems use the idea that the n-digit number $d_n d_{n-1} \cdots d_2 d_1$ has a check sum, S, calculated as

$$S = (w_n \times d_n + w_{n-1} \times d_{n-1} + \cdots + w_1 \times d_1) \bmod p$$

where p is a prime number and $w_n \cdots w_1$ are called *weights*. If d_i is incorrect by the amount δ, then the sum must change by $\delta \times w_i$, where $-9 \leq \delta \leq +9$. Provided $p > 9$ and w_i isn't divisible by p, then $(w_i \times \delta) \bmod p \neq 0$, and the check sum must change.

How does the simple '*Admission No* divides by 13' rule respond to errors? The set of weights here are the 'natural' weights: 10,000,000, 1,000,000 … 10, 1.[14] None of these divides by 13. From left to right, here are their remainders after division by 13:

10 1 4 3 12 9 10 1

---

[12] One theory is that we silently verbalise such sequences. We rely on the same short-term memory buffer that nature has given us to interpret speech.

[13] The argument here is a bit circular. If the admission numbers didn't include check digits, they would only have seven digits and would be easier to copy — especially since the first digit always seems to be a '2' and needn't be memorised.

[14] $d_n d_{n-1} \cdots d_2 d_1 = d_n \times 10^{n-1} + d_{n-1} \times 10^{n-2} + \cdots + d_2 \times 10^1 + d_1 \times 10^0$.

Owing to the following theorem of modulo arithmetic,

$$(w \times d) \mod p = ((w \mod p) \times (d \mod p)) \mod p$$

the remainders are equivalent to the natural weights themselves. Since none is zero, every *single* transcription error will be detected. Despite this, consider the admission number '25825566', which is evenly divisible by 13. Suppose, owing to poor handwriting, both '2' digits are misinterpreted as '7'. The result would be the number '75875566', which is also divisible by 13. In this example, the check failed because the sum of w_7 (10) and w_4 (3) is 13. In general, a multiple transcription error will be missed if the sum of the weights involved is divisible by p.

Now consider a transposition error, where '2**5**82556**6**' becomes '2**6**82556**5**'. This time, the error escapes detection because $w_6 = w_0$. In general, no two weights should be equal modulo p. In this particular case, such an error is unlikely to occur, as most transposition errors involve adjacent digits.

Real check digit schemes are more sophisticated than this example, which was deliberately chosen to illustrate potential problems.

8.3.2.2 Recursion

> 'We interrupt this programme to bring you a newsflash. But first, a word from our sponsor!'

How many levels of interruption can we process? Compare the following sentences:

1. The letter, in which his marks were detailed, was read by the candidate.
2. The letter, in which his marks, which the program printed, were detailed, was read by the candidate.
3. The letter that the program that the systems analyst specified printed, was read by the candidate.
4. The letter, in which his marks, which the program that the systems analyst specified printed, were detailed, was read by the candidate.
5. The systems analyst specified the program that printed the marks detailed in the letter that the candidate read.
6. The candidate read the letter detailing the marks printed by the program the systems analyst specified.

All are valid English sentences. No one has trouble with the first, which contains one nested clause. The second and third contain two levels of nesting, but the pile-up of three successive verbs causes some difficulty in remembering their contexts. The fourth contains three levels of nesting and strains most readers' abilities. The final two contain the same information as the fourth, but iteration replaces nesting.

What does this mean? Short-term memory can handle about three levels of *embedded* recursion, although it can handle an almost unlimited amount of head or tail recursion.

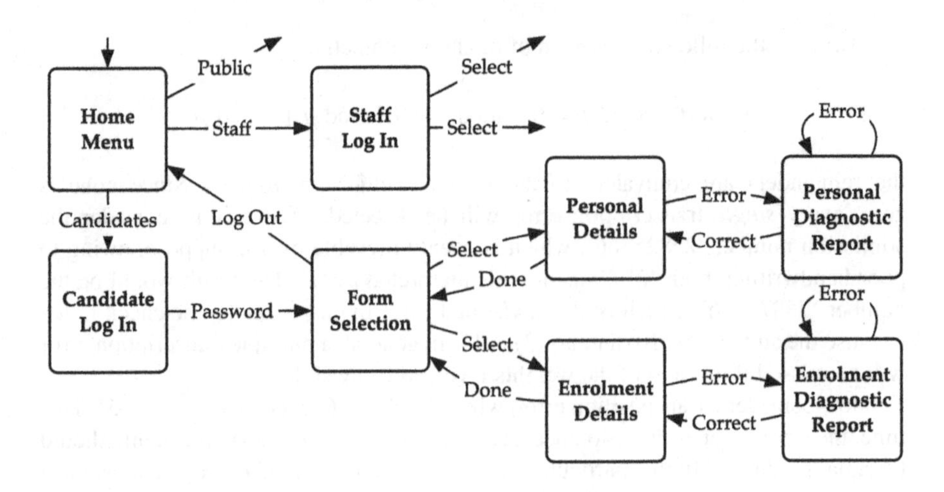

FIGURE 8.7

A finite-state automaton describing part of the story board for the NUTS web site. (Once registered, a smartphone app would go directly to the *Form Selection* state.)

Therefore, if the computer must interrupt an interactive task with an error message, which causes the candidate to seek help, then that had better be the limit.[15] In this instance, the computer is remembering the context and will not err, but clearly it is better, when the help is dismissed, to return back to the level where the error can be corrected, rather than to the error message. Better still, *the error message should provide the needed help.*

We need to apply these ideas when we construct a story board (Figure 8.7). Imagine a candidate wishing to enrol online:

Find Home Page: The candidate first uses a search engine to find the NUTS web site.

Home Menu: The candidate is presented with several useful links, one of which offers an opportunity to access the Academic Records System.

Log In: The candidate is asked to log in, by providing an *Admission No* and password.

Form Selection: The candidate is given a choice between amending personal details, amending enrolment data, or logging out and returning to the *Home Menu.*

[15] The same phenomenon is often apparent to computer programmers. Procedure P fails, possibly because of an error where it calls procedure Q. An examination of Q suggests that the fault may be in procedure R. The programmer inspects procedure R, but has by now forgotten what the problem was.

Enrolment Details: The candidate then chooses to amend enrolment data and completes a form — perhaps similar to that of Figure 8.3 — or chooses to return to the *Form Selection* level.

Diagnostic Report: The candidate is shown an error report or is given confirmation of success, and then given the chance to return to the *Enrolment Details* form.

The only time when the candidate must remember a context is when reading the *Diagnostic Report*. The candidate's memory can be aided by the report displaying the *Enrolment Details* form again, highlighting the errors. Even so, consider the caution in the following sub-section.

8.3.2.3 Stability

Screen displays should be stable. Form layouts should be standardised. The same information should be in the same place on each form. In moving between web pages or forms, *unchanged information should not move*.

Conversely, new data should not silently take the place of old. If the candidate submits a form that contains errors, it is wise to display a dramatically different-looking form in response, calling for acknowledgement. An error message added to the existing form can simply pass unnoticed — even if it is brightly coloured. If it is made to *flash*, it will certainly be noticed, but might be hard to read. A better option is a flashing icon adjacent to the text.

A more subtle cognitive factor is the **pop-out effect**: an extra feature is easier to spot than a missing feature. (See Table 8.1.) For example, although it is better to reward correct behaviour than to draw attention to errors, it would be most unwise to indicate an error by the *absence* of a tick-mark.

8.3.2.4 Muscle memory

It is useful to exploit **muscle memory**. Frequently repeated actions, even complex ones, become automatic and require no conscious effort. It is best for gestures, keystrokes, keywords, or commands to retain the same meanings independently of their context (called modeless operation). Otherwise, the user is likely to request an action in the wrong context and be surprised by its effect. Likewise, when menu items are unavailable, they should be disabled, *but the menu layout should remain the same*. Fortunately, using typical development tools, these factors have already been decided, and the implementor is left with few opportunities to do things badly.

Sometimes it is useful to break the rhythm of muscle memory. Normally, it is bad practice to mix pointing and typing activity. It requires movements that can quickly tire the arms and shoulders. Graphical interfaces often offer alternatives so that natural rhythm doesn't need to be broken. For example, arrow keys may be used to position the cursor, instead of a mouse, trackpad, or touchscreen; the *return* or *enter* key may be used to click a button. Despite this, if it is necessary to force the operator to pause, for example, to read and acknowledge an error message, then a switch between typing and pointing can be justified.

Table 8.1 The pop-out effect. The character that doesn't belong on the left-hand side of the table is usually obvious immediately. We have to look harder for the one that doesn't belong on the right-hand side

C	C	C	C	C	C	C	C	C	C	G	G	G	G	G	G	G	G	G	G
C	C	C	C	C	C	C	C	C	C	G	G	G	G	G	G	G	G	G	G
C	C	C	C	C	C	C	C	C	C	G	G	G	G	G	G	G	G	G	G
C	C	C	C	C	C	C	C	C	C	G	G	G	G	G	G	G	G	G	G
C	C	C	C	C	C	C	C	C	C	G	G	G	G	G	G	G	G	G	G
C	C	C	C	C	C	C	C	C	C	G	G	G	G	G	G	G	G	G	G
C	C	C	C	G	C	C	C	C	C	G	G	G	G	G	G	G	G	G	G
C	C	C	C	C	C	C	C	C	C	G	G	G	G	G	G	C	G	G	G
C	C	C	C	C	C	C	C	C	C	G	G	G	G	G	G	G	G	G	G
C	C	C	C	C	C	C	C	C	C	G	G	G	G	G	G	G	G	G	G

Consider a situation where Enrolment Office staff are populating the Academic Records System database with historical data recorded on file cards.[16] The staff use the form illustrated in Figure 8.8. Items in upright bold face are *boiler-plate*, which appear whenever the form is displayed. Items in italics are supplied interactively. The operators type the data you see in the boxes. The subject names are retrieved from the *Subjects* table as soon as each subject code has been entered.

When enrolment staff use the form, their eyes are focussed on the file cards, not the form. Muscle memory guides their fingers. As they complete the form, the computer validates each datum: the *Admission No* is checked for divisibility by 13, the *Date of Birth* is checked to see if it forms a valid date (and the month converted to text), *Phone Number* is checked to be all digits, each *Subject* is checked to see if it is the key of a row in the *Subjects* table, and *Mark* is checked to make sure it is in the range 0–100. If these tests all succeed, the operator is supposed to check that the display matches the file card. To encourage this, the operator must break rhythm and click a button to update the database.

Consequently, as soon as an error is detected, a warning is sounded, and keyboard input is ignored until the operator clicks a button that is *differently* positioned. (See Figure 8.9.) (This prevents the operator simply leaving the cursor positioned over the button and clicking without looking.)

---

[16] Yes, it is hard to believe that an institution that teaches technical subjects would still have a manual system, but I want you to consider the worst possible scenario.

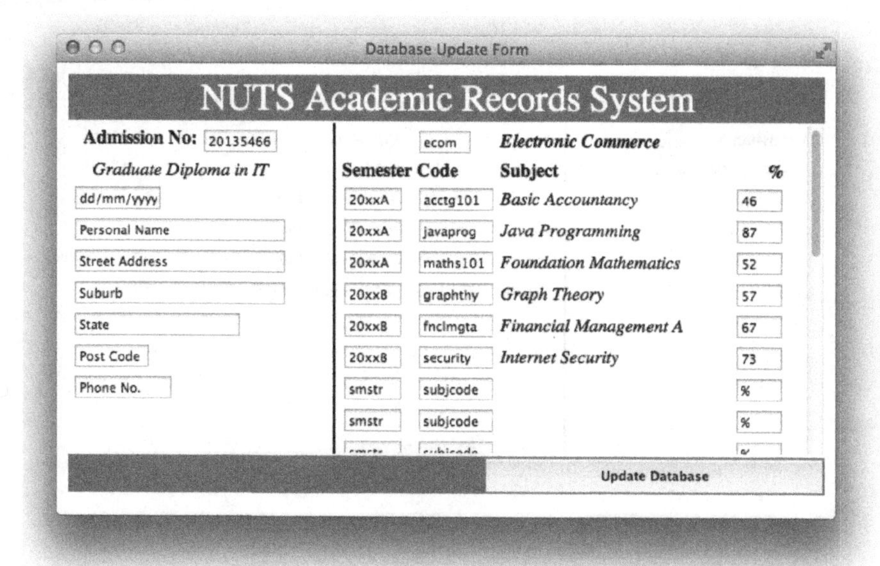

FIGURE 8.8

A data-entry form. This is how the form that enrolment staff use to populate the database looks once it is ready to be submitted. At this stage the Enrolment Office is concerned with preparing the examination timetable, and the candidates' personal details have been left blank. The operator is meant to check the form before clicking the *Update Database* button. The layout closely resembles the form that candidates use.

8.3.3 BEHAVIOURAL FACTORS

8.3.3.1 Operant conditioning

The reader is probably familiar with **classical conditioning**, in which Pavlov, by sounding a bell before giving food to dogs, caused them to salivate at the sound of a bell. On the other hand, the reader may be less familiar with Skinner's **operant conditioning**. The difference is that whereas classical conditioning links existing behaviour to a new *stimulus*, operant conditioning teaches new *behaviour*.

Operant conditioning exploits the following property of the nervous system:

- When an action is immediately followed by a reward, it becomes more likely;
- When an action is immediately followed by a punishment, it becomes less likely.

Importantly, the principle does *not* state that undesired behaviour followed by punishment will be replaced by desired behaviour. It may well be replaced by even less desirable behaviour.

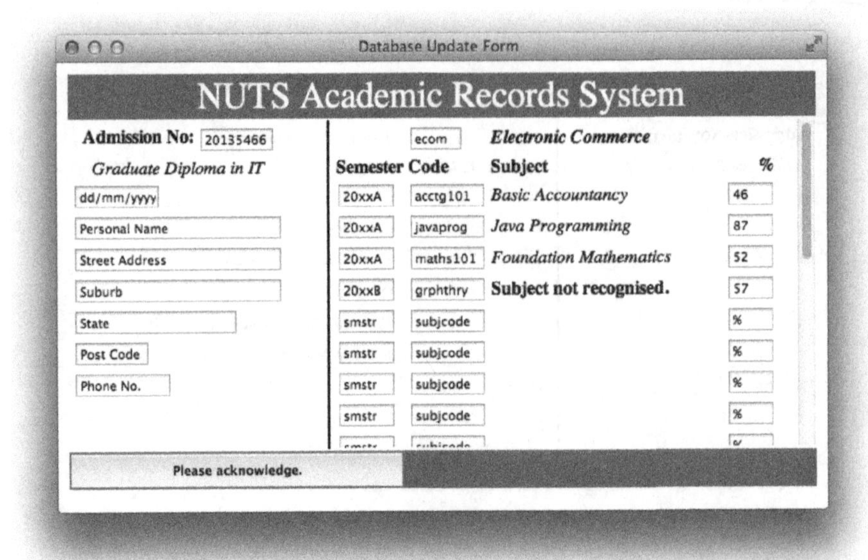

FIGURE 8.9

A form that highlights an error. How the form of Figure 8.8 looks when an error is detected. The operator is supposed to acknowledge it by clicking the button. The button is now on the left. This serves two goals: the button cannot be clicked blindly, and its apparent movement is intended to attract the corner of the eye.

The words 'immediately followed' mean 'within about one second' — optimally, after about 0.8 seconds. The reward is then seen as a direct consequence of the action. When operant conditioning is used to train animals, the reward is usually food. Nonetheless, if you have ever seen a dolphin perform, you will have noticed that the trainer blows a whistle to reward a good trick. The whistle is a Pavlovian promise of fish to come. It is more important that the reward is prompt than that it is real.

As a teaching technique, operant conditioning does *not* punish undesired behaviour, but rewards desired behaviour, or anything that approximates desired behaviour. To teach a dolphin to jump through a hoop, the trainer cannot wait until the dolphin accidentally jumps through the hoop, then reward it. The dolphin is rewarded for approaching the hoop, gradually needing to get closer and closer to get the reward. Then the dolphin is rewarded for jumping out of the water close to the hoop, and so on. This process is called **shaping**.

Although dolphin behaviour is shaped through hunger, human behaviour is more easily shaped by simply saying what is needed. So, instead of trying to guess what the operator has done wrong, a system should state the rules the operator should follow.

For example, consider the input '13-7-93' and the computer response, 'Day is not in month'. The operator may try '13-07-93', '13-Jul-93', '13-July-93', etc., without hitting on a valid input. The alternative response, 'Please enter the candidate's date of birth as yy-mm-dd', reveals the problem. In the first case, the program is trying to guess what the operator has done wrong, believing that '13' is the year, and '93' is supposed to be a day in July; in the second it is stating its own rules. Not only is the second more helpful, it is also easier for a computer program to do. The operator is flexible; the program isn't.[17]

An important property of operant conditioning is that if *every* instance of desired behaviour is rewarded, as soon as the reward ceases, so does the desired behaviour. But if the desired behaviour is rewarded only *part* of the time, at random, the behaviour will become persistent. (The reason for 'random' is that any underlying pattern is soon detected. If the trainer rewards a dolphin only at 3 o'clock, the dolphin will only perform at 3 o'clock.)

One might argue that humans are more complicated creatures than dolphins, which is certainly true. But consider problem gamblers, who are willing to gamble away their savings in return for random reinforcement from a poker machine. The loss takes time, but the rewards, when they do come, are immediate.[18]

Operators have to learn the rules the computer dictates, and how to use them to solve the problem at hand. How this happens should be part of system design; it shouldn't be an afterthought.

> ***Every interactive system is a teaching system, even if that is not its purpose.***

It turns out that the protocol used above to enter data from file cards has unfortunate side effects. Operators using the form become stressed, request frequent rest breaks, and complain of uncomfortable working conditions. The reason, perhaps surprisingly, is that they regard the warning sound as a punishment. Although the sound is neutral in itself, it soon becomes, Pavlovian-fashion, associated with making an error and the extra work involved in correcting it. The operators become averse to the sound. The undesired behaviour of making an error is replaced by the even more undesirable behaviour of finding excuses to avoid the situation where the perceived punishment occurs. *The system is teaching the operators to avoid it.*

How can the punishment be replaced by rewards?

Clearly, there has to be *some* way of attracting the operator's attention when an error occurs. (It doesn't have to be a sound; the whole display screen could be made to flash.) Whatever means is chosen will soon be perceived as a punishment. Punishments quickly follow actions; rewards, which only occur when a form is

---

[17] Some programmers see users as adversaries who are trying to bamboozle their programs with faulty data. They therefore adopt an aggressive attitude when writing error messages. It is better to accept humbly that the user was probably trying to do something perfectly sensible, but the program failed to understand it. Error messages should be respectful.

[18] Computer programmers shouldn't assume that other people are like them. Writing and debugging a computer program is mostly pain. Rewards are few and far between. Not many people enjoy such masochistic pursuits.

completed successfully, follow slowly. As for a problem gambler, short term will dominate long term. The remedy here must therefore be to reward quickly and often.

One possibility is to use two sounds: a pleasant one to reward desired actions, and a less pleasant one to punish undesired actions. The pleasant sound could be used every time a datum is entered but no error is detected.

Unfortunately, experiments suggest that expert operators will eventually find this annoying. A better approach is to reward in proportion to the error rate. This means that a beginner will receive plenty of encouragement, but an expert operator won't be irritated. For example, after an error has been detected, when the datum is finally submitted correctly, a reassuring message should be shown, such as, "Input accepted. Please continue." As a further refinement, several such rewards could be issued at random times, because we know that leads to persistent behaviour. A simple reward can also be given every time a complete form is submitted correctly. For reasons we shall shortly discuss, a suitable reward would be to display the number of forms successfully completed. The number of wrong attempts shouldn't be mentioned.

An alternative approach is to make the 'punishments' less prompt. This could be done by waiting until the operator has completed the entire form before checking for errors. This is common practice with forms that are transmitted over the Internet: checking doesn't occur until the form is submitted. The danger is that errors may cascade. For example, if the operator types too many or too few lines in the *Postal Address*, then every datum that follows might be entered into the wrong box.

8.3.3.2 Motivation

Maslow's Hierarchy of Needs is the psychological theory that we can only pursue higher-level needs when lower-level needs are already satisfied. Those who get adequate food and sleep and who are in no immediate danger will try to satisfy their *social needs* by interacting with others. Once they have established a circle of friends, they will try to win their *esteem*. Ultimately, they will pursue *self-realisation*, trying to become the person they want to be.[19]

Herzberg's Theory of Motivation is based on Maslow's hierarchy. Herzberg divides needs into two categories: **hygiene factors** and **motivation factors**. Hygiene factors include adequate salaries, comfortable working conditions, and so on. Motivation factors include social interaction, self-esteem, and self-realisation. The theory is that *hygiene factors no longer have the power to motivate*, although their inadequacy can cause dissatisfaction. In fact, it is possible for a worker to feel satisfaction with his or her work while feeling dissatisfaction with his or her working conditions. Motivation can only be effective on the social and esteem levels. This means that a good computer interface should increase its user's esteem and not interfere with social needs. A system that removes its user from social interaction and makes embarrassing noises isn't rewarding.

---

[19] Only programmers see computers as the road to self-realisation.

In general, how can a system reward correct behaviour? Humans see progress with the task as rewarding in itself, provided they understand what they are trying to achieve, regard it as useful, and can assess their own performance.

Rewards therefore lie outside the scope of a computer program, but lie in the context in which the program is used. A worker should be given a meaningful job involving social interaction, in which progress eventually leads to the satisfaction of some goal. The goal can be an artificial one. For example, at the end of entering a batch of enrolment records, the operator can compare the total number of records entered against an externally calculated control figure. This is a safeguard against items being overlooked. Paradoxically, obtaining a correct total will *not* be seen as a chore, but as satisfying.

Attainment of longer-term goals should be rewarded by the achievement of sub-goals: **closure** is the achievement of a self-contained step in a task after which *the amount of mental information carried forward is reduced.*

In however small a way, a closure is a reward. In reading, the end of each sentence is a minor closure, and the end of a paragraph is a major closure. In each case, the meaning of the words can be condensed into a single idea. In using a screen-based form, completing a text field or selecting a radio button is a minor closure, and submitting a form is a major closure.

Closures should give users the chance to correct errors. If data are entered incorrectly, it shouldn't be difficult to go back and correct them, e.g., there should be an 'undo' function. Major closures should give the user the chance to take a break. A visual cue should acknowledge each major closure, for example, "131 candidates and 254 enrolments recorded."

The computer should give constant feedback. When the computer is engaged in a long task, it shouldn't lapse into sulky silence. It should display a progress bar, or at least a message saying, 'Please wait...'.

A system's output is only as good as its input: the quality of input depends on the user's work environment. One approach to quality is through *external control*, e.g., by comparing totals with independently derived figures, using check digits, or by entering the same datum twice.[20] When controls cannot be used, it is important that the person responsible for entering data is *motivated* to enter data both quickly and correctly. This often means re-specifying the task.

To apply motivational ideas to the design of the NUTS Academic Records System we first need to understand how existing the paper-based system works:

- First, beginning candidates are sent a letter advising them of their success in gaining an admission.
- Second, each candidate is asked to attend on certain dates to complete an enrolment form. (See Figure 8.10.) This is basically the same form for both new and re-enrolling candidates. Their choices are signed off by a programme advisor.
- Third, *from the forms*, enrolment staff either create a new record card or amend an existing record card for each candidate, which lists their enrolment choices and

[20] A typical strategy used when creating passwords.

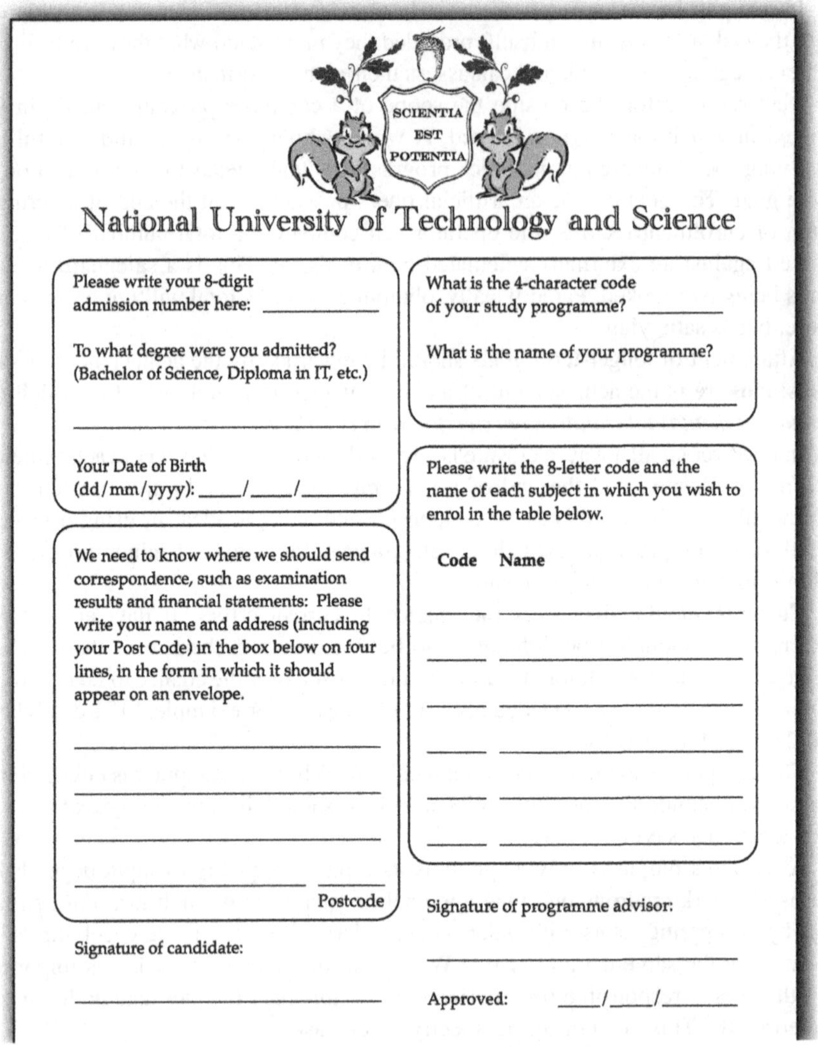

SCIENTIA
EST
POTENTIA

National University of Technology and Science

Please write your 8-digit admission number here: ________

To what degree were you admitted?
(Bachelor of Science, Diploma in IT, etc.)

____

Your Date of Birth
(dd/mm/yyyy): ___/___/_____

We need to know where we should send correspondence, such as examination results and financial statements: Please write your name and address (including your Post Code) in the box below on four lines, in the form in which it should appear on an envelope.

____

____

____

________ ____
Postcode

Signature of candidate:

____

What is the 4-character code of your study programme? ________

What is the name of your programme?

____

Please write the 8-letter code and the name of each subject in which you wish to enrol in the table below.

Code	Name
____	________
____	________
____	________
____	________
____	________
____	________

Signature of programme advisor:

____

Approved: ___/___/_____

FIGURE 8.10

The existing paper enrolment form. The left side of the form concerns candidates, and the right concerns study programmes. Placing *Degree Name* on the left is intended to motivate candidates by reminding them of their goals.

any previous marks they may have obtained. If candidates wish to change their choices, they complete a special form (which again must be signed off) and the record cards are amended.

- Fourth, after enrolments are frozen, the tedious work begins: A sheet is printed for each subject, headed with the name of the subject. Then, enrolment staff take the record cards for each candidate and mark off the candidate's *complete list* of enrolments onto every subject sheet in which the candidate is enrolled. At the end of this process, each subject sheet should have every clashing subject marked off. These 'clash lists' form the basis of the timetabling process. The process is similar to one we described on page 192, except that the final scheduling is done by amending the previous year's timetable to remove clashes. A provisional timetable is published, and candidates then report any clashes that affect them personally. With luck, these are corrected in time for the examinations.

What are the Enrolment Office's motives for replacing the existing manual system? There are two: First, the existing method of preparing examination timetables is long-winded, labour-intensive, and error-prone. Second, it is hoped that most of the enrolment data will *eventually* be entered by the candidates themselves using the Internet, saving clerical work. Despite this, the Enrolment Office isn't planning to introduce Internet enrolments until any teething problems have been solved internally. Unfortunately, management is neither willing to risk everything on the new system being successful, nor to allocate more staff during the changeover. We therefore have to solve the staffing problems and motivational problems for both stages of implementation.

We shall have no problem motivating *candidates*. Every candidate will want to ensure that they are enrolled in the subjects they wish to study, otherwise they will be unable to sit their examinations. Likewise, they will want to ensure that they aren't enrolled in subjects they don't wish to study, otherwise they will pay fees for them. All that is needed is to make sure they have the necessary information in time to correct their choices.

The Enrolment Office *staff* may perhaps prove to be a problem. Although they will be relieved of some chores, especially examination timetabling, they may be worried about their futures: If they aren't going to do the work that they do now, what work will they do? In addition, management has deemed that the initial trial of the computer system will be done in parallel with the existing system. This means additional work for them during the trial.[21] On top of that, existing data for re-enrolling candidates will need to be transferred to the new system. Currently, these data are held on hand-written cards, copied from hand-written forms completed by candidates. One factor that will help the transition to the new system is that, before the start of the enrolment period, the *Tertiary Education Admissions Centre (University Panel)* assigns beginning candidates to various educational institutions — including the NUTS. Currently *TEACUP* sends each institution the list of beginning candidates as a computer print-out, but they are willing to send it in machine-readable form. This means that the *new* rows of the *Candidates* table can be in place before enrolments begin.

[21] In effect, they are being asked to work harder to make themselves redundant. Assurances will need to be made!

Here is a possible plan, which depends on convincing the enrolment staff that the computer can prepare the examination timetable and therefore save them much trouble:

- First, well ahead of the enrolment period, some *Information Technology* candidates will be hired to enter the data into the *Programmes*, *Subjects*, and *Components* tables. This information is available from existing publications.
- Second, the machine-readable *TEACUP* data will be converted to a spreadsheet, and an extra column will be added to give each candidate an admission number. Using a database utility program, the spreadsheet data will be used to populate the *Candidates* table.
- Third, beginning candidates will complete a new-style enrolment form. This form will have pre-printed details derived from the *Candidates* table and will invite the candidates to make their enrolment choices and any necessary corrections to their personal details. The data from the enrolment forms will be entered into the database by the enrolment staff. On the other hand, because continuing candidates' details have not yet been recorded in the database, they will be dealt with exactly as in the past and have their existing record cards updated.
- Fourth, a trial run of the timetabling program will be demonstrated at the earliest possible opportunity — with some fanfare — using only the data for beginning candidates. This should convince the enrolment staff that it will be wiser to enter the enrolment data for continuing candidates than to continue to use the existing timetabling method.
- Fifth, assuming a successful outcome, the enrolment choices for continuing candidates will be entered, permitting the complete examination timetable to be planned. (It is not necessary to enter all their personal data: the *Admission No* is enough for the present purpose. This information can be collected from their record cards.)
- Sixth, as time permits, the personal details of continuing candidates can be entered into the database.
- Because candidates still submit written forms, it is always possible to fall back to the old system.

What are the motivations that should encourage staff to accept the new system?

- First, enrolling a beginning candidate no longer requires staff to copy personal details from a form onto a record card, because the *Candidates* table already contains them. Even copying enrolment choices becomes easier, *provided the computer form is well designed.*
- Second, when the computer timetabling system is demonstrated, the workers who made it possible should feel a great sense of achievement. Management should lavishly praise the contribution that the staff have made. With luck, this will provide the encouragement to enter the current enrolment choices for continuing candidates.

- Third, the database will provide better means of recording examination results, being able to print both mark sheets and academic transcripts from one set of data. (These are two separate jobs in the old system.)
- Fourth, when the time comes to enter examination results for continuing candidates onto file cards, staff are likely to believe that it would be better to copy the data into the database instead.

Although I cannot claim that this plan is perfect, it does illustrate the kinds of issues that are involved in any change from an old system to a new one. Learning to use a new system is unsettling, and unless the staff making the change enjoy some personal benefit from it, they won't fully cooperate. I regret that I cannot offer the reader a tractable algorithm for deriving such plans. Even so, considering the motivational and behavioural aspects of design is a key ingredient and must never be overlooked.

8.4 PSYCHOLOGY EXPERIMENTS

Suppose that we have designed two forms, *Form A* and *Form B*, that allow candidates to record enrolments, and we want to know which is better. One way to measure 'better' is to ask which form the candidates prefer to use, either by simple choice, or by rating their qualities on a scale of one to ten.[22] An alternative is to measure the speed with which candidates are able to complete the two forms.

The main purpose of this section is to warn the reader of some of the pitfalls involved in such experiments. Although it makes use of some well-known statistical methods, it isn't an attempt to teach statistics, and the reader who wants more detail is advised to study a statistics textbook.

Suppose that ten candidates are recruited as **experimental subjects**, and seven prefer *Form A* to *Form B* or work faster with *Form A* than *Form B*. Does this prove that *Form A* is superior?

We first need to ask if the experiment was conducted fairly:

- Five of the subjects — chosen at random — should have been tested with *Form A* before *Form B* and the other five tested in the reverse order. The experiment with *Form A* may have given them an experience that helped them to use *Form B*, or conversely, it may have confused them. This is called a **learning effect**. To reduce the effects of learning, we *balance* two sets of trials: *A* then *B* and *B* then *A*.
- Were the experimenters supervising the trials aware of their purpose? If they believed the purpose was to prove that *Form A* was better than *Form B*, they may have influenced the subjects through conscious or unconscious suggestion. If the supervisors don't know the purpose of the trial, it is called a **double-blind experiment**.

[22] Because of the *Rule of Seven*, we may prefer to use a 7-point scale such as, 'Appalling, Bad, Fairly Bad, Average, Fairly Good, Good, Brilliant'.

Assuming these conditions were met, then if the ten candidates tested are the *only* candidates that will ever use the forms, yes, *Form A* is better. But if other candidates will use the forms, *we have to ask whether such results could have arisen by chance.*

All the candidates who *potentially* might use the forms, we call the **population**. The *individuals* whom we have tested form a **sample** of that population. Perhaps *Forms A* and *B* are equally good, but we accidentally selected a sample of individuals who mostly happened to prefer *Form A*. If there truly were no difference, the odds of obtaining this result would then be exactly the same as if we had tossed a coin and found that it came up heads 7 out of 10 times. We therefore need to paddle in the murky waters of sample statistics.

8.4.1 THE BINOMIAL DISTRIBUTION

Suppose we record the outcome of a trial as AABAAABBAA, meaning that the first and second candidates preferred *Form A*, the third preferred *Form B*, etc. This outcome is essentially the same as the outcome AAAAAAABBB — the way we number the experimental subjects is immaterial.

How many sequences contain seven As and three Bs? Imagine that we have ten cards numbered from 1 to 10. The first three cards we pick up will determine the positions of the Bs, and the rest will represent the positions of the As. There are 10 choices for the first card, 9 for the second, and 8 for the third, making $10 \times 9 \times 8 = 720$ possible sequences. If we happened to choose 3, 7, and 8, we would get the outcome AABAAABBAA above. But the sequence [8,7,3] would have given the same result, as would four other sequences. It is the *set* of choices that matters, so there are only $720 \div 6 = 120$ sequences containing 7 As and 3 Bs. In general, the number of distinct ways of choosing k items from n items is given by

$$\binom{n}{k} = \frac{n!}{k!(n-k)!}$$

Let us suppose that the proportion of all potential candidates (the *population*) who will prefer *Form A* is a, and of those who prefer *Form B* is b, where $b = (1 - a)$. When we randomly choose ten subjects to take part in our experiment (the *sample*), the probability that any one of them will prefer *Form A* is a.[23]

The probability of the particular sequence AABAAABBAA is $aabaaabbaa = a^7b^3 = a^7(1-a)^3$. Therefore the probability of any sequence with exactly seven As and three Bs is $\binom{10}{7}a^7b^3 = 120a^7b^3$. Since $\binom{10}{7}$ is the coefficient of a^7b^3 in the binomial expansion of $(a+b)^{10}$, the **probability distribution** of possible outcomes in this experiment is called a **binomial distribution**. Figure 8.11 shows this distribution in the case that 70% of the population prefer *Form A*, i.e., $a = 0.7$. In a series of random samples, although 27% of the time 7 out of the 10 subjects will be observed to prefer *Form A*, observing 8 out of 10 or 6 out of 10 is almost as likely.

[23] This assumes that the population is very large, because, having chosen a candidate who happens to prefer *Form A*, we slightly increase the fraction of those who prefer *Form B* among those who remain unchosen.

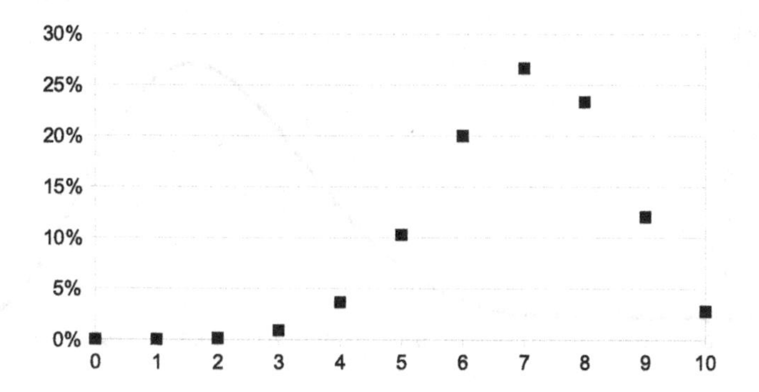

FIGURE 8.11

A binomial distribution. The horizontal axis shows how many subjects from a sample of 10 are observed to prefer *Form A*. The vertical axis shows the probability of such an observation if the underlying population probability is 0.7.

Conversely, given the result of an experiment involving a sample, how can we use this to estimate the proportion of the population who prefer *Form A*?

Let us assume that the number of people in a *sample* of size n who prefer *Form A* is s. How does the probability of observing s vary as a function of a? For $s = 7$ and $n = 10$,

$$P(a) = \binom{10}{7} a^7(1-a)^3 = 120(a^7 - 3a^8 + 3a^9 - a^{10}) \tag{8.4.1}$$

Figure 8.12 shows the graph of this function, from which we can see that if the underlying population probability is truly 70%, we have only a 27% chance of actually observing that 7 out of 10 subjects prefer *Form A* to *Form B*. On the other hand, if the underlying population probability were really only 50%, we would still have about 11.7% chance of observing the same result.

Although the shape of Figure 8.12 resembles that of Figure 8.11, and both peak at 27%, they aren't the same curve. In particular, the rightmost point of Figure 8.11 shows that there is a finite chance that all 10 subjects in a sample will prefer *Form A*, whereas the rightmost point of Figure 8.12 shows that it is impossible for only 7 out of 10 subjects to prefer *Form A* if *everybody* prefers it.

By integrating Equation 8.4.1, we find the **cumulative probability function**:

$$\int_0^a P(x)\,\mathrm{d}x = 120\left(\frac{a^8}{8} - \frac{3a^9}{9} + \frac{3a^{10}}{10} - \frac{a^{11}}{11}\right) \tag{8.4.2}$$

If we substitute $a = 1$ in Equation 8.4.2, we find that $\int_0^1 P(x)\mathrm{d}x = \frac{1}{11}$. Indeed, such integrals prove to be $\frac{1}{11}$ for each of the 11 possible values of s from 0 to 10. Figure 8.13 shows the *normalised* cumulative probability, scaled to 100%, i.e., $11 \times \int_0^a P(x)\mathrm{d}x$.

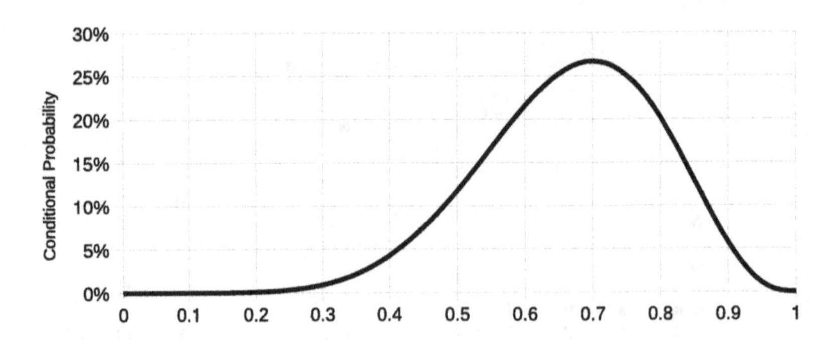

FIGURE 8.12

The probability that 7 out of 10 subjects will prefer *Form A*, if the underlying population probability (horizontal axis) has a given value.

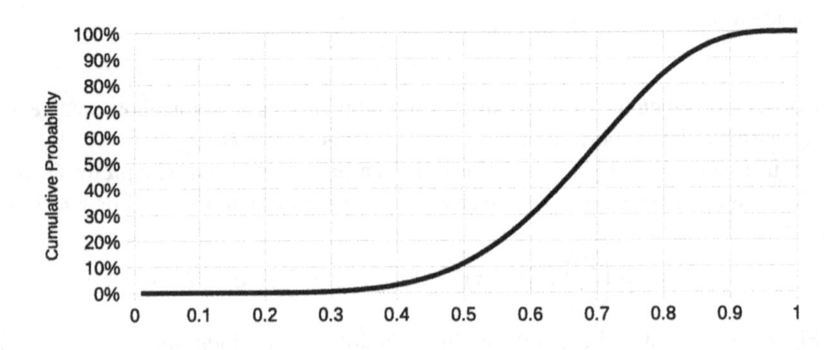

FIGURE 8.13

The normalised cumulative probability that 7 out of 10 subjects will prefer *Form A*, if the underlying population probability (horizontal axis) has a given value.

8.4.2 UNBIASED ESTIMATES

There are two ways we can interpret these graphs:

- The first is to compare our result with a **null hypothesis**: in this case, that *Form A* is no better than *Form B*. We learn from Figure 8.13 that populations with $a \le 0.5$ account for about 12% of the total, while those for $a > 0.5$ account for the remaining 88%. If we assume that we have no prior knowledge of the true value of a, and that any value is therefore equally likely, we may say that *Form A* is actually better than *Form B* at the 88% **level of significance**.
- The second is to establish a **confidence interval**: in this case, to observe that a sample of 10 subjects from populations with $a < 0.43$ or $a > 0.86$ would each have less than a 5% chance of containing 7 subjects who preferred *Form A*.

Again, assuming we have no prior knowledge of a, we may say that we have 90% **confidence** that a lies in the interval $0.43 \leq a \leq 0.86$.

Suppose that we aren't convinced that these results prove that *Form A* is better, so we repeat the experiment. This time, we observe that only 6 out of 10 subjects prefer *Form A*. Provided the sampling process didn't favour subjects from the first experiment, we may safely combine these results with our earlier ones, meaning that 13 out of 20 subjects prefer *Form A*. How does this affect the null hypothesis and the confidence interval?

Using a similar analysis to that above, despite the weaker preference, the first and second experiments together reject the *null hypothesis* with close to 90% *significance*, and the 90% *confidence interval* is narrowed at both ends to become $0.46 \leq a \leq 0.81$.

Since the second experiment showed a weaker preference than the first, we decide to repeat the experiment a third time, again scoring 7 out of 10, making a total of 20 out of 30 subjects who prefer *Form A*. Not surprisingly, the third experiment sharpens the result still further, rejecting the null hypothesis at the 96% level of significance and narrowing the 90% confidence interval to $0.51 \leq a \leq 0.78$. In general, the more subjects we sample, the narrower the confidence interval will become,[24] and the greater the statistical significance of the result.

Figure 8.12 shows $P(s|a)$, the **conditional probability** that s members of the *sample* will prefer *Form A*, given that the proportion of the *population* who prefer *Form A* is a. But what we really want to know is $P(a|s)$, the probability that the proportion of the *population* who prefer *Form A* is a, given that s members of the *sample* preferred *Form A*. These quantities are related by **Bayes' Theorem**.

$$P(s \wedge a) = P(s) \times P(a|s) = P(a) \times P(s|a) \tag{8.4.3}$$

From Equation 8.4.3, we conclude that $P(a|s) = P(s|a) \times P(a) \div P(s)$.

The problem with this equation is that the probability $P(a)$, that a will have any particular value, is zero. We can only talk sensibly about the probability of a lying in an infinitesimal *range* from x to $x + \mathrm{d}x$. $P(x)$ is called a **probability density function**. If we then assume that any value of x is equally likely, the probability that the proportion of the population lies in the range $\mathrm{d}x$ is simply $\mathrm{d}x$. Likewise, if we assume that any value of s is equally likely, $P(s) = \frac{1}{11}$, and we obtain $P(a|s) = 11 \times P(s|a)$. The most likely value for a in Figure 8.12 is therefore 0.7, coinciding with the maximum value of $P(s|a)$. This is called an **unbiased estimate** of a.

It is certainly possible for an estimate to be *biased*. For example, an unbiased estimate from the second experiment, taken alone, would suggest that $a = 0.6$. But following the first experiment, all values of a were no longer equally likely, so that the best estimate becomes $a = 0.65$.

---

[24] For a sample size of n, the confidence interval is proportional to $\frac{1}{\sqrt{n}}$. To halve the confidence interval, we need to increase the sample size four times.

What we have seen above is a simple example of **meta-analysis**: we have combined the results of three experiments to produce a more precise result. Meta-analysis is straightforward given sample sizes and confidence intervals. It is hard to combine results based on significance testing.[25]

A final word of caution: 'significant' does not mean 'important'; a result can be *important* without being *statistically significant.* A high level of *significance* is usually needed for the results of an experiment to be published. Consequently, many suggestive but potentially *important* results remain unpublished. For example, the finding that some expensive cancer treatment isn't 'significantly' better than a placebo could be *important* in allocating limited funds, but such a result is unlikely to be published unless it contradicts some previous announcement that claimed that the treatment gave a statistically significant benefit. Nowadays, partly because of common misunderstanding of the word 'significance', it is considered better practice to publish sample sizes and confidence intervals.

8.4.3 INDIVIDUAL DIFFERENCES

We have already seen that sampling and learning effects can influence the results of experiments. We also know that the experimental supervisors can unconsciously suggest the desired outcome of an experiment, and people like to please those in charge. Finally, we need to consider the great differences in our subjects' abilities and experiences. Such factors depend on a great many small influences, and in such a case, we may expect them to vary according to a **normal distribution.**[26]

Like *any* distribution, a normal distribution has a **mean** given by

$$\mu = \frac{1}{n}\sum_{i=1}^{n} x_i$$

and a **sample variance**

$$\sigma^2 = \frac{1}{(n-1)}\sum_{i=1}^{n} (x_i - \mu)^2$$

where the x_i are sample measurements, and n is the total number of measurements. The square root of the variance, σ, which is measured in the same units as μ, is called the **sample standard deviation.**[27]

[25] We have been guilty of a subtle error of procedure here: if our motive was to prove that *Form A* is better than *Form B*, we have biased our results by continuing the experiment until a significant difference was found, and then *stopping*. Perhaps, if a fourth sample had been tested, it might have led to a disappointing result. Procedurally, it is better to choose a suitable sample size before starting the experiment, and stick to it.

[26] The normal distribution is the limiting case of a binomial distribution as n approaches infinity.

[27] For example, the heights of Australian males closely follow a normal distribution with μ = 175 cm and σ = 7 cm. *Theoretically*, this allows someone to have a negative height, but the predicted probability is microscopic.

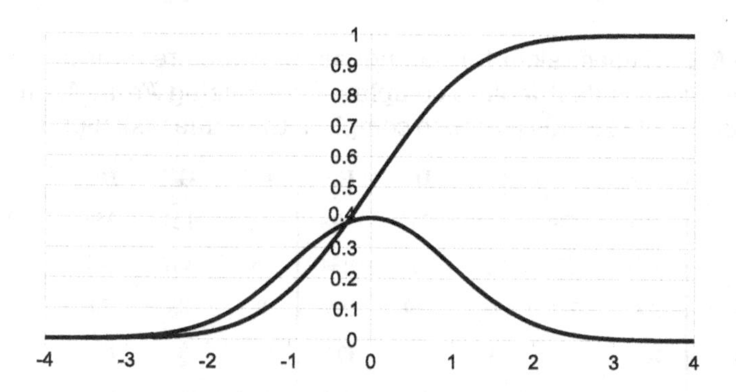

FIGURE 8.14

The normal distribution with $\mu = 0$ and $\sigma = 1$. The bell-shaped curve shows the probability density; the sigmoid curve shows the cumulative probability.

A *normal* distribution has a *probability density function* given by the following formula:

$$P(x) = \frac{1}{\sigma\sqrt{2\pi}} e^{-\frac{(x-\mu)^2}{2\sigma^2}}$$

where μ and σ are its mean and standard deviation, respectively. Figure 8.14 shows the graphs of the probability density function and the cumulative probability of the normal distribution with $\mu = 0$ and $\sigma = 1$.

A normal distribution is symmetrical about its mean, and its probability density falls away exponentially with the square of the number of standard deviations from the mean.[28] It has the property that 68% of samples will lie within the range $(\mu - \sigma)$ to $(\mu + \sigma)$, over 95% will lie within $(\mu - 2\sigma)$ to $(\mu + 2\sigma)$, and fewer than 3 samples in a thousand will lie outside the range $(\mu - 3\sigma)$ to $(\mu + 3\sigma)$.

Suppose now, instead of asking our subjects which form they prefer, we measure the time it takes for them to complete a certain task. We have 20 subjects: 10 complete the task using *Form A* then *Form B*, and 10 complete the task using *Form B* then *Form A*. The results are shown in Table 8.2. It is clear that there is great variation between individuals. Subject *B* recorded the longest time, 358 seconds, and Subject *R* recorded the shortest, 58 seconds.

A simple way to analyse the results is to find the average times subjects took with the two forms: we find $\mu(t_A) \approx 204$ seconds and $\mu(t_B) \approx 246$ seconds, where $\mu(t_A)$ and $\mu(t_B)$ are the average times using *Form A* and *Form B*, respectively. It seems that using *Form A* proved 42 seconds quicker to use than *Form B*, but is this difference significant?

[28] The factor $\frac{1}{\sigma\sqrt{2\pi}}$ ensures that the cumulative probability sums to 100%.

Table 8.2 Times taken for individuals to complete a form. t_A is the number of seconds it took to complete a task using *Form A*, and t_B is the number of seconds it took to complete the same task using *Form B*

Subject	A	B	C	D	E	F	G	H	I	J
t_A	195	278	146	247	198	209	121	256	262	131
t_B	250	358	187	313	234	254	137	347	326	181
$t_A \div t_B$	.78	.78	.78	.79	.85	.82	.88	.74	.80	.72
Subject	**K**	**L**	**M**	**N**	**O**	**P**	**Q**	**R**	**S**	**T**
t_A	229	271	243	222	262	168	98	58	242	238
t_B	287	308	272	260	304	195	104	61	257	290
$t_A \div t_B$	.80	.88	.89	.85	.86	.86	.94	.95	.94	.82

Clearly, the individual values vary so much that this difference might be due to chance. We consider a *null hypothesis*, that all the observations belong to a single normal distribution: if we combine both sets of results, we obtain a distribution with $\mu = 225$ and $\sigma = 80.5$. In fact, only 5% of the observations (both due to *Subject R*) lie outside the 95% confidence interval, $64 \leq x \leq 359$, and only 28% lie outside the range $\mu \pm \sigma$ — much as we should expect. The reader can see that the variations among individuals easily outweigh the 42 second difference between the forms, rendering the difference suggestive, but not statistically significant.[29]

In contrast, if we compare the *ratios* between individual performances (the third row in Table 8.2), we find that, on average, subjects completed the task using *Form A* in 0.84 of the time they took using *Form B*. In this case, $\sigma(t_A/t_B) = 0.07$, which means that *Form A* is faster than *Form B* with better than 98% significance.

How is it that the *ratios* of the times prove more significant than the difference in their averages? Quite simply because the averages lose track of the individual experimental subjects. Individual differences smear the times, giving a large standard deviation, but when we look at ratios, we compare two times recorded by the same individual. We expect to see two slow times, or two fast times, but not one of each. For the same reason, we find that the average of the differences, $\mu(t_A - t_B)$, also proves significant, even though the difference of the averages, $\mu(t_A) - \mu(t_B)$, doesn't.

The points in Figure 8.15 plot the pairs of test results for each individual. As expected, the two times correlate. The trend line shows the best estimate of the relationship between them, assuming it is linear.[30] Its slope is 0.77, which is within one standard deviation of our previous estimate of $\mu(t_A/t_B)$.

[29] A proper analysis of these results would use **ANOVA (analysis of variance)** — beyond the scope of this section.

[30] But, unlike our use of the ratio t_A/t_B, the best estimate doesn't assume that the trend line passes through the origin.

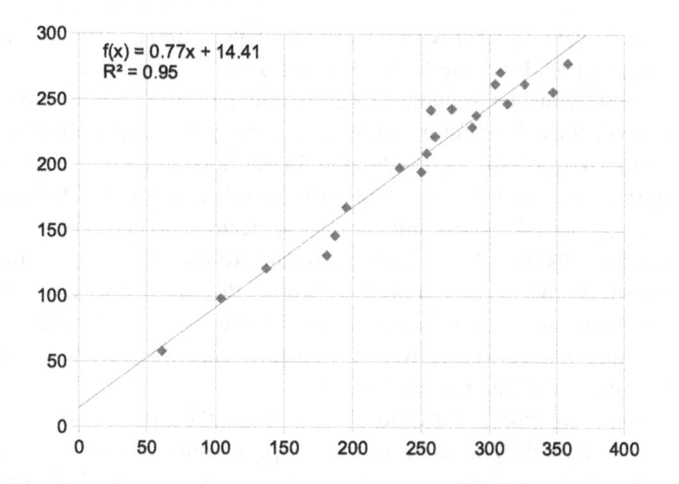

FIGURE 8.15

The individual results for t_A (vertical axis) and t_B (horizontal axis). The trend line accounts for 95% of the observed variance and suggests that $t_A = 0.77t_B + 14.41$.

If we now use our further knowledge that the first 10 subjects were tested with *Form B* first, and the last 10 with *Form A* first, we may estimate the *learning effect*. In the first group, we find that $\mu(t_A \div t_B) = 0.79$, and in the second, that $\mu(t_A \div t_B) = 0.88$. (In both cases $\sigma(t_A \div t_B) \approx 0.05$.) We discover that whichever form was tested first put the second form at about a 5% advantage.[31]

In the above analysis, we made the reasonable assumption that an individual experimental subject is consistently fast or consistently slow. Only by acknowledging such individual differences could we assert that the difference between the forms was significant.

To sum up, in conducting psychological experiments, we have to ensure that the experimenters don't unconsciously influence the results, we must compensate properly for learning effects, and we must take account of individual differences. We then have to turn our measurements of a suitably large sample into inferences about the population. There are many perils, and any reader who is interested in conducting such trials would be well advised to read a specialist statistics textbook.

8.5 FORMS DESIGN

Forms are the means by which an information system asks for a human decision. We may classify forms into three categories: interactive forms, which ask for an instant decision, reports, which allow more thought, and data-entry forms, which record the

[31] What we have achieved here is a crude form of **factor analysis** — again beyond the scope of this section.

results of such thought. In other words, we regard reports and data-entry forms as *two parts of an interaction.* It is wise for them to use a similar style and layout.

As an example, consider a candidate's academic transcript. The candidate will use this to make decisions about future enrolments, which will later be entered onto an enrolment form. As a result, the candidate will be motivated to read the report and to complete the form carefully. The written form subsequently has to be copied by enrolment staff, who have less motivation to be accurate. Therefore some external control is needed. One possibility is for candidates to receive a report similar to the statement issued by the Bursary (Figure 7.9) *after* the enrolment staff have entered the data, but *before* subject enrolments become frozen. The dual threats of missing a needed examination or paying for an unwanted enrolment should motivate the candidate to see that the database is corrected.

The long-term solution is for candidates to make their own enrolments interactively. Unfortunately, this ignores the role of the programme advisors, who help them make good choices. One solution is for the advisors to review the choices candidates have made, but a more interesting one is to encapsulate their knowledge in an *expert system*, a possibility we explore in the next chapter.

We now see that the main purpose of an academic transcript is to help a candidate plan future enrolments. To this end, the transcript should not only display all the candidate's earlier results, but the list of subjects that remain to be studied. We have seen a hint of this in Figure 8.3 (page 259), but that only showed subjects relevant to the current semester. A similar layout could be used for the academic transcript, showing all the subjects. Displaying the fee for each subject would also help the candidate plan. A further improvement would be to list subjects in a logical order, rather than alphabetically by subject code, but the database schema we have proposed doesn't contain enough information to make this possible. Although subjects that have already been studied can be ordered by semester, those remaining to be studied would need to be assigned priorities: essentially, the order in which components are normally studied as part of a programme. The interactive enrolment form should be modified to match.

Regrettably, if a form similar to Figure 8.3 were used for data entry by the Enrolment Office staff, it would prove error-prone. Assuming that staff don't watch the display while transcribing a candidate's form, they would need to mentally count how many subjects to skip between marking the check boxes for those in which the candidate wishes to enrol. One mistake in counting, and the result could be totally wrong — in a way the system might be unable to detect. Therefore, the written form, while resembling Figure 8.3, should also list subject codes, which would then be entered explicitly, as in Figure 8.8 (page 273).

As this example shows, form design begins with examining the decision-making process of a human. It is important to present information relevant to a decision in an intelligible way: too little information, and the decision cannot be an informed one; too much, and the reader might be overwhelmed. In some cases, only experiment will find the right balance. It is also wise to regard every report as one half of an interaction. If it does not lead to a decision, or is not required by law, it is useless.

Having decided on the purpose of a form, the next consideration is whether the person using it will be motivated to use it accurately. Assuming that this sometimes difficult problem can be solved, the physical layout of the form needs to take account of the psychological principles we have discussed. Last, but not least, are the technical problems: the interface requires that we design a *story board* linking the various forms and reports we need and that we create an infrastructure that allows the forms to invoke the kernel event procedures. These include matters of data flow, and whether additions need to be made to the database. We have already seen that *Candidates* rows ought to include a password and that *Components* should have priorities. These additions were not part of the kernel specification, but resulted from decisions about implementation. Sad to say, systems synthesis is often an iterative process.

8.6 SUMMARY

After a discussion of the technical issues concerning the design of reports and forms, we turned to the capabilities and limitations of typical humans. We found that several limitations of the human visual system and short-term memory ought to limit our ambitions in the design of forms and dialogues.

On a larger scale, we considered the importance of operant conditioning in making the work environment rewarding and stimulating, thus avoiding technophobia.

We have pointed out the many pitfalls of psychological experiments, especially concerning the interpretation and significance of their results.

Finally, we reviewed the relationship between forms, reports, and decision-making in the light of what we have learned about human behaviour.

8.7 FURTHER READING

Miller's Law was originally described in George A. Miller's 1956 article, 'The Magical Number Seven, Plus or Minus Two: Some Limits on Our Capacity for Processing Information', *Psychological Review* 63 (2): 81–97.

Operant conditioning was first described by Burrhus Frederic Skinner in his 1938 book, 'The Behavior of Organisms: An Experimental Analysis', *Appleton-Century-Croft*.

Frederick Herzberg's theory of motivation was originally proposed in his 1959 book, 'The Motivation to Work' (ISBN 978-1-56000-634-3).

Human interface design is discussed in the textbook 'Designing the User Interface: Strategies for Effective Human-Computer Interaction' by Ben Shneiderman and Catherine Plaisant (2009, ISBN: 0-321-53735-1).

No one does statistical calculations in the laborious way I have done here, but instead uses *SPSS*, an acronym that originally stood for 'Statistical Package for the Social Sciences', but no longer does. A useful *SPSS* textbook is 'SPSS for

Psychologists', by Nicola Brace, Richard Kemp, and Rosemary Snelgar (2012, ISBN: 1848726007). A cheaper alternative is 'SPSS for Applied Sciences: Basic Statistical Testing', by Cole Davis (2013, ISBN: 978-0-643-10710-6).

For a deeper understanding of why confidence intervals should be preferred to statistical significance, the reader should consult 'Understanding The New Statistics: Effect Sizes, Confidence Intervals, and Meta-Analysis' by Geoff Cumming (2011, ISBN: 978-0-415-87968-2).

For more information about the technology of speech recognition, see 'Spoken Language Processing: A Guide to Theory, Algorithm, and System Development' by Xuedong Huang, Alejandro Acero, and Hsiao-Wuen Hon (2001, ISBN: 0130226165).

8.8 EXERCISES

1. Consider the merits of a check digit scheme that requires an *Admission No* to be divisible by 17.
2. What does our inability to nest more than three interruptions tell us about the process of understanding the text of a computer program? (Consider, for example, the implementation of the *Enrol* event in Section 7.4.2.)
3. *HTML* provides a **<blink>** directive that causes text to flash. Unfortunately, the feature is non-standard, and not all browsers support it. When do you think this feature would be useful? (See Section 8.3.2.3.)
   Can you think of a good alternative, if **<blink>** is not supported?
4. The NUTS has a purchasing department and a store room, used mainly for stationery supplies. In theory, when any academic or administrative department (the *originating department*) needs stationery or equipment, it should send an order to the purchasing department. The purchasing department will record the order in a database, thereby printing a purchase order to be sent to a preferred supplier. It also sends an electronic copy of the order to the store room.
   When the goods arrive, they are delivered to the store room. The store hand informs the purchasing department of their arrival, which in turn informs the department that originated the order, recording the number of items *received* in the database. When the store room issues the goods to the originating department, the store hand records the number of items *issued* in the database.
   In some cases, e.g., paper-clips and most other small stationery items, the purchasing department makes sure that there is always a reserve supply available for immediate use.
   The purchasing manager receives a computer-generated report from the store hand listing the numbers of items of various kinds that are in the store. Unfortunately, the report is often wildly inaccurate, sometimes displaying a negative number of items in stock. This report has made both the store hand and the computer programming team the objects of scorn. What do you think causes the problem?

Apart from this, it is known that some departments bypass the purchasing system when they need something in a hurry. They send orders directly to the supplier, and inform the purchasing department later.
What can be done about this problem? Is a behavioural issue involved?

5. In an experiment, eight out of ten subjects preferred Form A to Form B. Is Form A better than Form B with 95% significance?

CHAPTER

Rules

9

CHAPTER CONTENTS

INTRODUCTION

The rules that an organisation uses to make decisions are a valuable resource. These rules are often understood by only a few experts. Computer programs that incorporate such knowledge are called **expert systems**. Too often in the past, such expert knowledge has been captured by procedures written in a programming language that has then become passé. For example, many businesses have had to convert *Cobol* code into more up-to-date languages. This has been a painful process. It is difficult to deduce the specification of a program from procedural code. The systems analyst often faces a similar problem when interviewing staff: they know *what* they do, but may not know *why*. Consequently, many organisations have sought ways of expressing knowledge in a way that is less dependent on the technology of the implementation.

Traditionally, converting rules to procedures has been the function of the computer programmer, but more and more, businesses are turning to 'intelligent' computer tools. These tools can function by generating program code or by interpreting rules at run-time. The first choice leads to faster programs, but the power of modern computers makes the second choice increasingly attractive.

9.1 BUSINESS RULES

We saw in Sections 7.4.2 and 7.5.3 (pages 224 and 238) that a simple event procedure, such as that for *Enrol*, can become segmented into several methods within different program objects. Such segmentation is essential in batch systems, such as that of Figure 7.10 (page 233). On the other hand, as we saw in Algorithm 7.3 (page 229), it can sometimes be avoided in an object-oriented design by organising the objects as a hierarchy, with the most subordinate containing the business rules.[1]

Segmentation presents two challenges to the systems analyst:

- Discovering existing business rules from the procedures that implement them, and
- Designing the infrastructure of a new system so that it will best preserve the original business rules.

Thus we can see that the division of labour between the Enrolment Office and the Bursary isn't a *requirement* of the *Enrol* event, but a historical decision made in implementation. It isn't *necessary* to involve two separate departments. The analyst is free, in principle, to make the Enrolment Office responsible for collecting fees. That is why we suggest that the analyst should initially concentrate on *what* is done and ignore *where* it is done. For similar reasons, kernel specifications ignore error processing. Errors are not a business requirement, and the ways in which they are handled depend on the implementation. For example, contrast the way errors have to be handled in a batch system (a postmortem report) with the more interactive way they are handled over the Internet.

As a result, the analyst is freed from the pain of specifying details that can best be filled in later by a programmer — even if the analyst will later be that programmer. The analyst can describe procedures more conveniently: by specifying rules.

[1] Another defence against segmentation is to use a literate programming tool. Among its other advantages, rather than allow the programming language to control the order in which code must be presented, it allows the programmer to document a program in the order best suited to a human reader. Thus all the code for the *Enrol* event could be written as a group, then distributed to necessary code modules by the tool.

9.2 RULE-BASED SYSTEMS

Consider the kernel specification of the *Enrol* event:

$$
\begin{aligned}
Enrol(c,\ s) \Leftarrow\ & Candidate\ State(c) = Admitted\ ;\\
& Subject\ State(s) = Offered\ ;\\
& Filled(s) < Quota(s)\ ;\\
& p = Programme(c)\ ;\\
& Component\ State(p,\ s) = Offered\ ;\\
& e = (Current\ Semester,\ c,\ s)\ ;\\
& Enrolment\ State(e) = Null\ ;\\
& Enrolment\ State(e)' = Enrolled\ ;\\
& Mark(e)' = \mathbf{null}\ ;\\
& Filled(s)' = Filled(s) + 1\ ;\\
& Balance\ Owing(c)' = Balance\ Owing(c) + Fee(s).
\end{aligned}
$$

Now consider some further rules concerning possible errors. First, *Candidate State* might also be *Graduated* or *Null*:

$$
\begin{aligned}
Enrol(c,\ s) \Leftarrow\ & Candidate\ State(c) = Null\ ;\\
& Error.No\ Such\ Admission(c).
\end{aligned}
$$

$$
\begin{aligned}
Enrol(c,\ s) \Leftarrow\ & Candidate\ State(c) = Graduated\ ;\\
& Error.Graduated(c).
\end{aligned}
$$

where *Error* is a program object that reports errors, and *No Such Admission*(c) is a method that displays 'No existing candidate record matches Admission Number c', or words to that effect.

Second, *Subject State* might also be *Inactive, Frozen, Obsolete*, or *Null*:

$$
\begin{aligned}
Enrol(c,\ s) \Leftarrow\ & Subject\ State(s) = Inactive\ ;\\
& Error.Subject\ Inactive(s).
\end{aligned}
$$

$$
\begin{aligned}
Enrol(c,\ s) \Leftarrow\ & Subject\ State(s) = Frozen\ ;\\
& Error.Subject\ Frozen(s).
\end{aligned}
$$

$$
\begin{aligned}
Enrol(c,\ s) \Leftarrow\ & Subject\ State(s) = Obsolete\ ;\\
& Error.Subject\ Obsolete(s).
\end{aligned}
$$

$$
\begin{aligned}
Enrol(c,\ s) \Leftarrow\ & Subject\ State(s) = Null\ ;\\
& Error.No\ Such\ Subject(s).
\end{aligned}
$$

Third, there might be no free place in the subject:

$$\begin{aligned} Enrol(c,\ s) \Leftarrow\ & Subject\ State(s) = Offered\,; \\ & Filled(s) \geq Quota(s)\,; \\ & Error.Quota\ Filled(s). \end{aligned}$$

Fourth, the subject might not be part of the candidate's study programme:

$$\begin{aligned} Enrol(c,\ s) \Leftarrow\ & Candidate\ State(c) = Admitted\,; \\ & Subject\ State(s) = Offered\,; \\ & p = Programme(c)\,; \\ & Component\ State(p,\ s) = Null\,; \\ & Error.Not\ In\ Programme(p,\ c,\ s). \end{aligned}$$

There are programming languages that support the use of rules. *Prolog* has much the same form as our specification language: each of the rules that we have just described for the *Enrol* event could be directly expressed as a *Prolog* rule.

OPS5, *CLIPS*, and similar **production system** languages consist of rules that modify a knowledge base of *facts*. If the knowledge base contains a set of facts that satisfy a rule, the rule modifies the knowledge base. Then the system iterates. To save testing each rule against each fact on every iteration, such languages use the **rete algorithm**.[2] Basically, the rete algorithm keeps track of which rules currently depend on each fact. When a new fact is asserted (e.g., an event occurs) or an existing fact retracted, the algorithm updates the status of all the rules that depend on it.[3]

We saw in Algorithm Algorithm 1.1 (page 10) that a set of rules can be systematically converted into a procedure in an imperative language such as *Java* or *C*. Unfortunately, we also saw that the resulting procedure depends on the order in which preconditions are tested. For example, we can test *Subject State*(s) either before or after testing *Candidate State*(c). On the other hand, the *Determines* relation forces the test $Filled(s) < Quota(s)$ to come after checking *Subject State*, and checking *Component State* must come after checking both *Subject State* and *Candidate State*. The remaining flexibility can be valuable to the implementor. In the batch system of Figure 7.12 (page 241), the infrastructure virtually forces the tests to be made in one particular order.

9.2.1 DECISION TABLES

A set of rules can often be expressed more concisely in the form of a **decision table**. Each row of Table 9.1 expresses one of the nine rules needed to define the *Enrol* event and its error processing. (For reasons of space, there is some informality here: for example, *Create Enrolment*(c, s) needs further explanation.)

[2] 'Rete' is a Latin word meaning 'net' and is pronounced to rhyme with 'treaty'.

[3] This can be seen as an application of use-definition analysis, discussed in Chapter 7.

Table 9.1 A decision table for the *Enrol* event

Candidate State(c)	Subject State(s)	Filled(s) < Quota(s)	p = Programme(c)	Component State(p,s)	Create Enrolment(c,s)	Error Message
Admitted	Offered	Yes	Yes	Offered	Yes	None
Null	–	–	–	–	No	No such admission
Graduated	–	–	–	–	No	Graduated
–	Inactive	–	–	–	No	Subject inactive
–	Frozen	–	–	–	No	Subject frozen
–	Obsolete	–	–	–	No	Subject obsolete
–	Null	–	–	–	No	No such subject
–	Offered	No	–	–	No	Quota filled
Admitted	Offered	–	Yes	Null	No	Not in programme

There are three kinds of entries in the table:

- When the column heading is an expression that can be either true or false, or an action that can either be performed or not, the entries in that column are either 'Yes' or 'No'. These are called **limited entries.**
- When the column heading invites several possible parameter values, the entries in that column provide the values. These are called **extended entries.**
- When an entry is left blank (–), it indicates 'inapplicable', 'irrelevant', or 'other'. These are called **don't cares**. For example, the condition *Filled*(s) < *Quota*(s) is *inapplicable* unless *Subject State*(s) = *Offered*, but in the bottom row, which considers the case that s is not part of the candidate's programme, it is *irrelevant*. If we wanted to be unhelpful, we could replace the error messages in rules four to seven by a single rule having the message: 'Subject s is not offered'. The entry for *Subject State*(s) would then be a dash, meaning 'other' (i.e., other than *Offered*).

The virtue of don't care entries is that they allow rules to be expressed more concisely. For example, the second rule of Table 9.1 applies to five different values of *Subject State*(s) and two values each of its following three conditions. If all the combinations of conditions had to be spelt out, forty rules would be needed to replace it.

Table 9.1, like the rules it expresses, is somewhat ambiguous: If more than one rule applies, what should be done? For example, if both *Candidate State*(c) = *Graduated* and *Subject State*(s) = *Inactive*, should one error message be displayed, or two? And if only one, should it be *Graduated* or *Subject Inactive*?

There are three widely accepted methods of **conflict resolution**:

- Execute *all* the applicable rules.
- Execute the *first* executable rule in the list.
- Execute the most *specific* executable rule in the list.

To understand this last point, suppose that we added a tenth rule to deal with the case that both *Candidate State*(c) = *Graduated* and *Subject State*(s) = *Inactive* are true. This would be more specific than either the second or fourth existing rule and would override both.[4]

Software tools exist that can convert such tables into procedures. The simplest is called the **rule-mask technique**. This evaluates the conditions at the head of the table, then matches their values with the rows of a *data structure* representing the decision table. This data structure can then be modified at any time to reflect later changes to the business rules. This method has the advantage that no specialised tool is needed to create the necessary software. It also means that the table is easily reusable if the implementation technology is changed.[5]

The rule-mask technique can deal with any of the three ways of interpreting ambiguity that we have described. The first two interpretations are straightforward. To select the most **specific** rule, we first sort the rows according to their entries, giving don't care entries the lowest priority. If we sort Table 9.1 in this way, we obtain Table 9.2. As a result of the sorting process, the case when both *Candidate State*(c) = *Graduated* and *Subject State*(s) = *Inactive* would be treated according to the third rule in Table 9.2 rather than the sixth. This is because, in sorting, the first column has a bigger effect than the second.

Decision tables can also be converted into procedures using an extension of Algorithm Algorithm 1.1 (page 10).[6] Since this generates structured code, it can result in long, but readable, programs. For example, if a rule requires four conditions to be satisfied, one branch of the code handles the case when they are all satisfied, but four other branches are needed to deal with the cases where the conditions aren't satisfied.

A third approach is to treat the rules as the specification of an NFA. This can then be converted into a DFA, as described in Section 3.2. The resulting program is then compact, but almost unintelligible.

Processors that convert decision tables to procedural code must necessarily check the rules for completeness and ambiguity, thus detecting potential errors at compile time, instead of leaving them latent.

---

[4] Production systems sometimes offer a fourth means of conflict resolution, based on how recently a rule has been activated.

[5] The process can be made more efficient by using bit maps (page 66). Each conditional outcome is represented by a bit, and each rule is represented by a bit map showing which outcomes need to be true. Evaluating the conditions creates a run-time bit map that can be quickly matched against the bit maps for the rules.

[6] Extended entries are first converted to a series of limited (Yes/No) entries.

Table 9.2 The result of sorting rules into priority order

Candidate State(c)	Subject State(s)	Filled(s) < Quota(s)	p = Programme(c)	Component State(p,s)	Create Enrolment(c,s)	Error Message
Admitted	Offered	Yes	Yes	Offered	Yes	None
Admitted	Offered	–	Yes	Null	No	Not in programme
Graduated	–	–	–	–	No	Graduated
Null	–	–	–	–	No	No such admission
–	Frozen	–	–	–	No	Subject frozen
–	Inactive	–	–	–	No	Subject inactive
–	Null	–	–	–	No	No such subject
–	Obsolete	–	–	–	No	Subject obsolete
–	Offered	No	–	–	No	Quota filled

9.2.2 CHECKING STUDY PROGRAMMES

As part of an interactive system that candidates can use to enrol, the NUTS propose to check that students have the prerequisite knowledge to study the subjects they have chosen. This information is found in the documents that describe the syllabus of each subject and is a resource that programme advisors consult when they help a candidate choose a study plan:

> **Project Management A (projmgta):**
>
> ***Prerequisite knowledge*:** a pass (50%) in *Financial Management A* (fnclmgta) and a pass (50%) in *Graph Theory* (graphthy).
>
> ***Syllabus*:** Scheduling activities using the Critical Path Method, Management by Objectives, *and so on . . .*

Consider the *Graduate Diploma in Information Technology (Electronic Commerce) (ecom)* programme that was shown in Figure 4.2 (page 119). The prerequisite requirements for this programme can be summarised in the form of a decision table, shown in Table 9.3. The first column of the table lists the subjects that are part of the *ecom* programme, and the top row lists the prerequisites on which they might depend. Each row following the first represents a rule: a candidate wishing to enrol in the subject listed in the leftmost column must have obtained at least the marks indicated in the grid for each corresponding subject listed in the top row. For example, the twelfth rule states that a candidate who wishes to enrol in *Project Management A* must already have scored a mark of at least 50% in both *Financial Management A* and *Graph Theory*.

Table 9.3 A prerequisite relation between subjects. Each row shows the prerequisite knowledge assumed before a candidate can enrol in the subject in the leftmost column. Most subjects require a pass (50%), but *graphthy* requires only a 40% score in *maths101*. The last row is an *else rule*, and applies only if no other rule applies

Subject	acctg101	fnclmgta	graphthy	javaprog	maths101	security	dataanal	eeconomy	hmnbhvra	marketec	netmgt1	projmgta	qtheory	sqlprgrm	dbadmin1	htmlprog	Allow?
acctg101	–	–	–	–	–	–	–	–	–	–	–	–	–	–	–	–	Yes
fnclmgta	50	–	–	–	–	–	–	–	–	–	–	–	–	–	–	–	Yes
graphthy	–	–	–	–	40	–	–	–	–	–	–	–	–	–	–	–	Yes
javaprog	–	–	–	–	–	–	–	–	–	–	–	–	–	–	–	–	Yes
maths101	–	–	–	–	–	–	–	–	–	–	–	–	–	–	–	–	Yes
security	–	–	–	50	–	–	–	–	–	–	–	–	–	–	–	–	Yes
dataanal	–	–	50	–	–	–	–	–	–	–	–	–	–	–	–	–	Yes
eeconomy	–	50	–	–	–	50	–	–	–	–	–	–	–	–	–	–	Yes
hmnbhvra	–	–	–	–	–	–	–	–	–	–	–	–	–	–	–	–	Yes
marketec	–	–	–	–	–	–	–	–	50	–	–	–	–	–	–	–	Yes
netmgt1	–	–	50	–	–	–	–	–	–	–	–	–	50	–	–	–	Yes
projmgta	–	50	50	–	–	–	–	–	–	–	–	–	–	–	–	–	Yes
qtheory	–	–	–	–	50	–	–	–	–	–	–	–	–	–	–	–	Yes
sqlprgrm	–	–	–	–	–	–	50	–	–	–	–	–	–	–	–	–	Yes
dbadmin1	–	–	–	–	–	–	–	–	–	–	–	–	–	50	–	–	Yes
fnclmgtb	–	50	–	–	–	–	–	50	–	–	–	–	–	–	–	–	Yes
htmlprog	–	–	–	–	–	50	–	–	–	–	–	–	–	50	–	–	Yes
intrface	–	–	–	–	–	–	–	–	50	–	–	–	–	–	–	50	Yes
mrktrsch	–	–	–	–	–	–	–	–	–	50	–	–	–	–	–	–	Yes
projmgtb	–	–	–	–	–	–	–	–	–	–	–	50	–	–	–	–	Yes
services	–	–	–	–	–	–	–	–	–	–	–	–	–	–	50	–	Yes
socmedia	–	–	–	–	–	–	–	–	–	50	–	–	–	–	–	–	Yes
–	–	–	–	–	–	–	–	–	–	–	–	–	–	–	–	–	No

The last row is called an **else rule**. Because it contains so many don't care entries, it has the lowest priority and is invoked only when no other rule succeeds.

Depending on the design of the interface,[7] this table might be used in two possible ways:

- To check that a subject a candidate has already chosen is acceptable, or
- To offer the candidate a choice from only those subjects that are acceptable.

In addition, if the table is stored as a data object and interpreted by software, a program can discover what the requirements of any given subject are and give intelligent feedback. For example, it might report:

> You are unable to enrol in *Project Management A* because you failed to score 50% in *Financial Management A*.

Apart from conciseness, decision tables have other advantages:

- Compared with procedural code, they make it easier for a programmer to discover what conditions can lead to a particular outcome. For example, if the *Enrol* event procedure displays the *Subject Frozen* message, rather than searching its code for occurrences of the method call, inspection of Table 9.2 would quickly tell us that this can only happen if $Subject\ State(s) = Frozen$ — and rather more subtly, only if $Candidate\ State(c) = Admitted$.
- Decision tables can express an underlying problem more directly. For example, Table 9.3 is essentially the matrix representation of the graph of Figure 9.1, which captures the overall prerequisite structure of the *ecom* programme.
- Decision tables are more easily understood by non-programmers. Similar tables are often seen in government publications, etc.
- A decision table can be stored as a data structure. Expressed as a matrix, Table 9.3 would occupy about 1,000 bytes of storage. A sufficiently general procedure could be written to interpret a similar table for *any* study programme. Writing this procedure would surely be much easier than writing — and maintaining — a hundred different procedures specific to each programme. The matrices themselves could be prepared by a spreadsheet program.

9.3 EXPERT SYSTEMS

A key distinction can be made between **declarative knowledge** and **procedural knowledge**. A familiar example of this distinction is provided by a GPS navigation system in a car. The navigator takes *declarative knowledge* in the form of a map and plans a route that is relayed to the driver as series of *procedural* steps: 'turn left', 'turn right', and so on. In this way, the navigator takes a finite amount of knowledge and can plan an almost infinite number of possible routes.

The important part of this process is some kind of **inference engine**, or planner. This is what gives life to the **knowledge base**.

Let us suppose that the National University of Technology and Science wants to take the knowledge of the programme advisors and replace it by an expert system.

[7] See Figure 8.3, for example.

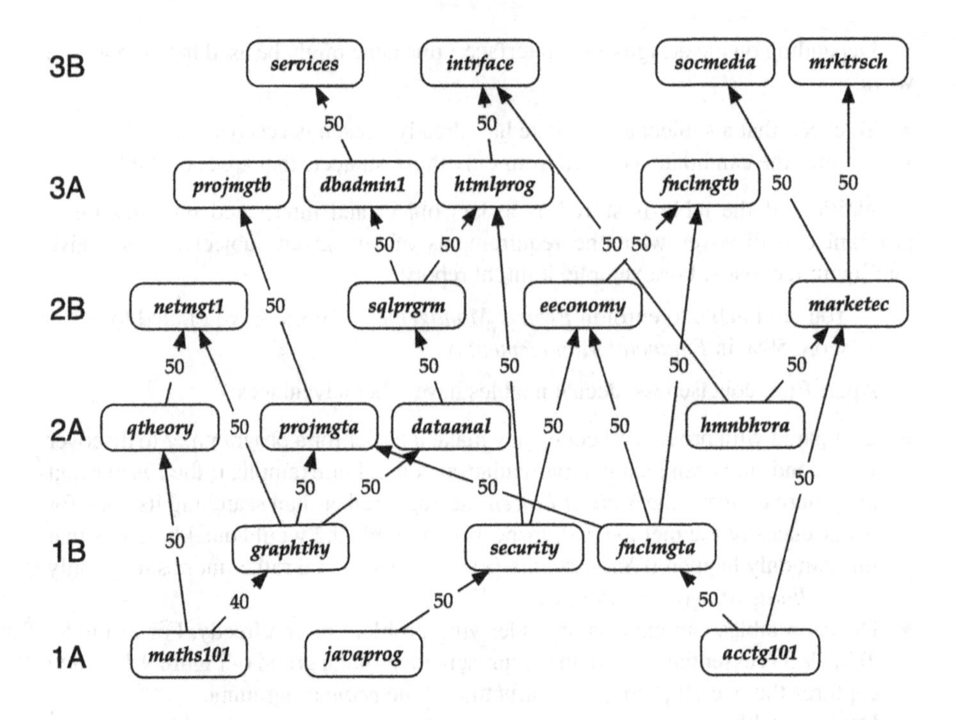

FIGURE 9.1

The structure of the *Graduate Diploma in Electronic Commerce*. The codes on the left indicate the year level and the semester in which the subjects are meant to be taught. (For the full names of the subjects, see Figure 4.2.)

For the moment, we shall assume that this knowledge can be modelled by directed acyclic graphs (DAGs), such as that shown in Figure 9.1 for the *Graduate Diploma in Information Technology (Electronic Commerce)* (*ecom*) study programme.

In the figure, an edge $S \underset{m}{\rightarrow} T$ means that a candidate must obtain at least m marks in *Subject S* before enrolling in *Subject T*. Figure 9.1 defines the *Prerequisite* relation for the *ecom* study programme. The relation is transitive:

$$(S\ Prerequisite\ T \wedge T\ Prerequisite\ U) \Rightarrow S\ Prerequisite\ U.$$

The graph can be modelled by a database table of the form, *Prerequisites*(*Programme, Subject, Prerequisite*, *Minimum*). For example, the row (*ecom*, *graphthy*, *maths101*, 40), would mean that a candidate wishing to enrol in *graphthy* as part of the *ecom* programme would need to have obtained at least 40% in *maths101*. Ideally, we would like to be able to store and maintain such rules in a database and interpret them using a suitable inference engine.

9.3.1 FORWARD AND BACKWARD CHAINING

Given a set of subjects that the candidate has already passed, say {*maths101*, *javaprog*, *acctg101*, *graphthy*, *security*, *fnclmgta*}, the graph tells us that the candidate may enrol in any subset of {*qtheory*, *projmgta*, *dataanal*, *hmnbhvra*} in Semester 2A, and in *eeconomy* in Semester 2B. If we assume the candidate chooses {*projmgta*, *dataanal*, *hmnbhvra*} in Semester 2A, then the set of available choices in Semester 2B is {*sqlprgrm*, *eeconomy*, *marketec*}, but cannot include *netmgtl*. Working in this direction, from knowns to unknowns, is called **forward chaining**. Forward chaining will find every subject that can eventually be studied given the subjects already passed. Essentially, in this case, forward chaining finds the transitive closure of the *Prerequisite* relation.

A more interesting question than 'What subjects can I study?' is, 'I want to take *Interface Design* and *Market Research Methods* in my final year. What subjects do I *need* to study?' Figure 9.1 tells us that to take *mrktrsch*, the candidate will need to study *marketec*. This will in turn need the candidate to study *hmnbhvra* and *acctg101*. To enrol in *intrface*, the candidate will need to study *htmlprog* and *hmnbhvra*, but *htmlprog* requires *sqlprgrm*, which, in turn, requires *dataanal*. Assuming the candidate has passed the same set of subjects as before, it would be necessary to enrol in *dataanal* and *hmnbhvra* in Semester 2A. Working in this direction, from questions or hypotheses to known facts, is called **backward chaining**. Backward chaining is usually preferred to forward chaining because it is goal-directed and typically needs to explore only a subset of the graph.

Both forward and backward chaining systems are able to offer **how-explanations**. In forward chaining, the how-explanation is a trace of the path from the givens to a particular goal. In backward chaining it is the reverse of the path from the goal to its antecedents. A how-explanation for *intrface* might read, 'Because you have passed *graphthy* you may study *dataanal*. This will allow you to study *sqlprgrm*. In conjunction with your pass in *security* this will enable you to study *htmlprog*. In conjunction with *hmnbhvra*, which has no prerequisites, you may then study *intrface*'.

Backward chaining systems often elicit facts interactively, asking questions like, 'Have you passed *security*?' The candidate may wonder why this matters and might ask, 'Why?' Backward chaining can then offer a **why-explanation**, as follows, 'You need to pass *security* so that you can study *htmlprog*, which you need to study *intrface*'.

An interactive system of this kind only needs to ask relevant questions: it doesn't need to ask if the candidate has passed *javaprog*; it is implied by success in *security*; nor does it need to ask if the candidate has passed *acctg101*, as this is not a prerequisite of *intrface*, either directly or indirectly. If the candidate later asks what subjects are needed to study *mrktrsch*, *acctg101* does become relevant, but the candidate would be irritated if the inference engine enquired again about *hmnbhvra*, because this question would already have been asked in connection with *intrface*. Consequently, an interactive system should remember the facts that it has already elicited.

At this point, the reader may wonder if there is any difference between backward chaining, and forward chaining using the reverse of the original graph — but this would be to miss the point. The key difference is that forward chaining works from knowns to unknowns, whereas backward chaining works from unknowns to knowns.

9.3.2 LOGIC TREES

The *Prerequisite* relation as we have described it is called an **and-tree** — even though it is not a tree. Enrolment in *intrface* is possible if *htmlprog* **and** *hmnbhvra* have been passed. The reverse relation, *Prerequisite*$^{-1}$ forms an **or-tree**; passing *graphthy* enables a candidate to enrol in *netmgtl* **or** *projmgta* **or** *dataanal*.

Now consider a situation in which new candidates often lack the mathematical background to study *maths101*. For the benefit of these candidates, it is decided to offer two additional subjects: *Remedial Mathematics A* (*remathsa*) and *Remedial Mathematics B* (*remathsb*), offered in Semesters 1A and 1B respectively. The content of *remathsa* is chosen so that it remains possible to study *graphthy* in Semester 1B, but *remathsb* is needed in order to study *qtheory* in Semester 2A. Thus, *graphthy* may be studied after a pass in either *maths101* **or** *remathsa*, and *qtheory* may be studied after a pass in either *maths101* **or** *remathsb*. The resulting structure, which is no longer a homogeneous relation, is called an **and-or tree**. An and-or tree can be drawn as a bi-partite graph, with alternating and-nodes and or-nodes. It is acceptable for either kind of node to have only one antecedent.

And-or trees can be explored by either forward or backward chaining, but it is important for backward chaining to discover *all* ways of reaching a goal, otherwise it might insist that a candidate should pass *maths101*, without considering the alternative of *remathsb*. On the other hand, if the inference engine has already discovered that the candidate has passed *maths101*, it would be stupid then to ask the candidate about *remathsb*.

No logic system is complete without it, but *negation* must be treated with care. Unless explicitly stated to be false, a proposition may be *assumed* to be false if it cannot be proved to be true. This is called **negation by failure** and is valid if all relevant facts are known and the universe of discourse contains a finite number of provable propositions, which is known as the **closed world assumption**. For example, a candidate may be assumed not to have passed *maths101* if no instance of a pass can be found. Negation would also be useful to express the fact that the two subjects, *remathsa* and *remathsb*, which together are only equivalent to *maths101*, cannot both be counted towards a degree. (For simplicity, in the following, we shall take the view that *remathsa* is a prerequisite of *remathsb*, but doesn't itself count towards the degree.)

A forward-chaining system can only deduce that a proposition is *false* after it has finished finding everything that is true. In contrast, a goal-directed, backward-chaining system can test if a hypothesis is false by immediately attempting to prove it is true, and failing.

Although neither type of inference engine has a problem with *this* example, negation can be a major source of difficulty: Suppose we tell a logic system that 1 is an odd number, and that if n is an odd number, then $(n + 2)$ is also an odd number. Then, to prove that 2 is *not* odd, the system would have to evaluate *all* the odd numbers to prove that 2 is not among them. Clearly, the closed world assumption is invalid, and this is a situation where the logic engine could use some help, for example, that $(n \bmod 2) = 1$ if n is odd.

Perhaps surprisingly, **and**, **or**, and **not** operations can all be expressed within a homogeneous graph,[8] by using the **nor** operator, where

$$\mathbf{nor}(a,\ b,\ \ldots) = \neg(a \vee b \vee \cdots),\ \text{and}\ \mathbf{nor}(a) = \neg a.$$

For example, given the expression

$$(a \vee b) \wedge (c \vee \neg d) \tag{9.3.1}$$

we may derive

$$\neg(\neg(a \vee b) \vee \neg(c \vee \neg(d)))$$

using De Morgan's Theorem (Equation 2.1.22 on page 27). The resulting expression may be expressed as

$$\mathbf{nor}(\mathbf{nor}(a,\ b), \mathbf{nor}(c,\ \mathbf{nor}(d))).$$

If we chose to use **nor trees** to express the prerequisite structures of programmes, we would be wise to use a tool to create them from some more convenient notation. In the case of prerequisites, the form of expression 9.3.1 would be convenient. The expression is in conjunctive normal form (CNF). (See Section 2.1.3.) If prerequisite requirements are expressed in CNF, the transformation to **nor** is trivial: *all* operators are replaced by **nor**. For example.

$$(a \vee b) \wedge (c \vee \neg d) = \mathbf{and}(\mathbf{or}(a,\ b),\ \mathbf{or}(c,\ \mathbf{not}(d))) = \mathbf{nor}(\mathbf{nor}(a,\ b),\ \mathbf{nor}(c,\ \mathbf{nor}(d)))$$

Although **and-or-not trees** and **nor** trees are flexible, they aren't always the most convenient way to represent a situation. The *ecom* programme offers 24 subjects (including *remathsa* and *remathsb*), but only 18 of them need to be passed to obtain a degree. At Level 3, candidates must choose at least six from the eight offered. There are $\binom{8}{6} = 28$ ways of choosing six subjects, eight ways of choosing seven subjects, and one way of choosing all eight. In order to express the requirements of a degree, we would need 37 rules of the kind we have described — not an attractive proposition. For that reason, an inference engine should also be capable of arithmetic.

A further type of tree can conveniently express the constraint that to pass the *ecom* programme, a candidate must pass at least six Level-3 subjects. For this purpose, we introduce a **from operator**, with which we write

(≥ 6) **from** $\{$*Passed(services)*, *Passed(intrface)*, *Passed(socmedia)*,

[8] Homogeneous, that is, except for its leaves. (Nor would this surprise an electronics engineer.)

$$Passed(mrktrsch),\ Passed(projmgtb),\ Passed(dbadmin1),$$
$$Passed(htmlprog),\ Passed(fnclmgtb)\}.$$

The **from** operator counts the number of its right-hand operands that are *True*. If this number satisfies the condition on its left, **from** returns *True*, otherwise it returns *False*.

Ignoring *remathsa* and *remathsb* for the moment, a similar **from** expression can specify that at least 18 subjects must be passed in all:

$$(\geq 18)\ \mathbf{from}\ \{Passed(services),\ Passed(intrface),\ \ldots,\ Passed(acctg101)\} \quad (9.3.2)$$

where the list contains all 24 subjects except *remathsa* and *remathsb*. We can therefore combine these requirements into the single expression

$$(=2)\ \mathbf{from}\ \{(\geq 6)\ \mathbf{from}\ \{Passed(services),\ \ldots,\ Passed(fnclmgtb)\},$$
$$(\geq 18)\ \mathbf{from}\ \{Passed(services),\ \ldots,\ Passed(acctg101)\}\}.$$

Since *remathsa* and *remathsb* together are only the equivalent of *maths101*, we need a rule to make sure that only *remathsb* counts. We replace *Passed(maths101)* with the expression

$$(\geq 1)\ \mathbf{from}\ \{Passed(maths101),\ Passed(remathsb)\}$$

whose effect is that, if the candidate has passed *maths101* **or** *remathsb*, it returns *True*. If by chance the candidate has passed both subjects, the expression still returns a single *True* value, which therefore counts as only one subject. In this way, a **from-tree** can be constructed to represent all subject combinations that satisfy the requirements of the programme.

The **from** operator is general enough to subsume $\wedge$, $\vee$, $\neg$, and **nor**:

$$(a \wedge b \wedge c) \Leftrightarrow (=3)\ \mathbf{from}\ \{a,\ b,\ c\}$$
$$(a \vee b \vee c) \Leftrightarrow (>0)\ \mathbf{from}\ \{a,\ b,\ c\}$$
$$\neg a \Leftrightarrow (=0)\ \mathbf{from}\ \{a\}$$
$$\mathbf{nor}(a,\ b,\ c) \Leftrightarrow (=0)\ \mathbf{from}\ \{a,\ b,\ c\}.$$

Consequently **from trees** can express everything that **and-or-not** trees can, but more concisely. As a result, we can also construct **from** trees to express the *Prerequisite* graph. For example, we can express the prerequisites for *qtheory* as

$$qtheory\ \mathbf{needs}\ (\geq 1)\ \mathbf{from}\ \{Passed(maths101), Passed(remathsb)\}$$

where **needs** is essentially the inverse of the *Prerequisite* relation.

If we assume that the prerequisite requirements for *graphthy* are rather more subtle, requiring either a pass in *remathsa* or at least 40% in *maths101*, then after introducing a suitable **in** operator, we may write

$$graphthy\ \mathbf{needs}\ (\geq 1)\ \mathbf{from}\ \{40\ \mathbf{in}\ maths101, 50\ \mathbf{in}\ remathsa\}$$

where $Passed(s) = 50\ \mathbf{in}\ s$.[9]

[9] Although 40% in *maths101* suffices to study *graphthy*, only a pass counts towards graduation.

A series of such definitions makes up the **needs** relation, and in combination with the **from**-tree of Figure 9.2 on the following page, is sufficient to express the *knowledge base* the programme advisors use.

9.3.3 A PRACTICAL INFERENCE ENGINE

To give advice, an inference engine must interpret the data structures just described. If we assume that the *Electronic Commerce* (*ecom*) programme is typical of those that the National University of Technology and Science offers, then an inference engine that works well for *ecom* should work well for other courses — provided that it doesn't take advantage of knowledge specific to *ecom*.

In essence, programme advice consists of a subset of the components of a programme, forming a **study plan** that satisfies whatever constraints the candidate imposes. Assuming that we add *remathsa* and *remathsb* to the 22 subjects in Figure 9.1, the *ecom* programme comprises a set of 24 components, so it has 2^{24} subsets — over 16 million of them. This is our **search space**.

One approach would be to generate all 2^{24} subsets and test which ones satisfy the requirements for a valid plan. Although such an **exhaustive search** is possible, we know that there must be fewer than 190,050 valid subsets,[10] because at least 18 subjects are required to graduate.

At first sight, it may seem that conditions a candidate places on the subjects to be studied — 'I want to take *Interface Design* at Level 3', 'I don't want to have to study *Financial Management A*', and so on — make the problem more complex, but they actually simplify it. If a candidate insists on studying a particular subject, or has already passed it, at most $\binom{23}{17} = 100{,}947$ options remain. If a candidate rules out a subject, then there remain at most $\binom{23}{18} = 33{,}649$ combinations with 18 subjects. The worst situation is when the candidate makes no conditions at all, because that is when the number of options is the greatest. If we can deal with that situation, we can deal with *any* situation.

To simplify the following discussion, we introduce two new definitions:

- If a set of subjects satisfies the conditions needed to graduate, we shall say it is **sufficient**.
- If a set of subjects satisfies all the prerequisite requirements, we shall say it is **valid**.

To construct an inference engine, we may then choose from three basic strategies:

1. Use the **from** tree of Figure 9.2 to generate sets of subjects that are *sufficient*, and select those that are *valid*.
2. Use the **from** tree corresponding to the prerequisite graph of Figure 9.1 to generate sets of subjects that are valid, and select those that are sufficient.
3. To generate all feasible sets and select those that are both valid and sufficient.

We shall define what we mean by *feasible* shortly.

[10] $190{,}050 = \binom{24}{18} + \binom{24}{19} + \binom{24}{20} + \binom{24}{21} + \binom{24}{22} + \binom{24}{23} + \binom{24}{24}$.

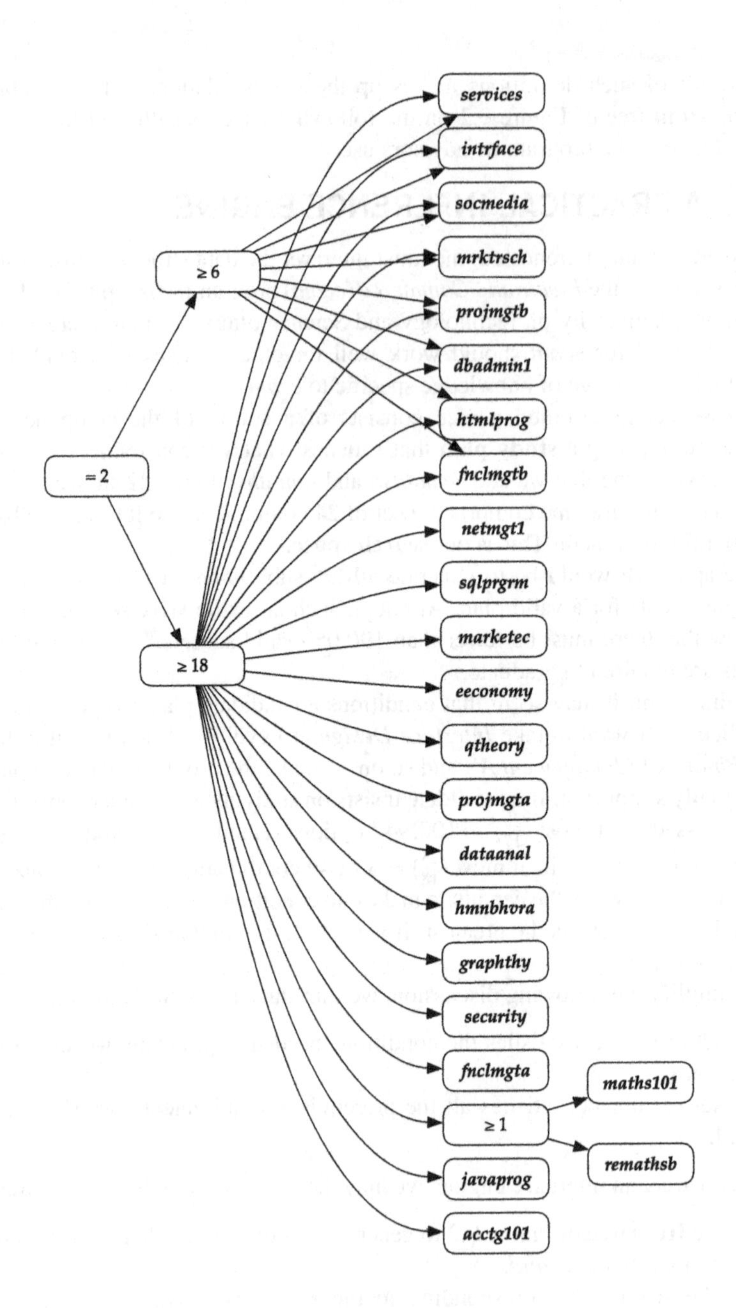

FIGURE 9.2

The from-tree that specifies the graduation requirements of the *ecom* study programme.

Although it is certainly possible to use a **from** tree to generate all *sufficient* or all *valid* sets, it is complicated to do and complicated to explain. So we adopt the third approach — if only for tutorial reasons.

It is often convenient to consider a search space as a *tree* resulting from a series of binary decisions. We might start by considering *acctg101*. We then have two branches: one in which *acctg101* is included, and one in which it isn't. If the next consideration is *javaprog*, we have four branches:

- One that includes both *acctg101* and *javaprog*,
- One that includes *acctg101* but not *javaprog*,
- One that includes *javaprog* but not *acctg101*,
- One that includes neither.

We can continue in this way until we reach the 2^{24} leaves of the **search tree**. Clearly, this approach is exponential in the number of components in the programme. Although searching the entire tree is within the reach of modern computers, it is clearly faster and more efficient to **prune** the tree, so that fewer subsets need to be considered. For example, if we can prove that no subset that omits both *acctg101* and *javaprog* can satisfy the constraints of a study plan, we can chop off that entire sub-tree. The secret of efficient pruning is to check the feasibility of each alternative as soon as it is generated — or even before.

Fortunately, many subject combinations can be ruled out, some of them quite quickly:

- Any subject that a candidate excludes also excludes the subjects of which it is a prerequisite. Ruling out *Financial Management A (fnclmgta)* would rule out *projmgta, eeconomy*, and *fnclmgtb*. The only choice left to the candidate is whether to study *maths101* or *remathsa* and *remathsb*.
- On the other hand, a candidate's decision to include *Interface Design* (*intrface*) forces *htmlprog, sqlprgrm, dataanal, graphthy, security, javaprog*, and *hmnbhvra* to be included. Ignoring the choice between *maths101* or *remathsa* and *remathsb*, it remains necessary only to choose nine subjects from a total of 13 — 715 combinations.
- Certain subjects have to be a part of *any* plan. For example, *graphthy* is a prerequisite, directly or indirectly, of nine other subjects. Any plan that omits *graphthy* can contain at most 14 subjects and can't be viable. As another example, omitting *javaprog* leaves 18 possible subjects, but no viable study plan. Only five third-year topics remain, and six are needed to graduate.

We could choose to make inferences such as these in advance and store them as part of the database for each programme. We shall ignore this possibility here, because, although such knowledge makes the search space smaller, it doesn't change the essential problem. Indeed, since we are constructing an inference engine that deals with all programmes, not just *ecom*, we cannot use the fact that *graphthy* and *javaprog* must be included in a plan, because it is only by running the inference engine that we know this is true.

Fortunately, *every* study plan must have two important and useful properties:

- We cannot forbid a candidate from taking unnecessary subjects; if a set of subjects S is *sufficient*, then every superset of S will be *sufficient*. Conversely, if a set of subjects S is *insufficient*, then every subset of S will be *insufficient*.
- Every *Prerequisite* relation must be acyclic and have at least one topological sort. (See Section 2.5.1 on page 47.) We can see this must be so, otherwise there could be no *valid* sequence in which a candidate could complete a programme. The initial vertices of the *Prerequisite* graph include those subjects taught in Semester 1A, and its final vertices include those taught in Semester 3B.

The order in which we build a search tree is usually important. Here, we can choose both the order in which subjects are considered, and whether we consider including a subject before or after we consider excluding it. There are therefore four different search orders we might choose:

1. Inclusion before exclusion, from initial to final vertices.
2. Inclusion before exclusion, from final to initial vertices.
3. Exclusion before inclusion, from initial to final vertices.
4. Exclusion before inclusion, from final to initial vertices.

Before we discuss which of these is best, let us specify the data structures we propose to use to represent sets of subjects and the search tree. We refer the reader to Section 2.6.6 (page 66), where we discussed representing a small set of integers using a bit map. Each integer maps onto a corresponding bit: 1 if the integer is a member of the set, and 0 if it is not. This bit map represents the characteristic function of the set. We can easily put the components of a course into correspondence with integers by using their positions in a topological sort. For example, if *hmnbhvra* is the ninth element of the sorted sequence, then it corresponds to the ninth bit. As a result, each possible subset of the components corresponds to a different binary number. In the case of *ecom*, a 24-bit number will suffice. Other programmes might need more bits, but we may be confident that a 64-bit word can represent every programme that is offered.

The characteristic function also serves to describe a path in the search tree: '1,0,0,1,...' represents the decision to include the first component, to omit the second and third, to include the fourth, and so on. With it, we can explore every possible path simply by counting from zero to $(2^{24} - 1)$, or *vice versa*.

A useful property of the characteristic function is that all the subsets of S can be derived by changing various 1 bits to 0. This means that a pattern such as '$s_1 \ldots, s_n, 1,0,1,1,1$' is a superset of '$s_1 \ldots, s_n, 1,0,1,0,1$', and so on. Consequently, if '$s_1 \ldots, s_n, 1,0,1,1,1$' is not *sufficient*, neither is '$s_1 \ldots, s_n, 1,0,1,0,1$'. This further implies that, if we count *down* towards zero and discover that '$s_1, \ldots, s_n, 1,0,1,1,1$' is *insufficient*, we can immediately jump the count to '$s_1, \ldots, s_n, 0,1,1,1,1$', skipping many of the subsets of '$s_1, \ldots, s_n, 1,0,1,1,1$'. (The other subsets would involve changing the prefix that we denoted by '$s_1, \ldots, s_n$'.) Therefore, we should consider big sets before small sets and restrict our attention to the first two search orders.

Similar considerations apply to deciding whether the least significant bits should represent initial vertices or final vertices. We choose the second of the four search orders, in which they represent initial vertices. The reason is this: if a sequence such as '$s_1, \ldots, s_n, 1, 0, 1, 1, 1$' is not valid, that is because it doesn't contain enough prerequisite subjects to meet the requirements of later subjects. For example, assume that '1,0,1,1,1' fails to meet the requirements of '$s_1 \ldots, s_n$'. Then no subset of '1,0,1,1,1' can be valid, so as before, we can jump immediately from '$s_1, \ldots, s_n, 1, 0, 1, 1, 1$' to '$s_1, \ldots, s_n, 0, 1, 1, 1, 1$'.

We can incorporate both these ideas in the following heuristic:

> If a characteristic function represents a set of components that is either insufficient or invalid, split it into a prefix and a suffix, where the suffix comprises the longest possible trailing sequence of **1** bits. Decrement the prefix by one, but leave the suffix unchanged. (This process has the property that it always extends the suffix by at least one bit.)

The sets that we generate in this way, we call **feasible**, because we don't yet know whether they are sufficient and valid. On the other hand, the sets we exclude are definitely either invalid or insufficient.

This pruning heuristic proves surprisingly effective in practice. It turns out that the *ecom* programme has 512 possible variations. (Many of them exceed the minimum requirements for graduation.) The pruning heuristic reduces the 16,777,216 *possible* subsets to a mere 2,816 *feasible* subsets, of which 2,123 are sufficient, and 512 are also valid.[11] A personal computer can find all these variations in a fraction of a second and offer them for scrutiny in any desired order, for example, the cheapest first. For the reader's interest, Table 9.4 shows the cheapest variation that the programme offers. Unfortunately, this variation chooses four Level 3 subjects in Semester A and only two in Semester B.

The method we have described is easily modified to deal with a candidate's special wishes. The simplest method is to generate plans as we have described and simply discard plans that include undesired subjects or that omit desired subjects.

Dealing with subjects that the candidate has already passed is a little more complicated. A simple approach is to consider them only after the set of valid plans has been found. The costs of subjects that have already gained a sufficient mark can be subtracted from the total cost of each plan. Sorting the set of programmes in order by their remaining costs will favour those plans that make best use of subjects already studied.

Our expert system for finding the cheapest study plans is presented as Algorithm 9.1.

---

[11] Many of these plans differ only in their various combinations of *maths101*, *remathsa*, and *remathsb*. If *remathsa* and *remathsb* were omitted, the corresponding figures would be 1,186, 846, and 101.

Table 9.4 One of several *ecom* programmes that have the least cost

Code	Name	Level	Fee
acctg101	Basic Accountancy	1a	$500
javaprog	Java Programming	1a	$700
maths101	Foundation Mathematics	1a	$500
fnclmgta	Financial Management A	1b	$500
graphthy	Graph Theory	1b	$500
security	Internet Security	1b	$600
dataanal	Data Analysis	2a	$550
hmnbhvra	Human Behaviour A	2b	$500
projmgta	Project Management A	2a	$650
qtheory	Queueing Theory & Simulation	2a	$550
eeconomy	The Electronic Economy	2b	$600
sqlprgrm	SQL Programming	2b	$600
dbadmin1	Database Administration 1	3a	$850
fnclmgtb	Financial Management B	3a	$550
htmlprog	HTML & PHP Programming	3a	$600
projmgtb	Project Management B	3a	$650
intrface	Interface Design	3b	$750
services	Online Services & Cloud Comp.	3b	$700
Total			$10,850

9.3.4 BRANCH AND BOUND SEARCH

Completing a study programme in the shortest *time* can be more important than minimising cost. Assuming that a graduate will earn substantially more than an undergraduate, teaching fees must be comparatively unimportant — otherwise, why study at all? Indeed, much care went into designing the structure of the *ecom* programme to ensure that someone studying full-time (three subjects per semester) could complete the course within six semesters. Sadly, it isn't so well designed for anyone who is studying part-time or who has failed a prerequisite subject.

Designing an optimal study *schedule* is a **planning problem**. Many such problems can be solved using **search techniques**. Search problems are characterised by having a **starting state** and a **goal** or objective. To get from the starting point to the goal, the planner can choose from several possible **operators**. Each operator transforms the state. The goal is reached by passing through a series of intermediate states, called the **path**. The sequence of operators linking the states is called the **plan**.

Algorithm 9.1 Finding all valid and sufficient study plans in order of increasing cost.

1. Find a topological sort of the vertices of the prerequisite graph.
2. Use the sort to assign bits to subjects such that the bit representing each subject is more significant than all the bits representing its prerequisites.
3. Modify the **from**-trees for *sufficient* and *valid* to use bit map representation.
4. Create a bit map M representing the whole set of subjects, and a list L, initially empty.
5. Until M is zero
   - If the value of M represents a sufficient and valid set of subjects,
     a. If M includes all desired subjects and omits all undesired subjects, add M to L.
     b. Decrement M by 1.
   - Otherwise, use the pruning heuristic to decrement M, thus skipping infeasible plans.
6. For each member P of L, find the total fees needed by plan P, ignoring any subjects in which the candidate has scored sufficient marks.
7. Sort the members of L into order by increasing cost.

There is usually some criterion associated with each possible path, called its **cost**. A **search problem** is one of finding a least cost path from the starting state to the goal.

Search problems can be represented by graphs. The vertices represent **states**, and the edges represent the transitions or operators between states and can be labelled with their costs.

In the case of a GPS navigator, the starting state is your current position and direction, and the goal is your planned destination. The intermediate states are typically the road junctions along the route. The operators can include, 'Take the third exit from the roundabout', 'Continue ahead for 5 Km', and so on. Their costs are expected journey times or delays incurred making difficult turns at intersections.

To solve such problems, we divide the states into three categories:

- States that have been visited as part of the search,
- States that are adjacent to visited states, and
- The remaining unexplored states.

Initially only the starting state is considered to be visited. The states that are adjacent to visited states, we call the **frontier** or **fringe** of the search. The next step is to visit one of the **fringe states**. The key question is, which fringe state should be chosen?

To aid the choice, we may use two measures: the **accumulated cost** of reaching the fringe state from the starting state, and the estimated **future cost** of reaching the goal from the fringe state. In the case of the navigator, the accumulated cost would be the sum of the journey times along the path from the starting point to the fringe state, and the future cost would be the *estimated* time from the fringe state to the goal. The estimate could be based on the line-of-flight distance to the goal. Typically, as roads often don't follow the shortest path, this estimate might underestimate the true cost.

The obvious fringe vertex to choose is the one for which the sum of the known past cost and the estimated future cost is minimum. This sum is an estimate of the total cost from starting point to goal via that vertex. If the estimated future cost were totally accurate, this strategy would lead straight to the goal. In practice, it is rarely the case that an accurate estimate is known — the only way to have a perfect estimate is to have already solved the problem.

Now consider the problem of planning a study schedule for a candidate who only has time to study at most two subjects per semester. We can pose this as a search problem. The starting state is the set of subjects that the candidate has passed. The operators are to enrol in, and pass, up to two subjects that are offered in the current semester, thus updating the state. The goal isn't just one specific state, but *any* state that satisfies the requirements for graduation.

The obvious way to measure the accumulated cost is to find the sum of the fees for the subjects that are members of the fringe state. The obvious way to estimate the future cost is to find the minimum cost of any set of subjects that will complete the degree.

There are two objections to this scheme: First, we can only find the set of subjects needed to complete the degree by a search like the one we described in the previous section. Second, it would fail to differentiate between states: all the sums of accumulated and future costs would tend to be equal. In particular, the empty set of subjects would always be a cheapest option. This would encourage a strategy of doing nothing. We need to introduce a cost element that puts a premium on completing a degree in the shortest time. This cost is caused both by foregoing earnings in order to study and by delaying the potential increase in the candidate's earnings after graduation.

Suppose that every semester's delay costs the candidate $10,000. By including this opportunity cost, we favour plans that minimise the time needed to graduate, even if they incur higher fees. We can estimate the cost to reach the goal by counting the number of subjects still needed to graduate, dividing — in this case — by two to give the minimum number of remaining semesters, and multiplying by $10,000. The cost estimate of a plan is therefore the sum of the fees for its subjects, the opportunity cost for each semester's delay, and an estimate of the remaining opportunity cost. This cannot overestimate the true cost and will favour plans that take the least time.

Since graduation requires passes in a total of 18 subjects, we can find a lower bound on the number of subjects needed to complete the degree: we simply subtract the number of subjects in the fringe state from 18.

One problem remains: although *we* know that 18 subjects are needed, it is hard to deduce that from the information available to the planner. The **from**-tree doesn't make this explicit; indeed, *we* can only deduce it after realising that the eight Level 3 subjects are a subset of the 23 subjects from which 18 must be chosen. Luckily, it makes no difference whether the total number is 18 or 20 or any other number. Subtracting the number of subjects completed from any number still puts a premium on completing subjects as quickly as possible. In fact, the effect is the same if we subtract the number of subjects in each fringe state from zero. Accordingly, we assign each subject in a fringe state a cost equal to its fee, *minus* $10,000. In effect, we no longer measure the cost to the candidate, but the complement of its *value*.

To implement such a strategy, the search algorithm must keep track of the set of fringe states in order by total estimated cost.[12] Each fringe state is represented by its estimated cost, the semester from which the next set of subjects must be chosen, and its path: a list of the subjects in the state.

The algorithm picks the first, and cheapest, fringe state, removes it from the fringe, and replaces it with all its possible successor states. Each successor is found by appending zero, one, or two subjects to the parent state's path. The chosen subjects must be offered in the right semester and have prerequisites that are already members of the parent state's path. The cost of each successor's path is estimated, and the semester is changed from *A* to *B* or *vice versa*. The algorithm then iterates using the extended fringe.

As soon as the subjects in the path of the cheapest fringe state satisfy the graduation requirements, an optimum path has been discovered. This becomes the candidate's subject schedule.

We can prove that the first plan to satisfy the goal is optimal, provided that estimated future costs are lower bounds on true costs. When the first goal state has been found, its future estimate must be zero, and its total cost must equal its true cost. Since the goal state chosen was the cheapest state in the fringe, all the remaining states in the fringe have estimates that are at least as great, and their true costs may be even greater. (That is why we mustn't allow the estimates to be overestimates; it would then be possible for an apparently more costly fringe state to prove even cheaper.)

This algorithm is called **Branch and Bound** (see Algorithm 9.2).

Unfortunately, in the form we have described it, *Branch and Bound* proves to take so much space to store the fringe, and so long to explore it, as to be practically useless. Despite this, its combination with the fast pruned search we described earlier works beautifully. First, we use the pruned search to discover a list of plans, then we use *Branch and Bound* to schedule them.

There are several reasons why *Branch and Bound* works much better in this case.

- It doesn't need to consider subject fees. The total fees for a plan are known; the order in which they are scheduled doesn't affect this cost. The only cost to be

[12] The heap representation of a priority queue is a good data structure for this purpose. (See Section 2.7.3.)

Algorithm 9.2 Branch and Bound search.

1. Initialise the fringe to include only the starting state, its path (empty), its accumulated cost (zero), and a lower bound on its future cost.
2. Remove the state S with the lowest estimated total cost from the fringe.
   (As a further refinement, we may order plans with equal *total* cost in order of *future* cost, thus favouring plans that appear to be closer to the goal.)
3. If S satisfies the goal, exit the algorithm. Its path is the optimum plan, and its cost is the true cost of reaching the goal.
4. Otherwise, for each operator O applicable to S,
   a. Find the state T resulting from applying O to S.
   b. Append O to the path leading to S to give the path leading to T.
   c. Add the cost of O to the accumulated cost of S, giving the accumulated cost of T.
   d. Estimate a lower bound on the future cost of reaching the goal from state T.
   e. Add T, its path, and its accumulated and estimated future costs to the fringe.
5. Repeat the algorithm from Step 2.

considered is the opportunity cost. Its objective is therefore to find the minimum number of semesters needed to graduate.

- Provided that the plans have already been sorted by their total fees, the *first* schedule that takes only S semesters must be the *cheapest* schedule that takes only S semesters.
- There is never any point in deferring an available subject. Consequently, this reduces the number of sensible schedules. For example, if a candidate is prepared to study up to three subjects per semester and four subjects are available, then only *four* options are sensible (which subject to omit) out of a potential 15 (all subsets of four subjects that contain three subjects or less).
- As soon as the cheapest path remaining in the fringe can take no fewer semesters than the best schedule so far, *Branch and Bound* can reject it.
- Already knowing exactly which subjects are to be studied, we can estimate the fewest number of semesters that a schedule can take. If this is no fewer than the number of semesters in the cheapest schedule found so far, it cannot be optimal, and *Branch and Bound* doesn't need to be invoked.

The more accurately we estimate the remaining number of semesters, the more quickly *Branch and Bound* will find an optimum subject schedule. An obvious estimate is to divide the remaining number of subjects by the number of subjects that a candidate is willing to study per semester. As it happens, we can improve on this.

Suppose a candidate intends to study three subjects per semester and hasn't yet studied any. If the study plan contains nine subjects in Semester A and nine subjects in Semester B, it can be completed in six semesters. Unfortunately, if the plan contains ten subjects in Semester A but only eight in Semester B, it will take at least seven semesters. But when the situation is reversed, and the plan contains ten subjects in Semester B but only eight in Semester A, it will take *eight* semesters, leaving the candidate idle in the seventh semester.

From this, it might seem that only balanced plans can be useful, but the situation is different for part-time candidates. A candidate willing to study only two subjects per semester can graduate in nine semesters, provided that ten of the subjects fall in Semester A and only eight fall in Semester B.[13] Table 9.5 shows the cheapest of ten such schedules.

Algorithm 9.3 summarises how the expert system works. When *Branch and Bound* is used in this more directed way, it is much more effective, and optimum schedules are found in a fraction of a second.

Algorithm 9.3 An expert system for scheduling tailored study programmes.

1. Use Algorithm 9.1 to form a list L of acceptable study plans in ascending order by total fee.
2. Set the value of *Shortest* to an impossibly large value.
3. For each element P of L,
   - Use Algorithm 9.2 to schedule P.
   - If at any stage of Algorithm 9.2, the estimated number of semesters needed by the currently most promising state is *no fewer* than *Shortest*, abandon the search.
   - Otherwise, set the value of *Shortest* to the number of semesters required by P, and set *Best* to the path assigned to P.
4. Present *Best* as the optimum schedule.

9.3.5 SEARCH AND PLANNING

Branch and Bound exploits the **Dynamic Programming Principle**:

> *The best way to get from A to C via B consists of the best way to get from A to B followed by the best way to get from B to C.*

There are two important points here: First, if we have found a best path from A to B, we have no need to consider alternative paths from A to B when extending the path to

[13] The same 10:8 distribution also proves optimal for an exceptionally gifted candidate willing to take four subjects per semester, who can then graduate in only five semesters.

Table 9.5 The least cost schedule for a part-time *ecom* candidate willing to study two subjects per semester. The total cost includes an opportunity cost of $10,000 per semester

Code	Name	Level	Fee
acctg101	Basic Accountancy	1a	$500
maths101	Foundation Mathematics	1a	$500
fnclmgta	Financial Management A	1b	$500
graphthy	Graph Theory	1b	$500
javaprog	Java Programming	1a	$700
dataanal	Data Analysis	2a	$550
security	Internet Security	1b	$600
sqlprgrm	SQL Programming	2b	$600
hmnbhvra	Human Behaviour A	2a	$500
projmgta	Project Management A	2a	$650
eeconomy	The Electronic Economy	2b	$600
marketec	Marketing for e-Commerce	2b	$650
dbadmin1	Database Administration 1	3a	$850
fnclmgtb	Financial Management B	3a	$550
services	Online Services & Cloud Comp.	3b	$700
socmedia	Exploiting Social Media	3b	$650
htmlprog	HTML & PHP Programming	3a	$600
projmgtb	Project Management B	3a	$650
Total			$100,850

C. Second, the principle does not specify the choice of *B*. So although we may have to consider many intermediate states, *B*, we only have to remember one path to each.

Branch and Bound makes passive use of this principle, in that sub-optimal paths are never favoured over optimal paths. A variant of *Branch and Bound*, called **A* Search** (*A-star Search*), uses it more aggressively, by checking if a newly developed path reaches an already visited state. As an example, consider the case of a part-time *ecom* candidate studying two subjects per semester. One possible schedule begins with the sequence [*maths101*, *graphthy*, *javaprog*, *security*, *acctg101*, *hmnbhvra*, *fnclmgta*, *marketec*], another with the sequence [*acctg101*, *hmnbhvra*, *fnclmgta*, *marketec*, *maths101*, *graphthy*, *javaprog*, *security*]. After four semesters, both these sequences reach the same state for the same cost, so there is little point in extending both. As it turns out, none of the nine-semester schedules begins this way, so the true future cost of the first sequence turns out to be more than its estimate. Therefore

Branch and Bound will turn to the second sequence, which still has its original underestimate. This must prove a waste of effort. The best extension of the second sequence will cost exactly the same as the best extension of the first. On the other hand, *A* Search* will avoid the waste because it actively checks each new state to see if it has been previously reached, and, if so, any new paths that lead to it are ignored.

Branch and Bound can be applied to many problems, but, as we have seen, it works best when we have a sharp estimate of future costs. If it has only a poor estimate, it can consume too much memory and too much time to be useful.

In some cases, there is no way to estimate future cost, in which case, a search is called a **blind search**. If an estimate is possible, it is called an **informed search**.

- **Beam Search** is a variant of *Branch and Bound* in which only the best w states are added to the fringe at each step, where w is the *width* of the beam. It therefore needs less storage and less time than *Branch and Bound*. When a solution is found, it will be good, but it may be less than optimal.
- **Best-First Search** is a blind variant of *Branch and Bound* in which future costs are all assumed to be equal. When a solution is found, it is the cheapest, but, for reasons recently discussed, such a search is rarely practical.
- **Greedy Search** is a variant of *Branch and Bound* in which all *past* costs are ignored. The intention is to find a solution as quickly as possible. The solution isn't necessarily optimal.
- **Breadth-First Search** is a variant of *Best-First Search* in which all operators have unit cost. It finds a solution that has the fewest number of edges from the start state to the goal.
- **Depth-First Search** is a blind variant of *Best-First Search* in which all operators have a negative cost. It mistakes activity for productivity. Its compensating feature, as explained in Section 2.7.4 (page 70), is that the fringe only contains the successors of the states already visited. It is therefore very economical in its use of storage.
- **Iterative Deepening** is a variant of *Depth-First Search* in which the search is truncated at a certain depth. Successive searches are conducted at depths of $1, 2, 3, \ldots$ until the goal is reached. Perhaps surprisingly, the early iterations don't add greatly to the time taken to complete the search; they are usually easily outweighed by the final one. If each edge has unit cost, the result is the same as for *Breadth-First Search*. If edges do not have equal costs, *Iterative Deepening* may return the solution with the fewest edges rather than the least cost.
- **Two-Way Search** is a variant of any of the above searches in which it is possible to search both forward from the start state and backwards from the (unique) goal. Once a state from the forward search is found among the states visited in the backward search (or *vice versa*), a solution has been found.

We have to admit that there are planning problems where none of these search techniques is adequate. A classic example of this kind is changing the battery in a TV remote controller. At the beginning of the process, the battery cover is on, and the remote contains a battery. At the end of the process, the battery cover is on, and the remote contains a battery. The first step, to remove the cover, is apparently a step *away* from the goal state, which a search method would be reluctant to do and would only stumble upon after an exhaustive search.

The point is that only one operator, *Insert New Battery*, can fit the remote with its new battery. This step has preconditions, *Cover Off* and *Battery Slot Empty*. Only one operator can achieve the state *Battery Slot Empty*, and that is *Remove Battery*. The precondition for this step is *Cover Off*. Having decided on the crucial *Insert New Battery* step, the planner must then formulate two sub-plans: a pre-plan (*Remove Cover*, *Remove Battery*) and a post-plan (*Replace Cover*). Thus, in contrast to the search methods we have discussed, the first thing such a **planner** decides to do (*Insert New Battery*) may not be the thing it decides to do first (*Remove Cover*).

The study programme scheduling problem is poorly suited to this planning approach: the goal is to pass all the subjects in the chosen plan. The operators are to study, and pass, subjects belonging to the plan. Their preconditions are the prerequisite subjects. Their effects are to add subjects to the goal state. A planner might (arbitrarily) first choose to add *projmgta* to the schedule. It would then need to formulate a pre-plan (comprising *fnclmgta*, *graphthy*, *acctg101*, and *maths101*) and a post-plan (comprising all the remaining subjects). Unfortunately, it is only after the completion of the schedule that its duration can be assessed, so this particular approach is *blind* or *uninformed*, and will do no more than find a possible topological sort.

A detailed discussion of planning methods is beyond the scope of this book.

9.3.6 FUZZY LOGIC

Fuzzy logic is a completely different way of capturing expert knowledge: it is characterised by situations in which several factors interact, and a human expert can specify what should happen in certain clear-cut situations. A fuzzy logic system is then able to interpolate between these exemplars, to deal with intermediate, less well-defined situations.

Suppose that the marks given to a candidate in a certain subject are primarily determined by examination, but the examiner may give a bonus of up to 5% to a borderline result if the candidate has scored good marks for assignment work. At the same time, the examiner may be suspicious about a candidate who has scored very well in one area without scoring well in the other. The examiner may therefore use the following rules of thumb:

- If the candidate's examination result is borderline and the candidate's assignment work is good, then bonus is maximum.
- If an assignment mark is very good but the examination mark is rather poor or the assignment mark is rather poor and the examination mark is very good, investigate the possibility of cheating.

It is not for us to question the wisdom of these rules, but to implement them.

The rules have the same form as logical statements, but the meanings of their terms are imprecise: What is meant by 'borderline'? Is it a mark of 50% exactly, or is it a mark of *about* 50%? What is meant by 'good' or 'poor'? What is meant by 'very good' or 'rather poor'?

All these concepts are **fuzzy**; they can be true, false, or *somewhat true*. We give such concepts a **degree of truth**: a number between zero and one. Zero represents 'definitely false', 1 represents 'definitely true', and intermediate values represent 'somewhat true but somewhat false'. We denote the degree of truth of proposition X as μX, where μX is called the **membership function** of X.

We then need operations that will implement the logical operations of $\wedge$, $\vee$, and $\neg$ in a sensible way. We would like these operations to work just like the classical logical operators when $\mu X = 0$ (false) or $\mu X = 1$ (true).

One option is to treat μX and μY as independent probabilities:

$$\begin{aligned}\mu(\neg X) &= 1 - \mu X\\ \mu(X \vee Y) &= \mu X + \mu Y - \mu X \times \mu Y\\ \mu(X \wedge Y) &= \mu X \times \mu Y.\end{aligned}$$

Predictably, this choice gives silly results in the case $Y = X$, where setting $\mu X = 0.5$ gives $\mu(X \wedge X) = 0.25$ and $\mu(X \vee X) = 0.75$.

An alternative is to treat μX and μY as so-called disjoint probabilities:

$$\begin{aligned}\mu(\neg X) &= 1 - \mu X\\ \mu(X \vee Y) &= \mu X + \mu Y\\ \mu(X \wedge Y) &= 1 - ((1 - \mu X) + (1 - \mu Y))\end{aligned}$$

where $\mu(X \wedge Y)$, which would always be zero if X and Y were truly disjoint, is instead derived from the equation for $\mu(X \vee Y)$ using De Morgan's law.

Again, this choice gives silly results in the case $Y = X$. Setting $\mu X = 0.5$ gives $\mu(X \wedge X) = 0$ and $\mu(X \vee X) = 1$.

As a result of these problems, most fuzzy logic systems use **min/max logic**:

$$\begin{aligned}\mu(\neg X) &= 1 - \mu X\\ \mu(X \vee Y) &= \max(\mu X, \mu Y)\\ \mu(X \wedge Y) &= \min(\mu X, \mu Y)\end{aligned}$$

which has the advantages of working perfectly in the classical case, being simple to implement, and not giving silly results when $Y = X$. The only points of difficulty are that, when $\mu X = 0.5$, $\mu(X \wedge \neg X) = 0.5$, and $\mu(X \vee \neg X) = 0.5$, whereas classical logic would give the values 0 and 1 respectively. This is similar to asking, 'Is the result good, or not good?' and getting the answer, 'Well, you can't say it's good, and you can't say it's not'.

As the reader can see, the mathematical foundations of fuzzy logic are shaky, and a proper statistical approach would be much more sound. But this hasn't stopped fuzzy logic being successful, partly because its applications don't rely much on the mathematics being consistent, and partly because the statistical information that a mathematically strict approach needs is hardly ever available.

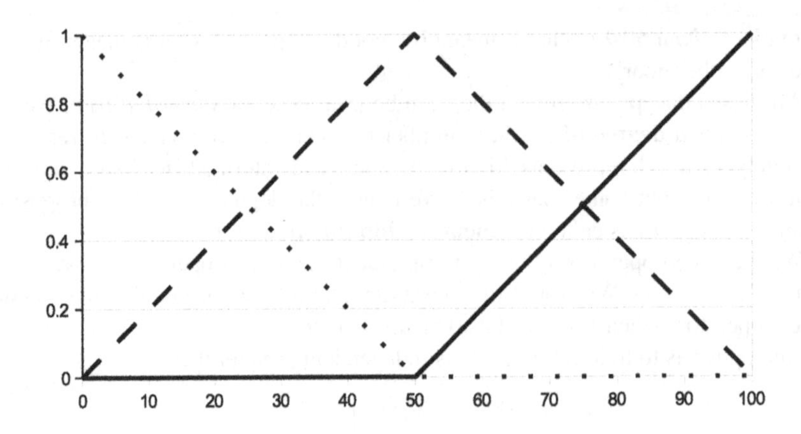

FIGURE 9.3

Possible membership functions for *'Poor'* (dotted line), *'Borderline'* (dashed line), and *'Good'* (continuous line), as a function of *Mark*. *'Poor'* peaks at 0%, *'Borderline'* peaks at 50%, and *'Good'* peaks at 100%.

Applying fuzzy logic requires four steps:

1. **Fuzzification**: converting **crisp values**, such as examination marks, into membership functions of fuzzy sets, such as '*Good*', '*Borderline*', or '*Poor*'.
2. **Fuzzy inference**: applying the rules of fuzzy logic to establish the degree of truth of possible consequents.
3. **Aggregation**: determining how the degree of truth of a potential result, such as '*Bonus* is maximum', varies with its crisp value.
4. **Defuzzification**: choosing a crisp value for each result, e.g., *Bonus* = 4. There are two distinct cases: the result is a continuous quantity, as in the case of *Bonus*, or a discrete value, such as *Investigate* = *Yes* or *Investigate* = *No*.

9.3.6.1 Fuzzification

Fuzzification begins with choosing suitable membership functions for the fuzzy sets. Figure 9.3 shows a typical choice, a series of straight line segments. (Other choices are possible, e.g., sigmoid curves.) Every possible mark is a member of at least one fuzzy set. From the figure, we can see that when *Mark* = 80, $\mu(Poor) = 0$, $\mu(Borderline) = 0.4$, $\mu(Good) = 0.6$.

We also need to quantify the terms 'very' and 'rather'. These linguistic hedges can be interpreted as follows:

$$\mu(very\ X) = (\mu X)^2$$

$$\mu(rather\ X) = \sqrt{(\mu X)}.$$

With these interpretations, a mark of 80% has $\mu(very\ Good) = 0.36$, but $\mu(rather\ Good) = 0.775$.

9.3.6.2 Fuzzy inference

We now need to express the informal statements as fuzzy logic statements:

> 'If the candidate's examination result is borderline and the candidate's assignment work is good, then bonus is maximum'.

$$\min(\mu(Examination = Borderline), \mu(Assignment = Good)) \Rightarrow (Bonus = Maximum) \quad (9.3.3)$$

To balance this inference, we derive its complement using De Morgan's Theorem, otherwise *Bonus* = *Maximum* would be the only conclusion we could draw:

$$\max(1 - \mu(Examination = Borderline), 1 - \mu(Assignment = Good)) \Rightarrow (Bonus = None). \quad (9.3.4)$$

In the particular case of a candidate who has scored 47% in the examination and 80% for assignments, $\mu(Examination = Borderline) = 0.94$, and $\mu(Assignment = Good) = 0.6$. As a result,

$$\min(\mu(Examination = Borderline), \mu(Assignment = Good)) = 0.6, \quad (9.3.5)$$

and,

$$\max(1 - \mu(Examination = Borderline), 1 - \mu(Assignment = Good)) = 0.4. \quad (9.3.6)$$

We note that the value of the expression 9.3.6 is one minus the value of expression 9.3.5, as we ought to expect.

> 'If an assignment mark is very good but the examination mark is rather poor, investigate the possibility of cheating'.

$$\min(\mu(Assignment = Good)^2, \sqrt{\mu(Examination = Poor)}) \Rightarrow (Investigate = Yes) \quad (9.3.7)$$

Again, this inference should be given a complement:

$$\max(1 - \mu(Assignment = Good)^2, 1 - \sqrt{\mu(Examination = Poor)}) \Rightarrow (Investigate = No). \quad (9.3.8)$$

Consider the case of a candidate who has scored 90% for assignments and only 30% in examinations, so $\mu(Assignment = Good) = 0.8$ and $\mu(Examination = Poor) = 0.4$. The left-hand side of inference 9.3.7 evaluates to $\min(0.64, 0.632) = 0.632$. Not surprisingly, the value of the left-hand side of inference 9.3.8 is $0.368 = (1 - 0.632)$.

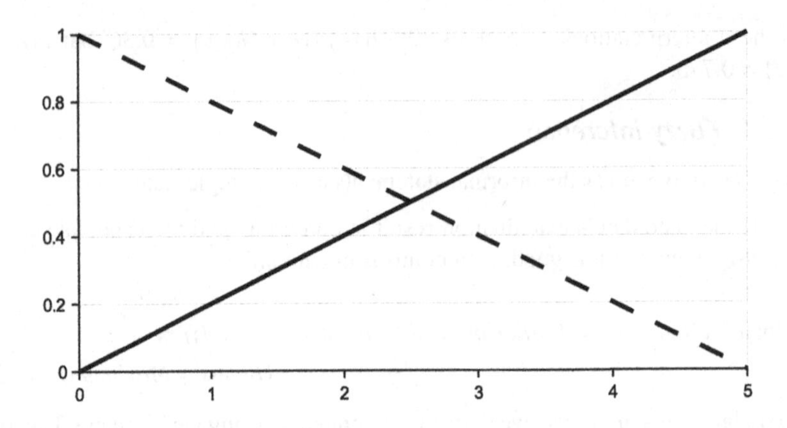

FIGURE 9.4

Possible membership functions for *'None'* (dotted line) and *'Maximum'* (continuous line) as a function of *Bonus* as *Bonus* ranges from zero to five.

Finally, we have the rule,

> 'If the assignment mark is rather poor and the examination mark is very good, investigate the possibility of cheating'.

$$\min(\sqrt{\mu(Assignment = Poor)},\ \mu(Examination = Good)^2) \Rightarrow (Investigate = Yes) \quad (9.3.9)$$

with complement,

$$\max(1 - \sqrt{\mu(Assignment = Poor)},\ 1 - \mu(Examination = Good)^2) \Rightarrow (Investigate = No). \quad (9.3.10)$$

Consider again the case of the candidate who scored 90% for assignments and only 30% in examinations, so $\mu(Assignment = Poor) = 0$ and $\mu(Examination = Good) = 0$. The left-hand side of inference 9.3.9 evaluates to $\min(0,0) = 0$, and the left-hand side of inference 9.3.10 evaluates to $\max(1,1) = 1$.

9.3.6.3 Aggregation

Aggregation is the act of determining how the membership function of an output, such as 'bonus mark', varies with its crisp value. We now have several rules that are satisfied to various extents. Equation 9.3.5 gives a weighting of 0.6 to *Bonus* = *Maximum* and equation 9.3.6 gives a weighting of 0.4 to *Bonus* = *None*. What crisp value should we assign to *Bonus*?

To answer this question, we must define membership functions for *Bonus* = *Maximum* and *Bonus* = *None*. A possible choice is shown in Figure 9.4. We see that setting *Bonus* = 3 gives $\mu(Maximum = 3) = 0.6$ and $\mu(None = 3) = 0.4$ exactly. So the crisp value of *Bonus* is 3. Therefore, our hypothetical candidate would receive a final mark of 50%.

The rules for *Investigate* really concern two different kinds of investigation, so it would be possible to consider rule 9.3.7 and rule 9.3.10 as applying to two different conclusions: *Investigate Assignment* and *Investigate Exam*. On the other hand, if we choose to combine them into a *single* conclusion, then rules 9.3.7 and 9.3.9 should be combined using $\vee$, and we should consider the maximum truth value of the two. Consequently, because of De Morgan's laws, we must then combine rules 9.3.8 and 9.3.10 using $\wedge$ and take the *minimum* of the two.

9.3.6.4 Defuzzification

Defuzzification is the act of turning fuzzy inferences into crisp values.

If we have several rules of the form $P_1 \Rightarrow Q, P_2 \Rightarrow Q, \ldots$ sharing the same consequent Q, and we know the values of $\mu P_1, \mu P_2, \ldots$, what crisp value should we assign to Q? Ideally, the value that makes $\mu Q = \mu P_1 = \mu P_2 = \cdots$, as were able to do for *Bonus*.[14] But what if no value of Q satisfies all the constraints? One answer is to find values of $Q_1, Q_2, \ldots$ that satisfy each constraint separately, then take their average.[15]

Although *Bonus* can be assigned a continuous set of values — albeit rounded to an integer in practice, *Investigate* must be either *Yes* or *No*. The left-hand side of rule 9.3.7 yields a degree of membership of 0.632 to the conclusion *Investigate* = *Yes*, whereas rule 9.3.10 yields a degree of membership of 1 to the conclusion *Investigate* = *No*. These conflicting inferences need to be reconciled: Should the examiner investigate or not?

Remembering that rules 9.3.8 and 9.3.10 should be combined using $\wedge$, we must take the *minimum* of their two truth values. After doing so, we have $\mu(Yes) = 0.632$ and $\mu(No) = 0.368$. As a result, we conclude that there is a stronger basis for investigating than not. Typically, faced with several discrete alternatives, we should choose the one with the strongest basis.

9.3.6.5 Interpretation

As we said earlier, fuzzy logic is a way of interpolating between clear-cut situations. Figure 9.5 lets us visualise *Bonus* as a function of *Assignment* and *Examination*. The values of *Bonus* lie on the upper surface of a skewed pyramid and on the horizontal plane in the foreground. We can see that *Bonus* is greatest for *Examination* = 50% and *Assignment* = 100%, and falls to zero everywhere for *Assignment* $\leq$ 50. This continuous surface has been interpolated from a few points by the rules of fuzzy inference. Although this surface is relatively simple, fuzzy logic is often applied to

---

[14] There was no problem here because rule 9.3.3 is the logical complement of rule 9.3.4.

[15] There are several other traditional ways of obtaining a crisp value, but we shan't discuss them here.

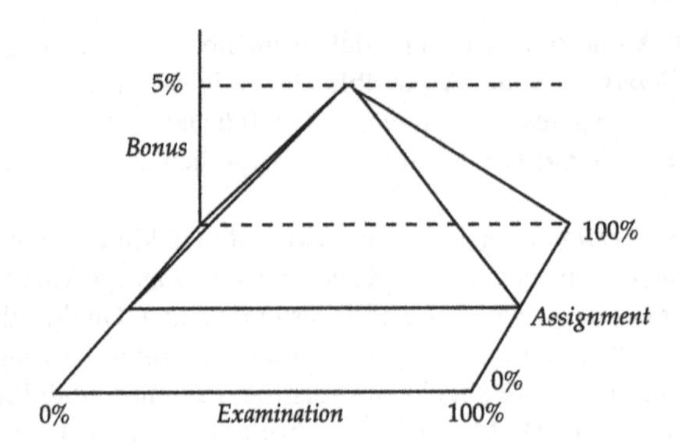

FIGURE 9.5

Bonus as a function of *Assignment* and *Examination*, drawn as a two-dimensional projection of a three-dimensional surface. The vertical axis represents the value of *Bonus*, the horizontal axis represents the value of *Examination*, and the diagonal axis represents the value of *Assignment*.

problems with many more variables, where the resulting multi-dimensional surface is much harder to visualise.

Fuzzy logic is used in many consumer products. The fuzzy logic controller of an automatic gearbox in a car may take into account the accelerator position, engine speed, road speed, manifold pressure, and brake pressure to determine the best gear ratio. Some driving situations are clear: engine speed low, road speed low, maximum accelerator setting, low negative manifold pressure — change to a lower gear to improve acceleration. Engine speed low, road speed high, high negative manifold pressure, foot on brake — again, change to a lower gear, this time to improve deceleration. Other situations are less clear, and it isn't easy to create a set of rules that will always make good choices. Fuzzy logic serves as a way of deriving continuous functions that join scattered sets of well-defined points — for which min/max logic usually proves adequate.

9.4 SUMMARY

We have discussed various means by which business rules can be expressed other than by embedding them in computer procedures. A collection of rules can be compactly expressed as a decision table. Such tables not only document what happens in a given situation, but also what situations can cause a given thing to happen. There are various software tools that can either interpret decision tables at run-time or translate tables into program code.

There are also ways of expressing knowledge *declaratively* in such a way that an inference engine can generate an appropriate sequence of actions to solve various related problems. Such knowledge is usually expressed in the form of a tree or graph.

We have seen that, armed only with a set of degree requirements and the prerequisite relations between subjects, an expert system can promptly offer meaningful advice to a candidate of any given study programme.

Offering the best advice is a theoretically intractable problem. It was solved in two stages: first by finding feasible sets of subjects, using an effective method of pruning, and second, by scheduling the subjects using a well-established and effective search method combined with an effective heuristic. The first stage greatly improved the efficiency of the second. Until finding the right tools, there was no guarantee that a workable algorithm existed. *It would be very risky to start implementing an online enrolment system until the intractable component at its core had been solved.* Such uncertainties should always be resolved before serious money is spent.

Finally, we have demonstrated a simple application of fuzzy logic to the adjustment of examination marks and the highlighting of inconsistent assessments. This relied on an expert (the examiner) stating what should happen in certain clear-cut situations, but didn't require the expert to offer a mathematical formula that would deal well with *every* situation.

9.5 FURTHER READING

Decision tables are unfashionable at present. This is a pity, because they are easily prepared and edited by spreadsheet programs. If you are interested in how a sophisticated processor can convert tables into program code, you might like to read the author's online document, 'Experience with the COPE Decision Table Processor'.

Many of the techniques in this chapter belong to the discipline called artificial intelligence (AI), an area of constant research. Roughly speaking, artificial intelligence researchers can be divided into 'neats' and 'scruffies'. Neats like techniques soundly based in logic and mathematics; scruffies are willing to use anything that works. In this respect, fuzzy logic has to be regarded as a scruffy technology. Although fuzzy inference *could* be dealt with more soundly using probability theory, it is usually impossible to know all the conditional probabilities needed to do it correctly. (Indeed, the whole idea of adjusting examination marks lacks any formal logical basis.) On the other hand, finding the optimum study programme for a candidate is a well-defined problem with a well-defined optimum, and therefore 'neat'.

In preparing the examples in this chapter, I have consulted three textbooks: The first, 'Artificial Intelligence: A Modern Approach', by Stuart Russell and Peter Norvig (3rd ed., 2009, ISBN: 0-136-04259-7), is strong on theory. Russell and Norvig may be classified as 'neats' and have no room in their book for fuzzy logic. Fuzzy logic (among many other things) is discussed in Michael Negnevitsky's 'Artificial Intelligence: A Guide to Intelligent Systems' (3rd ed., 2011, ISBN: 1-40822-574-3), and in 'Expert Systems: Principles and Programming' by Joseph C. Giarratano and Gary D. Riley (4th ed., 2004, ISBN: 0-534-95053-1), which also discusses

CLIPS. The rete algorithm used by such production systems was first described by Charles Forgy in his 1979 PhD thesis, 'On the Efficient Implementation of Production Systems' (Carnegie-Mellon University).

Prolog was first implemented around 1972 by Alain Colmerauer and Philippe Roussel. 'Prolog Programming for Artificial Intelligence' by Ivan Bratko (4th ed., 2012, ISBN: 0-321-41746-1) provides an excellent introduction to the *Prolog* language.

9.6 EXERCISES

1. If 50% in *remathsa* or 40% in *maths101* is enough to study *graphthy*, and 50% in *remathsa* is sufficient to study *remathsb*, it only seems fair that a candidate who scored 40% in *maths101* should be allowed to study *remathsb.* Assuming this change, express the **needs** tree for *remathsb* in the same style as the examples on page 308.
2. The decision table of Table 9.3 contains 23 rules, one of which is an *else* rule. If this table were implemented using structured (**if** ... **else**) code, how many branches would be needed to express the *else* rule?
You may assume that the conditions will be tested in order from left to right, and that *Subject* is already constrained to have one of the values listed in the first column.
Suppose, assuming that the prerequisites for subjects are the same for all programmes, that a giant two-dimensional matrix were prepared for all 500 subjects. Assuming an average of one rule per subject — even though subjects with alternative prerequisites might need two or more, and assuming that minimum marks are represented by one-byte integers, how much storage would it need?
Can you think of a more compact way of storing the table?
Is a giant matrix a good idea?
3. On page 313, we suggested that the programme planner should omit subjects that a candidate wishes to avoid from the search. There are at least three ways this can be achieved:
   - By deleting the unwanted subjects from the search tree.
   - By ignoring the candidate's wishes until a solution is generated, then discarding any programmes that happen to include unwanted subjects.
   - By ignoring any feasible bit maps that include the unwanted subjects.

   Comment on their relative merits.
4. The National University of Technology and Science owns many lecture rooms, both big and small. The number of seats in a lecture room must exceed or equal the *Quota* of every subject taught within it. In addition, each subject must be allocated a certain number of lecture hours per week.
Subjects *clash* if at least one candidate is enrolled in both of them. It is necessary to find an allocation of subjects to lecture rooms in such a way that the lecture rooms are big enough to hold them, and no room is busy outside the hours of 9

am–5 pm from Monday to Friday. In addition, no pair of clashing subjects may have lectures at the same time.
Suggest a heuristic that could be used to allocate lecture rooms to subjects.
Suggest a heuristic that could be used to schedule lecture times without causing clashes.
Can you foresee situations where these approaches might fail? What could be done in such cases?
There are two approaches we could take in designing an expert system to schedule the use of lecture rooms:

- Allocate lectures to rooms, then schedule lecture times, or,
- Schedule lecture times, then allocate lectures to rooms.

Can you see ways in which the two sub-problems might interact?
Which sub-problem should be tackled first?

5. Consider the fuzzy logic system of Section 9.3.6. Calculate how many bonus marks would be given to a candidate who scored 45% in the examination and 90% for assignments.
The lecturer concerned complains that the intention was for such a candidate to receive a pass mark of exactly 50%. In other respects, the system functions exactly as intended. What can be done to fix the problem?

CHAPTER

System dynamics

10

CHAPTER CONTENTS

INTRODUCTION

The *NUTS Academic Records System* that we have considered records external events, but only influences them indirectly. Among other things, its database helps candidates make decisions about their study programmes. If the system were enhanced to include *advising* candidates, we would need to be careful to make sure it chose wisely, by comparing its behaviour with the behaviour of experts. But we should not wait to see if the system performed well in practice, in case it proved to be a disaster. Instead, we should first **simulate** its behaviour to see how it would have performed on historical data. This simulation would best be done at an early stage of design, as feedback from the simulation might affect the algorithms chosen to be implemented, or the external interfaces, or even whether our plans should go ahead.

In this chapter, we present two examples where simple simulations (using only a spreadsheet program) might save a system from failure.

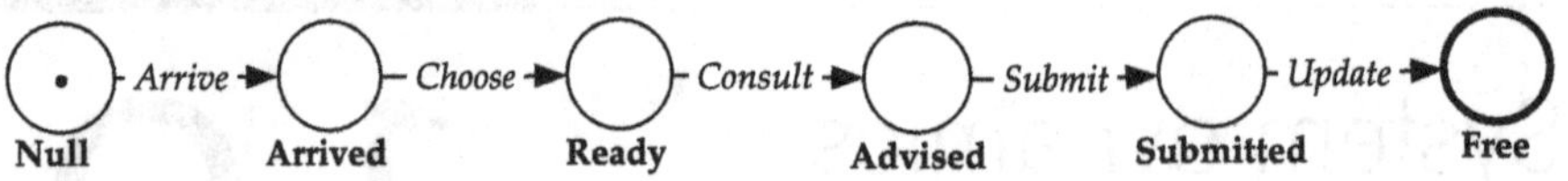

Arrive *Choose* *Consult* *Submit* *Update*

Null **Arrived** **Ready** **Advised** **Submitted** **Free**

FIGURE 10.1

States of *Candidates* enrolling for programmes.

10.1 QUEUEING MODELS

Queues arise whenever resources have to be shared. Consider the immediate plans for the NUTS Academic Records System, in which it is proposed that during the enrolment period, candidates will complete enrolment forms and then have them checked by programme advisers. There are four kinds of programme adviser, having differing expertise: *Engineering*, *Information Technology*, *Biological Sciences*, and *Physical Sciences*. Candidates must consult the correct kind of adviser. They will then take their completed forms to terminal operators in the Enrolment Office, who will record their choices in the database. The resources that candidates will share include the programme advisers, the terminal operators, the database, and some desks on which they fill in their forms. After completing a form, each candidate will need to queue to have it approved by a programme adviser, then queue again to submit the completed form to a terminal operator.

We assume that 10,000 active candidates will enrol during one 40-hour week, an average of 250 per hour. How many programme advisers and enrolment staff will be needed to make sure that things run smoothly? Will the database system handle the load, or will it prove to be a bottleneck?

We may draw a state diagram for *Candidates*, as in Figure 10.1. The *Null* state means that the candidate has not yet entered the system. The *Arrived* state means that the candidate has collected an enrolment form and is waiting for a desk at which to write. The *Ready* state means that the candidate has chosen a study programme, completed the form, and is waiting to speak to a suitable programme adviser. The *Advised* state means the candidate has received advice and is waiting for a terminal operator to submit the form to the database. The *Submitted* state means that the operator has finished entering the data, and the candidate is waiting to be sure that the update was successful. The *Free* state means that the candidate is free to leave.

In practice, some other transitions might occur, not shown in the figure. For example, an adviser might tell a candidate to return to a desk to correct a form, or a terminal operator may inform the candidate to make a different choice because a chosen subject has already filled its quota.

We could draw similar — but simpler — state diagrams for *Desks*, *Advisers*, *Operators*, and the *Database*. To show that the resources share certain events, we

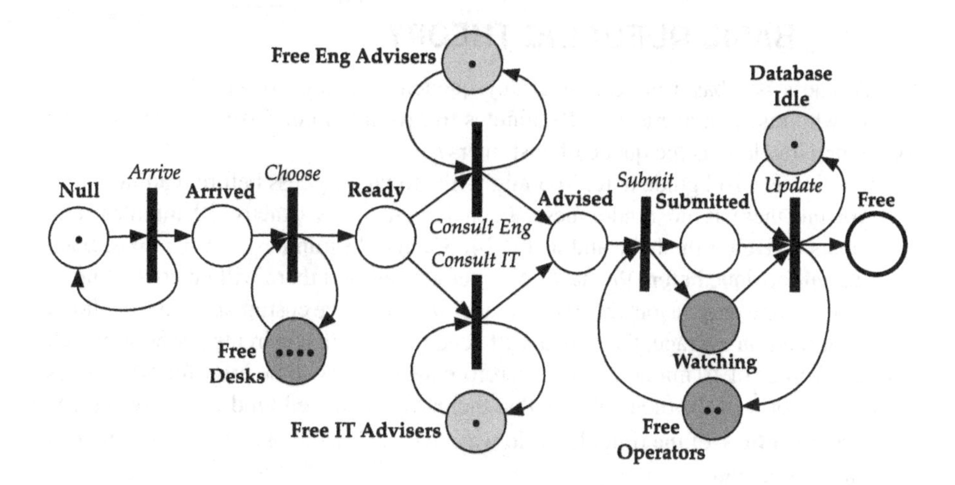

FIGURE 10.2

A bi-partite graph showing the states of *Candidates* (white circles), *Desks* (dark grey), *Advisers* (light grey), *Terminal Operators* (also dark grey), and the *Database* (also light grey) in an enrolment centre. Circles represent places (states), and lines represent transitions. For simplicity, the diagram shows only two kinds of programme advisers, although there are really four. The net is shown containing tokens in an initial state. The cycle between *Null* and *Arrive* is capable of putting an unlimited number of candidate tokens into the *Arrived* state. From there, provided at least one *Free Desk* is available, the candidate token can pass through the *Choose* transition to become *Ready*. A *Ready* token can then be absorbed by either the *Consult Eng* or *Consult IT* transition non-deterministically, to become *Advised*. Neither transition can fire unless there is a least one *Free Adviser*. Subsequently, the candidate must find a *Free Operator* to *Submit* a form and, once the database has been updated, is then *Free* to leave. Only one *Update* may proceed at a time and must wait for the database to be *Idle*. An operator in the *Watching* state is waiting for the update to succeed.

may then combine these state diagrams to form a bi-partite graph (Figure 10.2), similar in style to the Petri net of Figure 3.21 (page 104). In the figure, the circles (places) represent states where a resource is inactive and waiting to offer or receive a service; the solid bars (transitions) represent the services. Unlike a true Petri net, the transitions are not assumed to be atomic. Several candidates can be in the *Choose* state at the same time, and the first to enter the state is not necessarily the first to leave it. On the other hand, like a Petri net, for a transition to fire, it must have at least one token on all its input places, and once having fired, it places a token on each of its output places.

In states such as *Ready*, the candidate is idle and said to be *queueing* for service. We therefore need to introduce the reader to the basics of **queueing theory**.

10.1.1 BASIC QUEUEING THEORY

Randomness is a basic ingredient of any queueing model. To see why, consider a barber who takes an average of 10 minutes to give a haircut. Five customers arrive per hour. How long is the queue of customers?

Intuition is a bad guide here. You might like to take a guess before reading on.

Assume there is no randomness. Each haircut takes exactly 10 minutes, and customers arrive on the hour and at 12, 24, 36, and 48 minutes past the hour. Each haircut will be done before the next customer arrives, and there will be no waiting.

Now assume that customers arrive at random. Since five customers arrive per hour, it is clear that, on average, the barber will need to finish five haircuts per hour, which take an average of 10 minutes each. Therefore the barber will be **busy** for 50 minutes in the hour, or $\frac{5}{6}$ of the time. We say that the barber's **offered load** is $\frac{5}{6}$. Conversely, the barber is **idle** $\frac{1}{6}$ of the time. How do we calculate how long a customer will need to wait for a haircut?

The following is a simplistic argument, which is easy to remember, *but nonetheless gives the right answer*. Since the barber is idle only $\frac{1}{6}$ of the time, he can offer only $\frac{1}{6}$ of the service he offers when he has no customers. Consequently, a haircut now takes six times longer, an average of 60 minutes. Ten of these minutes are spent having a haircut, and 50 minutes are spent waiting. Since customers leave the queue to have a haircut every 10 minutes on average, the queue must contain an *average* of five customers.

This argument cannot tell us the probability distribution of the queue length. Even so, it is clear that although the average queue contains five customers, $\frac{1}{6}$ of the time no one is waiting. In fact, the standard deviation of the length (a measure of its variability) is always more than the length itself. We shouldn't be surprised to sometimes see a queue of ten customers or more.

The same reasoning applies to computer resources: a file server may have an average access time of 10 mS when it has zero offered load, but when its offered load is 90%, any given user will have to wait an average of 100 mS per access and may sometimes have to wait much longer.

In general, if the offered load is ρ, the probability that the queue has length n is $(1-\rho)\rho^n$, its average length is $\rho/(1-\rho)$, and the standard deviation of the length is $\sqrt{\rho}/(1-\rho)$. A queue is said to be **stable** if it doesn't grow without limit, and this requires that $\rho < 1$. As a result $\sqrt{\rho} > \rho$, so the standard deviation of the queue length always exceeds its average.

Now assume that there are *two* barbers, but customers arrive twice as often. How does this affect things? Since both barbers are just as busy as one was before, on the face of it, nothing changes. Indeed, that would be true if each barber had his own private set of customers, but not if either barber can serve any customer.

There are two ways we can analyse this situation:

- The activities of the two barbers are totally correlated. In effect, customers arrive in pairs, and both take the same time to have their hair cut. The result is the same

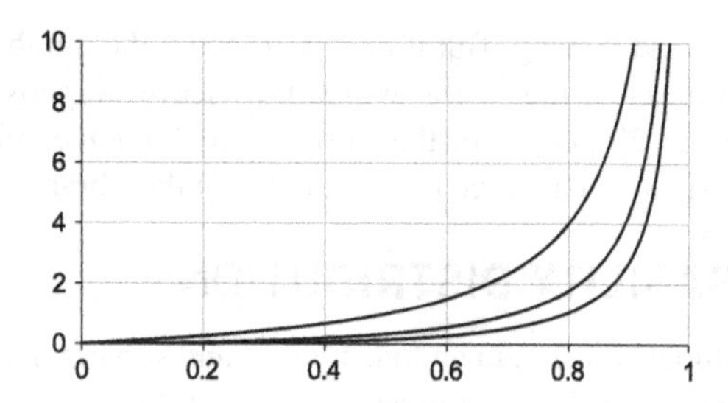

FIGURE 10.3

How queue length varies with offered load, when the activities of one, two, or three barbers are uncorrelated (upper, middle, and lower curves), and any barber can serve any customer.

as before: customers have to wait 50 minutes for a haircut, and the queue contains an average of ten (5×2) customers.

- The activities of the two barbers are completely uncorrelated. A customer will only fail to get service if both barbers are busy, which occurs $(\frac{5}{6})^2$ of the time. For the remaining $\frac{11}{36}$ of the time, at least one barber is free, so the rate of service is reduced by the same ratio. It now takes $\frac{360}{11}$, or about 33, minutes to get a haircut, of which 23 minutes are spent in the queue, which therefore has a much shorter average length of 2.3 customers. (See Figure 10.3.)

Both these arguments are too simplistic. The truth lies between the two estimates: when the first barber is busy, it becomes more likely (but not certain) that the second barber will become busy. The correct formula is surprisingly complicated and hard to remember.[1] In practice, almost all queueing problems are best solved by simulation.

Even so, from this one example, we may safely draw a general conclusion:

It always pays to pool resources.

As a further example, in the NUTS Academic Records System, it will be more efficient to allow a terminal operator to be able to serve any candidate rather than to allocate candidates to operators on the basis of personal name, study programme, or any other fixed basis. Similarly, with a given number of programme advisers, it would be better for them all to be able to deal with any candidate. Unfortunately, their specialised knowledge of particular programmes makes this impossible.

In the above argument, we used an obvious-seeming result known as **Little's Law**:

> *In any queueing system, the* number *of clients in the system is the product of their average* time *in the system and the* rate *at which they leave it.*

As a simple example of its use, suppose you want to know the average time people spend in a library. This is hard to measure directly, because you would have to identify

[1] See the reference to L. Kleinrock in Section 10.4.

each person entering and leaving. But it is easy to count the number of people in the library and easy to measure the rate at which they leave. Suppose you observe, on average, that there are 20 people in the library, and 60 leave per hour (one every minute). Then they spend an average of 20 minutes in the library.

10.1.2 PROBABILITY DISTRIBUTIONS

In most queueing studies it is assumed that sources and servers 'have no memory' — for example, the time when one customer arrives is unaffected by when the previous customer arrived — and that the probability that a customer will arrive at any given time is constant. We therefore talk about customers having a *uniform* arrival rate. Consequently, the distribution of **inter-arrival times** (times between new arrivals) follows an exponential or geometric distribution.

To illustrate, consider the probability of a giant asteroid annihilating all life on earth. We may assume that the chance that an asteroid will arrive at any given time is constant, say, once every 10 million years. But if we ask in which year life is most likely to be destroyed, the answer is *this one*, because, for it to be destroyed next year, we first have to survive not being destroyed this year. For every year that goes by, the chance that the *first* extinction event will occur in that year decreases exponentially by a factor of 0.9999999.

When we consider the number of events over a period of time, such as the number of life extinction events expected over the next billion years, we conclude that we should expect 100 of them. If life on earth is lucky, there will be none at all. If it is unlucky, there could be over 200 of them. The expected number of events in a given interval follows a **Poisson** distribution.

Similarly, if candidates arrive at an enrolment centre *uniformly*, the times *between* their arrivals will be distributed *exponentially*, and the number of candidates arriving each hour will follow a *Poisson distribution*. Because of these properties, we sometimes speak of events having a uniform, an exponential, or a Poisson distribution, with the same thing meant in each case. (See Figure 10.4.)

For a server, we may assume the time that will be taken to complete its service is independent of the time already spent. Therefore the chance the service will be completed at any given time is uniform, the time taken to give the service has an exponential distribution, and the number of clients served in an hour has a Poisson distribution. (Although it is perhaps unrealistic to assume, for example, that a programme advisor has a uniform chance of ending an interview at any time, in what follows, we shall adopt the assumption nonetheless.)

10.1.3 A SIMPLE QUEUEING MODEL

Let us assume, as earlier, that 10,000 candidates must enrol at the start of each semester. This will occur over the course of a 5-day week — an average of 2,000 candidates per 8-hour day.

Ideally, we would model the situation using a special simulation program, but we shall use a spreadsheet here. To keep matters simple, we snapshot the situation every

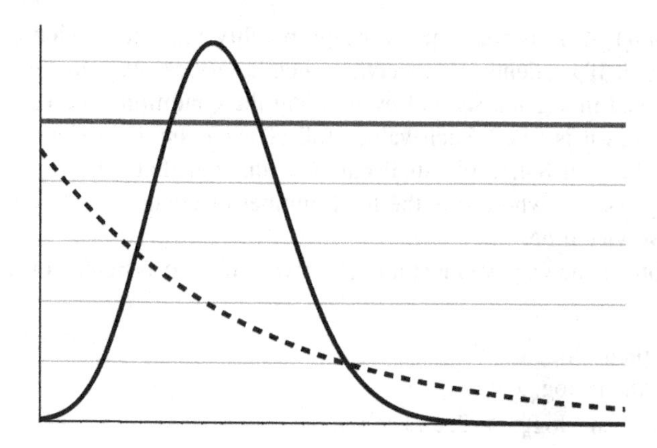

FIGURE 10.4

A *uniform distribution* (horizontal line), *exponential distribution* (dotted line), and a *Poisson distribution* (solid curve). *(Not to scale.)* The uniform distribution would represent a constant probability of an event happening at a given time, the exponential distribution would then represent the probability that the *first* such event would happen at the given time, and the Poisson distribution would then represent the expected number of events that would occur in a given interval of time. (Although the probability of *zero* events occurring in the interval is minuscule in this example, it is not zero.)

second. This could mean some unnecessary work, because in some seconds nothing interesting might happen. On the other hand, 1 second might be too coarse a time scale — maybe we ought to model the situation every $\frac{1}{10}$ second.

A proper simulation program doesn't work this way. It takes a snapshot only at the times when an event happens, while keeping a list of events that are yet to happen. For example, when a candidate starts to speak to a programme advisor, such a simulator decides how long the interview will last and places an 'end of interview' event in a priority queue. It then picks the earliest event from the head of the queue, sets its internal clock to the time of that event, processes it, and so on. Eventually, the 'end of interview' event will reach the head of the queue and be processed. There are efficient data structures that can maintain queues in priority order,[2] but they are difficult to implement using a spreadsheet, so we shall accept a certain amount of inefficiency and inaccuracy here. Our purpose is merely to demonstrate the basic idea.

For our simulation, we shall need to generate various random quantities. The basic tool will be the rand() function built into the spreadsheet program, which, each time it is invoked, returns a different random number, $0 < \text{rand}() \leq 1$, with uniform distribution. Consequently, $(\text{rand}() \leq p)$ is true with probability p. Equivalently, if the average time for a service to complete is s seconds, then the probability

[2] See Section 2.7.3.

that ($s \times \text{rand}() \leq 1$) is true equals the probability that the service completed in a given second. If n clients are receiving such a service, we can find the number of clients served in a given second by iterating the condition n times and counting how many times it is true.[3] Such values follow a *binomial distribution*, which we discussed in Section 8.4.1. We shall call the function that returns the number of clients 'sample(n,s)', where n is the total number of clients being served, and s is the average service time.

More subtle is the way we must model arrivals into the system: Consider $-\log_n x$, for $n > 1$.

- If $x \leq 1$, then $-\log_n x \geq 0$.
- If $x \leq \frac{1}{n}$, then $-\log_n x \geq 1$.
- If $x \leq \frac{1}{n^2}$, then $-\log_n x \geq 2$, and so on.

Consequently, the function $\lfloor -\log_n \text{rand}() \rfloor$, has the value 0 with probability $(1 - \frac{1}{n})$, the value 1 with probability $(\frac{1}{n} - \frac{1}{n^2})$, the value 2 with probability $(\frac{1}{n^2} - \frac{1}{n^3})$, and so on. Its expected value is $\frac{1}{n} + \frac{1}{n^2} + \frac{1}{n^4} + \frac{1}{n^4} + \cdots = \frac{1}{(n-1)}$. Therefore, to simulate the random number of arrivals in 1 second, when the average inter-arrival time is s seconds, we use the formula

$$\textit{Arrived In}_t = \lfloor (-\log_{s+1} \text{rand}() \rfloor \tag{10.1.1}$$

To find the queue lengths and offered loads in our model, we need to keep track of the numbers of candidates, desks, programme advisors, and terminal operators in each state. But we also need variables to represent the flows between states.

To see why these flow variables are necessary, consider the two outputs from *Submit*, which flow to *Submitted* and to *Watching*. If $\textit{Submit}_t$ is the number of forms being processed at time t, and therefore the number of candidates receiving a service, then the number of such services that complete in 1 second is

$$\textit{Submit Out}_t = \text{sample}(\textit{Submit}_{t-1}, \textit{Submit Time}) \tag{10.1.2}$$

where *Submit Time* is the average time for an operator to enter the data from an enrolment form into the database. We may then write

$$\textit{Submit}_t = \textit{Submit}_{t-1} + \textit{Submit In}_t - \textit{Submit Out}_t \tag{10.1.3}$$

$$\textit{Submitted}_t = \textit{Submitted}_{t-1} + \textit{Submit Out}_t - \textit{Submitted Out}_t$$

$$\textit{Watching}_t = \textit{Watching}_{t-1} + \textit{Submit Out}_t - \textit{Watching Out}_t \tag{10.1.4}$$

Unfortunately, if we were to substitute the expression 'sample($\textit{Submit}_{t-1}$,*Submit Time*)' for each occurrence of $\textit{Submit Out}_t$, because of the behaviour of the rand() function, each instance might return a different value. As a result, the numbers of tokens entering the *Submitted* and *Watching* states would rarely equal the number leaving the *Advised* state.

---

[3] This requires the definition of a simple spreadsheet macro.

Each state variable, whether it represents a queue or a transition, has the same form as Equations 10.1.3–10.1.4.

There are two kinds of flow equation: those leaving a transition, and those entering a transition. The flows that leave a transition tend to have the same form as Equation 10.1.2:

$$\begin{aligned}
\mathit{Choose\ Out}_t &= \text{sample}(\mathit{Choose}_{t-1}, \mathit{Choose\ Time}) \\
\mathit{Update\ Out}_t &= \text{sample}(\mathit{Update}_{t-1}, \mathit{Update\ Time}) \\
\mathit{Advised\ Eng}_t &= \text{sample}(\mathit{Consult\ Eng}_{t-1}, \mathit{Consult\ Time}) \qquad (10.1.5)
\end{aligned}$$

($\mathit{Advised\ IT}_t$, $\mathit{Advised\ Bio}_t$, and $\mathit{Advised\ Phys}_t$ have the same form.)

The flows entering a transition have a different form, because a transition cannot fire unless it has tokens on all its input places:

$$\begin{aligned}
\mathit{Choose\ In}_t &= \min(\mathit{Free\ Desks}_{t-1}, \mathit{Arrived}_{t-1}) \\
\mathit{Submit\ In}_t &= \min(\mathit{Free\ Operators}_{t-1}, \mathit{Advised}_{t-1}) \\
\mathit{Update\ In}_t &= \min(\mathit{Database\ Idle}_{t-1}, \mathit{Submitted}_{t-1}, \mathit{Watching}_{t-1}) \\
\mathit{Consult\ Eng\ In}_t &= \min(\mathit{Ready\ Eng}_{t-1}, \mathit{Free\ Eng\ Advisers}_{t-1}) \qquad (10.1.6)
\end{aligned}$$

($\mathit{Consult\ IT\ In}_t$, $\mathit{Consult\ Bio\ In}_t$, and $\mathit{Consult\ Phys\ In}_t$ have the same form as Equation 10.1.6.)

We also need a means of allocating candidates who have completed their enrolment forms into four categories, one for each kind of programme advisor. We aim to do this randomly, but fairly. We use the following equations:

$$\begin{aligned}
\mathit{Ready\ In\ Eng}_t &= \text{sample}(\mathit{Choose\ Out}_t, 4) \qquad (10.1.7) \\
\mathit{Ready\ In\ IT}_t &= \text{sample}(\mathit{Choose\ Out}_t - \mathit{Ready\ In\ Eng}_t, 3) \\
\mathit{Ready\ In\ Bio}_t &= \text{sample}(\mathit{Choose\ Out}_t - \mathit{Ready\ In\ Eng}_t - \mathit{Ready\ In\ IT}_t, 2) \\
\mathit{Ready\ In\ Phys}_t &= \mathit{Choose\ Out}_t - \mathit{Ready\ In\ Eng}_t - \mathit{Ready\ In\ IT}_t - \mathit{Ready\ In\ Bio}_t
\end{aligned}$$

(We cannot give all four equations the same form as Equation 10.1.7, because of the behaviour of the rand() function. Invoking sample($\mathit{Choose\ Out}_t$,4) four times cannot be expected to generate four random numbers whose total happens to equal $\mathit{Choose\ Out}_t$.)

After receiving advice, candidates join a single queue to have their enrolments recorded:

$$\begin{aligned}
\mathit{Advised}_t = {} & \mathit{Advised\ Eng}_t + \mathit{Advised\ IT}_t + \mathit{Advised\ Bio}_t + \mathit{Advised\ Phys}_t \\
& + \mathit{Advised}_{t-1} - \mathit{Submit\ In}_t
\end{aligned}$$

As a simple exercise, the reader is invited to complete the list of equations needed for the simulation. (Note that the value of a flow variable at time t is typically an expression in the values of the state variables at time $t-1$, or the values of other flow variables at time t. The value of a state variable at time t depends on its value at time

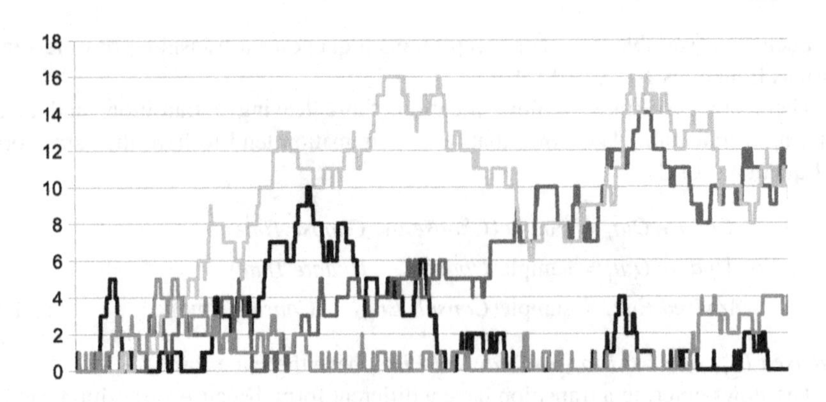

FIGURE 10.5

How the simulated numbers of candidates waiting for each adviser might vary during the first hour of activity of the enrolment centre. At first, the queue for the *Engineering* adviser (black) lengthens to a peak of five candidates, then that for the *IT* adviser (dark grey) grows to the same length. The queue for *Biological Sciences* (pale grey) then climbs to 16 candidates and remains long throughout the rest of the hour. Towards the end, it is overtaken by the queue for IT. The queue for *Physical Sciences* (medium grey) remains short throughout. This is despite the average demand for each adviser being almost equal. It also suggests how much one run of the same simulation can differ from the next.

$t-1$, plus its inward flow, minus its outward flow.) More ambitious readers may like to create their own spreadsheet simulation, which they should find very instructive.

Figure 10.5 shows some results from the spreadsheet. Based on previous experience, the conditions chosen for this simulation were:

- Candidates arrive at intervals of 12 seconds on average, a rate of 300 per hour.
- It takes an average of 40 seconds to consult a programme adviser. There are four advisers, one for each programme area. The offered load per adviser is therefore $\frac{40}{4\times 12} = \frac{5}{6}$.
- There are two terminal operators, who take an average of 20 seconds to enter a candidate's enrolment data, with an offered load also of $\frac{5}{6}$.
- There are four desks, and it takes 40 seconds to complete a form. The offered load is again $\frac{5}{6}$.
- The database takes 1 second to update, with an offered load of $\frac{1}{12}$.
- The simulation covers a period of 1 hour — 3,600 iterations.

In the particular run of the simulation shown in Figure 10.5 the average queue lengths for the *Biological Sciences*, *IT*, *Engineering*, and *Physical Sciences* advisers proved to be 9.3, 5.8, 2.1, and 1.2 candidates respectively. This is despite the offered loads for each adviser being almost equal. The differences are due to the random nature of arrivals and the large associated standard deviation (about 5.5) in the expected queue length (5).[4]

The average queue lengths for desks and terminal operators were 0.1 and 1.0 respectively. This is despite both these resources having the same offered load ($\frac{5}{6}$). There were four desks, which is a bigger pool than two terminal operators. As Figure 10.3 (page 337) showed, the bigger the pool, the shorter the queue. Although there were also four programme advisers with the same offered load, they do *not* constitute a pool, because a candidate can only be served by one particular adviser. Consequently, despite the four advisers having the same offered load as the four desks, their queue lengths were dramatically different.

The database had an offered load of $\frac{1}{12}$ and wasn't a bottleneck. This is because we assumed that the terminal operators complete a form, which they then submit. On the other hand, the situation would be radically different if the operators locked the database throughout their 20-second data-entry time. The database would then have an offered load of $\frac{20}{12}$ and the system would be incapable of handling more than 180 candidates per hour. Clearly, an implementation that locked the whole database during data entry would be impractical.[5]

Although this simulation merely counts candidates in various states and does not keep track of individuals, we can exploit Little's Law to estimate the average time a candidate spends in the system.

Take the case of the queue for the *Biological Sciences* adviser. Candidates entered the system at 12-second intervals, on average. Assuming one quarter of them chose *Biological Sciences*, they entered and left the queue at an average rate of one every 48 seconds. As a result, they spent an average time of 9.3×48 seconds in the queue, or about $7\frac{1}{2}$ minutes.[6] Considering the other times that they spent queueing or being served, the *Biological Sciences* candidates took an average of about 10 minutes to pass through the enrolment office, which is likely to be considered 'good enough'. Even so, since the queue for their adviser sometimes held as many as 16 candidates, some unfortunates took as long as 15 minutes.

The reader should note that although *Biological Sciences* candidates formed the longest queue in *this* simulation run, recalculating the spreadsheet would generate different values of the rand() function, probably causing a different queue to swell. Every time the simulation is run, the results will be different.

---

[4] We saw that in the case of the barber, if customers arrived at 12-minute intervals, no one had to wait. But when several customers arrive in a short time, a queue will build up, which then takes a long time to clear.

[5] A more practical alternative would be to lock only the candidate row and those subject rows in which the candidate enrolled. If we assume, pessimistically, that *every* programme required *Maths101* as one of the 11 subjects it currently offered and that candidates enrolled in an average of two subjects, then two out of 11 candidates would need to enrol in *Maths101* on average. Enrolments in *Maths101* would then occur once every $12 \times 11 \div 2 = 66$ seconds. If we assume that, on average, the *Maths101* row were locked for half the data-entry time, i.e., 10 seconds, then its offered load would be only $\frac{10}{66}$. Therefore an implementation that locked individual database rows would certainly perform adequately.

[6] A more accurate estimate can be derived by inspection of the spreadsheet. But queue lengths are so variable that the extra accuracy is meaningless.

An advantage of a simulation model is that it is easy to change its parameters. For example, one might ask how many candidates per hour the system could handle if there were more course advisors.[7] One might even make the arrival rate a function of time, to allow for a lunch-time rush hour, and so on.

Clearly, a systems analyst who is proposing a radical change to the way enrolments are processed would be well advised to construct such a model.

10.2 CONTROL SYSTEMS

In the previous section, we saw that random inputs can cause a system to exhibit complex behaviour, but in Section 3.7 (page 96), we suggested that systems like pendulums and aircraft can show interesting autonomous behaviour even in the absence of external input. When they interact with their environment, they can behave almost unintelligibly. The NUTS Academic Records System doesn't offer us a handy example of this kind, so we consider here the dynamic behaviour of a simple stock **control system**.

The senior management of **Transmute Ltd** has realised that its existing computer system, which keeps track of stocks of raw materials and finished goods, has all the information it needs to decide how many units of finished goods to manufacture and how many units of raw materials to buy. Discussions with the production manager reveal that *Transmute Ltd* maintains levels of finished goods that are sufficient to supply one week's sales, with an added margin of 20% in case of unexpected demand. A discussion with the purchasing officer reveals a similar strategy: the levels of raw materials should be sufficient to supply 1 week's production, with a margin of 20%.[8]

From this information, the systems analyst decides that it should be possible to automate the decisions they make and devises suitable rules for the computer to follow. But, being wise, the analyst decides to simulate the proposed system before implementing it.

Transmute Ltd actually manufactures many products from a mix of raw materials, but the analyst decides to keep things simple by considering a single, hypothetical product that consumes a single raw material — reasoning that if the system performs badly in this simple case, it cannot be expected to work well in a more complex situation or in practice.

10.2.1 THE FIRST SIMULATION

In the analyst's simplified simulation, *Transmute Ltd* purchases a single raw material, which it processes and sells. Exactly two units of raw material are needed to

[7] Clearly, the system is sensitive to the times we estimate each operation will take. If it actually took candidates 45 seconds to consult course advisors rather than 40, the advsiors' offered loads would increase to $\frac{15}{16}$, tripling the lengths of their queues.

[8] Rather than such specific rules, the analyst is more likely to hear fuzzy logic statements such as, 'If stock is very low, then order quantity is maximum', and so on.

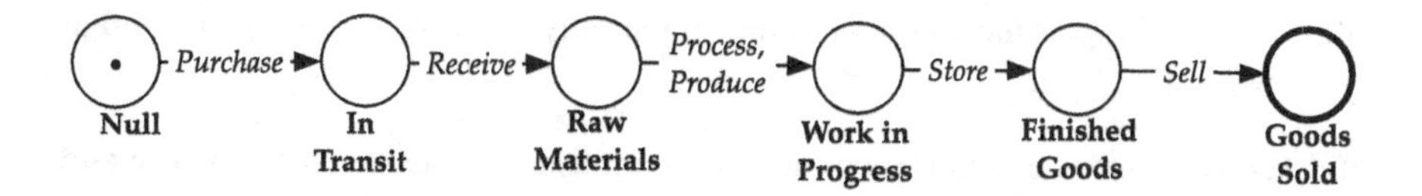

FIGURE 10.6

The state-transition diagram of units of materials. Once materials are *Purchased* they are *In Transit*. Once they are *Received*, they are added to the *Raw Materials* inventory. Once it is decided to *Process* them to *Produce* the product they become *Work in Progress*. On completion, they are *Stored* and added to the *Finished Goods* inventory. *Selling* the goods adds them to the total *Goods Sold*.

manufacture one unit of product.[9] Materials become of interest to *Transmute Ltd* when they are purchased, and they cease to be of interest when they are sold. As in the real world, materials are conserved and pass through several states, as shown in Figure 10.6.

The analyst defines the following *flow* variables:

$Purchases_w$ is the number of units of raw material purchased **during** week w.

$Receipts_w$ is the number of units of raw material received **during** week w.

$Process_w$ is the number of units of *raw material* processed **during** week w.

$Produce_w$ is the number of units of *finished goods Transmute Ltd* begins to make **during** week w.

$Stored_w$ is the number of units of finished goods completed and added to inventory **during** week w.

$Customer\ Orders_w$ is the number of finished units that customers order **during** week w.

$Sell_w$ is the number of finished units *Transmute Ltd* actually sells **during** week w.

The analyst also defines the following variables, which count the number of units in each *state*:

$In\ Transit_w$ is the number of units of raw material purchased but not yet delivered **at the end of** week w.

$Raw\ Materials_w$ is the number of units of raw material in inventory **at the end of** week w.

[9] Although the factor of 2 complicates things a little, it prevents lines colliding on the graphs that follow.

Work in Progress$_w$ is the number of units of partially processed goods **at the end of** week w.

Finished Goods$_w$ is the number of units in the finished goods inventory **at the end of** week w.

Goods Sold$_w$ is the total number of units ever delivered to customers **by the end of** week w.

Transmute Ltd has a weekly reporting cycle. Quantities at the *beginning* of week w equal the quantities at the *end* of week $(w - 1)$:

- If *Transmute Ltd Purchases* materials *during* week w, they are still *In Transit* at the *end* of week w and are *Received during* week $(w + 1)$, augmenting the *Raw Materials* inventory measured at the end of week $(w + 1)$. We assume that *Transmute Ltd's* supplier is totally reliable and prompt, so that $In\ Transit_w = Purchases_w$ and $Receipts_w = In\ Transit_{(w-1)}$.
- If *Transmute Ltd* chooses to *Produce* finished goods during week w, for every two units *Processed* and removed from the *Raw Materials* inventory measured at the *beginning* of week w (*end* of week $(w - 1)$), one unit is *Produced* and added to the *Work in Progress* measured at the end of week w and *Stored during* week $(w + 1)$, augmenting the *Finished Goods* inventory measured at the *end* of week $(w + 1)$.
- When *Transmute Ltd Sells* goods during week w, they are removed from *Finished Goods* measured at the *start* of week w (*end* of week $(w - 1)$) and augment the *Goods Sold* measured at the *end* of week w.

The following equations express the conservation of materials:

$$\begin{aligned} In\ Transit_w &= In\ Transit_{w-1} + Purchases_w - Receipts_w \\ Raw\ Materials_w &= Raw\ Materials_{w-1} + Receipts_w - Process_w \qquad (10.2.1) \\ Process_w &= 2 \times Produce_w \qquad (10.2.2) \\ Work\ in\ Progress_w &= Work\ in\ Progress_{w-1} + Produce_w - Stored_w \\ Finished\ Goods_w &= Finished\ Goods_{w-1} + Stored_w - Sell_w \\ Goods\ Sold_w &= Goods\ Sold_{w-1} + Sell_w \end{aligned}$$

Raw materials remain in transit and work remains in progress for only 1 week:

$$\begin{aligned} Receipts_w &= In\ Transit_{w-1} = Purchases_{w-1} \qquad (10.2.3) \\ Stored_w &= Work\ in\ Progress_{w-1} = Produce_{w-1} \qquad (10.2.4) \end{aligned}$$

The number of units *Transmute Ltd* can *Sell* during week w is determined by *Customer Orders* during week w, but is limited by the number of *Finished Goods* at the *start* of week w:

$$Sell_w = \min(Customer\ Orders_w, Finished\ Goods_{w-1})$$

Transmute Ltd has no control over *Customer Orders*. As a result, *Transmute Ltd* must forecast the orders that customers will make, which it does in the simplest possible way: it assumes that customers will order the same amount in week $(w + 1)$ as they did in week w.

$$Forecast\ Sales_w = Customer\ Orders_{w-1}$$

There are two decisions that *Transmute Ltd* makes at the beginning of week w (the *end* of week $(w - 1)$): how many units of material to *Purchase* during week w and how many units to *Process* during week w. At this time, *Transmute Ltd* knows the values of all variables for week $(w - 1)$, and, because of Equations 10.2.3 and 10.2.4, they also know $Receipts_w$ and $Stored_w$.

As mentioned above, the production manager tries to make sure that the number of units of *Finished Goods* is equal to the *Forecast Sales*, plus a safety margin of 20%. The analyst interprets this statement by estimating the future number of *Finished Goods* as

$$Finished\ Goods_w = Finished\ Goods_{w-1} + Work\ in\ Progress_w - Forecast\ Sales_w \tag{10.2.5}$$

But $Finished\ Goods_w$ should equal $1.2 \times Forecast\ Sales_w$. Combining Equations 10.2.4 and 10.2.5, the analyst obtains

$$Produce_w = Forecast\ Sales_w + (1.2 \times Forecast\ Sales_w - Finished\ Goods_{w-1})$$

where the first term represents the number of units needed to satisfy expected sales, and the second term represents the number of units needed to restore the finished goods inventory to its desired level.

In addition, there are some restrictions: *Transmute Ltd* doesn't have the capacity to process more than 300 units of raw material per week, it cannot process more units than are already in the raw materials inventory at the beginning of week w, and, trivially, the number produced cannot be fractional or negative. With these constraints, and using Equation 10.2.2,

$$Produce_w = \lfloor \max(0,\ \min(150, (2.2 \times Forecast\ Sales_w - Finished\ Goods_{w-1}),\ Raw\ Materials_{w-1}/2)) \rfloor$$

Using similar reasoning, the analyst predicts the future level of *Raw Materials* by taking into account the forecast number of units that will be consumed by *Processing*:

$$Raw\ Materials_w = Raw\ Materials_{w-1} + Purchases_{w-1} - Process_w \tag{10.2.6}$$

To determine how many units of material to *Purchase*, the purchasing officer aims also to restore the level of *Raw Materials* to equal *Process* plus a safety margin of 20%:

$$Purchases_w = Process_w + (1.2 \times Process_w - Raw\ Materials_{w-1}) \tag{10.2.7}$$

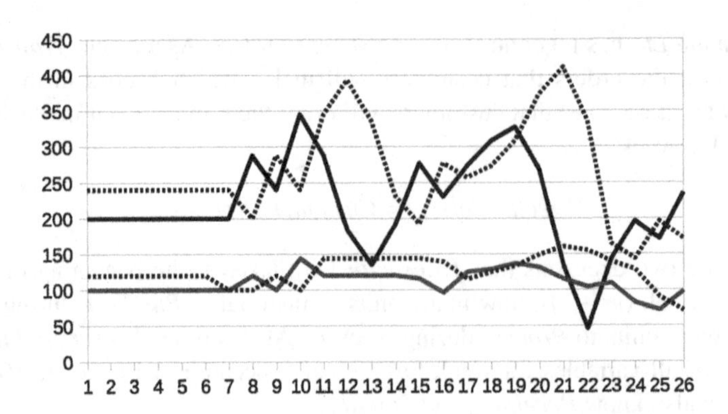

FIGURE 10.7

The first simulation of a proposed stock control system. Initially, there are 240 units of *Raw Materials*, *Purchases* are stable at 200 units, *Finished Goods* inventory is stable at 120 units, and *Produce* is steady at 100 units. All goes well until the system is disturbed by an increase in *Customer Orders* from 100 to 120 units. This causes a corresponding increase in *Produce* from 100 to 120 units, causing a dip in *Raw Materials* and a matching increase in *Purchases*. Meanwhile, because of delays in the supply of materials, *Finished Goods* dips below *Customer Orders* (120), resulting in lost sales. From then on, things only get worse. (The straight lines are merely a way of helping the reader connect the dots. The functions have values only at integer values of w and have no values in between.)

where the first term represents the number of units consumed in production, and the second term represents the number of units needed to restore the raw materials inventory to its desired level.

Finally, *Purchases* can never be fractional or negative.

$$Purchases_w = \lfloor \max(0, (2.2 \times Process_w - Raw\ Materials_{w-1})) \rfloor \qquad (10.2.8)$$

(The purchasing and production decisions are, of course, linked by Equation 10.2.2.)

At this point, I ask the reader to pause: Can you see anything wrong with the analyst's reasoning?[10]

The analyst now uses these equations in a spreadsheet simulation covering a period of 26 weeks. Its initial conditions are that customers order 100 finished units each week, finished goods inventory is 120, and 240 units of raw materials are in stock. The formulae correctly determine that 200 units of raw materials need to be processed and 200 units of raw materials need to be ordered each week. On the face of it, all is well.

Wisely, the analyst also decides to test how well the system responds to a sudden increase in sales, by changing *Customer Orders* from 100 to 120 in the seventh week onwards. At this point, as Figure 10.7 shows, the system goes haywire.

[10] Well, *can* you?

Such a system is unstable. It is like a pencil balanced on end: the slightest disturbance tips it over. After a disturbance, the proposed stock control system exhibits divergent behaviour. A peak in *Purchases* causes a peak in *Raw Materials*, which causes a trough in *Purchases*, which causes a trough in *Raw Materials*, which causes a peak in *Purchases*, in an endless cycle whose amplitude will ultimately be limited only by *Transmute Ltd's* inability to process either more than 300 or fewer than zero units of raw material.

Anyone who has showered in a hotel room will understand the problem. Having turned on the tap and waited for the hot water to come through, at first the shower is too hot, so you increase the flow from the cold tap. Nothing happens, so you increase it again. The result is a comfortable temperature, but soon afterwards the water becomes freezing cold. You hastily turn the cold water down. Nothing happens, so you do it again. The water is warm again, but soon afterwards, it is scalding. You soon realise that the changes you make take a long time to take effect, as if there were an enormous length of pipe between the taps and the showerhead. Consequently, you learn to allow for the delay and to wait for each adjustment to take effect before making the next one. Unfortunately, the *Transmute Ltd* stock control system is unable to learn.

Transmute Ltd's difficulty is that it takes 4 weeks for *Purchases* (the tap) to affect *Finished Goods* (the showerhead), but the system adjusts *Purchases* every week. We may expect similar behaviour in any system that tries to correct its behaviour on a shorter cycle than its internal delay. One answer would be to adjust *Purchases* more cautiously, by only a fraction of what seems to be needed, slowing its response to match the slowness of the system. But we can do better, as we shall see shortly.

10.2.2 THE SIMPLIFIED SIMULATION

What have we learned from the simulation? First, that if a computer system were to adopt the rules that the analyst has derived, *Transmute Ltd* would be ruined: sales would be lost and workers would alternate between being idle or being paid overtime penalty rates. Second, that since *Transmute Ltd* is actually still in business, and the existing control system does not oscillate in this alarming way, the rules the analyst has derived can't be those the production manager or the purchasing officer actually uses.

To understand what has happened, let us return to the example of the pendulum discussed in Section 3.7 (page 96). We observed that a real pendulum won't oscillate forever, but loses energy to the environment, mainly through air resistance. There are two forces acting on the pendulum; one (gravity) is proportional to its displacement from the vertical, the second (air resistance) is proportional to the speed with which it approaches the vertical. We say that the oscillation of the pendulum is **damped** by air resistance.

If we dangle a pendulum in water, it will be damped[11] more than in air and reach equilibrium more quickly. If it is immersed in treacle, it will certainly not oscillate,

[11] No pun intended.

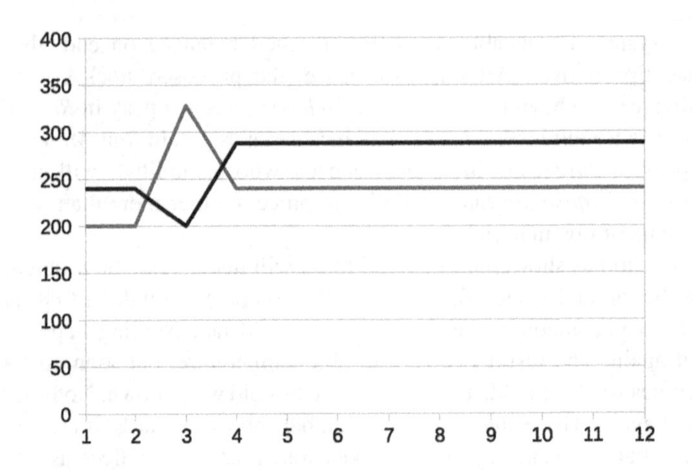

FIGURE 10.8

A simulation using the rule of Equation 10.2.9 with $\alpha = 1$ and $\beta = 1$. The purchasing system is critically damped and rapidly reaches a new equilibrium.

but will take much longer to reach a steady state. Somewhere between extremes lies a **critical damping factor** that lets the pendulum reach equilibrium soonest.

The analogy with the purchasing system is that ($1.2 \times Process_w - Raw\ Materials_{w-1}$) represents the displacement of *Raw Materials* from its desired value, and the expression ($Process_w - In\ Transit_{w-1}$) represents the speed with which it approaches the desired value. This second term can provide the damping needed to reduce the oscillation.

The analyst therefore considers adjusting α and β in the following equation:

$$Purchases_w = Process_w + \alpha(1.2 \times Process_w - Raw\ Materials_{w-1}) + \beta(Process_w - In\ Transit_{w-1}) \tag{10.2.9}$$

The analyst then constructs a much simpler simulation, shown in Figure 10.8 and Figure 10.9. This simulation isolates the purchasing control system from the fluctuating demands of the process control system by giving *Process* an initial value of 200 units and then a *constant* value of 240 units. Figure 10.8 shows that setting $\alpha = 1$ and $\beta = 1$ makes the system critically damped, and *Raw Materials* reaches its final value in the least possible time. In contrast, Figure 10.9 shows that removing the damping by setting $\beta = 0$ reproduces the oscillatory behaviour of the first simulation.[12]

This experiment clearly demonstrates that the term ($Process_w - In\ Transit_{w-1}$) stabilises the purchasing control system. Indeed, given the 2-week delay between

[12] In effect, by overlooking the goods already ordered, the system orders twice as much as needed — then overcompensates.

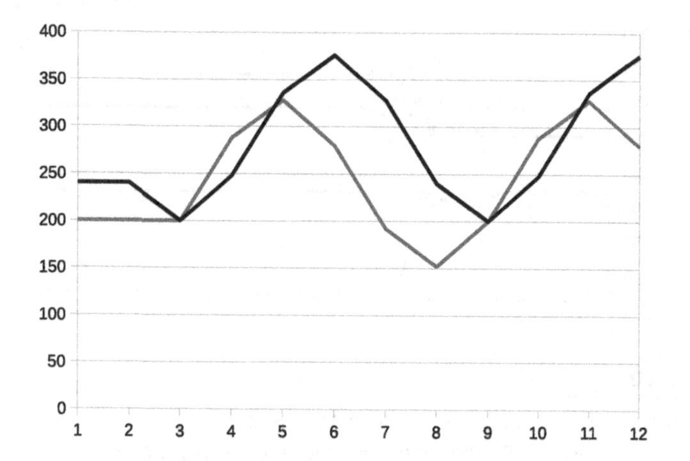

FIGURE 10.9

A simulation using the rule of Equation 10.2.9 with $\alpha = 1$ and $\beta = 0$. The purchasing system is on the edge of instability. Its oscillations neither grow nor shrink.

ordering raw materials and their availability for processing, no system could perform better: *Purchases* and *Raw Materials* both stabilise by the second week. This is indeed the optimum solution.

10.2.3 THE SECOND SIMULATION

Since the production control system obeys similar rules to the purchasing system, it can be made to behave in a similar way. Even so, the analysis still has a weakness: the need for more raw materials is recognised only after their stock has been depleted by the production control system. Since the production control system will itself take 2 weeks to react to increased demand, *Raw Materials* doesn't stabilise until 4 weeks after the increase in sales.

The problem is easily corrected: in discussion, the analyst learns that the purchasing officer doesn't consult the production manager to choose the correct level of *Purchases*. So, what *does* the purchasing officer do? The same thing as the production manager: the officer uses the value of $Forecast\ Sales_w$. But the analyst didn't use this figure, foolishly believing that $Process_w$ was the more realistic demand and would give a better result. The analyst decides to replace $Process_w$ (i.e., $2 \times Produce_w$) by $2 \times Forecast\ Sales_w$. Nonetheless, after experimentation, the following hybrid proves even better:

$$\begin{aligned} Purchases_w = {} & (1.2 \times 2 \times Forecast\ Sales_w - Raw\ Materials_{w-1}) \\ & + (2 \times Forecast\ Sales_w) + (Process_w - In\ Transit_{w-1}) \end{aligned} \tag{10.2.10}$$

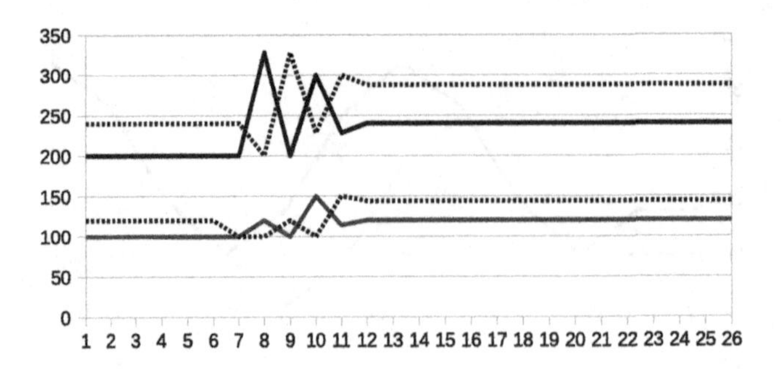

FIGURE 10.10

The second simulation of a proposed stock control system. The proposed system now settles to a steady state in week 12, and the peaks and troughs in *Purchases* and *Produce* are greatly reduced.

The first term on the right-hand side of the equation represents the number of units needed to bring the raw materials inventory to the desired level, the second term represents the *estimated* number of units that will be consumed per week, and the third term (which continues to use $Process_w$) is the *known* number of units that will be lost in production but not yet replaced by goods in transit.

The results of this second simulation are shown in Figure 10.10. It shows that both *Purchases* and *Produce* react to the increased demand in week 7 immediately and appropriately. Despite the increase in *Purchases* in week 8, because of the week's delay in transit, *Raw Materials* dips to 200 units in week 8, limiting *Finished Goods* to 100 units in week 10. As a result, potential sales of 20 units are lost in each of weeks 7, 8, and 10. Impressively, the system reaches its new equilibrium in week 12, only 6 weeks later than the increase in demand. It is clear that, although both the purchasing and production systems take only 2 weeks to react individually, their combination reacts more slowly, due partly to their interaction, but also to shortages.

This simulation cannot be considered the final word. The analyst needs to check how the system would respond to a change in *Customer Orders* other than 20 units. The analyst would also be wise to simulate a more complex situation, for example, one in which product P consumes raw materials R and S, and product Q consumes materials S and T. (P and Q share raw material S.) On the other hand, if the simulation becomes too complicated, there is a danger that the analyst will fail to learn anything from its behaviour.

10.2.4 LINEAR SAMPLED-DATA SYSTEMS

Can we predict the behaviour of a control system without the need for simulation? Yes and no: 'yes', when the system is **linear**, and 'no' when it is **non-linear**.

What do we mean by linear? Essentially, when the system is governed by an equation like 10.2.10, where the **control variable** (*Purchases*) is directly proportional to the terms on which it depends. What do we mean by non-linear? When the control variable depends on the terms in any other way, such as in Equation 10.2.8, which includes a non-linear max() function.

Additionally, we can usefully forecast the **small-signal behaviour** of non-linear systems by treating their non-linear behaviour as locally linear. For example, the purchasing system has two linear regions: one when the desired value of $Purchases_w$ is positive, and one when the desired value is negative, and the max() function substitutes a zero. The production control system has three linear regions: one where the desired value of $Produce_w$ is negative, one where it is in the range zero to 150, and one where it would exceed 150 units, which is the limit of *Transmute Ltd's* productive capacity. Typically, as in the analyst's first simulation, the usual (but not universal) effect of non-linearities is to prevent oscillations increasing without limit.

The value of the **delta calculus**, which we now explain, is that it can predict the behaviour of a linear system, can help the analyst stabilise an otherwise unstable system, and can even help to get the optimum performance from it.

We consider the case of linear systems that sample data at regular intervals, $t, t+\delta, t+2\delta$, and so on. The interval δ can be a week in the case of a stock control system, a millisecond in the case of an aircraft control system, or any other regular interval. The sampled states of the system variables then form a movie-like series of snapshots, $S_t, S_{t+\delta}, S_{t+2\delta}$, and so on, where S_t is the initial state of the system, and some set of linear equations derives $S_{t+\delta}$ from S_t.

We now introduce the **delta operator**, Δ, which corresponds to a time delay of δ, the time between two successive samples, so that $\Delta S_t = S_{t-\delta}$. We replace the notation $S_{t-2\delta}, S_{t-\delta}, S_t$ by $\Delta(\Delta S), \Delta S, S$, where S is the current state. Instead of $\Delta(\Delta S)$ we shall write $\Delta^2 S$, etc.

Consider a system in which an output variable $y(t)$ is simply the sum of the values of an input variable $x(t)$. We assume that in each interval, the new value of y is equal to the previous value of y, plus x. We write

$$y = \Delta y + x$$

We now solve for y:

$$y = \frac{x}{(1-\Delta)} \tag{10.2.11}$$

By long division or otherwise, we obtain the infinite series

$$y = (1 + \Delta + \Delta^2 + \Delta^3 + \Delta^4 + \Delta^5 \cdots)x$$

$$\therefore y = x + \Delta x + \Delta^2 x + \Delta^3 x + \Delta^4 x + \Delta^5 x \cdots \tag{10.2.12}$$

proving that y is the sum of all previous values of x.

There is a second way we can interpret this series: the term $\Delta^3 x$, for example, means that the current value of y depends on the value of x three time units earlier. Conversely, the current value of x affects the value of y three time units later. The coefficients of Equation 10.2.12 may therefore be interpreted as future values of y, given a **unit impulse** (the sequence 1, 0, 0, 0, ...) now. This is called the system's **impulse response**. In this case, the impulse response is a constant stream of unit impulses, called a **unit step function**.

If a system is linear, we can find the output resulting from an arbitrary input by regarding it as a stream of impulses of various sizes. In this case we simply sum their resulting step functions. Given the input sequence 1, 2, 3, 0, 0, ... the future values of y become

$$\begin{aligned} y &= 1, 1, 1, 1, 1, 1, 1, \ldots \\ &+ 0, 2, 2, 2, 2, 2, 2, \ldots \\ &+ 0, 0, 3, 3, 3, 3, 3 \ldots \\ &= 1, 3, 6, 6, 6, 6, 6 \ldots \end{aligned}$$

In other words, y takes the successive values 1, 3, 6, 6, 6 ..., and correctly sums the values of x.[13]

This series of coefficients is exactly what we obtain by treating the input as a polynomial in Δ, and multiplying it by the impulse response:

$$\begin{aligned} y &= (1 + 2\Delta + 3\Delta^2)(1 + \Delta + \Delta^2 + \Delta^3 + \Delta^4 + \Delta^5 \cdots) \\ \therefore y &= 1 + 3\Delta + 6\Delta^2 + 6\Delta^3 + 6\Delta^4 + \cdots \end{aligned}$$

The response to a unit impulse tells us all we need to know about a linear system. Any pattern of input can be considered as the sum of a sequence of impulses of various sizes, and (because it is linear) the system's output will simply be the sum of its responses to the individual impulses.

Let's apply the Δ calculus to the simplified purchasing system of Section 10.2.2: We begin by expressing Equation 10.2.6 — which relates the change in *Raw Materials* to *Purchases* and *Process* — using the Δ operator:

$$\textit{Raw Materials} = \Delta \textit{Raw Materials} + \Delta \textit{Purchases} - \textit{Process}$$

Next, we re-express the purchasing rule of Equation 10.2.9 in terms of Δ, initially setting $\alpha = 1$ and $\beta = 0$.

$$\textit{Purchases} = \textit{Process} + (1.2 \times \textit{Process} - \Delta \textit{Raw Materials})$$

We may then eliminate *Purchases* and solve for *Raw Materials* in terms of *Process*, giving

$$\textit{Raw Materials} = \frac{2.2\Delta - 1}{1 - \Delta + \Delta^2} \times \textit{Process}$$

[13] The attentive reader may well be questioning why we have dared to treat Δ as if it were an ordinary algebraic variable, when it is actually an operator. The answer is that our delta calculus is merely a variation of the well-known z-transform, widely used in signal processing. We are merely replacing z^{-1} by Δ here. Intuitively, we can see that an input at time $t + 2\delta$ subjected to an internal delay of 3δ will result in output at time $t + 5\delta$. Correspondingly, $\Delta^2 \times \Delta^3 = \Delta^5$.

By long division, we obtain

$$\textit{Raw Materials} = (-1 + 1.2\Delta + 2.2\Delta^2 + 1\Delta^3 - 1.2\Delta^4 - 2.2\Delta^5 - 1\Delta^6 + \cdots)\textit{Process}$$

where the pattern of coefficients repeats *ad infinitum*. The unit impulse response of the system is therefore an oscillation with a period of 6 weeks, on the borderline of instability. By accumulating terms from left to right, we obtain its response to a unit step function, $[-1, +0.2, 2.4, 3.4, 2.2, 0, -1 \ldots]$, identical (apart from scaling) to the behaviour shown in Figure 10.9.

If we now introduce damping by setting $\beta = 1$, the analogue of Equation 10.2.9 becomes

$$\textit{Purchases} = \textit{Process} + (1.2\textit{Process} - \Delta \textit{Raw Materials}) + (\textit{Process} - \Delta \textit{Purchases}).$$

After eliminating *Purchases* as before, we get

$$(1 - \Delta)(1 + \Delta)\textit{Raw Materials} = (2.2\Delta - 1)\textit{Process} - \Delta^2 \textit{Raw Materials}.$$

The terms in Δ^2 miraculously cancel, leaving

$$\textit{Raw Materials} = (2.2\Delta - 1)\textit{Process}.$$

The response of *Raw Materials* to a unit step in *Process* is therefore the infinite sequence $[-1, 1.2, 1.2, 1.2, \ldots]$, meaning that *Raw Materials* reaches its correct level (with the added 20% safety margin) in only 2 weeks.

The linearity property implies that if *Transmute Ltd* makes several products from several raw materials, their stock control systems will behave independently — as long as their behaviour remains linear. Different products will interact only when there is a shortage of labour or materials, in which case *Transmute's* management will have to make hard decisions: which products will get priority, which customers will be disappointed, and so on. Such decisions are beyond the scope of the control system.

10.3 SUMMARY

We have discussed two problems that might cause a proposed system to fail. In the case of the *NUTS* enrolment system, we estimated how long it might take for a candidate to pass through the enrolment centre and verified that the resources we planned to make available would be adequate. Basic queueing theory gave us some guide to where bottlenecks might exist, but the system as a whole was too complex for theory to be a reliable guide. We found that, even without special software, we could build a realistic model of the system that showed a range of its possible behaviours. No two runs of the model gave similar results. That is a lesson in itself.

In the case of *Transmute's* stock control system, although it wasn't immediately obvious, the exact rules for reordering materials and for planning the level of production are crucial to the survival of the company. A simple simulation was sufficient to reveal a serious problem. Although a systems analyst with a background in control theory might have foreseen the problem, *any* analyst would be wise to verify that the proposed rules will work as they should. An analyst should not trust that what experts seem to be saying is always the full story. We also learnt that when we replace a manual system with a rigid set of rules, we need to be careful that they will actually make an improvement. Computers are fast, but they have no common sense.

There is much more to inventory management than we have discussed: for example, maintaining high levels of inventory ties up money that could be better spent, so good management practice must balance costly stock levels against the risk of shortage. Also, there are recurrent administrative and transport costs involved whenever goods are ordered from a supplier. This results in the concept of an economic order quantity (EOQ): If the EOQ is too great, then inventory levels will often be higher than they should be. On the other hand, if the EOQ is too small, then recurrent costs will become excessive.

In order to give the reader a deeper understanding, we described the delta calculus, which makes it possible to predict the behaviour of linear sampled-data systems.

10.4 FURTHER READING

Queueing problems are rarely simulated in the crude way we have illustrated here. There are programming languages especially designed for the purpose. One of the earliest — and still popular — is *GPSS* (General Purpose Simulation System). Broadly speaking, building a *GPSS* model involves writing a script that describes the behaviour of a typical client entering the system. Such a client can join queues, seize resources (such as desks or course advisors), and release them again after a random length of time. (Various random distributions can be chosen.) *GPSS* suits the *NUTS* enrolment centre problem quite well and requires only a declarative style of programming — but it is rather inflexible. Other simulation languages have a wider range of applications. *Simula* and *SIMSCRIPT* are among the earliest of these, although there are now many alternatives.

One of the most important contributors to queueing theory has been Leonard Kleinrock, who was a pioneer of what became the Internet. In 1975 he published the first volume of a two-volume work, 'Queueing Systems: Volume I — Theory' (ISBN 978-0471491101), and in 1976 published its second volume, 'Queueing Systems: Volume II — Computer Applications' (ISBN 978-0471491118).

We observed that feedback control systems don't always behave in the way intuition might tell us.

With the aid of the delta calculus, we could predict the behaviour of linear sampled-data systems without the need for simulation. Unfortunately, using long division to predict system behaviour proved almost as laborious as simulation. Is there some way to predict how a system will behave that will take less effort?

Yes, there is. The impulse response R of a linear sampled-data system can always be described as the ratio of two polynomials in Δ. The values of Δ for which the numerator equals zero are — logically enough — called the **zeros** of R. The values of Δ for which the denominator equals zero, and R is therefore infinite, are called the **poles** of R. If any pole lies inside the unit circle on the complex plane the system will be unstable.[14] Conversely, careful manipulation of the poles and zeros allows us to *synthesise* any desired feasible response.[15]

If you are interested in this level of detail, read any book on sampled-data control systems, such as 'Optimal Sampled-data Control Systems' by Tongwen Chen and Bruce A. Francis (1995, ISBN 3-540-19949-7).

10.5 EXERCISES

1. Suppose enrolments for *Biological Sciences*, *IT*, *Engineering*, and *Physical Sciences* were decentralised, so that each course adviser had a separate enrolment centre with its own desk and computer operator. Construct a spreadsheet queueing model for one such centre. The average service times should be the same as in the model of Section 10.1.3, but the arrival rate for candidates should obviously be one quarter of the value used before. You can ignore the time taken to update the database.
2. A lathe produces axles whose diameter must be accurate to 0.01 mm. Every morning, a sample of the previous day's output is collected, and the average diameter of the sample is calculated. If this average is x mm too great, the setting on the lathe is changed by $-x$ mm on the following morning. Is this a good idea? If you think this is a bad quality control system, what should be done to improve things?

---

[14] If you don't know what is meant by the complex plane, ignore this paragraph!

[15] This process is closely related to the synthesis of an active linear filter, which we briefly referred to on page 9.

CHAPTER

Project management

11

CHAPTER CONTENTS

INTRODUCTION

A systems analyst must often act as the interface between senior management and the team that will implement a new system. (Indeed, the analyst may become the project leader or, in a small organisation, even *be* the team that implements the system.) The analyst must therefore understand both what motivates management and what motivates designers and programmers.

Project management is the art of producing systems that are delivered on time and within budget. There is no one-size-fits-all method of project management. Projects differ both in their scale and their degree of uncertainty. They also differ in the numbers, skills, and expectations of the teams who implement them. Even within the context of the NUTS Academic Records System, which we have discussed in earlier chapters, there is a contrast between the skill set needed to implement the enrolments database and that needed to implement the programme adviser expert system. Most of what we have discussed involved well-understood database schema synthesis and forms design, followed by routine implementation, using *SQL* and *PHP*, for example. Such work is easily completed by members of a team working in parallel, working on different forms, events, and reports. The design and management of the database itself is central, but even this needn't be finished before work can begin. Once routine work has been divided into its component tasks, it should be relatively easy to estimate the time and effort needed to complete them. On the other hand, there is no guarantee that an expert system can be created that will prove fast enough or clever enough to be useful.[1] The skill set needed for an expert systems project is more likely to involve *Prolog* or a similar logic-oriented language — at least at the prototype stage.

Another major contrast is between creating a new system and maintaining or enhancing an existing one. For example, the *NUTS* may call in a software services company to create a bespoke Academic Records System, decide to create it itself, or it may choose to purchase a licence to use a pre-written package from a company that tailors, maintains, and enhances an existing general-purpose academic software system. The nature of a project to create a system from scratch is likely to differ profoundly from that to adapt an existing system to a specific environment.

Rather than recommend some particular fashionable approach to project management, this chapter focusses on tools that are common to almost all approaches.

11.1 DISCOUNTED CASH FLOW

A project of any kind involves investing money in the hope of a future benefit. It might seem that, provided the future benefits exceed the initial investment, a project is viable. But it isn't as simple as that. The same money could have been invested in a different project that brought more benefit. The money that could have been earned in that way is called the **opportunity cost**, i.e., the cost of the missed opportunity. In the simplest case, the money could be invested in an interest bearing bank account. If such an account offers an interest rate of 5%, it is said to offer a 5% **return on investment**. To compete with the bank, a project would need to offer a return on investment of better than 5%.

Another way of looking at this is to recognise that businesses have to borrow money to make money. They may need to borrow from a bank or other financial institution or issue shares and pay the shareholders a dividend. One way or another, interest has to be paid on the money they have borrowed. This is called their **cost of capital**. The cost of capital is usually estimated by a business's chief financial officer. In many cases, especially in government and semi-government organisations, an artificial figure is agreed upon, say 10%.

What would you rather have, $1,000 now or $1,100 next year? Ignoring any short-term needs you may have, the logical way to answer this question is to consider what would happen if you took the $1,000 and invested it. If you could invest it at an

[1] Until I had finished writing Section 9.3, I wasn't sure either.

interest rate of 5% per annum, you would have $1,050 by next year, and you would probably benefit by waiting for the $1,100. In contrast, if you could invest the $1,000 at 15%, you would have $1,150 by next year. By waiting a year for the $1,100, you would lose $50 interest and incur a $50 *opportunity cost*. Clearly, you would be wiser to invest the $1,000 now — or maybe not, because by next year, owing to inflation, $1,150 may not buy you as much as $1,000 does now. (Inflation is tricky: Some goods and services — for example, wages — cost more from year to year. On the other hand, you may be able to pay less for the same or a better computer next year than the computer you can buy now. Different kinds of future expense need to be costed individually.)

Consider a project that requires us to invest $200,000 in a new system now, plus a further $185,000 at the end of this year, but which is expected to save us $60,000 every year for the next 10 years ($600,000 in total). How do we decide if this investment is worthwhile?

One commonly used approach is to *discount* future cash flows by the *cost of capital*. If a business's cost of capital is 10%, then $1.00 now is worth $1.10 next year. Conversely, $1.10 next year is worth $1.00 now. In general, if the interest rate is r per accounting period, an investment would grow exponentially by a factor of $(1 + r)$ in each period. Consequently, to make future cash flows comparable with current cash flows, we should discount them by a factor of $d = 1/(1 + r)$ per period, where, in this case, the **discount factor** is $d = \frac{10}{11}$. A flow n periods in the future should be discounted by a factor of $(\frac{10}{11})^n$ or $1/(1 + r)^n$. We may write

$$PV = \frac{FV}{(1 + r)^n} \text{ or, equivalently, } FV = PV(1 + r)^n$$

where PV is the **present value**, FV is its corresponding **future value** after n periods, and r is the interest rate per period, expressed as a fraction. (A rate of 10% means that $r = 0.1$.)

To find the present value of the proposed project, we add up all the flows — negative for outflows or expenses and positive for inflows or incomes.[2] The result is shown in Table 11.1.

From the table, we can conclude that the project has a positive present value of $492. This means that if the cost of capital is 10%, the project is just viable; if the cost of capital were less, it would become more viable.

An alternative way of expressing the value of a project is to calculate its **rate of return**. This is defined as that value of the cost of capital at which the project's present value is zero. Clearly, the rate of return for this project is slightly above 10%. The rate of return is a useful way of comparing projects of different scales: a big multi-million dollar project might have a greater present value than a smaller project, but the smaller project might have the higher rate of return.

---

[2] When we have a long series of equal flows — as in a mortgage or bank loan, there is a short-cut method for calculating their sum. First observe that an investment of PV now, at interest rate r, is equivalent to an *infinite* series of flows of value $PV \times r$. A *finite* series of equal flows is simply the difference between two infinite series. If the series begins now and terminates at the end of period n, its sum is $PV \times (1-(1+r)^{-n})$.

Table 11.1 A Discounted Cash Flow Table for a certain investment when the cost of capital is 10% per annum. The end of year 0 is the start of year 1, and so on. The second row of the table shows the discount factors for each year, the third row shows the un-discounted cash inflow, and the fourth shows the discounted cash flow. The bottom row displays the accumulated net present value for each year

Year	0	1	2	3	4	5	6	7	8	9	10
D.F.	1	0.91	0.83	0.75	0.68	0.62	0.56	0.51	0.47	0.42	0.39
U.C.F.	-200000	-125000	60000	60000	60000	60000	60000	60000	60000	60000	60000
D.C.F.	-200000	-113636	49587	45079	40981	37255	33868	30789	27990	25446	23133
N.P.V.	-200000	-313636	-264050	-218971	-177990	-140735	-106866	-76077	-48086	-22640	492

A third way of expressing the value is the **payback period**, which is usually defined as the reciprocal of the rate of return — in this case, 10 years. Loosely, the term is sometimes taken to mean the time it takes before the present value of the project exceeds zero, which is more properly called its **break-even period**. In this example, the break-even period and the payback period are equal, but they need not be.

In practice, a period of one year is too coarse a scale for a typical project, and cash flows are more likely to be monitored monthly. Although dividing the annual interest rate by 12 gives a good approximation to the monthly interest rate, a more accurate calculation (due to the compounding of interest) is $m = \sqrt[12]{(1 + a)} - 1$, where a is the annual interest rate, and m is the monthly interest rate.

The importance of discounted cash flow accounting to the systems analyst is to recognise the need to bring incomes forward and to defer expenses as much as possible. For example, one common use of a computer system is to shorten the billing cycle: to minimise the time between a payment becoming due and the payment being collected. The *NUTS* collects about $12,000,000 in fees each semester, so invoicing candidates 1 month earlier could earn roughly $50,000 in interest at 5% per annum.[3] This happens twice a year, so earlier billing would earn $100,000 per year, with a present value of $2,000,000 — easily enough to pay for a better system to be developed.[4] It would therefore make good sense for the NUTS Academic Records System to give a high priority to invoicing candidates as soon as enrolments become frozen.

In this example, and indeed, in most projects, there is a degree of risk: If candidates are invoiced earlier, will they actually pay earlier? There will be a range of possible benefits. Alternatively, there can be risk associated with the implementation: the programme adviser expert system is of this kind. There might be great uncertainty about how long it would take to develop a successful system. The present value of

[3] $12,000,000 × 0.05/12.

[4] $100,000 = $2,000,000 × 5%.

a project should therefore be considered probabilistically, and a decision to proceed should be based on its *expected value*.[5]

In practice, deciding what priorities should be given to projects is rarely the analyst's responsibility, but that of someone we shall refer to as the **client**. The client may be concerned less with things that are profitable and more with things that need urgent repair. In the case of the *NUTS Academic Records System*, one such issue is the existing examination scheduling system, which is slow and error-prone. From the client's point of view, this has a high perceived cost, although one that may be hard to quantify.

11.2 CRITICAL PATH ANALYSIS

Critical path analysis is a means of discovering the minimum time that a project can take, given

- The set of component activities that make up the project.
- How long each activity is expected to take.
- What *predecessor* activities each activity depends on.

For example, one activity might be *Write Components Update*. This might be estimated to take 4 days and to depend on the completion of the prior activities *Design Components Interface* and *Create Components Table*.

Because of the predecessor relation between activities, it is possible to draw an **activity graph** whose vertices represent activities and whose edges represent the predecessor relation between them. (See Figure 11.1.) If the division into activities is done correctly, the resulting graph must be acyclic.[6] Since any acyclic graph has a topological sort, the sort gives one feasible sequence in which the activities may be completed. (See Section 2.5.1.) Such a sequence might be useful when one person does all the work, but the underlying graph may offer opportunities for many workers to operate in parallel.

The **critical path** is the longest path in the activity graph, where the length is measured as the sum of the times taken by the activities along it. We associate three properties with each activity:

- Its *Duration*: the time it takes to do it.
- Its *Earliest Start*: the earliest time at which the activity can start.
- Its *Latest Start*: the latest time at which the activity can start.

From these properties, we can easily calculate,

$$\textit{Earliest Finish} = \textit{Earliest Start} + \textit{Duration} \qquad (11.2.1)$$

[5] If a project will have a present value of \$100,000 if it is successful, but has only 20% chance of succeeding, its **expected value** is only \$20,000.

[6] Any apparent cycles can be resolved by splitting activities. Suppose activity A seems to both precede and follow activity B, then, for example, A might decompose into A_1, which precedes B, and A_2, which follows B.

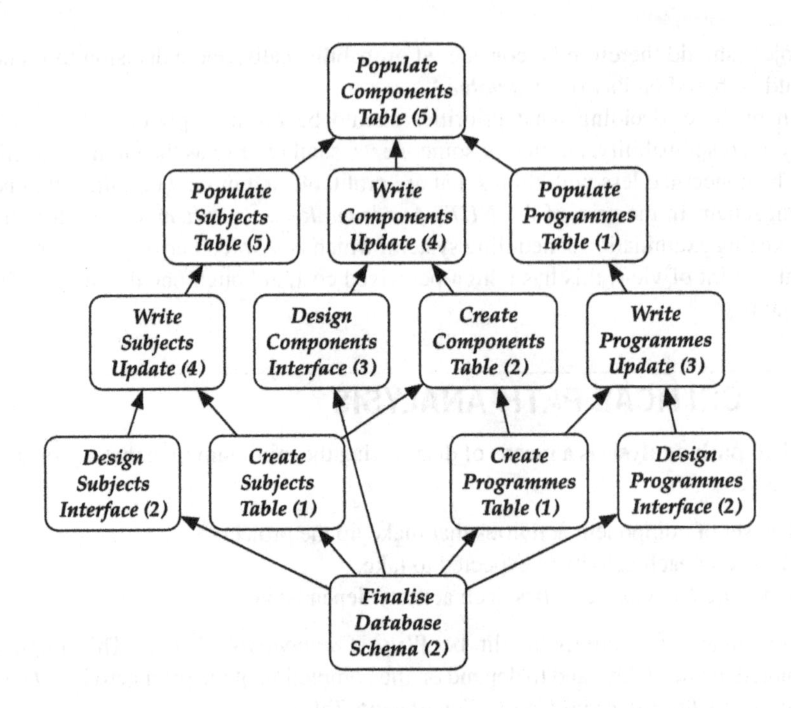

FIGURE 11.1

An activity graph. It shows the ordering of the activities needed to populate the *Subjects*, *Programmes*, and *Components* tables. The number of days allowed for each activity is shown in parentheses.

$$\textit{Latest Finish} = \textit{Latest Start} + \textit{Duration} \tag{11.2.2}$$

With this information, it is easy to find the duration and *Earliest Finish* of the project, using Algorithm 11.1.[7] If this is no later than the project deadline, all is well, and it may even be decided to bring the deadline forward. If not, we have to think again. Sometimes, we simply have to accept a later deadline, but at others we may be able to achieve the original deadline by restructuring the project into a set of activities that allow more people to work concurrently.

The value of the critical path method to the analyst is rarely in the calculation itself. Listing the activities and finding the predecessor relation is often enlightening enough for the analyst to want to devise a better plan.

Let us consider the implementation plan described on page 279.

1. Populate the *Programmes*, *Subjects*, and *Components* tables.
2. Populate the *Candidates* table with the *TEACUP* data for beginning candidates.

[7] The algorithm assumes that the duration of each activity is known precisely. Sometimes it is uncertain how long a task will take. Such situations can be modelled probabilistically, perhaps using spreadsheet techniques similar to those we used to model queues in Section 10.1.3.

Algorithm 11.1 Scheduling a project using critical path analysis.

1. Construct an activity graph, and establish the desired start time and deadline of the project as a whole.
2. Find a topological sort of the activity graph.
3. In ascending topological order, for each activity,
   a. Find the *Earliest Start* of the activity as the latest of the *Earliest Finish* times of all its predecessors. If it has no predecessors, its *Earliest Start* is the start time of the project.
   b. Find the *Earliest Finish* of the activity using equation 11.2.1.
4. Find the *Earliest Finish* of the *project* as the latest of the *Earliest Finish* times of all its activities.
5. If the *Earliest Finish* is later than the project deadline, think again!
6. In descending topological order, for each activity,
   a. Find the *Latest Finish* of the activity as the earliest of the *Latest Start* times of all its successors.
   If it has no successors, its *Latest Finish* is the project deadline.
   b. Find the *Latest Start* of the activity using equation 11.2.2.
7. Find the *Latest Start* of the *project* as the earliest of the *Latest Start* times of all its activities.

3. Populate the *Enrolments* table with the data for beginning candidates.
4. Demonstrate the timetabling program.
5. Populate the *Candidates* and *Enrolments* tables with skeleton data for continuing candidates, and produce the examination timetable.
6. Populate the *Candidates* table with the remaining data for continuing candidates.

The first thing to note about this plan is that each activity consists of several smaller activities, for example, the first step involves at least three sub-activities: one for each table. The second thing to note is that this plan is just part of a larger future plan, involving the interface with the Bursary, allowing candidates to make Internet enrolments, and developing the expert system for programme advice. The third thing to notice is that some important activities haven't been mentioned, the most obvious being to finalise and implement the database schema.

We shall assume that we are at a point in time when the kernel specifications of the atoms (Section 5.6) and events (Section 5.7) are complete. We now need to flesh out the specifications by designing the database schema and some interfaces. From these, we can proceed with the implementation of certain kernel events. We need these completed on schedule to reach our **project milestones** on time: the dates when candidates enrol and are examined are predetermined and not negotiable.[8]

[8] Milestones can be modelled as activities of zero duration.

Table 11.2 Finding a critical path. The *Duration*, *Earliest Start (E.S.)*, *Earliest Finish (E.F.)*, *Latest Start (L.S.)*, *Latest Finish (L.F.)*, and *Slack*, for each activity needed to populate the *Subjects*, *Programmes*, and *Components* tables. The rows are topologically ordered. The earliest times are calculated in a pass from top to bottom, and the latest times in a pass from bottom to top. The earliest start time for an activity is the maximum of the earliest finish times of its predecessors. The latest finish time of an activity is the minimum of the latest start times of its successors

Activity	Duration	E.S.	E.F.	L.S.	L.F.	Slack
Finalise Database Schema	2	0	2	0	2	0
Design Subjects Interface	2	2	4	2	4	0
Design Programmes Interface	2	2	4	7	9	5
Design Components Interface	3	2	5	6	9	4
Create Subjects Table	1	2	3	3	4	1
Create Programmes Table	1	2	3	8	9	6
Create Components Table	2	3	5	8	9	5
Write Subjects Update	4	4	8	4	8	0
Write Programmes Update	3	4	7	9	12	5
Write Components Update	4	5	9	9	13	4
Populate Subjects Table	5	8	13	8	13	0
Populate Programmes Table	1	7	8	12	13	5
Populate Components Table	5	13	18	13	18	0

Also, from a behavioural point of view, the time between Step 3 and Step 4 should be as short as possible, to reward enrolment staff for their efforts. This implies that work on the timetabling program should be begun immediately, even before its data are ready.[9]

To illustrate how the start and finish times of each activity are calculated using Algorithm 11.1, consider the graph of Figure 11.1, which shows the main activities needed to achieve the first objective above. Table 11.2 lists the activities in topological order, with an initial vertex at the top and a final vertex at the bottom. The earliest start for the initial vertex is zero, and its earliest finish is the end of day 2, which is zero plus its duration. Its earliest finish becomes the earliest start for each of its successors. The calculation continues down the table, row by row. Row 7 (*Create Components*

[9] Which implies in turn that the larger scale plan on page 279 should itself be expressed as an activity graph.

Table) has two predecessors (*Create Subjects Table* and *Create Programmes Table*), so it is given an earliest start equal to the *later* of their two finish times. Continuing in this way, *Populate Components Table* is assigned an earliest start of day 13, so its earliest finish is day 18.

To find the *latest* start and finish times, the process is reversed, by working upwards from the bottom of the table. We assume, for now, that the latest finish time for the project as a whole is day 18. The latest start for *Populate Components Table* is therefore day 13, which becomes a potential latest finish for each of its predecessors. When an activity has more than one successor, its latest finish is the *earliest* of their latest start times.

Once the earliest and latest times are known, the *slack* is simply their difference. Activities with zero slack are called **critical activities** and lie on the *critical path.* In this example, the critical path passes through *Finalise Database Schema, Design Subjects Interface, Write Subjects Update, Populate Subjects Table*, and *Populate Components Table.* If the project as a whole is to be completed in 18 days, no *critical* activity can be late. On the other hand, activities that have non-zero slack can be completed any time between their earliest and latest finish times.

Before we continue, consider what may be the greatest benefit of the critical path method: *No matter how many workers we throw at it, this development plan cannot be completed in fewer than 18 days.*

What if we don't have 18 days? Is there an alternative plan that will meet an earlier deadline?

In this case there is. The most noticeable defect in the plan is that the IT candidates cannot start work until day 9. Why not have them prepare the data by typing it into spreadsheets before the database tables are ready? A single programmer could then design the schema and define the tables within 6 days. The spreadsheet data could then be loaded into the database using a utility program. The *Subjects, Programmes,* and *Components* update programs would remain to be written, but they would no longer be urgent, and the unused programming resources could be focussed on writing the *Enrolments* update instead.

Nonetheless, in the following, we ignore this exciting possibility and return to the original scenario.

Often, the next stage in developing a plan is to draw a **Gantt Chart** (Figure 11.2). Although the Gantt Chart contains exactly the same information as Table 11.2, it makes it easier to see how slack time may be exploited. Remembering that the actual task of populating the tables is to be done by IT candidates, the programming work must occupy only 13 days. Even so, 24 days' worth of work must be done, so at least two programmers will be needed. As it happens, only one activity can be done on days 1 and 2, and only one programming activity remains on day 13. Consequently, 21 days' programming work has to be done in the 10 days between day 3 and day 12. We can resolve this issue either by adding a third programmer or by extending the deadline. A third possibility is for *both* programmers to work on the first task:

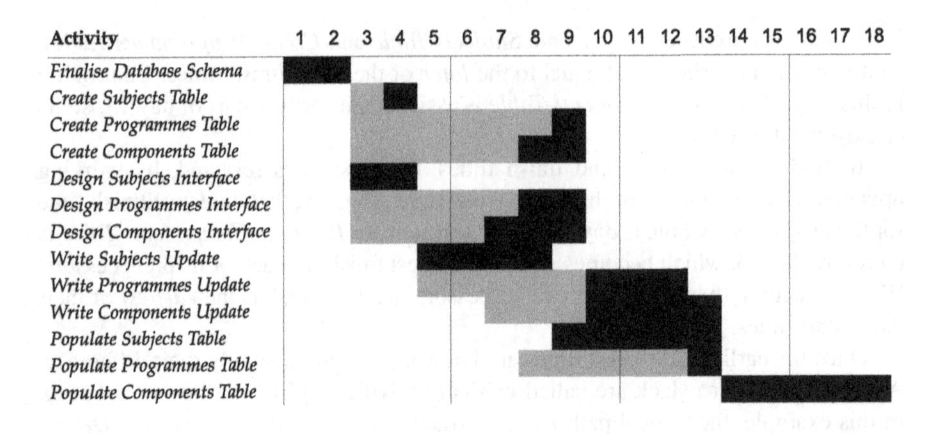

FIGURE 11.2

A Gantt Chart. The black bars show the latest durations of activities. The grey bars show the times when activities may be started before they become critical. The critical path comprises activities with zero slack.

Finalise Database Schema. Even so, there is no guarantee that two programmers can complete this task any sooner than one can.[10]

11.3 PROJECT CONTROL

To ensure that projects keep to their schedules, it is important to measure and review progress. This serves three purposes:

- To give an early warning that a project may be late unless extra resources are made available,
- To stimulate implementors to redouble their efforts, and
- To improve the process of estimation, which can help with later projects.

It is normal to review progress at regular meetings — typically weekly. This serves to make implementors constantly aware of deadlines, and also to discover if a change of plan is needed. It is also normal to deliver working software to the client on a regular basis — typically monthly. Ideally, the product should offer the client an immediate benefit, even if it is only a prototype that can be used for evaluation or training. Even if a product isn't immediately useful to the client, it should still be delivered in order to enforce discipline, motivate the development team, and give the client a tangible indication that progress is being made.

[10] One woman can produce a baby in 9 months, but that doesn't mean that two women can produce a baby in four-and-a-half.

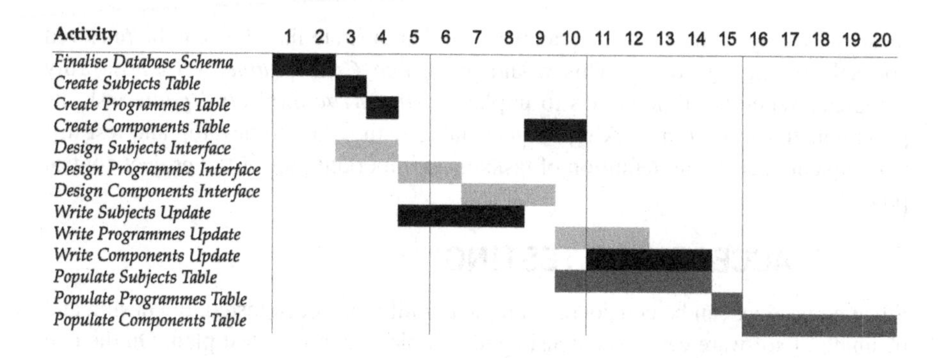

FIGURE 11.3

A project timetable. The dark grey bars in the lowest three rows show the data-entry tasks allocated to the IT candidates. The black bars show the tasks allocated to the first programmer, who is experienced in database implementation. The light grey bars show the tasks allocated to the second programmer, who is more experienced in interface design.

As a consequence, it would be reasonable in *this* project to stretch the deadline by 2 days, thus making it a period of four 5-day weeks. The additional slack time should make it possible to complete the programming work with only two programmers.

Figure 11.3 shows how two programmers can complete the work in 13 days in such a way that the IT candidates can start populating the tables on time. Unlike the problem of finding a critical path, the **task scheduling** problem is NP-complete. It is analogous to the problem of constructing a study plan discussed in Section 9.3, with the additional complication that, due to individual differences (page 286), the time taken to complete an activity depends on who it is assigned to.[11] It is well-suited to solution by the Branch and Bound or A* Search algorithms; a tight upper bound on the remaining cost of a plan can be estimated by a *critical path analysis* of its remaining tasks.

It is usually clear when a product delivered to the client is complete: The usual **definition of 'done'** is that the programmer has 100% confidence that absolutely no more work is needed before the product can be handed to the client. Nonetheless, confusion and delay can result from a poor definition of when an individual *task* is complete. The programmer may think that *Create Subjects Table* is complete once the *SQL* statements to define the *Subjects* table have been written (or perhaps, accepted by the database management system), whereas, what was really meant was that the programmer would also create a small test population of subjects (say

[11] The *Prerequisite* graph of Figure 9.1 contains two critical paths, from *maths101* to *services*, and from *maths101* to *interface*, which means that any programme containing these pairs of subjects must take at least six semesters. In comparing task scheduling with programme planning, tasks are analogous to subjects, skills (e.g., programming) are analogous to semesters *A* and *B*, and the number of workers per skill is analogous to the number of subjects per semester.

those belonging to the *ecom* programme) and demonstrate that they can be retrieved correctly by an *SQL* query. This would prove that *Create Subjects Table* is truly complete. When the time comes to implement the *Write Subjects Update* task, its programmer will need a working *Subjects* table with which to develop and test it.[12] Consequently, accurate definition of tasks is an important part of the project control process.

11.3.1 ACCEPTANCE TESTING

Since no product can be considered complete until it has been tested, the design and planning of software cannot be considered complete without a test plan. On the one hand, testing is essential; on the other, once a milestone has been reached, and the software delivered, the testing seems to be just so much wasted effort. The answer is not to save testing until the product seems to be complete — as a rule of thumb, the time it takes to find a bug seems to increase with the square of the length of the program. The answer is to find efficient ways of incorporating testing into the overall plan.

Some languages allow procedures to be tested by invoking them from the command line. Such languages encourage **bottom-up development**: starting by writing the lowest-level routines (those that make no procedure or function calls except to library routines) and then, in effect, adding these tested routines to the library and developing successively higher levels. Bottom-up development in other languages requires special test routines to be written.

Suppose we wanted to test a procedure that is meant to implement the decision table of Table 9.2. To test the first rule, we might use something similar to the following snippet of code, which carefully checks that the *Admit* procedure doesn't corrupt its input data:

```
Candidate_State = 'Admitted';
Subject_State = 'Offered';
Filled = 99; Quota = 100;
Programme = 'ecom'; p = 'ecom';
Component_State = 'Offered';
Admit(Candidate_State, Subject_State, Filled, Quota,
      Programme, p, Component_State, Create_Enrolment,
      Error_Message);
if Candidate_State == 'Admitted'
  and Subject_State == 'Offered'
  and Filled == 99 and Quota == 100
  and Programme == 'ecom' and p == 'ecom'
  and Component_State == 'Offered'
```

[12] This would suit the method of program development by *stepwise refinement* discussed on page 372.

```
    and Create_Enrolment == 'Yes'
    and Error_Message == ''
    then write 'OK'
  else
    write(Candidate_State, Subject_State, Filled, Quota,
          Programme, p, Component_State, Create_Enrolment,
          Error_Message);
```

Although there are only nine rules in the decision table, the rules share five preconditions, which have 32 combinations of truth values. We should therefore write 32 snippets of this kind. Fortunately, it isn't hard to generate all the possible combinations using a suitable test generator, a spreadsheet, or even an ordinary text editor. Generating the test cases is not a waste of effort. They become useful documentation, which tells a potential user exactly what the procedure does — and just as important, what it omits to do. If a procedure needs to be rewritten to make it faster, the test cases ensure that the new implementation is compatible with the old.

Testing should demonstrate that *all* the program code has been executed — that no bugs have escaped detection. Sometimes this leads to a program being simplified, because some branch of the code has been written to handle a combination of conditions that can never arise. Many software development systems include **program profilers**, which can measure exactly how often each branch of a program is executed.

A danger with bottom-up development is that modules might be written to provide more functionality than will ever be needed. For example, an expert system application might need a generalised *Graph* object class to represent several relations of interest. The temptation would be to implement methods for transitive closure, transitive reduction, and so on, which might never prove to be needed in practice.

The opposite approach to *bottom-up* is **top-down development**, in which the main, outer, procedure is written first — called procedures being replaced by dummy code. This code merely has to handle the specific cases that arise in testing the outer procedure, and might be of the form

```
procedure Admit(Candidate_State, Subject_State, Filled,
                Quota, Programme, p, Component_State,
                Create_Enrolment, Error_Message) is
  if Candidate_State == 'Admitted'
    and Subject_State == 'Offered' and Filled == 99
    and Quota == 100 and Programme == 'ecom'
    and p == 'ecom' and Component_State == 'Offered'
    then {Create_Enrolment = 'Yes'; Error_Message = ''}
  else ...
```

Clearly, the effort involved in writing all the possible test cases can exceed the effort involved in writing the procedure itself — unless the test cases are produced automatically by some kind of test generator.

One danger with top-down development is that programmers may defer solving difficult problems by pushing them down into lower and lower levels. For that reason, top-down development should not begin until the whole hierarchy of modules has been specified. A second danger of top-down development is that the implementor may fail to recognise that a particular problem is an example of a more general one, for example — and in contrast to bottom-up development, that several specific requirements could be implemented by a single *Graph* class.

A third approach is **stepwise refinement**, which is a type of evolutionary development (page 7). In this approach, a complete program is always tested, but it might not be fully functional. For example, the *Subject Update* program might initially merely display a form, with data values being supplied by built-in constants. As a second refinement, the interface might display a fixed row of the *Subjects* table. A third refinement might allow a subject code to be entered interactively and its matching row displayed. A fourth might allow some particular datum to be updated, and so on, until the final stage, which might be to implement the display of error messages that give the operator helpful feedback. The idea is that at each stage only a small section of new code needs to be tested, yet no special software has to be written to help with testing. In addition, development steps can be discussed with the operators who will eventually use the program. After the first step, the dummy form can be shown to its potential users, and their feedback can be used to refine the next step. As part of the final step, the operators could be asked how they would interpret the planned error messages, and so on.

A second strategy used in stepwise refinement is to give simplicity priority over efficiency: rather than develop the most efficient form of a procedure, replace it by a less efficient but *simpler* form.[13] Rather than implement an intricate linked data structure, replace it by a set of tables.

Stepwise refinement was used to develop the expert system of Section 9.3. The motive for developing the program was to remove uncertainty and prove that an on-line programme advice system could be a success.

The program was developed using the *Prolog* logic-programming language. (One advantage of *Prolog* is that merely by defining **from**, **needs**, etc., as operators, its built-in parser was able to convert a text file directly into internal tree structures similar to that of Figure 9.2.) The first iteration of development was to describe the structure of the *ecom* programme as a series of **needs** trees that could be read by the parser.[14]

The second step was to develop a procedure that could tell if a set of subjects satisfied the requirements of a given **needs** tree. This served to check whether a study plan was both *valid* and *sufficient*. At this stage, the program used exhaustive search, but proved too slow. It became clear that pruning was necessary, and the heuristic of checking for feasible bit maps was devised, as explained on page 312.

[13] Often referred to as **KISS**, or **Keep It Simple, Stupid**.

[14] A second advantage of *Prolog* is that its data structures are self-describing, so that a built-in procedure can display any acyclic structure in a readable form — saving a lot of time when debugging. (Cyclic structures can cause the built-in print routine to loop. *Prolog* programmers avoid them.)

The third step was therefore to find a topological sort of the subjects using depth-first search and use it to assign each subject to a particular binary digit, in preparation for the bit map representation. This coding scheme was then used to 'compile' the **needs** trees into bitmap form and to develop a decoder to turn binary representations back into subject codes. This step was an example of refinement by moving to a more sophisticated data structure.

The next step was to implement the *feasible* pruning heuristic. A subsequent refinement ordered the *valid* and *sufficient* plans by total fees.

The refinement of adding *Branch and Bound* search was initially implemented using a simplified future cost function. This cost function was then refined, in stages, as explained in Section 9.3, to eventually become that described on page 319.

A further refinement would have been to exploit the *dynamic programming principle* better by using the $A*$ algorithm, but the system already performed so well that this refinement was never implemented.

The refinement of adding user dialogue — allowing the user to specify desirable and undesirable subjects, and so on — wasn't implemented either, as it is clear that this would cause no technical difficulty.

11.4 PERSONNEL MANAGEMENT

One of the responsibilities of a project manager is to get the best performance out of the programming team. This is not done by bullying, but by exploiting the psychological phenomenon of **achievement motivation**:[15] programmers think of themselves as creative people, who set their own goals and take pride in achieving them. These goals may be short-term, as in, 'Complete *Write Subjects Update* in four working days', or they may be long-term, as in, 'Develop my knowledge of *HTML*'. So although we suggested that a project timetable (Figure 11.3) could be derived by a search algorithm, there is usually more to it than that. A project leader needs to be aware of the skill set of each team member and know when it is time for a member to acquire new skills. People differ in this respect. Some are content to constantly hone the skills they have; others are anxious to gain new ones.

The project leader can also exploit Herzberg's Theory of Motivation by making team members responsible for meaningful units of work, for example, by writing complete programs, rather than modules. It is important for the project leader to try to boost each member's self-esteem. This can be done by ensuring that the tasks that are set are not too difficult, so the worker will receive favourable feedback at the next project review meeting. On the other hand, the tasks should be challenging enough to give the worker a sense of achievement.

The project leader already had his team members in mind when constructing the activity graph of Figure 11.1 (page 364). Given the experience of the team members,

[15] Achievement motivation can be observed in children by asking them to throw balls at a target. Some stand so close they cannot fail, others stand so far away they cannot succeed. The third group, those who are motivated by achievement, stand at a distance where success is a realistic challenge.

the tasks they were assigned were chosen to be challenging, but not overwhelming. Goals should be such that team members can take pride in achieving them — without the project leader fearing that they might fail.

In constructing Figure 11.3 (page 369), the project leader would have been aware that one team member had good *SQL* skills, and the other was skilled at designing forms. This second member, however, having only had experience with forms in a local-area network, wanted to learn how to use *HTML*. The schedule of Figure 11.3 allowed the first member to complete the *Subjects* update. Since the *Programmes* update is logically similar, the second programmer can then modify the *Subjects* update, learning the basics of *HTML* in the process. Meanwhile, the first programmer will be working alongside, writing the *Components* update, and can be ready to answer questions. Such subtle interactions would prove hard for a planning algorithm to capture.

The times allocated were *agreed* between the project leader and the individuals involved. This is an example of **management by objectives** (**MBO**). People will strive to achieve a goal that they have already agreed is realistic. But if they are bullied into accepting someone else's unrealistic expectations, they will feel defeated before they begin. They will interpret failure as proof that their manager is ignorant and incompetent.

11.5 STANDARDS AND DOCUMENTATION

Some people regard program documentation as useless, because it is always out of date. Others (myself included) regard it as central to successful system development. The real questions are: *What* documentation should be produced? And *when* should it be produced?

To answer the second question first, documentation should *precede* programming. Answering the first question, we have already discussed the kinds of documents we need: event-state diagrams, FD graphs, mock-ups of forms and reports, kernel specifications, decision tables, story boards, process graphs, activity graphs, and the like. These are all *constructive*, in that they result in decisions that are steps towards writing working programs. (We may contrast these with charts that merely summarise intuitive decisions that have already been made.) Particular program development technologies may benefit from additional specialised planning tools, but these are the essentials.

Too often, programmers regard documentation as a waste of time and want to begin coding as soon as possible. But, without a clear objective, much time may be wasted. It is much easier to refine a story board for a user interface while it is still an FSA and a series of paper sketches or mock-ups than it is to revise a set of working programs. The ability to get things right the first time *without* prior planning only comes with much experience, and even then the lack of a documented plan is excusable only if the worker isn't part of a team. In the event that a key team member becomes sick, is reallocated, or finds new employment, his or her *mental* planning is completely wasted.

Another important aspect of early documentation is that it often reveals that one's first thoughts are misguided and that a better approach is possible. We can see examples of this in the Gantt Chart of Figure 11.2. Two of the three different data-entry forms proved to have so much in common that integrating them — or simply assigning them to the same implementor — would save much work. (Despite this, rather than give the two forms to the same person, the project manager thought it wiser to use one of them as a training exercise for a second team member.)

Most organisations have a set of **coding standards**. Coding standards make program debugging and repair easier by forcing team members to adopt a consistent style and a minimum standard of documentation. Suitable standards depend on the programming language used, the kind of application, and the length of time that the product is supposed to last. I regard only one standard as universally valuable: the names of variables and procedures should be correctly spelt English words that are self-explanatory.[16] Variables named x and y are permissible — as local variables only — if they have a short context and a consistent meaning. A good rule is that if a name is *not* a proper word, its meaning should be explained by a comment, and this meaning should hold globally.

Miller's Law (page 267) suggests that procedures should have at most seven parameters, because it would otherwise be hard to remember them all while debugging the code — but this should probably be a guideline rather than a rule. Similar limitations of short-term memory suggest that procedure calls shouldn't be nested more than three levels deep, but this is often unavoidable.[17]

It is important to distinguish between standards that aid communication or that accept the limitations of the human brain from those that merely express someone's personal aesthetics. For example, it is reasonable to insist that **if**-statements should be properly indented, but not quite so reasonable to insist that each level must be indented by exactly four spaces.

11.5.1 LITERATE PROGRAMMING

Should programs contain comments? In my view, only if the comments are written first. The programmer may reasonably begin by writing an explanation of what is planned, perhaps in the form of some imaginary higher-level language. Experiments have shown that writing a single block of comments is more useful than commenting individual lines of code.

In the context of what we have discussed earlier, a programmer may find it helpful to include a copy of the kernel specification of an event as a comment before coding its procedure. This would serve both as a reminder to the programmer of what is needed and a useful guide to someone reading the code. The only problem with this

[16] I once worked with a programmer who had a limited vocabulary where variable names were concerned. His programs were full of names like *total*, *total1*, *totalA*, and even *totle*. Having spent many hours failing to get one particular program working, I finally persuaded him to replace all the names with ones that were more explicit — and guess what?

[17] Function and method calls that are self-explanatory — and have trustworthy implementations — don't count.

suggestion is, as we have seen, event segmentation (Section 7.4.2) may cause an event procedure to be split between several program objects.

There are **literate programming** software tools that allow a program to be embedded in a sea of comments and then allow code segments to be extracted in order to be compiled. These tools allow program text to be presented and explained in a different sequence to that needed by the compiler and may even incorporate code written in several different languages. An important virtue of such tools is that when the logic of an event must be segmented (as in a batch system or object-oriented program), its segments can be collected together in the documentation, so that the whole may be more easily comprehended.

Such tools typically contain two functions: **tangle**, which shuffles the embedded code into source files, and **weave**, which typesets the documentation, perhaps inserting a table of contents. Some literate programming tools include macro-processing features that allow commonly used general-purpose code to be parameterised for a specific application. For example, the generalised logic of the sequential-update algorithm could be tailored to use a particular master file and transaction file.

A drawback of most literate programming tools is that they work in one direction only: if an error is found during debugging, it must be corrected by editing the source file used by the tool, then *tangling* it again.[18] A few tools exist that work in the opposite direction: comments are written within the program code. These comments can include mark-up. The comments can then be extracted and incorporated into a publication-quality document.

A potential drawback of either approach is that individual objects can be elegantly documented without any documentation being produced to describe the system as a whole. It is therefore useful to have tools that can extract information from a badly documented system — information that isn't already obvious by looking at the code. The following hypothetical tools would help extract useful information from an existing set of programs:

- With a tool to cross-reference procedures and database attributes, a programmer might find which events in the Academic Records System inspect or update *Subject State*, for example. From this information, a programmer could then sketch the event-state diagram of a *Subject* atom.
- It is easy to see what procedures a given procedure invokes, but not so easy to discover which procedures invoke it. A software tool could extract this information and display the relationships between procedures as a matrix or graph. If the tool then isolated the strongly connected components of the graph and presented them in topological order, so much the better. (Such a graph should once have been the foundation for a critical path analysis for a top-down or bottom-up implementation — and might also have revealed a development sandwich.)

[18] Many would argue that the object-oriented practice of gathering data definitions and methods together into a class definition makes literate programming unnecessary. Although this is partly true, the example of Section 7.4.2 shows that objects aren't always as self-contained as one might like.

- With software tools that can analyse data flows, the analyst could derive a state dependency graph. (This isn't as simple as it might seem because the names given to arguments within a procedure need not match the names of the arguments passed to the procedure when it is invoked.) In Chapter 7 we saw how to use a state dependency graph to make systems deadlock free and more efficient, and even to decide whether a system should have modes.
- With a tool that could analyse a database schema, an analyst could reconstruct its FD graph. (See Section 4.4.1.)
- Given a tool that could trace logical paths, a programmer could use it to express procedural code more concisely in the form of decision tables.

Such reverse-engineering tools can be expected to produce their output in the form of matrices or spreadsheets. Matrices have an advantage over linear text in that one can either scan their rows or their columns, revealing two contrasting aspects of the same information. In some cases, a matrix can be conveniently drawn as a graph, but in typical-sized projects, most such graphs would overwhelm the eye.

There are perhaps only two occasions when comments should be added *after* code is written:

- First, users of a program module may have difficulty understanding what it does or what its limitations are. Things that were obvious to the writer may be far less obvious to others — or even to the original author 6 months later.
- Second, the writer may need to add helpful comments during debugging. (Personally, I am a believer in **holistic debugging**: If your program doesn't work, it is because you don't understand it. If you don't understand it, it is because it is badly written. Make the names of procedures and variables more accurately reflect their purposes, and add explanatory comments until the bugs become obvious!)

11.6 FEASIBILITY STUDIES

There are two kinds of feasibility study: financial feasibility and technical feasibility. We have already dealt with financial feasibility to some extent when discussing *discounted cash flow* techniques (page 360). The key questions are what a project will cost, what its benefits are worth, and how long it will take to achieve them.

11.6.1 FINANCIAL FEASIBILITY

Financial feasibility is evaluated at an early stage and is largely the responsibility of the client. Even so, the systems analyst or project manager is heavily involved in estimating how much work must be done. With enough experience, it may be possible to assess the effort needed merely by counting how many database tables, events, forms, and reports have to be created, after perhaps assigning each one a degree of difficulty. Difficulty can be measured in terms of the numbers of attributes inspected and updated, numbers of rules, and so on. Such **metrics** can be combined with the project manager's assessment of each team member's performance for different kinds of task.

Some organisations keep detailed statistics of this kind to help with future estimates. In the context of a given fixed technology, the metrics can become quite accurate. An important use of metrics is to compare estimates with achievements at weekly or monthly project meetings. Realising that the original estimates for a project have proved over-optimistic can give early warning to the client that the project might overrun — unless given more resources. The client can then decide how to deal with this, bearing in mind that the money already invested is a sunk cost, which can't be recouped by cancelling the project.

Sadly, it is rarely the case that additional manpower will speed things up in the short term. **Brooks's Law** states that adding manpower to a late project will make it even later. Workers who are already fully loaded may need to spend time bringing the new helpers up to speed. If there are n workers, there are $O(n^2)$ potential inter-personal communication channels. Therefore, unless tasks have been structured to limit the number of channels, communication can become more time-consuming than productive activity. To make proper use of additional workers, the whole communication structure would need to be revised — if the task network can be revised to permit it. As within a computer, some tasks can be done in parallel, and some tasks must be done serially. Ultimately, however many resources are available, the serial tasks will dominate. Adding resources to the parallel tasks will show diminishing returns.

There *are* situations where speed is directly proportional to manpower, as in digging a ditch.[19] This gives rise to the concept of the man-hour as a unit of work. Clients who have worked in environments where the man-hour has some validity fail to understand the nature of software development, believing that all programmers are interchangeable, and that *more* means *faster*. In the activity graph of Figure 11.1, only the data-entry tasks are of the kind where more manpower means faster progress. If the client were to hire additional IT candidates, the project could be completed a few days earlier.

11.6.2 TECHNICAL FEASIBILITY

Technical feasibility concerns whether a project can meet its performance objectives. We have already seen some examples of this:

- Modelling the queues for the enrolment process (page 338) checked that the proposed system would cope with its expected load.
- Developing an expert system for programme advice checked that it would be feasible for candidates to make wise enrolment choices over the Internet.

The primary aim of a technical feasibility study is to remove uncertainty. Since the present value of a proposed project depends on the probability of its success, it is important to quantify the risk before committing money to it.

[19] Up to a point. Once the workers stand shoulder to shoulder, extra workers just get in each other's way.

For example, some uncertainty still surrounds the automatic timetabling of examinations: Will a computer system be able to fit the examination sessions into the available period, as the existing manual system has proved that it can?

A key problem here is that the input data for the algorithm won't be available until after enrolments are complete. If the timetabling application proves a failure, time and money will have been wasted, the enrolment staff will feel let down, and the reputation and morale of the project team will suffer. It is crucial that automatic timetabling succeeds. Unfortunately, even after the enrolment period, the data will only contain information for *new* candidates, and even a successful experiment won't guarantee similar success when a set of data for all candidates become available.

The information that is needed for input to the timetabling process is a list of subject clashes: two subjects clash if at least one candidate is enrolled in both of them, which implies that their examinations mustn't be held at the same time. To assume the worst case, that every subject clashes with every other, would lead to a timetable that only allowed one subject per session, which is clearly unrealistic.

A more realistic test could be made by using the fact that subjects can only clash with subjects that are common to the same study programme and the same semester. Each subject might be assumed to clash with all *Offered* subjects that share any programme with it. This scenario can be tested as soon as the *Components* table has been populated, *preceding* the enrolment period.[20] If the timetabling program works well with this set of data, it is certain to succeed in practice.[21]

But what if this test proves unsatisfactory? If the programme proves inadequate, it is well to know as soon as possible. With luck, the subject clashes that were used to schedule the previous semester's examination timetable might still be available. If they are, and a test using these data proves that the timetabling algorithm is adequate, all is well. All these considerations suggest a strategy for reducing the uncertainty about the project's success before too much effort is wasted.

Likewise, in *any* project, technical feasibility should be assessed before making any serious financial commitment. Unfortunately, in the eagerness to deliver a product on schedule, such experiments may seem like a waste of time — but if omitted, a project runs the risk of being an expensive failure.

11.7 METHODOLOGY

A **methodology** is essentially a series of procedures that an organisation uses to develop products. It is not the place of this book to suggest a universal methodology, if such a thing even exists. This book has presented a set of *tools* that will be

[20] In theory, Subjects S and T can't clash if S requires a previous pass in T. Therefore, if the *Prerequisites* table has already been populated in preparation for the programme advice expert system, it can immediately be put to good use.

[21] If *this* experiment is successful, it would be better for the project leader to remain quietly confident rather than to broadcast it. Letting the Enrolment Office staff know that its efforts to record enrolments are pointless would destroy its motivation to transcribe existing records. What is more, its efforts in constructing timetables in previous years will be revealed as having been pointless too.

common to many systems projects. In a given project, some may prove less useful than others — not everyone will need to worry about deadlock, for example. Also, many applications will need specialised tools of their own. For example, an aircraft control system has to concern itself with reliability, equipment failure, redundancy, and so on, but these are considerations that few computer applications need to worry about — except for routine data back-up.

Despite the above remarks, here is a sensible order in which the tools can be applied:

- Analyse the problem:
  - Discuss the objectives of the project with the client to determine which of its aspects are important and valuable.
  - Start by thinking about the kinds of atoms that are involved (Chapter 3).
  - Use *event-state analysis* to discover the states they can be in and what events can change their states (Section 3.5).
  - Consider when to distinguish individual atoms and when to merely count them (Section 3.6).
  - By thinking about shared events (Section 3.9) or by using entity-relationship analysis (Section 4.3), discover the nature of the relations between atoms, perhaps reifying relations to discover new kinds of atoms (Section 4.3.1).
  - Draw a functional dependency graph (Section 4.4.1) — whether you expect to use a DBMS or not.
  - Use the events and conceptual structures you have identified to write a *kernel specification* (Chapter 5).
    This step marks the important boundary between analysis and synthesis.
- Remove uncertainty:
  - If you have identified any mission-critical event for which you do not know a tractable algorithm, investigate its technical feasibility before committing resources to it, perhaps experimenting with expert system techniques like those in Section 9.3.
  - If the project is a part of a control system, analyse its dynamic behaviour (Section 10.2).
  - Use use-definition analysis to find the data flows induced by each event (Chapter 7) to see if they cause any deadlock problems (Section 6.5.3), perhaps resolving them by using modes (Section 7.6).
  - See where data flows interact with external interfaces, especially those involving humans.
  - Decide on the infrastructure that will be used. If there is any uncertainty about the capacity of the proposed infrastructure, simulate its queues and servers to determine whether additional resources will be needed (Section 10.1).
  - Estimate the time and money that will be needed to complete the project (Sections 11.1 and 11.2).
- Expand the specifications:
  - Decide how to structure your data (Sections 2.6–2.7, or Section 6.2).

- Create mock-ups of the human interfaces (including error messages), paying attention to psychological principles (Chapter 8), and discuss the interfaces with potential users.
- Document business rules and error-processing rules using decision tables or a rule-based specification language (Section 9.2.1).
- Segment kernel specifications into object or other program module specifications (Section 7.4.2).

- Implement the system:
  - Structure project development as best you can to satisfy a host of conflicting interests (Section 11.2).
  - Document task specifications and write test cases (Section 11.3.1).
  - Write the programs.

This list is in roughly the right order for most projects, always remembering that systems work is often iterative and that later steps can reveal omissions and errors in earlier ones that call for revision.[22] Even so, the importance of these steps will vary from project to project, and other steps, not listed, may prove even more important.

For example, consider the design of the controller for a set of lifts (elevators) in a large building. The business requirement is simple, 'Get me from Floor *A* to Floor *B*!' This requirement could be satisfied by several possible *infrastructures*: a flight of stairs, a series of escalators, an endless belt of ascending and descending cars, a set of lift shafts, or some combination of them.

Nonetheless, once we choose a conventional lift system infrastructure, we can easily identify its *atoms*: cars, floors, passengers, press-buttons, etc. We might then identify *events* — including *Call Lift* and *Go To Floor*, and *states*, such as *Stopped at Floor* 5. We can then do a *data-structure analysis* — which will prove almost trivial. *Data-flow analysis* will tell us that a passenger can either specify the destination while waiting for the lift or after entering it. The *interface design* is simple, and hardly needs any thought, although we should remember that passengers are likely to be irked by the car carrying them in the wrong direction or stopping unnecessarily. The *important* part of designing a lift system is therefore likely to be a simulation study:[23]

- How quickly can the system get people to their offices in the morning?
- How quickly can they leave after work?
- How much energy will the system consume?
- How long will passengers have to wait for service?
- How many lift shafts will be needed to satisfy the performance criteria?
- What is the most efficient algorithm for scheduling the movements of the lift cars?
- Does it help to know which floor a passenger wants to go to before assigning a particular car to the job?

---

[22] We saw an example of this when we designed a form for Internet enrolment and realised the need to include an encrypted password in the *Candidates* table.

[23] Assuming that the design parameters are not already predetermined by building codes and standards.

Although a lift system has little in common with an academic records system, and the *importance* of the various steps is quite different, the basic steps remain the same.

11.8 SUMMARY

A future benefit is not as valuable as a present benefit. The client's accountant will therefore use discounted cash flow to value a project. This puts pressure on the project manager to deliver useful software as early as possible. In addition, if a project involves risk, its expected value is reduced, and it may be uncertain whether the project will have a positive present value. Therefore a preliminary technical feasibility study is sometimes needed to reduce such uncertainty.

The system designer and project manager must find a development structure that satisfies several goals:

- It should allow the regular delivery of products to the client — ideally, useful ones.
- Each task should have a definition of when it is complete — and a means of proving it.
- The dependencies between tasks should be acyclic and should lead to an activity graph whose critical path is sufficiently short.
- The tasks should suit the capabilities and motivations of their implementors — concerning which we discussed achievement motivation and management by objectives.

There are fashions in project management: managers, especially those with no hands-on experience of software development — and blissfully ignorant of Brooks's Law — are constantly struggling to control what is often a fundamentally risky process. The tools of sound project management hardly change. The jargon changes constantly.

11.9 FURTHER READING

Discounted cash flow was first described by John Burr Williams in his 1938 Harvard University PhD thesis, 'The Theory of Investment Value', which was reprinted in 1997 (ISBN: 0-87034-126-X).

One of the earliest articles describing the critical path method is James Kelley's and Morgan Walker's 'Critical-path Planning and Scheduling', *Proceedings of the Eastern Joint Computer Conference* (1959) 160–173.

Brooks's Law and many other important software development ideas appeared in Frederick Brooks's 1975 book, 'The Mythical Man-Month', reprinted in 1995 (ISBN: 0-201-23595-9). The reprint contains the essay 'No Silver Bullet', in which Brooks argues that order-of-magnitude improvements in software productivity cannot happen in the short term. In other words, the investment in time and effort needed to develop better software tools usually far exceeds their benefit to any individual project. Their development costs have to be amortised over many projects.[24]

Achievement motivation was first described in 1953 by David Clarence McClelland in his book, 'The Achievement Motive' (*Appleton-Century-Crofts*).

Management by objectives was described by Peter F. Drucker in his 1954 book, 'The Practice of Management', (ISBN: 0-06-011095-3).

The virtues of literate programming are expounded in Donald E. Knuth's 1992 book, 'Literate Programming' (ISBN: 978-0-937073-80-3). Knuth's original literate programming tool, *Web*, was limited to one particular language and to TEX documentation. Since then, other tools have appeared for other languages. The open-source *FunnelWeb* literate programming tool supports any programming language — indeed a mixture of them — and can create beautiful documentation in both *HTML* and TEX format, without the programmer needing to know any *HTML* or TEX.

11.10 EXERCISES

1. Although a low cost of capital usually increases the present value of a project, it is not always the case.
   A small opencast mining project requires an initial investment, offers a good return in the first year, but is complete by the end of its second year. In its second year, it makes a loss, because the terms of the lease require the site to be filled and made good. The initial investment is \$10,000,000, the return at the end of Year 1 is \$21,500,000, and the loss at the end of Year 2 is \$11,550,000. Express the present value of the project as a quadratic equation in $1/(1+r)$, the discount factor. For what values of the cost of capital is the project worthwhile? (Hint: either solve the quadratic, or use a spreadsheet.)
   At what value of the cost of capital is the present value of the project greatest? What is the maximum value of the project?
   Would you advise the board of directors to proceed with this project?
2. Consider the project network of Figure 11.1. Because time is short, the project manager decides to personally assist the two programmers. The project manager is capable of performing any task. Compared with the Gantt Chart of Figure 11.3, how much time will this save?
   Is there some other way the project manager can help?
3. In connection with the project network of Figure 11.1, we suggested that it might be better to have the IT candidates type the data into spreadsheets while a single programmer designed and created the database tables. The activity graph would then consist of the following tasks:

[24] This is illustrated by software development kits that support application development for the operating system of a particular hardware platform. By developing the kits, hardware manufacturers encourage third-party developers to add value and appeal to their products.

- Finalise database schema.
- Create *Subjects* table.
- Create *Programmes* table.
- Create *Components* table.
- Populate *Subjects* spreadsheet.
- Populate *Programmes* spreadsheet.
- Populate *Components* spreadsheet.
- Use a utility program to load the data.

Allowing one day to use the utility program, and assuming that the remaining times and predecessor relationships are unchanged, arrange the activities in topological order, and find the length of the critical path. Model your answer on Table 11.2.

4. One particular member of the programming team continually takes twice as long to complete a task as he agrees with the project manager. Short of freeing him to pursue his career elsewhere, what can be done to correct the situation?

APPENDIX

Regular expressions and finite-state automata

CHAPTER CONTENTS

INTRODUCTION

In this section, we describe how a regular expression can be converted to a state-transition diagram. Although this isn't central to our purposes, it serves two functions here:

- First, it offers an excellent example of *synthesis by composition*, which, among other things, shows that a composition process tends to alter structure.
- Second, it shows that state-based and behaviour-based descriptions are equivalent and that neither is inherently more abstract than the other.

A.1 FROM REGULAR EXPRESSIONS TO STATE-TRANSITION DIAGRAMS

Initially, a given regular expression is converted by straightforward rules into a state-transition diagram. Unfortunately, this diagram is usually non-deterministic and not directly useful as it stands. It is then reduced to a deterministic form by *composing* similar states. It is possible to go further and find the (unique) **minimal DFA** that contains the fewest states.

We can convert a regular expression to a state-transition diagram using special constructions for each operator.

- If l is a letter, we construct a diagram for l by drawing an edge $u \underset{l}{\rightarrow} v$, where u and v are new states.

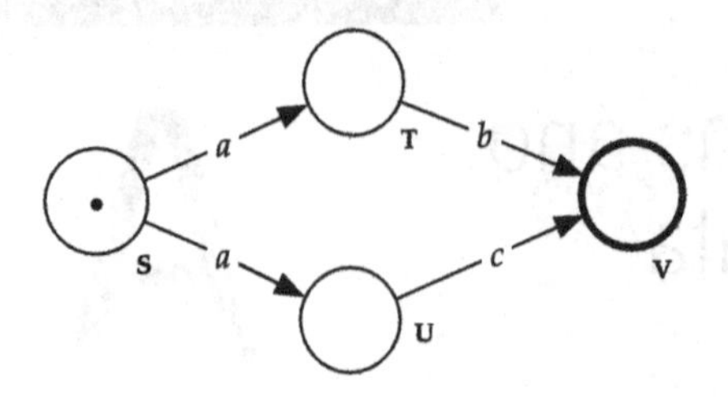

FIGURE A.1

The state-transition diagram of an NFA that accepts $(a\,;b) \cup (a\,;c)$. The NFA for $a\,;b$ is formed by making the final state for transition a (T) the initial state for transition b. The NFA for $a\,;c$ is formed in a similar way. Their union is formed by merging their initial states (S) and final states (V).

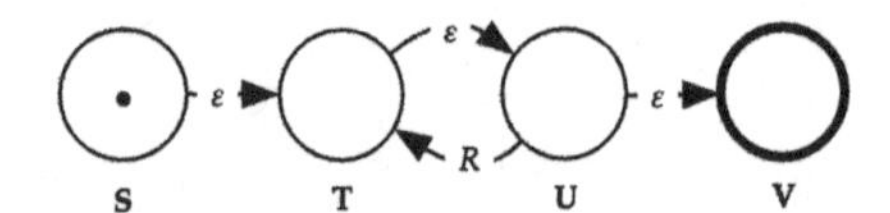

FIGURE A.2

The state-transition diagram of an NFA that accepts R^*.

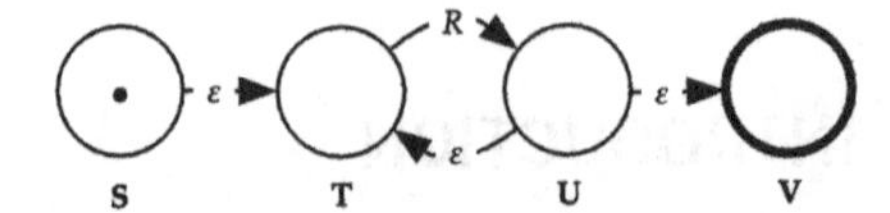

FIGURE A.3

The state-transition diagram of an NFA that accepts R^+.

- If R and S are regular expressions, we form the diagram for $R\,;S$ by merging the *final* state of the diagram for R with the *initial* state of the diagram for S. The initial state of R becomes the initial state of $R\,;S$, and the final state of S becomes the final state of $R\,;S$.
- If R and S are regular expressions, we form the diagram for $R \cup S$ by merging the *initial* state of the diagram for R with the *initial* state of the diagram for S and the *final* state of the diagram for R with the *final* state of the diagram for S.

Consider Figure A.1, which has been constructed using these rules from the expression $(a\,;b) \cup (a\,;c)$. It is non-deterministic: on input a, the initial state, S, has transitions to both state T and state U.

The constructions for the iterative operators are more complex and involve the use of ε transitions:

- If R is a regular expression, the diagram for R^* is that shown in Figure A.2.
- If R is a regular expression, the diagram for R^+ is that shown in Figure A.3.

FIGURE A.4

A faulty state-transition diagram of an NFA that equally accepts $R^* ; S^*$, $S^* ; R^*$, $(R \cup S)^*$, and $R^* \cup S^*$.

Without the ε transitions, the diagrams for $R^* ; S^*$, $S^* ; R^*$, $(R \cup S)^*$, and $R^* \cup S^*$ would all be that shown in Figure A.4. (Convince yourself that the expressions themselves are *not* all equivalent.)

Let us explore these constructions by creating a transition diagram: consider all sequences of binary digits whose penultimate (second-last) letter is a 1. A regular expression that defines such sentences is

$$(0 \cup 1)^* ; 1 ; (0 \cup 1)$$

that is, any number of 1s or 0s, followed by a 1, followed by exactly one 1 or 0. 10 and 10110011 are valid sentences; 111101 and 1 are invalid. We want to find a finite-state automaton that reaches a final state if and only if a given input matches the given expression. Using the constructions just described, the resulting diagram is shown in Figure A.5.[1]

This diagram is non-deterministic. In state S, even before any input, the automaton could be in any of the states $\{S,T,U,V\}$. (This is called the **ε-closure** of S: the set of all states Z such that $S \underset{\varepsilon}{\rightarrow}^* Z$.)

Now consider the situation following an initial input of 0. The NFA could only move from state U to the set $\{T,U,V\}$. Given an input of 1, it is impossible to know whether this will prove to be the second-to-last element of the input, so it is impossible to guess whether the automaton should move from U to $\{T,U,V\}$ or from V to $\{W\}$.

By pre-computing the possible sets of states, we can reduce the diagram to a more efficient, deterministic form. We do this by mapping each distinct *set of states* of the NFA onto a *single state* of the DFA. This process is the same as we described in Section 3.2 (page 83), except that, this time, the ε transitions have introduced additional states.

[1] In a computer algorithm, it is easier to *link* states using ε transitions than to merge them.

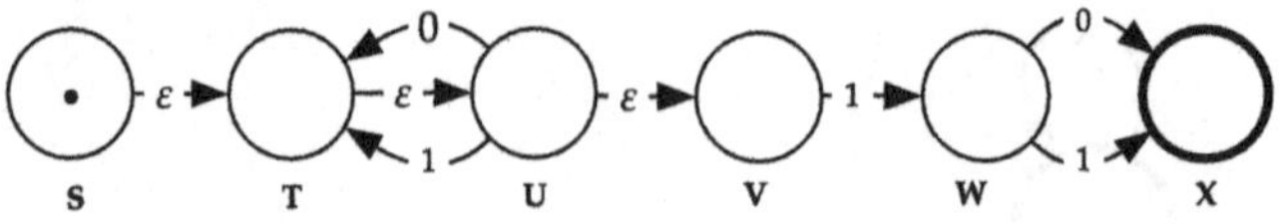

FIGURE A.5

The state-transition diagram of an NFA that detects if the penultimate letter of a binary sequence is a 1.

We therefore have two possible transitions between *sets* of states:

$$\{S,T,U,V\} \underset{0}{\rightarrow} \{T,U,V\}$$

$$\{S,T,U,V\} \underset{1}{\rightarrow} \{T,U,V,W\}$$

We have now found two new sets of states: $\{T,U,V\}$ and $\{T,U,V,W\}$. In similar style, we add four more transitions to our list:

$$\{T,U,V\} \underset{0}{\rightarrow} \{T,U,V\}$$

$$\{T,U,V\} \underset{1}{\rightarrow} \{T,U,V,W\}$$

$$\{T,U,V,W\} \underset{0}{\rightarrow} \{T,U,V,X\}$$

$$\{T,U,V,W\} \underset{1}{\rightarrow} \{T,U,V,W,X\}$$

We have again found two new sets of states: $\{T,U,V,X\}$ and $\{T,U,V,W,X\}$. As before, we consider the new transitions that arise from them:

$$\{T,U,V,X\} \underset{0}{\rightarrow} \{T,U,V\}$$

$$\{T,U,V,X\} \underset{1}{\rightarrow} \{T,U,V,W,X\}$$

$$\{T,U,V,W,X\} \underset{0}{\rightarrow} \{T,U,V,X\}$$

$$\{T,U,V,W,X\} \underset{1}{\rightarrow} \{T,U,V,W,X\}$$

Fortunately, no new sets of states have appeared, so we can draw the state transition matrix of the resulting DFA, which has five states in all. (See Table A.1.) Any set of states that includes X, the final state of the NFA, becomes a final state of the DFA and is marked with an asterisk.

In fact, the DFA really needs only four states. We know this is true, because one way to detect if the penultimate bit of a sequence is a 1 is to store the most recent 2 bits of the input in a shift register initialised with zeros and check that its leading bit is 1 once the input is complete. A 2-bit shift register has precisely four states. It turns

Table A.1 A state-transition matrix of a deterministic finite-state automaton that detects if the second-to-last letter of a binary sequence is a '1'

	$\{S,T,U,V\}$	$\{T,U,V\}$	$\{T,U,V,W\}$	$\{T,U,V,X\}^*$	$\{T,U,V,W,X\}^*$
0	$\{T,U,V\}$	$\{T,U,V\}$	$\{T,U,V,X\}$	$\{T,U,V\}$	$\{T,U,V,X\}$
1	$\{T,U,V,W\}$	$\{T,U,V,W\}$	$\{T,U,V,W,X\}$	$\{T,U,V,W,X\}$	$\{T,U,V,W,X\}$

Table A.2 The transition matrix of the unique minimal DFA that detects if the penultimate letter of a binary sequence is a '1'

	00	01	10	11
0	00	10	00	10
1	01	11	01	11

Table A.3 A state-transition matrix derived from Table A.1 that differentiates only between final and non-final states

	N	N	N	F	F
0	N	N	F	N	F
1	N	N	F	N	F

out that states $\{S,T,U,V\}$ and $\{T,U,V\}$ in Table 9.2 are equivalent, because both make the same transitions, and neither are final states. So we can actually combine $\{S,T,U,V\}$ and $\{T,U,V\}$ into a single initial state. Table A.2 shows the result of this simplification, with the states renamed to match the contents of the imaginary shift register.

In general, we can *minimise* the number of states of a DFA by assuming initially that only final and non-final states are distinct. We then differentiate between states only if they make transitions to distinct states. In Table A.3, the states are marked either F (final) or N (non-final).

We see that the first two columns are similar but the remaining columns are distinct. We therefore need to differentiate between the last two columns, because although they are both labelled F, they make different transitions. Likewise, the first two states are different to the third.

As Table A.4 shows, we have no reason to split the two columns labelled N_1; they are identical. We therefore have four states, as before, corresponding to those we previously labelled 00, 01, 10, and 11. The state diagram of this minimal DFA is shown in Figure A.6.

Table A.4 A state-transition matrix that differentiates between states on the basis of their transitions to final and non-final states

	N_1	N_1	N_2	F_1	F_2
0	N_1	N_1	F_1	N_1	F_1
1	N_2	N_2	F_2	N_2	F_2

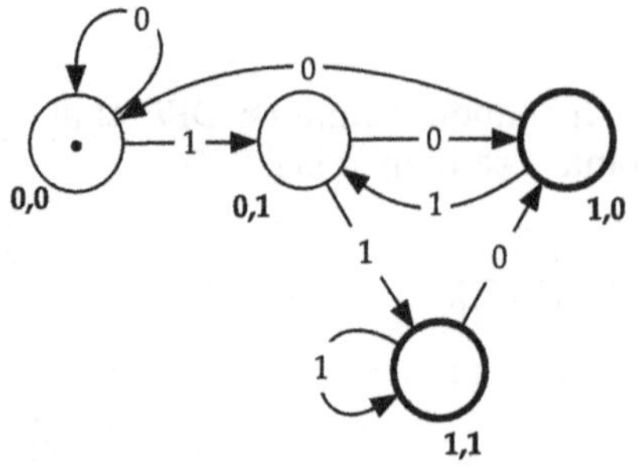

FIGURE A.6

The state-transition diagram of the minimal DFA that recognises if the penultimate bit of a binary sequence is a '1'.

The number of states of the *non-deterministic* automaton is always of the same order as the length of the regular expression from which it derives, but the number of states of the *deterministic* automaton can grow exponentially. If we consider detecting whether the last bit but *seven* of an input is a 1, we would need an 8-bit shift register, so we ought to expect the corresponding deterministic automaton to have 256 states. The example we have just analysed was chosen precisely to illustrate this possibility, but in practice, such exponential growth is rare.

A.2 FROM FINITE-STATE AUTOMATA TO REGULAR EXPRESSIONS

Kleene's Algorithm allows us to discover a regular expression for the sentences that a given FSA will accept. It is similar to Warshall's Algorithm (discussed in Section 2.5.5) in that it iterates through each intermediate state in turn. It extends the idea of establishing which paths between states exist, to finding the expressions that the paths spell.

Let us consider a simple FSA that checks that the parity of a sequence of binary digits is even. It has two states, *Even* and *Odd*. For this purpose, we set out the state-transition matrix with rows representing the source state, columns representing the target state, and cells representing the expressions that link them:

	Even	*Odd*
Even	0	1
Odd	1	0

The first step is adjust the diagonal terms to recognise that each state moves to itself on an empty input, ε:

	Even	*Odd*
Even	$\varepsilon \cup 0$	1
Odd	1	$\varepsilon \cup 0$

For each value of k, we take into account all the paths that pass through the state k one or more times. For a given pair of states i and j, if the existing expression is $E_{i \to j}$, we compute its new value as follows:[2]

$$E'_{i \to j} = E_{i \to j} \cup (E_{i \to k}\,;\, E_{k \to k}{}^{*}\,;\, E_{k \to j})$$

We begin by setting $k = Even$, after noting that $E^{*}_{Even \to Even}$ simplifies to 0^{*}:

	Even	*Odd*
Even	$\varepsilon \cup 0 \cup (\varepsilon \cup 0 ; 0^{*} ; \varepsilon \cup 0)$	$1 \cup (\varepsilon \cup 0 ; 0^{*} ; 1)$
Odd	$1 \cup (1 ; 0^{*} ; \varepsilon \cup 0)$	$\varepsilon \cup 0 \cup (1 ; 0^{*} ; 1)$

Before continuing, we simplify the path expressions:

	Even	*Odd*
Even	0^{*}	$0^{*} ; 1$
Odd	$1 ; 0^{*}$	$\varepsilon \cup 0 \cup (1 ; 0^{*} ; 1)$

We now repeat the process, setting $k = Odd$ and simplifying $(\varepsilon \cup 0 \cup (1 ; 0^{*} ; 1))^{*}$ to $(0 \cup (1 ; 0^{*} ; 1))^{*}$:

[2] As an alternative to the known path from i to j, there is also a path from i to k, followed by any number of cycles through k, followed by a path from k to j.

	Even	*Odd*
Even	$(0^* \cup 1;(0 \cup (1;0^*;1)))^*;1$	$0^*;1 \cup (1;(0 \cup (1;0^*;1))^*;0$ $\cup (1;0^*;1))$
Odd	$1;0^*$ $\cup ((0 \cup (1;0^*;1);(1;0^*;1))^*;0^*;1)$	$\varepsilon \cup 0 \cup (1;0^*;1)$ $\cup (\varepsilon \cup 0 \cup (1;0^*;1));(0 \cup (1;0^*;1))^*\varepsilon$ $\cup 0 \cup (1;0^*;1)$

Finally, we simplify the path expressions again:

	Even	*Odd*
Even	$(0 \cup (1;0^*;1))^*$	$(0 \cup (1;0^*;1))^*;1;(0 \cup (1;0^*;1))^*$
Odd	$(0 \cup (1;0^*;1))^*;1;(0 \cup (1;0^*;1))^*$	$(0 \cup (1;0^*;1))^*$

from which we conclude that, since the *Even* state is both the initial and final state, the expression $(0 \cup (1;0^*;1))^*$ describes all binary sequences of even parity.

It is clear from this example that without intermediate simplification the lengths of the path expressions grow by a factor of four at almost every step, so the algorithm is of little practical use. But its *existence* establishes that every FSA corresponds to a regular expression.

A.3 SUMMARY

The above suggests that, even though behaviour-based descriptions make no mention of states, they are no more abstract than state-based descriptions, since every regular expression directly corresponds to a non-deterministic automaton and *vice versa*. Indeed, a surefire way to prove that two regular expressions describe the same language is to derive their minimal DFAs. If the DFAs are identical (apart from the labels given to states), the expressions must be equal. Therefore, I make no apology for using state-based rather than behaviour-based descriptions.

Reflecting on what we have just seen, it is straightforward to convert a specification (the regular expression) into a workable but inefficient system (the NFA). The structure of the NFA is essentially the same as the structure of the expression. Unfortunately, to obtain an *efficient* implementation, we must optimise the NFA by *composing* sets of states and *minimising* them (to give the DFA). Consequently, such *optimisation can cause a profound change in the structure of a system.* There is often little resemblance between a regular expression and its minimal DFA.

An interesting point is that when we want to minimise a DFA, we don't merge states; we start with the states already merged, and split them. But on the other hand, as we have seen elsewhere, if we want to divide a system into components, we don't split, we start with the smallest parts and merge them. So, paradoxically, when we want to merge, we split; when we want to split, we merge. This isn't a general rule, but it does suggest we should always look at both possibilities.

A.4 FURTHER READING

Chapter 10 of 'Foundations of Computer Science: C Edition' by Alfred V. Aho and Jeffrey D. Ullman (1994, ISBN: 0-7167-8284-7) describes most of the material in this appendix. 'Introduction to Automata Theory, Languages, and Computation' by John E. Hopcroft, Rajeev Motwani, and Jeffrey D. Ullman (2001, ISBN 0-201-44124-1) also describes both Kleene's Algorithm and the state-minimisation problem.

Michael A. Jackson used behavioural descriptions in his influential 1975 book, 'Principles of Program Design' (ISBN: 0-123-79050-6). Jackson showed how program structures could be derived from the data structures they manipulate by drawing structure diagrams, a direct analogue of regular expressions. The expression or diagram describing the overall structure of the program should embed the expressions that describe its inputs and outputs. Unfortunately, some programs manipulate incompatible data structures, so in such cases he resorted to a process called program inversion, which is essentially the transformation from a regular expression to an FSA.

In his 1985 book, 'Communicating Sequential Processes' (ISBN: 0-131-53271-5), Tony Hoare also made a forceful case for behavioural descriptions of processes, strongly rejecting the need for states. Such descriptions seem natural in a programming context, because the program structures of sequence, selection, and iteration are closely related to regular expressions. They seem less natural in contexts in which systems are, effectively, victims of their environment.

A.5 EXERCISES

1. Using the rules of Appendix A.1 sketch an FSA for the expression $(a\,;c) \cup (b\,;c)$. Then find the minimal DFA for the expression, and sketch it.
2. Sketch an FSA for the expression $(a \cup b)\,;c$. Thence, or otherwise, prove that $(a\,;c) \cup (b\,;c)$ and $(a \cup b)\,;c$ define the same language.

APPENDIX

Normalisation

B

CHAPTER CONTENTS

INTRODUCTION

In this section, I shall present the ideas behind normalisation with the aid of FD graphs. I stated in Section 4.4 that tables should be formed from a single domain and its outgoing edges. If in the process of view integration we look at a report, we may see attributes from several tables. These attributes might be plucked almost arbitrarily from the database's underlying FD graph. We can't expect such collections of attributes to form a sensible table. I believe that by drawing the subgraphs that contain these sets of attributes I shall help the reader understand what might otherwise seem a rather obscure algebraic definition or decomposition technique. I hope the reader will agree.

B.1 JOINS AND VIEWS

A join is a combination of attributes from different tables. We may form *any* join in three steps:

- Form the (enormous) Cartesian product of the tables. That is, if we have tables *A*, *B*, and *C*, the rows of the product contain values from a row of *A*, a row of *B*, and a row of *C* in every possible combination.
- Select a subset of the resulting rows.
- Project the result onto a subset of their attributes.

Often we require that attributes from different tables should be equal. We call the result an **equi-join** and drop one of the equal attributes from the result. For example, we can join

$$Candidates(\underline{Candidate}, Programme, Personal\ Name)$$

$$Programmes(\underline{Programme}, Degree\ Name)$$

by matching the value of *Programme* in the *Candidates* table with a value of *Programme* in the *Programmes* table, effectively creating the table

$$Candidate\ Programmes(\underline{Candidate}, Personal\ Name, Programme, Degree\ Name)$$

Among other things, this table maps the *Admission No* of a candidate onto the name of the degree the candidate hopes to qualify for.

Although a typical DBMS will use a much more efficient method, we can imagine that we take the Cartesian product of the *Candidates* and *Programmes* tables, select only those rows where *Programme* matches, and then project the result onto *Candidate*, *Programme*, *Personal Name*, *Programme Name*, and *Degree Name*.

Clearly, the common attributes must belong to the same domain. Typically, one is a primary key, and the other is a foreign key that references it. Here, we shall call such a foreign key-primary key join a **natural join.**[1]

The ability to join tables means that the tables themselves can be simple, yet complex reports can be constructed by joining several tables. These reports are called views of the database. View integration is therefore a process of working backwards from complex views to simple tables.

B.2 THE DATABASE KEY

We have already noted that the non-terminal vertices of a canonical FD graph represent candidate keys. Extending this idea, the set of the *initial* vertices of the

[1] Some texts require that the two attributes have the same *name*, but this restriction is pointless.

graph constitute the **database key**. The FD graph of Figure 4.7 has two initial vertices, those for *Enrolments* and *Components*.

It is instructive to examine separately the subgraphs whose domains are in the transitive closure of these two vertices: the subgraph rooted at *Components* is shown in Figure B.1, and the subgraph rooted at *Enrolments* is shown in Figure B.2.

The two subgraphs describe different aspects of the Academic Records System: the first is concerned with the rules that limit what subjects a candidate may choose, and the second concerns the choices that candidates actually make. They support subsystems that are almost independent. *Components* would be updated during the vacations, and *Enrolments* would be updated during the enrolment and examination periods. The vertices for *Programmes* and *Subjects* are present in both subgraphs. (If the subgraphs were truly disjoint, we would have two separate databases and two separate systems.)

B.3 THE UNIVERSAL RELATION

Decomposition algorithms usually assume the worst possible scenario: a single table containing every possible attribute. Such a table is called a universal relation.

From Figure B.1, we can construct a table that contains everything it is possible to know about *Components*.

Component Details (*Programme, Subject, Component State,*
Subject Name, Subject State, Fee, Quota, Filled,
Programme Name, Programme State, Degree Name)

(The initial vertex of Figure B.1 forms the determinant, and all the vertices in its transitive closure become dependent attributes.)

It may be useful to *display* all this information, for example, as a report describing the options available in each programme, but that is no reason to store all this information in one table. The same values of *Programme Name*, *Programme State*, and *Degree Name* would be stored for every subject that is part of the programme identified by *Programme*. Similarly, the same values of *Subject Name*, *Subject State*, *Fee*, *Quota*, and *Filled* would be stored for every programme that teaches the subject identified by *Subject*. This redundancy means that if the *Fee* for a subject is changed, many rows of the table need to be updated. Careless programming might allow the attributes of subjects in different rows to become inconsistent. This is called an **update anomaly**. Worse, whenever a programme ceases to use a subject, one row containing the subject attributes is lost, so when the last such row disappears, *all* the subject data disappear. This is called a **deletion anomaly**. Clearly, it makes sense to store subject information in a *Subjects* table, programme information in a *Programmes* table, and the relationship between them in a third table.

The *Component Details* table is called the universal relation for the subgraph of Figure B.1.

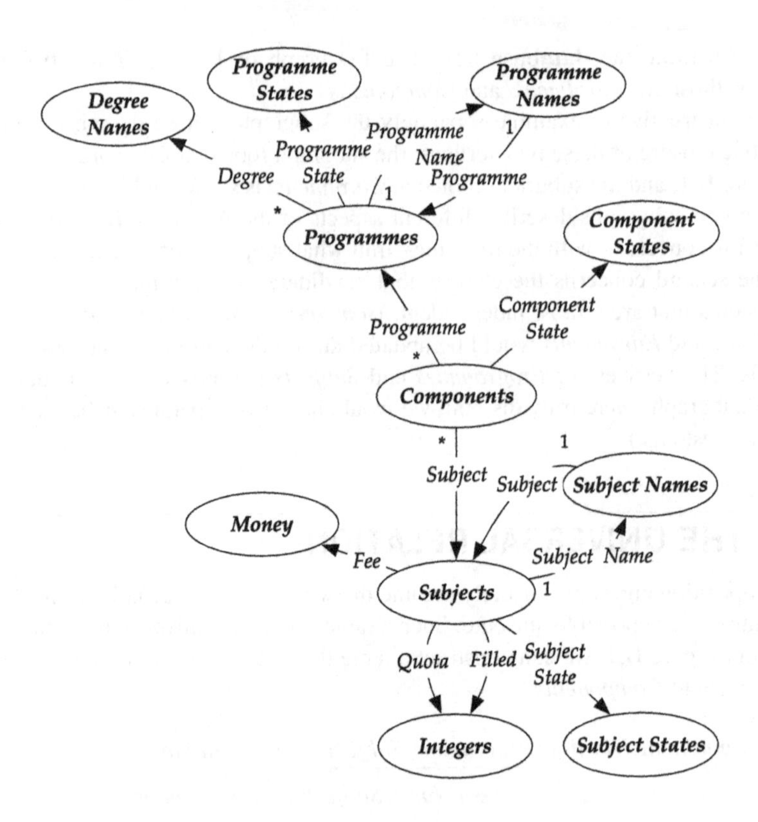

FIGURE B.1

The subgraph rooted at *Programmes* × *Subjects*.

Similarly, we can form the universal relation for Figure B.2:

Enrolment Details (*Candidate, Subject, Semester, Enrolment State, Mark,*
Personal Name, Postal Address, Phone, Date of Birth,
Balance Owing, Candidate State,
Programme, Programme Name, Programme State, Degree,
Subject Name, Subject State, Fee, Quota, Filled)

This table schema might be the basis for a detailed list of *Enrolments*. Organised by *Subject*, it would be a set of class lists, or organised by *Candidate*, it would be a set of academic transcripts. As in the case of *Component Details*, the table is highly redundant and susceptible to update and deletion anomalies.

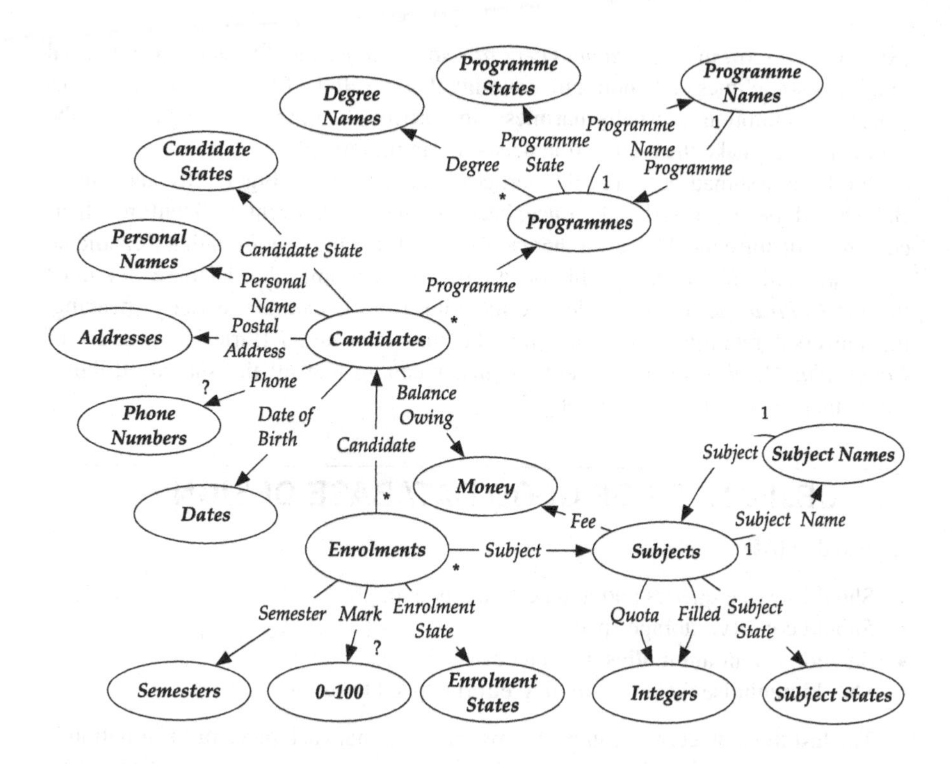

FIGURE B.2

The subgraph rooted at *Candidates* × *Subjects* × *Semesters*.

We may combine the two tables into one giant universal relation for the entire database:

Academic Records (<u>*Candidate, Programme, Subject, Semester*</u>,
Enrolment State, Mark,
Personal Name, Postal Address, Phone, Date of Birth,
Balance Owing, Candidate State,
Programme, Programme Name, Programme State, Degree,
Subject Name, Subject State, Fee, Quota, Filled,
Component State)

Given the database key (*Candidate, Programme, Subject, Semester*), all the remaining attributes are in its closure. Unfortunately, this does not imply that we can access every value of every attribute. For example, if a programme has no candidates, we are unable to supply a value of *Admission No* that gives access to the details of that programme. (This is an example of a deletion anomaly: although the programme may exist, it has no row in the *Academic Records* table.) Consequently, using the universal relation is sometimes tantamount to assuming that the all the FDs are onto. As it turns out, this assumption is usually harmless; the process of normalisation removes the anomalies that make the assumption necessary in the first place.

We have assumed here that the vertices shared between Figure B.1 and Figure B.2 should be expressed only once because the *Enrol* event will enforce their equality. But since the database schema fails to enforce the condition that a candidate may enrol only in subjects in his or her own programme, the database key must include *Programme*. It cannot deduce its value from *Candidate*. Strictly speaking, the universal relation should therefore be the (enormous) Cartesian product of *Component Details* and *Enrolment Details*, meaning that all the shared attributes would then appear in two different guises.

B.4 OBJECTIVES OF GOOD DATABASE DESIGN

A good database:

- Should service queries and updates quickly enough,
- Should conserve storage space,
- Should prevent **anomalies**, i.e., inconsistent data,
- Should minimise the programming effort needed to use it.

The last three objectives can be met by ensuring that each piece of information is stored exactly once. If a datum is stored twice, not only is one copy wasting space, but the programmer must ensure that both copies are always kept in step, otherwise an anomaly will result. The designer should therefore aim to eliminate redundancy.

Foreign keys are an exception. Although a foreign key duplicates the value of the primary key it references, it is vital to the structure of the database, and the DBMS enforces referential integrity to ensure that anomalies cannot arise.

Eliminating redundancy usually does no harm to the first objective, as it will always take less time to update one copy of a datum than two. Nonetheless, exceptions do occur. If the database is used mainly to answer queries that would take a long time to serve without redundancy, then the additional time and space needed to insert or update the extra data may be justified.[2] For example, in the Academic Records System, we decided to keep track of the total number of places in a subject using the *Quota* attribute and the number of places already filled in the *Filled* attribute. *Filled* is redundant because the number of places filled could also be found by counting the number of enrolments in the subject. Even so, although *Filled* needs to be adjusted whenever *Enrol* or *Withdraw* events occur, the overall effect on performance is likely to be positive.

The storage of redundant statistical information is heavily exploited in data warehousing. **Data warehousing** is a technique of taking snapshots of a database so that it can be analysed for strategic planning. Redundant information is added by

[2] It is reasonable to assume that this is very much the case for Internet search engines.

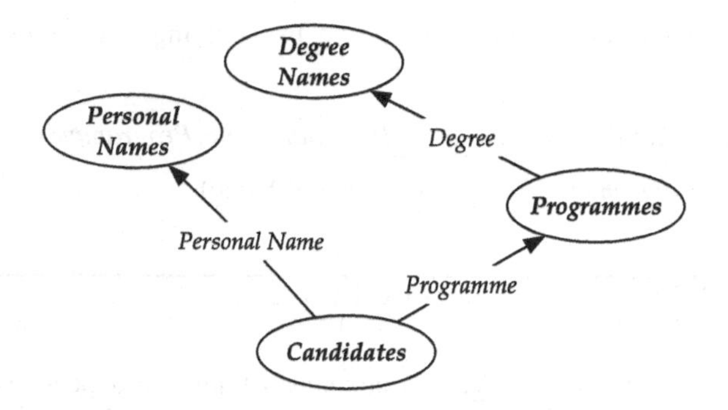

FIGURE B.3

A transitive dependency from *Candidates* to *Degree Names* via *Programmes*.

forming totals and other summary statistics. The redundancy is harmless because the snapshots are never updated. They are eventually replaced by new snapshots.

We now discuss how redundancy can arise, and present some successful ways of dealing with it.

B.4.1 TRANSITIVE DEPENDENCIES

Consider a table based on the subgraph of Figure B.3:

$$\textit{Candidate Degrees}\ (\underline{\textit{Candidate}}, \textit{Personal Name}, \textit{Programme}, \textit{Degree Name})$$

This table would contain the attributes of the relational product of two FDs:

$$\textit{Programme} : \textit{Candidates} \rightarrow \textit{Programmes}$$
$$\textit{Degree} : \textit{Programmes} \rightarrow \textit{Degree Names}$$

As a result, the FD *Degree*: *Programmes* → *Degree Names* would be expressed many times, once for each candidate admitted to a programme. This would be redundant and allow anomalies.

Degree Name is said to be transitively dependent on *Candidate*. A *transitive dependency* occurs when one non-prime attribute is dependent on another non-prime attribute.

> If a table schema $T(\underline{X},Y,Z)$ contains a dependency $f : Y \rightarrow Z$, where Y and Z are non-prime attributes, we say it contains a **transitive dependency** from X to Z.

We can see the same situation more clearly in Figure B.3. The path from *Candidates* to *Degree Names* is compound, with length two in this case. In general, a transitive dependency can be decomposed by applying Algorithm B.1. In this example, the correct decomposition is into

$$\begin{gathered}\mathit{Candidates}(\underline{\mathit{Candidate}}, \mathit{Personal\ Name}, \mathit{Programme}),\\ \mathit{Programmes}(\underline{\mathit{Programme}}, \mathit{Degree\ Name}).\end{gathered}$$

Algorithm B.1 Decomposing a transitive dependency.

If a table of the form $T(\underline{X}, Y, Z)$ contains a transitive dependency $f : Y \rightarrow Z$, where Y is a non-prime attribute, decompose it into two tables, $T_1(\underline{X}, Y)$ and $T_2(\underline{Y}, Z)$.

Although the transitive dependency is obvious in this example, it is possible for a transitive dependency to be obscured. If we had started with the schema

$$\begin{gathered}\mathit{Candidates}(\underline{\mathit{Candidate}}, \mathit{Personal\ Name}, \mathit{Degree\ Name}),\\ \mathit{Programmes}(\underline{\mathit{Programme}}, \mathit{Degree\ Name}),\end{gathered}$$

neither table would have contained a transitive dependency, but the problem would still have been present — so it couldn't be eliminated by Algorithm B.1.[3]

B.4.2 PARTIAL DEPENDENCIES

Consider a table based on the subgraph of Figure B.4:

$$\begin{aligned}\mathit{Enrolment\ Details}(&\underline{\mathit{Candidate}, \mathit{Subject}, \mathit{Semester}},\\ &\mathit{Enrolment\ State}, \mathit{Mark}, \mathit{Subject\ Name}),\end{aligned}$$

which contains the attributes of the relational product of two FDs:

$$\begin{gathered}\mathit{Subject} : \mathit{Candidates} \times \mathit{Subjects} \times \mathit{Semesters} \rightarrow \mathit{Subjects},\\ \mathit{Subject\ Name} : \mathit{Subjects} \rightarrow \mathit{Subject\ Names}\end{gathered}$$

As a result, the FD *Subject Name* : *Subjects* → *Subject Names* is expressed many times, once for each enrolment in a subject. This is redundant and allows anomalies. The fault is that the *Subject Name* attribute depends only on part of the primary key (*Subject*) of *Enrolment Details* and not the whole primary key. This is called a *partial dependency*. A partial dependency is a special case of a transitive dependency where the first FD is trivial.

---

[3] The Academic Records System contains a similar error: in the interview with the domain expert that we described earlier, the analyst became aware that *Programmes* and *Degrees* are different, but later chose to ignore this and to make *Degree Name* a dependent attribute of *Programmes*.

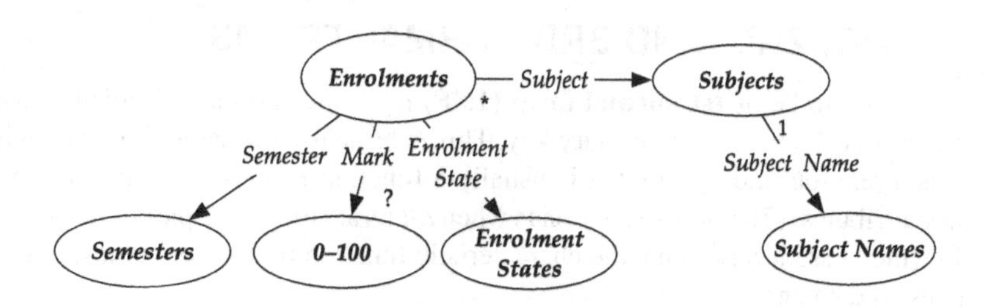

FIGURE B.4

A partial dependency from (*Candidates*, *Subjects*, *Semesters*) to *Subject Names*.

> If a table schema $T(W,\underline{X},Y,Z)$ contains a dependency $f : X \to Y$, where X is part of a composite key, we say it contains a **partial dependency** from X to Y.

The subgraph of Figure B.4 is similar to Figure B.3. Again there is a compound path of length two, this time from *Candidates×Subjects×Semesters* to *Subject Names*, but in this case its first edge is a projection function, *Subject*. A partial dependency may be removed using Algorithm B.2. In this case, the correct decomposition is

$$Subject\ Details(\underline{Subject}, Subject\ Name),$$
$$Enrolment\ Details(\underline{Candidate, Subject, Semester}, Enrolment\ State, Mark)$$

Algorithm B.2 Decomposing a partial dependency.

> If a table of the form $T(\underline{W,X},Y,Z)$ contains a partial dependency $f : X \to Y$, where Y is a non-prime attribute, decompose it into two tables, $T_1(\underline{X},Y)$ and $T_2(\underline{W,X},Z)$.

It is also possible for a partial dependency to obscured: if we had started with the schema

$$Subject\ Details(\underline{Subject}, Subject\ Name),$$
$$Enrolment\ Details(\underline{Candidate, Subject\ Name, Semester}, Enrolment\ State, Mark),$$

the same problem would be present, but neither table would contain a partial dependency.

B.4.3 1ST, 2ND, AND 3RD NORMAL FORMS

A table is said to be in **1st normal form (1NF)** if its attributes are atomic (no lists, arrays, or tables) and it has a primary key. This is basically the same thing as saying that it is a relation; indeed, a table is usually a function from its primary key to its non-key attributes. (The only exception is when *all* its attributes are part of its primary key, in which case it represents the characteristic function of a set of values, or pairs of values, and so on.)

A table is in **2nd normal form (2NF)** if it is in 1NF and all non-prime attributes are fully dependent on its primary key (no partial dependencies).

A table is in **3rd normal form (3NF)** if it is in 2NF and it is free of transitive dependencies involving non-prime attributes.

A database schema is in 1NF (2NF, 3NF) normal form if all its tables are in 1NF (2NF, 3NF).

Prime attributes are excluded from the definitions to avoid technical difficulties. Otherwise, every trivial dependency would also be a partial dependency, and any table having more than one candidate key would contain a transitive dependency. For example, because *Programme Name* is a candidate key,

$$Programmes(\underline{Programme}, Programme\ Name, Programme\ State, Degree\ Name)$$

contains the cyclic identity function

$$Programme\ Name\,;Programme : Programmes \rightarrow Programmes$$

Putting a schema into 3rd normal form is easy if it has only one candidate key. We simply group FDs together according to their determinants. A table is created for each group, with the determinant as the primary key, and a column for each right-hand side attribute. Grouping FDs with the same determinants together saves space; their common domain needs to be stored only once. Such a schema is always free of partial and transitive dependencies.

The Academic Records database contains the following set of minimal, simple FDs:

$$Programme\ Name : Programmes \rightarrow Programme\ Names$$
$$Programme\ State : Programmes \rightarrow Programme\ States$$
$$Degree\ Name : Programmes \rightarrow Degree\ Names$$
$$Subject\ Name : Subjects \rightarrow Subject\ Names$$
$$Subject\ State : Subjects \rightarrow Subject\ States$$
$$Fee : Subjects \rightarrow Money$$
$$Quota : Subjects \rightarrow Integers$$
$$Filled : Subjects \rightarrow Integers$$
$$Programme : Programmes \times Subjects \rightarrow Programmes$$
$$Component\ State : Programmes \times Subjects \rightarrow Component\ States$$

$$
\begin{aligned}
\mathit{Subject} &: \mathit{Programmes} \times \mathit{Subjects} \rightarrow \mathit{Subjects} \\
\mathit{Personal\ Name} &: \mathit{Candidates} \rightarrow \mathit{Personal\ Names} \\
\mathit{Postal\ Address} &: \mathit{Candidates} \rightarrow \mathit{Addresses} \\
\mathit{Phone} &: \mathit{Candidates} \rightarrow \mathit{Phone\ Numbers} \\
\mathit{Date\ of\ Birth} &: \mathit{Candidates} \rightarrow \mathit{Dates} \\
\mathit{Balance\ Owing} &: \mathit{Candidates} \rightarrow \mathit{Money} \\
\mathit{Candidate\ State} &: \mathit{Candidates} \rightarrow \mathit{Candidate\ States} \\
\mathit{Enrolment\ State} &: \mathit{Candidates} \times \mathit{Subjects} \times \mathit{Semesters} \rightarrow \mathit{Enrolment\ States} \\
\mathit{Mark} &: \mathit{Candidates} \times \mathit{Subjects} \times \mathit{Semesters} \rightarrow 0 \ldots 100 \\
\mathit{Candidate} &: \mathit{Candidates} \times \mathit{Subjects} \times \mathit{Semesters} \rightarrow \mathit{Candidates} \\
\mathit{Subject} &: \mathit{Candidates} \times \mathit{Subjects} \times \mathit{Semesters} \rightarrow \mathit{Subjects} \\
\mathit{Semester} &: \mathit{Candidates} \times \mathit{Subjects} \times \mathit{Semesters} \rightarrow \mathit{Semesters} \\
\mathit{Subject} &: \mathit{Subject\ Names} \rightarrow \mathit{Subjects} \\
\mathit{Programme} &: \mathit{Programme\ Names} \rightarrow \mathit{Programmes}
\end{aligned}
$$

Ignoring the last two FDs, $\mathit{Subject} : \mathit{Subject\ Names} \rightarrow \mathit{Subjects}$ and $\mathit{Programme}: \mathit{Programme\ Names} \rightarrow \mathit{Programmes}$, simply grouping the FDs together by determinant yields the following (unique) conceptual schema:

$$
\begin{aligned}
&\mathit{Subjects}(\underline{\mathit{Subject}}, \mathit{Subject\ Name}, \mathit{Subject\ State}) \\
&\mathit{Candidates}(\underline{\mathit{Candidate}}, \mathit{Personal\ Name}, \mathit{Postal\ Address}, \mathit{Phone}, \\
&\qquad \mathit{Date\ of\ Birth}, \mathit{Balance\ Owing}, \mathit{Candidate\ State}) \\
&\mathit{Programmes}(\underline{\mathit{Programme}}, \mathit{Programme\ Name}, \mathit{Programme\ State}, \mathit{Degree\ Name}) \\
&\mathit{Components}(\underline{\mathit{Programme}, \mathit{Subject}}, \mathit{Component\ State}) \\
&\mathit{Enrolments}(\underline{\mathit{Candidate}, \mathit{Subject}, \mathit{Semester}}, \mathit{Enrolment\ State}, \mathit{Mark})
\end{aligned}
$$

As it happens, the *Subjects* table will be able to enforce the *Subject* FD by assigning *Subject Name* the **unique** attribute. Similarly, the *Programmes* table will be able to enforce *Programme*, so *all* the FDs can be enforced. Consequently this schema is completely satisfactory. Once transitive and partial dependencies have been eliminated, this compositional approach is essentially the last step of the earlier derivation from the FD graph.

B.4.4 NORMALISATION ALGORITHMS

We now explain some algebraic normalisation algorithms, which decompose unsatisfactory schemas by removing **redundancies**. But first, we need to introduce some underlying criteria:

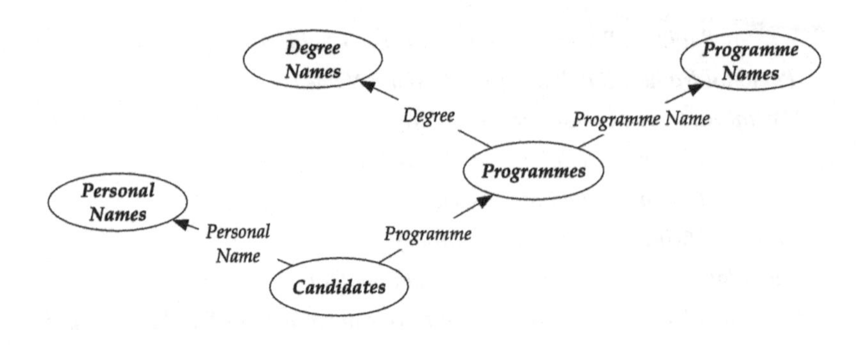

FIGURE B.5

The FD subgraph containing the attributes needed to produce the *Candidate Programmes* report.

B.4.4.1 The dependency-preserving property

As part of a process of view integration, the analyst is considering a report that shows the personal name, programme, and degree of each candidate. It expresses the following relation:

$$\textit{Candidate Programmes}(\textit{Candidate}, \textit{Personal Name}, \textit{Programme}, \textit{Programme Name}, \textit{Degree}),$$

with non-trivial FDs:

$$\begin{aligned} \textit{Personal Name} &: \textit{Candidates} \rightarrow \textit{Personal Names}, \\ \textit{Programme} &: \textit{Candidates} \rightarrow \textit{Programmes}, \\ \textit{Programme Name} &: \textit{Programmes} \rightarrow \textit{Programme Names}, \\ \textit{Degree} &: \textit{Programmes} \rightarrow \textit{Degree Names} \end{aligned}$$

Consider the (rather silly) decomposition:

$$\textit{Candidates}(\underline{\textit{Candidate}}, \textit{Personal Name}, \textit{Degree}),$$
$$\textit{Programmes}(\underline{\textit{Programme}}, \textit{Programme Name})$$

As Figure B.5 shows, the FD *Degree* : *Candidates* → *Degree Names* is really a transitive dependency via *Programmes*, but since *Programme* isn't an attribute of *Candidates*, both tables are technically in 3NF. Unfortunately, neither of the FDs *Programme* : *Candidates* → *Programmes* and *Degree* : *Programmes* → *Degree Names* is preserved.

A table is said to **preserve an FD** if both its determinant and dependent attribute lie in the same table. The above decomposition is not **dependency-preserving**. (A dependency is not actually *enforced* unless its determinant is a key of a relation, but once a dependency is *lost* by being split across two tables, it cannot be enforced by any future decomposition.)

B.4.4.2 The lossless join property

After studying a report on the list of enrolments within each programme component (again as part of view integration), an analyst might consider the schema:

$$\textit{Component Enrolments}(\underline{\textit{Programme}, \textit{Subject}, \textit{Candidate}, \textit{Semester}},\ \textit{Programme Name}, \textit{Subject Name}, \textit{Personal Name}),$$

with non-trivial FDs:

$$\begin{aligned} \textit{Personal Name} &: \textit{Candidates} \to \textit{Personal Names},\\ \textit{Programme Name} &: \textit{Programmes} \to \textit{Programme Names},\\ \textit{Subject Name} &: \textit{Subjects} \to \textit{Subject Names} \end{aligned}$$

Consider the dependency-preserving decomposition,

$$\textit{Enrolments}(\underline{\textit{Candidate}, \textit{Subject}, \textit{Semester}}, \textit{Personal Name})$$

$$\textit{Components}(\underline{\textit{Programme}, \textit{Subject}}, \textit{Programme Name}, \textit{Subject Name})$$

The *Enrolments* table preserves the FD *Name* : *Candidates* → *Personal Names*, and the *Components* table preserves the remaining two FDs. The only possible join of the tables is by *Subject*, but this is insufficient to ensure that the correct set of enrolments appears for each programme. Their join is *lossy*. This is despite every FD (including the trivial FDs) being preserved. (See Figure B.6.)

There are two solutions to this problem:

- First, if the *Enrolments* table had also included *Programme*, then *Enrolments* could be joined successfully with *Components* using its primary key, *Programme* × *Subject*. This leads to the following theorem:

  If we decompose table $T(X,Y,Z)$ into tables $T_1(X,Z)$ and $T_2(Y,Z)$, the decomposition is lossless-join if and only if the common attribute(s), Z, is either a key of T_1 or a key of T_2. (In other words, it is a primary key in one table, and a foreign key in the other.)

- Second, even without adding *Programme* to the *Enrolments* table, the join could still be made successfully if this third table were created:

  $$\textit{Component Enrolments}(\underline{Programme, Subject, Candidate, Semester})$$

  that is, a table containing the primary key of the original table. The correct join can then be made using its projection functions onto *Enrolments* and *Components*.

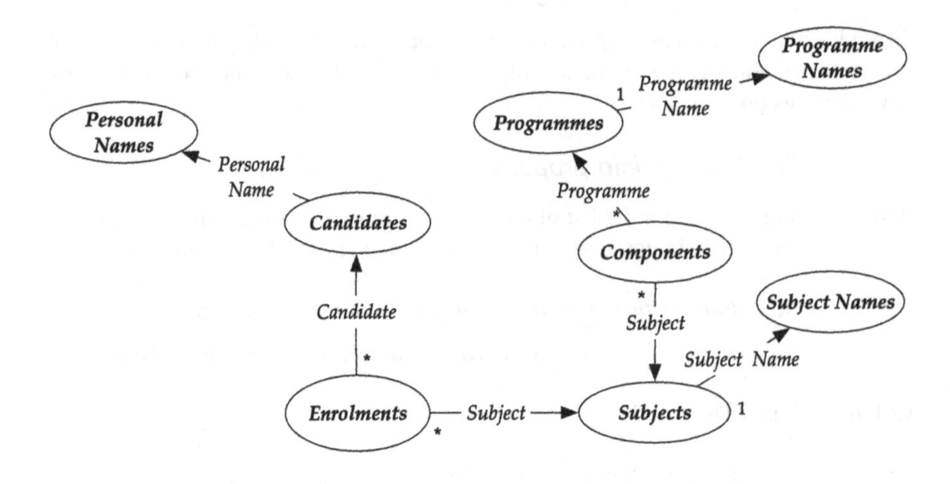

FIGURE B.6

The FD subgraph needed to produce the *Component Enrolments* report.

> In general, the decomposition of $T(\underline{A,B},C,D)$ with FDs $f : A \to C$ and $g : B \to D$ into $T_1(\underline{A},C)$ and $T_2(\underline{B},D)$ is lossless-join only if we also include $T_3(\underline{A,B})$.

B.4.5 BOYCE-CODD DEPENDENCIES

In order to preserve an FD, both its domain and codomain must be attributes of the same table. Inevitably, the table must also preserve any other FDs that happen to exist between those attributes.

If we use the FD *Subject Name* : *Subjects* → *Subject Names* to construct the table

$$Subjects(\underline{Subject}, Subject\ Name, \ldots),$$

then, whether we like it or not, the table also expresses the FD *Subject* : *Subject Names* → *Subjects*. This results from the one-to-one correspondence between *Subjects* and *Subject Names*, which makes them both candidate keys. Technically speaking, the table contains a transitive dependency, (*Subject Name* ; *Subject*) : *Subjects* → *Subjects*, which is an identity. This technicality explains why we exclude prime attributes from the normalisation rules. In the case of alternative candidate keys, as here, this does little harm, but in the case of a **Boyce-Codd dependency**, the analyst is faced with the choice of either expressing an FD redundantly or failing to express an FD at all.

The Academic Records System does not illustrate this particular difficulty. Instead, consider a small schema concerning a lecture timetable. In the process of view integration, the analyst discovers that a report is needed to show in which rooms

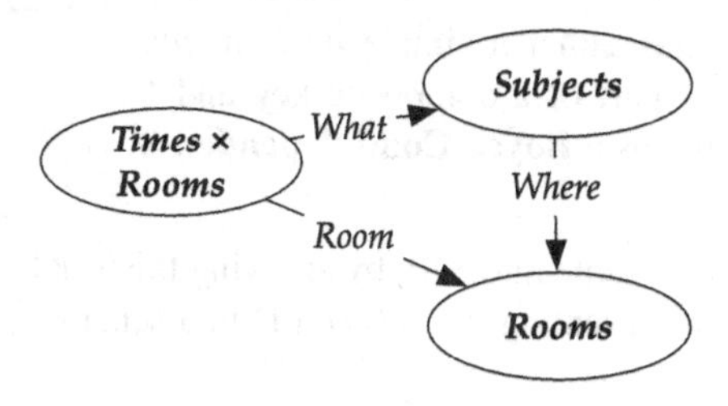

FIGURE B.7

A Boyce-Codd dependency. The trivial dependency *Times* × *Rooms* → *Rooms* is also a transitive dependency.

and at what times lectures will be held. Only one lecture can be held in a given room at a given time, which is captured by the FD:

$$What : Times \times Rooms \rightarrow Subjects$$

So far so good, but now imagine that in addition, in order to reduce confusion, it is a rule that all lectures in a given subject are held in the same room:

$$Where : Subjects \rightarrow Rooms$$

The corresponding FD graph is shown in Figure B.7.

Suppose we form a table to express the *What* FD:

$$Room\ Allocations(\underline{Time,\ Room}, What)$$

The dependency *Times* × *Rooms* → *Rooms* is now both trivial and transitive (*Times* × *Rooms* → *Subjects* → *Rooms*), i.e., it is both π_{Rooms} and *What* ; *Where*. But *What* is *not* a prime attribute, so the *Room Allocations* table is *not* in third normal form; the same combination of *What*, *Room* will appear redundantly for every lecture in a given subject.

Such an FD from a dependent attribute to *part* of the primary key is called a Boyce-Codd dependency.[4] As Figure B.7 shows, it occurs whenever a trivial dependency is also a product of FDs. Despite our earlier advice that tables should avoid containing attributes linked by compound paths, it seems unavoidable here. *Any* table capable of expressing the *What* FD must contain the projection *Times* × *Rooms* → *Rooms*, which, by virtue of the real-world situation, corresponds to the product *What* ; *Where*.

[4] If the FD maps from a dependent attribute to the *whole* of the primary key, the dependent attribute must be a candidate key. Such an FD can be enforced by using a **unique** constraint.

> If a table schema $T(\underline{W, X}, Y, Z)$ contains either a simple *or transitive* dependency $f : Y \rightarrow X$, where X is part of a composite key and Y is a non-prime attribute, we say it contains a **Boyce-Codd dependency** from Y to X.

We can attempt to put the *Room Allocations* table into 3NF by applying the usual decomposition for a transitive dependency, representing the *Where* FD in a separate table:

$$Lecture\ Times(\underline{Time},\ What)$$

$$Lecture\ Locations(\underline{Subject},\ Where)$$

Unfortunately, *Lecture Times* has no valid primary key, since many lectures can be scheduled at a given time. We may therefore be tempted to retain *Room* as part of its primary key. But this makes it identical to the *Room Allocations* table that we started with. Even so, we may regard this decomposition as a positive step, because *Lecture Locations* enforces the *Where* FD. So although the *Room Allocations* table contains the same combination of *Subject, Room* for every lecture in a given subject, *Lecture Locations* can be used to check that it is the correct one.

Luckily, precisely because of the *Where* FD, we may use *Subjects* as a proxy for *Rooms*, giving the schema

$$Lecture\ Times(\underline{Time,\ Subject})$$

$$Lecture\ Locations(\underline{Subject},\ Where)$$

which neatly avoids the Boyce-Codd dependency and captures the original *What* FD as the natural join of *Lecture Times* and *Lecture Locations* on *Subject Code.*

We can see the validity of this transformation by considering augmented forms of the *Where* and *What* FDs, *+Where* and *+What*:

$$+What : Times \times Rooms \rightarrow Times \times Subjects$$

$$+Where : Times \times Subjects \rightarrow Times \times Rooms$$

which make *Times* × *Rooms* and *Times* × *Subjects* a correspondence, as illustrated in Figure B.8.

An unfortunate side effect of this transformation is that the key of the *Room Allocations* table, *Times* × *Rooms*, automatically prevented a room from being double-booked, but the new schema can only check this constraint by forming the join of *Lecture Times* and *Lecture Locations*. Therefore, for most purposes it would be wise to include a third table to enforce this constraint. Correcting this problem, we obtain the following 3NF decomposition:

$$Lecture\ Locations(\underline{Subject},\ Where)$$

$$Lecture\ Times(\underline{Time,\ Subject})$$

$$Room\ Bookings(\underline{Time,\ Room})$$

We may generalise this process as Algorithm B.3.

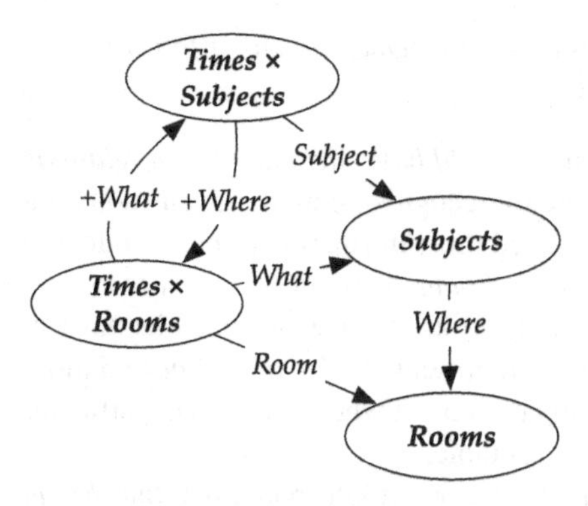

FIGURE B.8

The augmented FD graph of a Boyce-Codd dependency, showing that an alternative primary key is possible. The FDs +*What* and +*Where* are augmented forms of *What* and *Where*.

Algorithm B.3 Decomposing a Boyce-Codd dependency.

> While any table of the form $T(\underline{W,X},Y,Z)$ contains a simple or transitive Boyce-Codd dependency $Y \rightarrow X$, decompose it into three tables, $T_1(\underline{W,Y})$, $T_2(\underline{Y},X)$, and $T_3(\underline{W,X},Z)$.
> (If Z is empty, this preserves the characteristic function $T_3(\underline{W,X})$.)

This decomposition eliminates the transitive dependency, and the FD *Where* : *Subjects* → *Rooms* is no longer repeated redundantly. But it has failed to preserve the FD *What* : *Times* × *Rooms* → *Subjects*, because no table contains all three attributes. Consequently, a relational DBMS will be unable to check this FD automatically. In addition, the *Room Bookings* table seems redundant, because its contents can be deduced from *Lecture Times* and *Lecture Locations* — provided that the *Where* FD is *onto*. But without *Room Bookings*, a DBMS won't automatically prevent rooms from being double-booked.

> A table is said to be in **Boyce-Codd normal form** (**BCNF**) if every non-trivial FD is a key constraint, i.e., its determinant is a candidate key.

The *Room Allocations* table is not in BCNF because the *Where* FD is not a key constraint. In contrast, the decomposition into *Lecture Times*, *Lecture Locations*, and *Room Bookings is* in BCNF, but only because the FD *What* : *Times* × *Rooms* → *Subjects* has been discarded.

A Boyce-Codd dependency always results in a schema that either fails to preserve a dependency or is not in BCNF.

A more disciplined approach to the problem would have yielded a set of relations in BCNF from the outset. This would be done by recognising that a given room at a given time can be either *Free* or *Booked*. We need this information to prevent double booking. We therefore need an FD, *Room State* : *Rooms* × *Times* → *Room States*.

From the point of view of someone wanting to know at what times lectures in particular subjects are held, we need a table to represent the *Lecture Times* relation, and from the point of view of someone wanting to find out where lectures in particular subjects are held, we need the *Lecture Locations* table.

As a later optimisation during implementation, we might recognise that *Room States* has only two values, so that (as in the earlier case of *Component States*) we can use the existence or non-existence of a row in the *Room Bookings* table to represent *Room State*. But as we stressed in Algorithm 4.4.3 (page 133) and Section 4.4.7 (page 139), this optimisation must be left until normalisation is complete.[5]

At no time in *this* analysis did we imagine that there was a need for the *What* FD, so we shouldn't be concerned about whether it is preserved. The original *Room Allocations* table containing the *What* FD can be derived by joining the *Lecture Locations* and *Room Bookings* tables. If the *What* FD were added to the schema, it would not enforce any constraint that isn't already enforced. In short, the *What* FD is (and always was) redundant.

This example again illustrates the importance of considering tables that seem to have no dependent attributes as FDs that represent characteristic functions. If we fail to do this, the determinant of the associated relation won't be part of any FD, so it can never be derived from the set of FDs. Personally, I would hazard that all Boyce-Codd dependencies result from similar muddled thinking — in this case, by prematurely optimising *Room State* out of existence. After making that mistake, but realising that a table with the key *Rooms* × *Times* was needed to prevent double bookings, an alternative FD with the *Rooms* × *Times* domain had to be found, so the redundant *What* FD simply *had* to be invented. I would suggest that it is usually wiser to decompose a Boyce-Codd dependency to BCNF than to preserve a potentially redundant FD.

B.4.6 MULTI-VALUED DEPENDENCIES

A **multi-valued dependency**, such as $R : X \xrightarrow{1\ \ *} Y$, occurs when R maps $x \in X$ to more than one value $y \in Y$. Multi-valued dependencies can arise either from many-to-many relations or as converses of FDs.

---

[5] We are now left with a seeming paradox. If *Room State* isn't represented explicitly, the *Room Bookings* table remains a convenient way of preventing double booking, but, strictly speaking, it is redundant because we can deduce its contents from *Room Allocations* and *Subjects*. On the other hand, if other attributes were associated with bookings, such as who made them, and when, then the *Room Bookings* table would be essential, and there would be no paradox.

Table B.1 Some rows from a proposed *Enrolment Options* table

Admission No	Subject Code	Programme Code
...	...	...
26312091	acctg101	ecom
26312091	maths101	ecom
26312091	javaprog	ecom
26313755	acctg101	ecom
...	...	...

Trivial multi-valued dependencies are common. They exist from every non-prime attribute to each candidate key of a table. For example, the table *Programmes*($\underline{\textit{Programme}}$*, Degree Name*) contains the trivial multi-valued dependency

$$\textit{Programme} : \textit{Degree Names} \xrightarrow{1 \quad +} \textit{Programmes},$$

which is simply the one-to-many converse of the many-to-one FD

$$\textit{Degree Name} : \textit{Programmes} \xrightarrow{+ \quad 1} \textit{Degree Names}.$$

A **non-trivial multi-valued dependency** arises from a set of pairs that expresses a many-to-many relation. For example, the schema

$$\textit{Enrolment Options}(\underline{\textit{Candidate}, \textit{Subject}}, \textit{Programme})$$

would pair every candidate admitted to a programme with every subject that is a component of that programme. Potentially, several values of *Subject* would be associated with a given *Candidate*, and many values from *Candidate* with a given *Subject*. As a result, both attributes must form part of the key.

Suppose the proposed *Enrolment Options* table includes the rows shown in Table B.1. Since the table pairs every candidate in the *ecom* programme with every subject in the *ecom* programme (many more than three in reality) irrespective of whether the candidate is enrolled in the subject, it is easy to guess the contents of two further rows in Table B.1; candidate *26313755* must be paired with the same three subjects as candidate *26312091*.

Clearly, the *Enrolment Options* table contains redundancies. We say the table contains two non-trivial multi-valued dependencies:

$$\textit{Admissions} : \textit{Programmes} \xrightarrow{1 \quad *} \textit{Candidates},$$

and

$$\textit{Components} : \textit{Programmes} \xrightarrow{1 \quad +} \textit{Subjects}.$$

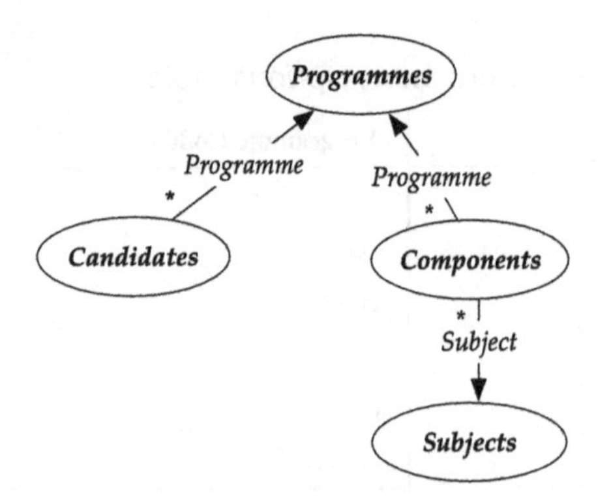

FIGURE B.9

The proposed *Enrolment Options* table creates multiple multi-valued dependencies.

The *Enrolment Options* table also contains two partial dependencies: *Programmes* × *Subjects* → *Programmes*, and *Programmes* × *Candidates* → *Programmes*. If we use the method of removing the partial dependencies explained in Appendix B.4.2, we obtain three tables:

$$\begin{gathered} \textit{Enrolment Options}(\underline{\textit{Candidate}, \textit{Subject}}), \\ \textit{Components}(\underline{\textit{Subject}, \textit{Programme}}), \\ \textit{Programmes}(\underline{\textit{Candidate}, \textit{Programme}}). \end{gathered}$$

Of these, *Enrolment Options* is simply the Cartesian product *Candidates* × *Subjects*, contains no useful information, and can safely be discarded.

Figure B.9 shows the part of the FD graph corresponding to the *Enrolment Options* table. It includes two FDs with *different domains*. To avoid non-trivial multi-valued dependencies, we propose the following simple rule:

A table should never contain FDs with different determinants.

Specifically, as we did in the graphical approach, we should always choose a set of FDs with a common domain to group into a table, and *never group FDs with different domains*.

A schema is in **4th normal form** (**4NF**) if no table contains more than one non-trivial multi-valued dependency. (Effectively, all domains are represented in separate tables.)

Finally, consider the relation that defines which subjects may be counted towards each degree programme. If *any* subject may be counted towards *any* programme, then this relation is simply the Cartesian product *Subjects* × *Programmes*. When the rules restrict which subjects may be counted, then the relation is some subset of the Cartesian product. It is a non-trivial multi-valued dependency

$$Subjects \xrightarrow{+\ *} Programmes$$

(or its converse). This relation would normally be expressed by a table, such as *Components*(*Programme*, *Subject*), which is in 4NF. (In general, *any* table that has a composite primary key may be said to have multi-valued dependencies, but they are inevitable and don't need to be resolved.)

A database is in **5th normal form** (**5NF**) if it is in 4NF and cannot be decomposed into simpler tables (except for trivial decompositions, such as decomposing $X \to Y \times Z$ into $X \to Y$ and $X \to Z$).

B.4.7 DECOMPOSITION ALGORITHMS

In Algorithm B.4, X, Y and Z may be simple or composite domains. The algorithm assumes that we know a cover of the set of FDs, not necessarily a minimal cover, that is, some FDs may be redundant, as a result of view integration. The algorithm is designed to remove them.

Algorithm B.4 A lossless-join (but not necessarily dependency preserving) decomposition to Boyce-Codd normal form.

Initialisation: Begin with the universal relation and a cover of its FDs.

Eliminate Cycles: While there is a cycle of FDs, $f_1 : X_1 \to X_2$, $f_2 : X_2 \to X_3, \ldots f_n : X_n \to X_1$, choose any X_j $(1 \leq j \leq n)$ as determinant, and replace each f_i by the corresponding product g_i: $X_j \to X_i$. (For example, if $j = 1$, we replace f_2 by $g_2 = f_1 ; f_2$, etc.) For all $i \neq j$, replace X_i by X_j in all other FDs. In other words, from a set of candidate keys, choose one to be the primary key.

Eliminate Boyce-Codd Dependencies: While any table contains a Boyce-Codd dependency, apply Algorithm B.3.

Eliminate Transitive Dependencies: While any table contains a transitive dependency, apply Algorithm B.1.

Eliminate Partial Dependencies: While any table contains a partial dependency, apply Algorithm B.2.

Aggregate: Combine all tables with the same keys, e.g., combine $R_y(\underline{X}, Y)$ and $R_z(\underline{X}, Z)$ to give $R_{yz}(\underline{X}, Y, Z)$.

By omitting the step *Eliminate Boyce-Codd Dependencies*, we obtain a second algorithm that guarantees that the decomposition is dependency-preserving and in

3NF (but not necessarily in BCNF). (As we saw in Section B.4.5, if a Boyce-Codd dependency is present, a schema cannot be both dependency-preserving and in BCNF.)

Consider applying Algorithm B.4 to the timetable schema of Section B.4.5, with universal relation

$$\mathit{Room\ Allocations}(\underline{\mathit{Time}, \mathit{Room}}, \mathit{What})$$

and FDs

$$\mathit{What} : \mathit{Times} \times \mathit{Rooms} \rightarrow \mathit{Subjects}$$
$$\mathit{Where} : \mathit{Subjects} \rightarrow \mathit{Rooms}.$$

Because of the Boyce-Codd dependency $\mathit{Where} : \mathit{Subjects} \rightarrow \mathit{Rooms}$, Algorithm B.4 requires us to decompose *Room Allocations* into

$$\mathit{Lecture\ Times}(\underline{\mathit{Time}, \mathit{Subject}})$$
$$\mathit{Lecture\ Locations}(\underline{\mathit{Subject}}, \mathit{Where})$$
$$\mathit{Room\ Bookings}(\underline{\mathit{Time}, \mathit{Room}})$$

which is in BCNF, but the dependency $\mathit{What} : \mathit{Times} \times \mathit{Rooms} \rightarrow \mathit{Subjects}$ is lost.[6]

If the step *Eliminate Boyce-Codd Dependencies* is omitted, *Room Bookings* $(\underline{\mathit{Time}}, \underline{\mathit{Room}}, \mathit{What})$ may still be considered to contain a *transitive* dependency, $\overline{\mathit{What}}; \overline{\mathit{Where}} : \mathit{Times} \times \mathit{Rooms} \rightarrow \mathit{Rooms}$. As a result, Algorithm B.1 suggests that we separate the role of *Room* as a dependent attribute from its role as part of the determinant,[7] resulting in the decomposition

$$\mathit{Lecture\ Times}(\underline{\mathit{Time}, \mathit{Room}}, \mathit{What})$$
$$\mathit{Lecture\ Locations}(\underline{\mathit{Subject}}, \mathit{Where}).$$

This schema is dependency-preserving, in 3NF, but not in BCNF.

In the case of the Academic Records database, with FDs

$$\begin{aligned}
\mathit{Personal\ Name} &: \mathit{Candidates} \rightarrow \mathit{Personal\ Names},\\
\mathit{Postal\ Address} &: \mathit{Candidates} \rightarrow \mathit{Addresses},\\
\mathit{Phone} &: \mathit{Candidates} \rightarrow \mathit{Phone\ Numbers},\\
\mathit{Date\ of\ Birth} &: \mathit{Candidates} \rightarrow \mathit{Dates},\\
\mathit{Balance\ Owing} &: \mathit{Candidates} \rightarrow \mathit{Money},\\
\mathit{Candidate\ State} &: \mathit{Candidates} \rightarrow \mathit{Candidate\ States},\\
\mathit{Enrolment\ State} &: \mathit{Candidates} \times \mathit{Subjects} \times \mathit{Semesters} \rightarrow \mathit{Enrolment\ States},
\end{aligned}$$

[6] As we noted earlier, even if it *were* preserved, it wouldn't enforce any constraint that isn't already enforced.

[7] In treating *Room* both as part of the determinant and as a dependent attribute, this textbook departs company with many other textbooks, which would have omitted this decomposition.

$$
\begin{aligned}
\textit{Candidate} &: \textit{Candidates} \times \textit{Subjects} \times \textit{Semesters} \to \textit{Candidates},\\
\textit{Subject} &: \textit{Candidates} \times \textit{Subjects} \times \textit{Semesters} \to \textit{Subjects},\\
\textit{Semester} &: \textit{Candidates} \times \textit{Subjects} \times \textit{Semesters} \to \textit{Semesters},\\
\textit{Mark} &: \textit{Candidates} \times \textit{Subjects} \times \textit{Semesters} \to 0 \ldots 100,\\
\textit{Programme Name} &: \textit{Programmes} \to \textit{Programme Names},\\
\textit{Programme State} &: \textit{Programmes} \to \textit{Programme States},\\
\textit{Degree Name} &: \textit{Programmes} \to \textit{Degree Names},\\
\textit{Component State} &: \textit{Programmes} \times \textit{Subjects} \to \textit{Component States},\\
\textit{Programme} &: \textit{Programmes} \times \textit{Subjects} \to \textit{Programmes},\\
\textit{Subject} &: \textit{Programmes} \times \textit{Subjects} \to \textit{Subjects},\\
\textit{Subject Name} &: \textit{Subjects} \to \textit{Subject Names},\\
\textit{Subject State} &: \textit{Subjects} \to \textit{Subject States},\\
\textit{Fee} &: \textit{Subjects} \to \textit{Money},\\
\textit{Quota} &: \textit{Subjects} \to \textit{Integers},\\
\textit{Filled} &: \textit{Subjects} \to \textit{Integers},\\
\textit{Subject} &: \textit{Subject Names} \to \textit{Subjects},\\
\textit{Programme} &: \textit{Programme Names} \to \textit{Programmes}.
\end{aligned}
$$

and universal relation

$$
\begin{aligned}
\textit{Academic Records}(&\underline{\textit{Candidate}, \textit{Programme}, \textit{Subject}, \textit{Semester}},\\
&\textit{Component State}, \textit{Enrolment State}, \textit{Mark},\\
&\textit{Personal Name}, \textit{Postal Address}, \textit{Phone}, \textit{Date of Birth},\\
&\textit{Balance Owing}, \textit{Candidate State},\\
&\textit{Programme}, \textit{Programme Name}, \textit{Programme State},\\
&\textit{Degree Name}, \textit{Subject Name}, \textit{Subject State}, \textit{Fee}, \textit{Quota}, \textit{Filled}).
\end{aligned}
$$

- *Subject* and *Subject Name* form a cycle, so one of them, say *Subject*, is elected as the primary key. The FD *Subject Name* : *Subjects* → *Subject Names* remains unchanged, but the FD *Subject* : *Subject Names* → *Subjects* is replaced by the product *Subject Name* ; *Subject*, which is an identity function and therefore redundant. A similar treatment eliminates the cycle *Programme Name* ; *Programme* by choosing *Programme* as the primary key.
- There is no Boyce-Codd dependency.
- Several partial dependencies exist, essentially one for each FD. After eliminating unnecessary components of their keys, we obtain the following decomposition:

$$
\begin{aligned}
&\textit{Subjects}_1(\underline{\textit{Subject}}, \textit{Subject Name}),\\
&\textit{Subjects}_2(\underline{\textit{Subject}}, \textit{Subject State}),
\end{aligned}
$$

$$\textit{Subjects}_3(\underline{\textit{Subject}}, \textit{Fee}),$$
$$\textit{Subjects}_4(\underline{\textit{Subject}}, \textit{Quota}),$$
$$\textit{Subjects}_5(\underline{\textit{Subject}}, \textit{Filled}),$$
$$\textit{Candidates}_1(\underline{\textit{Candidate}}, \textit{Name}),$$
$$\textit{Candidates}_2(\underline{\textit{Candidate}}, \textit{Postal Address}),$$
$$\textit{Candidates}_3(\underline{\textit{Candidate}}, \textit{Phone}),$$
$$\textit{Candidates}_4(\underline{\textit{Candidate}}, \textit{Date of Birth}),$$
$$\textit{Candidates}_5(\underline{\textit{Candidate}}, \textit{Balance Owing}),$$
$$\textit{Candidates}_6(\underline{\textit{Candidate}}, \textit{Candidate State}),$$
$$\textit{Programmes}_1(\underline{\textit{Programme}}, \textit{Programme Name}),$$
$$\textit{Programmes}_2(\underline{\textit{Programme}}, \textit{Programme State}),$$
$$\textit{Programmes}_3(\underline{\textit{Programme}}, \textit{Degree Name}),$$
$$\textit{Components}(\underline{\textit{Programme}, \textit{Subject}}, \textit{Component State}),$$
$$\textit{Enrolments}_1(\underline{\textit{Candidate}, \textit{Subject}, \textit{Semester}}, \textit{Enrolment State}),$$
$$\textit{Enrolments}_2(\underline{\textit{Candidate}, \textit{Subject}, \textit{Semester}}, \textit{Mark}),$$
$$\textit{Academic Records}_1(\underline{\textit{Candidate}, \textit{Programme}, \textit{Subject}, \textit{Semester}})$$

- All the FDs have determinants that are part of the database key, so there can be no transitive dependencies present.
- After combining tables with common keys, we obtain:

$$\textit{Subjects}(\underline{\textit{Subject}}, \textit{Subject Name}, \textit{Subject State}, \textit{Fee}, \textit{Quota}, \textit{Filled}),$$
$$\textit{Candidates}(\underline{\textit{Candidate}}, \textit{Personal Name}, \textit{Postal Address}, \textit{Phone},$$
$$\textit{Candidate State}, \textit{Date of Birth}, \textit{Balance Owing})$$
$$\textit{Programmes}(\underline{\textit{Programme}}, \textit{Programme Name}, \textit{Programme State},$$
$$\textit{Degree Name}),$$
$$\textit{Components}(\underline{\textit{Programme}, \textit{Subject}}, \textit{Component State}),$$
$$\textit{Enrolments}(\underline{\textit{Candidate}, \textit{Subject}, \textit{Semester}}, \textit{Enrolment State}, \textit{Mark}),$$
$$\textit{Academic Records}(\underline{\textit{Candidate}, \textit{Programme}, \textit{Subject}, \textit{Semester}})$$

This is similar to the schema we derived in Section 4.4.3 using the FD graph, with two exceptions.

First, there are no tables here corresponding to *Subject Index* and *Programme Index*. Fortunately, the dependencies they contain have been preserved, and therefore they are easily enforced (as in Section 4.4.3) by a **unique** constraint.

Second, the *Academic Records* table was not present in Section 4.4.3. It is present here because no step in Algorithm B.4 can remove the database key. We discussed the derivation of the key in Section B.3. In particular, the inclusion of *Programme* as part of the key was problematic. We, as analysts, are aware that the correct values of *Programme* can be derived from *Candidate* via the FD *Programme* : *Candidates* → *Programmes*, but the decomposition algorithm cannot discover this. The sensible thing is to exclude *Programme* from the key at Step 1. Once that is done, the remaining attributes of the *Academic Records* table become subsumed by the *Enrolments* table.

B.5 DIFFICULTIES WITH NORMALISATION

There are three difficulties with normalisation when compared with drawing FD graphs:

- First, view integration can pose problems that simply do not arise when using graphs. Consequently, not all possible decompositions will prove successful: we have to be concerned with preserving dependencies and avoiding lossy joins. As a result, individual steps can be rather complicated, especially ones that remove a Boyce-Codd dependency. It is also important to make decompositions in the right order.
  In most cases, what appear to be rather complex algebraic definitions and algorithms are easily understood when illustrated by FD graphs, as we have seen in the foregoing text. Although cycles and parallel paths are easily detected in an FD graph — either by eye or by finding the transitive root — they seem to crop up somewhat arbitrarily during decomposition.
- Second, the approach doesn't include any information about domains. Normalisation assumes that each FD has a unique name (something we have not done here), so that FD names — in a sense — *imply* their domains and codomains. The algebraic approach therefore has a problem in showing when two paths in the FD graph lead to the same codomain. This, in turn, makes it impossible to recognise when they are equivalent. In the case of the *NUTS* Academic Records system, both *Quota* and *Filled* map *Subjects* onto integers. The algebraic method cannot give both FDs the same name because that would imply that they are the same FD and one would be eliminated. If it gave them different names, it would fail to imply that they have the same codomain. There is a similar case with respect to *Balance Owing* and *Fee*. More importantly, there are two paths from an enrolment to a study programme, which should agree. The algebraic approach must express this by ensuring that both *Programme* FDs have the same name. The FD graph makes such parallel paths obvious even when FDs have different names. Although none of these parallel FDs proves redundant in the case of the *NUTS* Academic Records database, all three examples ought to alert an analyst to give them careful attention: e.g., *Filled* should never exceed *Quota*.
- Third, it treats every FD as total and onto, and makes no allowance for partial or into FDs. In practice, these difficulties are rarely serious.

Fortunately, Algorithm B.4 is independent of whether FDs are total and onto, except in the second step, when it assumes that *any* attribute that appears in a cycle is a candidate key. An example of this difficulty is that, because of the FD *Where* : *Subjects* → *Rooms*, the pseudo-transitivity rule would allow the database key of the *Room Bookings* table to be *Times*×*Subjects*, rather than *Times*×*Rooms*. But this choice of key would not give access to all the values of the (implicit) FD *Room State* : *Rooms* × *Times* → *Room States*, because the *Where* FD is only into. As a consequence, rather than being able to directly query whether a room is free at a given time, we would need to check the times for every subject that used the room.

In Section 4.4.5, we pointed out some other problems that can be caused by cycles involving partial or into functions. In particular, in the case that no suitable candidate key exists, we suggested that it might sometimes be necessary to introduce a *new* key that maps *onto* all the other candidate keys.

In addition, we note that although Algorithm B.4 requires us to deal with *cycles*, it does not explain how to find them.

- Fourth, there are very many ways we can select a set of attributes from an FD graph. It is not clear that normalisation can deal correctly with every arbitrary choice.

B.6 SUMMARY

The decompositional approach typically begins with a series of business documents that present views of the database. These views become potential table schemas. The potential schemas are examined for redundancies, such as transitive dependencies, and then decomposed to remove them.

We have encountered some difficulties, such as Boyce-Codd dependencies, inherent in certain sets of FDs, that decomposition can't resolve. They typically result from the analyst having omitted a characteristic function from the set of FDs. Even so, we shouldn't take difficulties too seriously; careful programming can enforce absolutely any constraint.

B.7 FURTHER READING

The book 'Databases and Transaction Processing' by Philip M. Lewis, Arthur Bernstein, and Michael Kifer (ISBN: 0-201-070872-8) discusses normalisation in more depth than we have space for here.

Algorithm B.4 is the author's fusion of algorithms described in Chapter 7 of 'An Introduction to Database Systems' by Bipin. C. Desai (ISBN: 0-314-6671-7). In my version, I have promoted the step of eliminating Boyce-Codd dependencies ahead of the others in order to deal with the rare case of a transitive (composite) Boyce-Codd dependency. I have also taken the view that, when making a dependency-preserving decomposition into 3NF, it is wise to treat a Boyce-Codd dependency as if it were a transitive dependency. To see why, consider Exercise 1 in the next section.

B.8 EXERCISE

1. The *National University of Technology and Science* owns several tutorial rooms. Only one tutorial may be held in a room at a given time. The tutorial concerns a particular subject. All tutorials for a subject are conducted by the same tutor. A tutor always occupies the same room, irrespective of the subject. We therefore have the universal relation

$$Tutorials : (\underline{Time, Room}, What, Who)$$

with the following FDs:

$$\begin{aligned} What &: Times \times Rooms \rightarrow Subjects, \\ Who &: Subjects \rightarrow Tutors, \\ Where &: Tutors \rightarrow Rooms \end{aligned}$$

Use Algorithm B.4 to find a BCNF schema.
By omitting the step *Eliminate Boyce-Codd Dependencies*, find a 3NF schema.

APPENDIX

Answers to the exercises

CHAPTER CONTENTS

C.1 CHAPTER 1

1. Apply Algorithm 1.1 to the set of rules on page 10 again, this time picking the other condition first. Verify that it produces the same result as the 'intuitive' solution.

   Step 1 The last actions differ; nothing happens.

   Step 2 We pick the condition (*status* = 'ok'):

   Group 1 contains

   Rule 1: If *balance* ≥ *debit* then
   set *approve* to 'yes' and *error* to 'no'

   Rule 2: If *balance* < *debit* then
   set *approve* to 'no' and *error* to 'no'

   Group 2 contains

   Rule 3: Set *error* to 'yes'.

Step 3: We generate

```
if (status == 'ok') then
    {Group 1}
else
    {Group 2};
```

Step 4: Applying the same process recursively to Group 1 we have

Step 1 Set *error* to 'no' is placed last and removed from consideration.

Step 2 The only condition is (*balance* < *debit*)

Group 1.1 contains

Rule 2: Set *approve* to 'no'.

Group 1.2 contains

Rule 3: Set *approve* to 'yes'.

Step 3 We generate

```
if (balance < debit) then
    {Group 1.1}
else
    {Group 1.2};
error = 'no';
```

Step 4 Applied recursively to Group 1.1 we have

Step 1 Since there is only one rule, trivially, the last actions agree. We generate

```
approve = 'no'
```

Step 2 Nothing to do.

Step 3 Nothing to do.

Step 4 Exit the recursion.

The remaining steps are similarly trivial.

At the end of this process we have,

```
if (status == 'ok') then {
  if (balance < debit) then
    approve = 'no'
  else
    approve = 'yes';
  error = 'no' }
else
  error = 'yes';
```

2. Suppose you have to sort a list of length l containing integers in the range $1 - n$. Is it possible to sort them in time with complexity less than $\mathbf{O}(l \log l)$?

   Yes. Create an array $A(1 \ldots n)$ and initialise it to zeros. Step through the list sequentially from 1 to l, and if the current element of the list is

x, add 1 to $A(x)$. Then step through the array sequentially, and if $A(x)$ has value y, append y occurrences of x to the result. Each of these steps is $\mathbf{O}(l)$ or $\mathbf{O}(n)$, so the overall complexity is $\mathbf{O}(l + n)$. On the other hand, if n is much larger than l, an $\mathbf{O}(l \log l)$ sorting algorithm may prove quicker.

3. A computer has a word length of n bits. Is the problem of listing each possible word tractable or intractable?

Strictly speaking, *intractable*. The number of possible words is 2^n, so the time taken to enumerate them grows exponentially with n. If a computer processed one value per nanosecond, it would take over 500 years to enumerate all the values of a 64-bit word.

This example shows that we need to take care in understanding what is meant by the *size* of a problem. If you had understood that the size was 2^n, then the problem would have been considered tractable. If you had understood that the size was the *length* of the binary representation of n, then it would have been considered $\mathbf{O}(2^{2^n})$.

C.2 CHAPTER 2

1. Complete the proof of Equation 2.1.27 on page 28.

   Using Axioms 2.1.8, we get

$$(P \Rightarrow Q) \wedge (P \Leftarrow Q) = (\neg P \vee Q) \wedge (\neg Q \vee P).$$

   We can then use Theorem 2.1.13 to reverse the order of the terms, giving

$$(\neg P \vee Q) \wedge (\neg Q \vee P) = (P \vee \neg Q) \wedge (Q \vee \neg P),$$

   which is what we needed to prove.

2. Prove Equation 2.1.11 on page 26. (Hint: Use Axiom 2.1.7 and the resolution theorem (page 29).)

   We use Axiom 2.1.7 to get the propositions into CNF, then use resolution: Q and $\neg Q$ cancel, leaving $\neg P \vee R$ as the resolvent, and then Axiom 2.1.7 is used to put the result into the required form.

$$\begin{array}{c} P \Rightarrow Q \\ Q \Rightarrow R \\ \neg P \vee Q \\ \neg Q \vee R \\ \hline \neg P \vee R \\ P \Rightarrow R \end{array}$$

   We can also use an Euler diagram to illustrate the theorem. Draw three concentric circles. Label the innermost one P, the intermediate one Q, and the outermost one R.

3. Use Axiom 2.1.7 to reduce $P \Rightarrow Q$ to CNF, then use De Morgan's Theorem to find its logical complement. How do you interpret your result?

   $\neg(P \Rightarrow Q) = \neg(\neg P \vee Q) = P \wedge \neg Q$. The resulting conjunction is the only case where $P \Rightarrow Q$ is defined as false in the truth table of page 25.

4. Why does a set of cardinality n have 2^n possible subsets? (See Section 2.3.1.)

   Each member of the set may either belong, or not belong, to a given subset. There are therefore n independent binary choices, giving 2^n possible subsets.

5. In what special circumstances does $R^1 ; R^{-1} = R^0$? (See inequality 2.4.26 on page 43.) (Hint: Consider the difference between the *Child* relation in connection

with family relationships, and the *Child* relation in rooted trees such as Figure 2.19.)

In the case that R^{-1} is a function. R may map a member x of domain X to any number of members $y_1, y_2 \ldots$ in codomain Y, but R^{-1} maps a given member of Y to only one member of X. Thus each value $y_1, y_2 \ldots$ is associated with x and only x. (Although $R^1 ; R^{-1}$ may map x to itself as many times as there are distinct values $y_1, y_2 \ldots$, because relations are *sets*, this is immaterial.)

6. By analogy with Equation 2.4.22 on page 43, define the *Has Uncle* relation.

$$Has\ Uncle = (Has\ Grand\ Parent\,;Has\ Son) \setminus Has\ Father$$

7. Represent the labelled graph of Figure 2.21 on page 53 as a matrix, with entries '0', '1', '0,1', or blank.

	S	T	U
S	0,1	1	–
T	–	–	0,1
U	–	–	–

8. Sketch a graph data structure corresponding to Figure 2.21 in the style of Figure 2.34 on page 71, labelling its edges '0' or '1'.

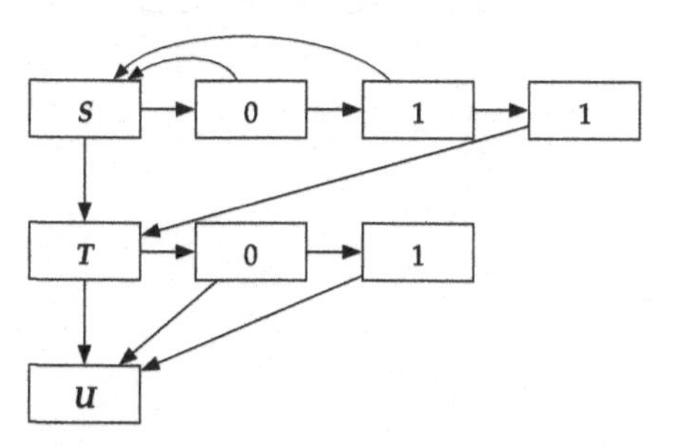

9. Graph G has vertices V, W, X, Y, and Z, and edges $V \rightarrow W$, $V \rightarrow X$, $W \rightarrow V$, $W \rightarrow X$, $X \rightarrow Y$, $X \rightarrow Z$, $X \rightarrow Y$, $Y \rightarrow Z$, and $Z \rightarrow Y$. Find the strongly connected components of G and sketch its reduced graph. (See Section 2.5.5.)

If the answer isn't immediately obvious, we begin by forming an adjacency matrix of the graph:

	V	W	X	Y	Z
V	–	X	X	–	–
W	X	–	X	–	–
X	–	–	–	X	X
Y	–	–	–	–	X
Z	–	–	–	X	–

The next step is to find the reflexive closure:

	V	W	X	Y	Z
V	X	X	X	–	–
W	X	X	X	–	–
X	–	–	X	X	X
Y	–	–	–	X	X
Z	–	–	–	X	X

We then use Warshall's Algorithm to find its reflexive transitive closure, but this doesn't introduce any new entries in the matrix. The rows are already neatly sorted, so we conclude that the strongly connected components are $\{V,W\}$, $\{X\}$, and $\{Y,Z\}$, with edges $\{V,W\} \rightarrow \{X\} \rightarrow \{Y,Z\}$.

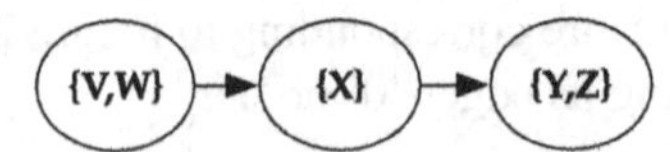

C.3 CHAPTER 3

1. Consider Figure 3.9 on page 91, which shows an FSA for books in a library. Sketch the figure and modify it to include an additional state, *On Hold*, and two additional events. Books can be placed *On Hold* by a *Hold* event when they are in any of the states *On Loan*, *Shelved*, or *Archived*. A book in the *On Hold* state can become *On Loan* as the result of a *Lend* event.

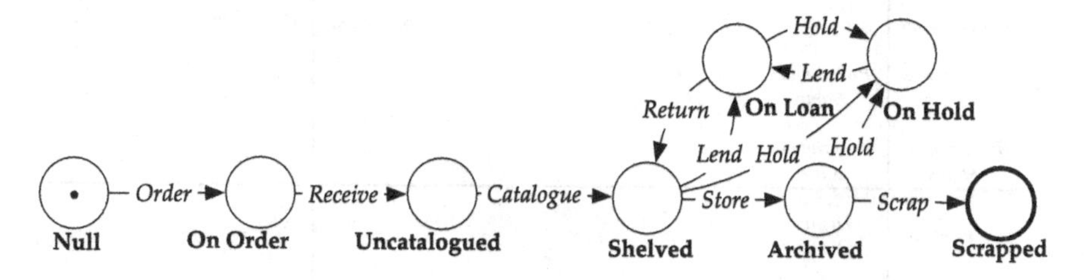

2. Consider the DFA of Figure 3.7 on page 86, which detects if the last letter but one of a binary sequence is '1'. By analogy, how many states would you expect a DFA that detects if the last letter but *two* of a binary sequence is '1'? How many states would you expect the DFA to have in order to detect if the last letter but n of a binary sequence is '1'? What does this tell you about the tractability of converting an NFA to a DFA?

   Using the shift register analogy, it would then need three stages, so it would have 2^3 states. Likewise, to detect if the last letter but n of a binary sequence is '1', a DFA would need 2^{n+1} states. Consequently, converting an NFA to a DFA is technically *intractable*, although rarely a problem in practice.[1]

3. What regular expressions describe the FSAs of Figure 3.1 and Figure 3.2?

$$[Open\,;(Deposit \cup Withdraw)^*\,;Close], \text{ and}$$

$$[Open\,;(Deposit \cup Withdraw \cup (Withdraw\,;Deposit^+))^*\,;Close].$$

   Are they equivalent?

   The expressions must be different, because after *Open*, Figure 3.2 sometimes rejects *Withdraw* and *Close* events, but Figure 3.1 always allows them until after the final *Close* event. Whereas a system based on Figure 3.1 might refuse to honour a *Withdraw* event once an account was overdrawn, one based on Figure 3.2 wouldn't accept the event at all.

[1] To prevent this ever becoming a problem, it is possible to convert the NFA to a DFA dynamically, constructing only those states actually encountered in analysing the input. The complexity is then bounded by the length of the input.

C.4 CHAPTER 4

1. Take the FD graph of Figure 4.6 on page 134 and express it as a matrix whose columns are domains and whose rows are codomains.

	Programmes	*Programme Names*	*Subjects*	*Subject Names*	*Components*	*Candidates*	*Enrolments*
Programmes	–	*Programme*	–	–	*Programme*	*Programme*	–
Programme Names	*Programme Name*	–	–	–	–	–	–
Programme States	*Programme State*	–	–	–	–	–	–
Degree Names	*Degree Name*	–	–	–	–	–	–
Subjects	–	–	–	*Subject*	*Subject*	–	*Subject*
Subject Names	–	–	*Subject Name*	–	–	–	–
Subject States	–	–	*Subject State*	–	–	–	–
Integers	–	–	*Filled, Quota*	–	–	–	–
Money	–	–	*Fee*	–	–	*Balance Owing*	–
Component States	–	–	–	–	*Component State*	–	–
Candidates	–	–	–	–	–	–	*Candidate*
Candidate States	–	–	–	–	–	*Candidate State*	–
Personal Names	–	–	–	–	–	*Personal Name*	–
Addresses	–	–	–	–	–	*Postal Address*	–
Phone Numbers	–	–	–	–	–	*Phone Number*	–
Dates	–	–	–	–	–	*Date of Birth*	–
0–100	–	–	–	–	–	–	*Mark*
Enrolment States	–	–	–	–	–	–	*Enrolment State*
Semesters	–	–	–	–	–	–	*Semester*

2. In Figure 4.4 on page 126 we considered *Enrolments* as a relation between *Admissions* and *Components*. What set of FDs would have resulted from this idea?

$$
\begin{aligned}
Programme\ Name &: Programmes \rightarrow Programme\ Names\\
Programme\ State &: Programmes \rightarrow Programme\ States\\
Degree\ Name &: Programmes \rightarrow Degree\ Names\\
Subject\ Name &: Subjects \rightarrow Subject\ Names\\
Subject\ State &: Subjects \rightarrow Subject\ States\\
Fee &: Subjects \rightarrow Money\\
Quota &: Subjects \rightarrow Integers\\
Filled &: Subjects \rightarrow Integers\\
Component\ State &: Programmes \times Subjects \rightarrow Component\ States\\
Programme &: Programmes \times Subjects \rightarrow Programmes\\
Subject &: Programmes \times Subjects \rightarrow Subjects\\
Personal\ Name &: Candidates \rightarrow Personal\ Names\\
Postal\ Address &: Candidates \rightarrow Addresses\\
Phone &: Candidates \rightarrow Phone\ Numbers\\
Date\ of\ Birth &: Candidates \rightarrow Dates\\
Balance\ Owing &: Candidates \rightarrow Money\\
Candidate\ State &: Candidates \rightarrow Candidate\ States\\
Enrolment\ State &: Candidates \times Programmes \times Subjects \times Semesters\\
&\quad \rightarrow Enrolment\ States\\
Mark &: Candidates \times Programmes \times Subjects \times Semesters\\
&\quad \rightarrow 0 \ldots 100\\
Candidate &: Candidates \times Programmes \times Subjects \times Semesters\\
&\quad \rightarrow Candidates\\
Component &: Candidates \times Programmes \times Subjects \times Semesters\\
&\quad \rightarrow Programmes \times Subjects\\
Semester &: Candidates \times Programmes \times Subjects \times Semesters\\
&\quad \rightarrow Semesters\\
Subject &: Subject\ Names \rightarrow Subjects\\
Programme &: Programme\ Names \rightarrow Programmes.
\end{aligned}
$$

(If you chose to use the name *Admissions* instead of *Candidates*, your answer is equally correct.)

C.5 CHAPTER 5

1. It is usual to provide a separate update event for every attribute in a database, with the possible exception of those that can change a primary or foreign key. In the case of the Academic Records System, which attributes do you think should be allowed to be updated, and which should not?

- *Candidates*:
  - *Candidate* should only be set by an *Admit* event.
  - *Candidate State* should only be updated by *Admit* and *Graduate* events.
  - *Personal Name* can be changed arbitrarily, e.g., when a candidate marries.
  - *Postal Address* can be changed arbitrarily, e.g., when a candidate moves house or wants mail redirected.
  - *Phone* can be changed arbitrarily, e.g., when a candidate moves house, loses a mobile phone, etc.
  - *Date of Birth* shouldn't need to be changed, except after a mistake.
  - *Programme* cannot be changed. Changing it might leave a candidate enrolled in subjects for the wrong programme. In any case, the candidate ought to seek readmission.
  - *Balance Owing* shouldn't be updated except by *Enrol*, *Withdraw*, and *Pay* events, otherwise bookkeeping errors will result.
- *Programmes*:
  - *Programme* is set by *Create Programme*.
  - *Programme State* is set by *Create Programme*. There is currently no provision to remove a programme, because it might cause candidates to belong to non-existent programmes. It could be done, even so, by first checking that no one has been admitted to it.
  - *Programme Name* can safely be updated, to correct a misspelling, say.
  - *Degree* likewise can safely be updated.
- *Subjects*:
  - *Subject State* is updated by *Create Subject*, *Offer*, *Freeze*, and *De-activate* events in order to control when *Enrol*, *Withdraw*, and *Assess* events can occur. Arbitrary updates shouldn't be permitted.
  - *Subject Name* can safely be updated, to correct a misspelling, say.
  - *Quota* is carefully changed by the *Set Quota* event in such a way that a subject cannot become overfull.
  - *Filled* is updated by the *Enrol* and *Withdraw* events. It needs to agree with the number of *Enrolments* rows in the subject concerned.
  - *Fee* is updated by the *Set Fee* event, which first ensures that enrolments aren't currently in progress, which could otherwise lead to bookkeeping errors.
- *Components*:
  - *Programme* is set when a component is created. Changing it could lead to existing enrolments becoming forbidden for the candidate's programme.

- *Subject* is dealt with like *Programme*.
- *Component State* is already updated by *Permit* and *Forbid* events.
- *Enrolments*:
  - *Enrolment State* is already updated by *Enrol*, *Withdraw*, and *Assess* events.
  - *Semester* cannot be changed without creating a bookkeeping error.
  - *Candidate* cannot be changed without creating a bookkeeping error.
  - *Subject* cannot be changed without creating a bookkeeping error.
  - *Mark* is already updated by *Assess* events. The existing specification only allows such an event to occur once. There should be provision for reassessing or correcting a mark.

2. Specify an event that will set the value of *Phone*, and a second event that will allow it to be restored to **null**.

$$Set\ Phone(c, ph) \Leftarrow Candidate\ State(c) \neq Null\,; Phone(c)' = ph.$$

$$Null\ Phone(c) \Leftarrow Candidate\ State(c) \neq Null\,; Phone(c)' = \mathbf{null}.$$

3. It is proposed to allow candidates to protect their data using passwords. Passwords are usually stored in encrypted form, in this case, as an integer.
Specify an FD that allows each candidate to store an encrypted password.

$$Password : Candidates \rightarrow Integers.$$

Assuming that you already have an *encrypt* function available, specify an event to check the correctness of a password, and a second event to change a password. The change should only be allowed if the existing password is known and the new password isn't the same as the old one.

$$\begin{aligned} Check\ Password(c, pw) \Leftarrow\ & Candidate\ State(c) \neq Null\,; \\ & Password(c) = encrypt(pw). \end{aligned}$$

$$\begin{aligned} Change\ Password(c, oldpw, newpw) \Leftarrow\ & Candidate\ State(c) \neq Null\,; \\ & Password(c) = encrypt(oldpw)\,; \\ & Password(c)' = encrypt(newpw). \end{aligned}$$

(Note that both events fail if the password is incorrect.)

Which existing procedure will need to be modified?

Admit will need to initialise *Password*. A suitable value would be *encrypt*(c).

C.6 CHAPTER 6

1. On page 192 we discussed a query that could produce a list of timetable clashes. It was suggested that only a very clever optimiser would spot that a cross join was being made between two identical sets of data.
Suppose that the DBMS's query optimiser proves inadequate and the resulting query takes too long. It is decided to write an embedded *SQL* procedure to create the cross join from a single instance of the data. This would involve fetching a list of enrolments for one candidate at a time, storing them in an array — ten locations would be more than enough — and inserting all the required pairs into a new table. Write a query that will produce the required lists of enrolments.

```
select e.Candidate, e.Subject
    from Enrolments e
    where e.Semester = '20xxA'
    order by asc e.Candidate;
```

or

```
select e.Candidate, e.Subject
    from Enrolments e
    where e.Semester =
        (select Current_Semester
            from Globals
            where The_Key = 1)
    order by asc e.Candidate;
```

2. Candidate *C* is enrolling in subjects *S* and *T* and has locked the corresponding rows of the *Subjects* table for updating. Concurrently, candidate *D* is updating subject *U*. Candidates *E* and *F* are currently reading subject *V*, but have no intention of updating it. Candidate *E* wants to read subject *T*, but must wait for candidate *C* to commit. In addition, candidate *F* wants to update subject *S* but must also wait for *C* to commit. At the same time, candidate *C* is waiting to read subject *U*.
Is the database deadlocked? If not, in what order can the candidates' transactions commit?

The situation is exactly that depicted in Figure 6.3 on page 196 with the labels *A*, *B*, *C*, and *D* replaced by *C*, *D*, *E*, and *F* and *R*, *S*, *T*, and *U* replaced by *S*, *T*, *U*, and *V*. *D* must commit first, followed by *C*.

Suppose that the DBMS detects that a candidate's transaction is about to create a cycle in the graph. Can the potential deadlock be avoided without killing one of the transactions?

Yes and no: If candidate *D's* next action were to try to enrol in subject *V* a deadlock would result. If this were reported back to candidate *D's* event procedure, it could choose one of at least three options:

- To loop, waiting for the condition to clear,
- To **rollback**
- To **commit**

The first isn't really an option. The DBMS has detected a potential cycle passing through *D*. If the cycle passes through *D*, then *D* must be locking a row that completes the cycle. Until *D's* transaction releases the row it must continue to wait and therefore cannot release the row. The event procedure is doing exactly what the DBMS is trying to avoid. Waiting solves nothing.

The second option certainly removes the deadlock, but it is effectively the same as killing candidate *D's* transaction. *D* may feel frustrated at having wasted time and effort.

If the third option is possible, it may be best. For example, the candidate's enrolment in *U* could be committed, and then the enrolment in *V* could be reattempted. Since *D* will have released the lock on *U*, the other candidates' transactions should have been able to complete, so that *D's* second attempt to enrol in *V* may well succeed.

In reality, if this third option exists, there is a better way to do things and that is to make each enrolment a separate transaction, i.e., to **commit** immediately after every update. That way, enrolling can never cause deadlock.

3. Consider the query on page 177, to find all candidates who scored better than average in *Maths101*.

```
select c.Personal_Name, e1.Mark
    from Candidates c, Enrolments e1
    where c.Candidate = e1.Candidate
    and e1.Subject = 'Maths101'
    and e1.Mark >
        (select avg(e2.Mark)
            from Enrolments e2
            where e2.Subject = 'Maths101')
    order by asc c.Personal_Name;
```

What sequence of *select*, *project*, *join*, and *sort* operations would you expect a query optimiser to choose to answer the query?

This is an example of an uncorrelated subquery:

a. *Enrolments* rows would be selected with *Subject* = 'Maths101', projected onto *Mark*. At the same time, the average mark would be calculated.

b. The *Enrolments* rows with *Subject* = 'Maths101' and a value of *Mark* above the average would be *selected* and *projected* onto *Mark* and *Candidate*.

c. The selected rows and columns would be joined with rows from the *Candidates* table on the foreign key *Candidate*, and a final projection would be used to generate the result.

Which join algorithms should be used?

The query — perhaps erroneously — doesn't specify a particular semester. If, pessimistically, *Maths101* is a component of every programme, about 1/20th of all enrolments are in *Maths101*. We are therefore joining 50,000 *Enrolments* rows with 100,000 *Candidates* rows. This is a foreign key-primary key join, so the Sort-Merge and Table Look-Up algorithms are applicable. Using the rule of thumb on page 187, *Sort-Merge* will prove faster provided the blocking factor of the *Candidates* table is more than two, which in practice it certainly will be.

Can you suggest an improvement?

Yes. The first step could produce a selection of the *Enrolments* table containing just the rows for *Maths101* that could be passed to the second step, removing the need for the second selection process. Alternatively, the programmer could first create a new table, derived from the *Enrolments* table, containing just the enrolments in *Maths101*, and substitute this for *Enrolments* in a simplified version of the above query. (A *view*, being a procedure, wouldn't help.)

C.7 CHAPTER 7

1. Consider the *Withdraw* event, which undoes the work of *Enrol*, provided that the subject has not yet been frozen:

$$
\begin{aligned}
Withdraw(c,s) \Leftarrow\ & e = (Current\ Semester, c, s)\,;\\
& Enrolment\ State(e) = Enrolled\,;\\
& Subject\ State(s) = Offered\,;\\
& Enrolment\ State(e)' = Null\,;\\
& Filled(s)' = Filled(s) - 1\,;\\
& Balance\ Owing(c)' = Balance\ Owing(c) - Fee(s).
\end{aligned}
$$

Construct a dependency matrix for the *Withdraw* event in the style of Table 7.1, marking the cells with the relevant rule or rules. (Use the same abbreviations as Table 7.1.)

	Balance Owing	*Balance Owing(c)*	*Balance Owing(c)′*	*Balance Owing′*	*c*	*e*	*Enrolment State*	*Enrolment State(e)*	*Enrolment State(e)′*	*Enrolment State′*	*Fee*	*Fee(s)*	*Filled*	*Filled(s)*	*Filled(s)′*	*Filled′*	*s*	*Subject State*	*Subject State(s)*	*Current Semester*
Balance Owing	–	s	–	–	–	–	–	–	–	–	–	–	–	–	–	–	–	–	–	–
Balance Owing(c)	–	–	a	–	–	–	–	–	–	–	–	–	–	–	–	–	–	–	–	–
Balance Owing(c)′	–	–	–	u	–	–	–	–	–	–	–	–	–	–	–	–	–	–	–	–
Balance Owing′	–	–	–	–	–	–	–	–	–	–	–	–	–	–	–	–	–	–	–	–
c	q	s	–	u	–	a	–	–	–	–	–	–	–	–	–	–	–	–	–	–
e	–	–	–	–	–	–	q	s	–	u	–	–	–	–	–	–	–	–	–	–
Enrolment State	–	–	–	–	–	–	–	s	–	–	–	–	–	–	–	–	–	–	–	–
Enrolment State(e)	–	–	c	–	–	–	–	–	c	–	–	–	–	–	–	–	–	–	–	–
Enrolment State(e)′	–	–	–	–	–	–	–	–	–	u	–	–	–	–	–	–	–	–	–	–
Enrolment State′	–	–	–	–	–	–	–	–	–	–	–	–	–	–	–	–	–	–	–	–
Fee	–	–	–	–	–	–	–	–	–	–	–	s	–	–	–	–	–	–	–	–
Fee(s)	–	–	a	–	–	–	–	–	–	–	–	–	–	–	–	–	–	–	–	–
Filled	–	–	–	–	–	–	–	–	–	–	–	–	–	s	–	–	–	–	–	–
Filled(s)	–	–	–	–	–	–	–	–	–	–	–	–	–	–	a	–	–	–	–	–
Filled(s)′	–	–	–	–	–	–	–	–	–	–	–	–	–	–	–	u	–	–	–	–
Filled′	–	–	–	–	–	–	–	–	–	–	–	–	–	–	–	–	–	–	–	–
s	–	–	–	–	–	a	–	–	–	–	q	s	q	s	–	u	–	q	s	–
Subject State	–	–	–	–	–	–	–	–	–	–	–	–	–	–	–	–	–	–	s	–
Subject State(s)	–	–	c	–	–	–	–	–	c	–	–	–	–	–	–	–	–	–	–	–
Current Semester	–	–	–	–	–	a	–	–	–	–	–	–	–	–	–	–	–	–	–	–

2. Draw the state dependency graph for the *Withdraw* event in the style of Figure 7.3.

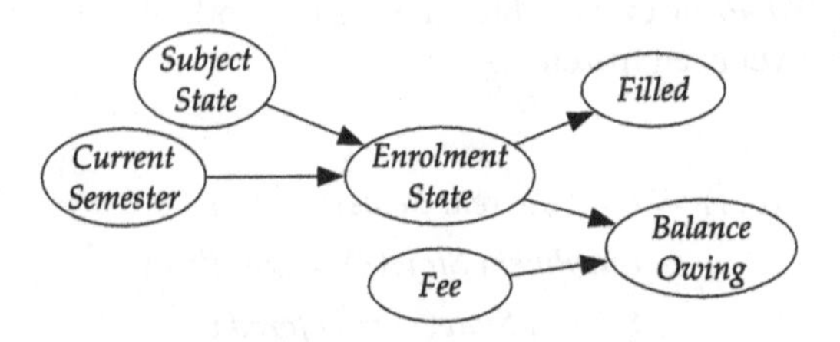

Is the graph you obtained canonical?

It contains no loops or multiple paths, so it is a minimal transitive reduction of itself. (With the proviso that the vertex labelled *Subject State* should be relabelled as the set {*Subject State*}, etc.)

3. Use Algorithm 7.1 on page 221 to find an optimised process graph for the *Withdraw* event.

Subject State and *Fee* share the same row of the *Subjects* table and can share the same read access. *Filled* also shares the same row, but to combine it with the others would create a cycle.

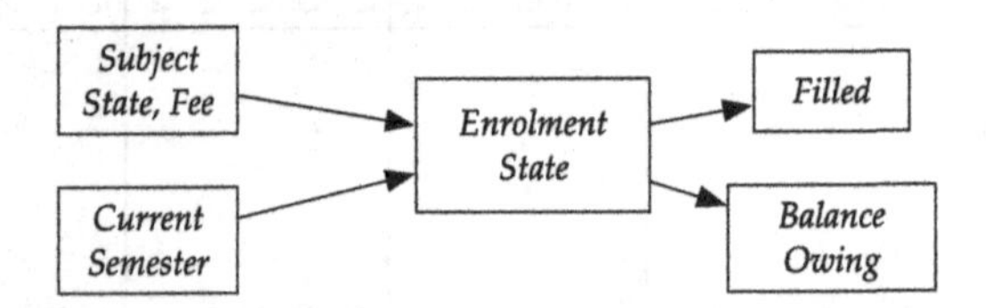

Ignoring all other types of event, is it possible to use a batch system to process *Withdraw* events?

Yes. It takes four stages.

a. The first process can check to see that the subject s is *Offered*. If so, it passes the values of *Fee*(s), s, and c to the second stage.
b. The second process uses *Current Semester*, s, and c to find e and check the state of *Enrolment State*(e). If all is well, it passes the values of *Fee*(s) and c to the third stage and passes the value of s to the fourth.
c. The third process decreases the value of *Balance Owing*(c) by *Fee*(s).
d. The fourth process decreases the value of *Filled*(s) by one.

The last two stages can be executed in either order or concurrently. The first and fourth processes cannot be merged.

4. On page 225, it was stated that although the object-oriented hierarchy of processes has no global **commit**, it is still safe, except in the event of a power or system failure. Why?

   The correctness of the *two-phase protocol* depends on none of the resources being unlocked until after the last one has been locked. It doesn't rely on all the unlocks being made atomically. In contrast, if they aren't unlocked atomically, during the time that they are being unlocked the consistency of the system faces three dangers:

   - A decision to **rollback**,
   - A power failure,
   - Any system failure that prevents the unlock sequence from completing.

   If resources are allocated hierarchically, as suggested, the system cannot deadlock, so the first danger is eliminated, but the last two remain.

5. In Section 7.4.2, we discussed an example of object-oriented design that used a particular series of classes: *Academic Records*, *Candidate*, *Subject*, *Enrolment*, *Ledger*. Assuming there is a separate object instance for each row of each class, synthesise the segmented specifications for the *Withdraw* event.

$$
\begin{aligned}
\mathit{Academic\ Records.Withdraw}(c,s) \Leftarrow\ & \mathit{Global.Read\ Lock}\,;\\
& \mathit{sem} = \mathit{Global.Get\ Semester}\,;\\
& \mathit{Candidate}(c).\mathit{Withdraw}(\mathit{sem},s)\,;\\
& \mathit{Global.Unlock}.
\end{aligned}
$$

$$
\begin{aligned}
\mathit{Candidate}(c).\mathit{Withdraw}(s,\mathit{sem}) \Leftarrow\ & c.\mathit{Read\ Lock}\,;\\
& \mathit{Subject}(s).\mathit{Withdraw}(c,\mathit{sem})\,;\\
& c.\mathit{Unlock}.
\end{aligned}
$$

$$
\begin{aligned}
& \mathit{Subject}(s).\mathit{Withdraw}(c,\mathit{sem}) \Leftarrow\\
& \qquad s.\mathit{Update\ Lock}\,;\\
& \qquad \mathit{Enrolment}((\mathit{sem},s,c)).\mathit{Withdraw}(s.\mathit{Fee},s.\mathit{Filled},s.\mathit{Filled}')\,;\\
& \qquad s.\mathit{Unlock}.
\end{aligned}
$$

$$
\begin{aligned}
\mathit{Enrolment}(e).\mathit{Withdraw}(\mathit{fee},\mathit{filled},\mathit{filled}') \Leftarrow\ & e.\mathit{Update\ Lock}\,;\\
& (\mathit{Enrolment\ State}(e) = \text{'Enrolled'}\,;\\
& \mathit{Enrolment\ State}(e)' = \text{'Null'}\,;\\
& \mathit{filled}' = \mathit{filled} - 1\,;\\
& c = \mathit{Candidate}(e)\,;\\
& \mathit{Ledger}(c).\mathit{Withdraw}(\mathit{fee}))\,;
\end{aligned}
$$

$\cup$ (*Enrolment State*(*e*) ≠ 'Enrolled' ;
Error.Not Enrolled(*e*) ;
filled′ = *filled* + 1 ;)
e.Unlock.

Ledger(*c*).*Withdraw*(*fee*) ⇐ *c.Update Lock* ;
Balance Owing(*c*)′ = *Balance Owing*(*c*) − *fee* ;
c.Unlock.

6. At the end of Section 7.8, we suggested that a candidate could submit a form that made several enrolments in one hit. To prevent deadlock, is it necessary to prioritise subject updates, e.g., in order by *Subject Code*?

Yes and no:

If the form is treated as a single transaction, it would be possible for candidate *C* to specify subject *S* followed by subject *T*, and candidate *D* to specify subject *T* then subject *S*. This might lead to a situation where *C* has locked *S* and *D* has locked *T*, and both are deadlocked. To avoid the possibility of deadlock, it would be necessary to ensure that both candidates locked the subjects in the same order.

Luckily, a single form doesn't necessarily imply a single transaction. As each enrolment is completed, its updates can be committed. (This has the minor disadvantage that the relevant *Candidates* row will need to be locked and unlocked each time — not a serious issue.)

C.8 CHAPTER 8

1. Consider the merits of a check digit scheme that requires an *Admission No* to be divisible by 17.

   The natural weights are

   5 9 6 4 14 15 10 1

   They are all non-zero, and no two weights sum to 17. Therefore all single and double transcription errors will be detected. (Also, most triple transcription errors, but weights 6, 10, and 1 sum to 17, for example.)

   All the weights are different, so all single transposition errors will be detected. No adjacent pairs have the same sum, so transpositions of pairs of digits will be detected too.

2. What does our inability to nest more than three interruptions tell us about the process of understanding the text of a computer program? (Consider, for example, the implementation of the *Enrol* event in Section 7.4.2 on page 224.)

   If we read the outer level of the program text and encounter a procedure call, we may need to interrupt our reading to see what the procedure does. In turn, the second procedure may call a third procedure, and so on. After three or four levels of such interruptions, when we return to the top level, we shall have forgotten what it does.

   The example of Section 7.4.2 is therefore hard for the uninitiated to understand, although if we know the method used to construct it, it becomes more intelligible.

   A good strategy in many cases is to first map out the procedure hierarchy, and then study the procedures from bottom to top. This exploits our better capacity for head and tail recursion.

3. *HTML* provides a **<blink>** directive that causes text to flash. Unfortunately, the feature is non-standard, and not all browsers support it. When do you think this feature would be useful? (See Section 8.3.2.3 on page 271.)

   This would be useful to draw the eye from its current focus to a different area, as when displaying an error diagnostic. Rather than flash the message itself, making it harder to read, it might prove better to flash an icon next to the message.

   Can you think of a good alternative, if **<blink>** is not supported?

   All browsers support animated GIFs. A flashing or otherwise irritating animated GIF placed next to the diagnostic should get the operator's attention.

4. The NUTS has a purchasing department and a store room, used mainly for stationery supplies. In theory, when any academic or administrative department

(the *originating department*) needs stationery or equipment, it should send an order to the purchasing department. The purchasing department will record the order in a database, thereby printing a purchase order to be sent to a preferred supplier. It also sends an electronic copy of the order to the store room.
When the goods arrive, they are delivered to the store room. The store hand informs the purchasing department of their arrival, which in turn informs the department that originated the order, recording the number of items *received* in the database. When the store room issues the goods to the originating department, the store hand records the number of items *issued* in the database.
In some cases, e.g., paper-clips and most other small stationery items, the purchasing department makes sure that there is always a reserve supply available for immediate use.
The purchasing manager receives a computer-generated report from the store hand listing the numbers of items of various kinds that are in the store. Unfortunately, the report is often wildly inaccurate, sometimes displaying a negative number of items in stock. This report has made both the store hand and the computer programming team the objects of scorn. What do you think causes the problem?

> One way that a negative number of items could seem to be in stock is if they have been recorded as issued but not recorded as having been received. This suggests that the store hand is more prompt in recording stock issues than the purchasing department is in recording receipts.
>
> We could solve this by making one department carry the responsibility for recording both receipts and issues, and the logical department to choose is the store room. The store hand should then be given full responsibility for producing the stock report. This will motivate the store hand to record deliveries promptly and to take pride in delivering a sensible report.

Apart from this, it is known that some departments bypass the purchasing system when they need something in a hurry. They send orders directly to the supplier, and inform the purchasing department later.
What can be done about this problem? Is a behavioural issue involved?

> There are two ways we could tackle this problem: try to prevent the originating departments from making unauthorised orders, or allow it to happen and adapt the system to cope.
>
> Both alternatives seem to have the same solution: The store hand already records the receipts of deliveries. Any receipts unmatched by a purchasing department order can then be investigated. This can be facilitated by a reconciliation report that compares purchase orders with store room receipts. The purchasing department already has the motivation to record orders promptly, otherwise the computer can't print the purchase order to be sent to the supplier. Exceptions on the reconciliation report will indicate either slow, overdue deliveries or

unauthorised purchases made directly by originating departments — provided the supplier sent the goods to the store room. Suppliers could be encouraged to comply by threatening to rescind their preferred status if they don't.

In the long term, the only way to discipline the originating departments — if that is deemed wise — is by senior management auditing their expenditure accounts, perhaps even going so far as to revise their budgets if they don't use the system properly.

5. In an experiment, eight out of ten subjects preferred Form *A* to Form *B*. Is Form *A* better than Form *B* with 95% significance?

The null hypothesis is that there is no difference between the forms. We need to calculate the probability that eight or more subjects out of ten would choose Form *A* by chance — but it is easier to count the ways that zero, one, or two subjects would choose Form *B*.

There is only one way zero subjects can choose Form *B*. There are ten distinct ways one subject can choose Form *B*. There are $(10 \times 9)/2 = 45$ ways that two subjects can choose Form *B*, making 56 ways in all. There are 2^{10} possible voting patterns. The odds are therefore 56/1,024 that two or fewer subjects would choose Form *B* by chance: 5.47%. The result therefore has 94.53% significance and just fails to reach the desired criterion.

C.9 CHAPTER 9

1. If 50% in *remathsa* or 40% in *maths101* is enough to study *graphthy*, and 50% in *remathsa* is sufficient to study *remathsb*, it only seems fair that a candidate who scored 40% in *maths101* should be allowed to study *remathsb*. Assuming this change, express the **needs** tree for *remathsb* in the same style as the examples on page 308.

   remathsb **needs** ($\geq$ 1) **from** {40 **in** *maths101*, 50 **in** *remathsa*}

2. The decision table of Table 9.3 on page 302 contains 23 rules, one of which is an *else* rule. If this table were implemented using structured (**if** ... **else**) code, how many branches would be needed to express the *else* rule?
   You may assume that the conditions will be tested in order from left to right, and that *Subject* is already constrained to have one of the values listed in the first column.

   Every one of the 23 numerical conditions in the table would need to be implemented by a separate branch in the code.

   Suppose, assuming that the prerequisites for subjects are the same for all programmes, that a giant two-dimensional matrix were prepared for all 500 subjects. Assuming an average of one rule per subject — even though subjects with alternative prerequisites might need two or more, and assuming that minimum marks are represented by one-byte integers, how much storage would it need?

   About 250,000 bytes for the matrix and 4,000 bytes for the subject codes.

   Can you think of a more compact way of storing the table?

   Use any of the sparse matrix representations described in Section 2.7. There will be just over 500 non-blank entries in the table.

   Is a giant matrix a good idea?

   Roughly 500 entries in a single table seems to be less of a maintenance problem than about 2,500 entries spread over 100 tables, although the table itself will need to be supplemented by lists of *Components* for each *Programme*. Fortunately, this information is already stored in the NUTS Academic Records System database.

3. On page 313, we suggested that the programme planner should omit subjects that a candidate wishes to avoid from the search. There are at least three ways this can be achieved:

   - By deleting the unwanted subjects from the search tree.
   - By ignoring the candidate's wishes until a solution is generated, then discarding any programmes that happen to include unwanted subjects.
   - By ignoring any feasible bit maps that include the unwanted subjects.

Comment on their relative merits.

Although the first option reduces the search space, we know that the algorithm is fast enough even without reducing it. The question then becomes one of choosing the easiest implementation. Modifying the search tree will prove more complex than creating a bit map of the unwanted subjects. The second and third options are both easy to implement. Of the two, the latter is slightly more efficient, as it avoids having to check if the unwanted bit maps are *valid* or *sufficient*.

4. The National University of Technology and Science owns many lecture rooms, both big and small. The number of seats in a lecture room must exceed or equal the *Quota* of every subject taught within it. In addition, each subject must be allocated a certain number of lecture hours per week.
Subjects *clash* if at least one candidate is enrolled in both of them. It is necessary to find an allocation of subjects to lecture rooms in such a way that the lecture rooms are big enough to hold them, and no room is busy outside the hours of 9 am–5 pm from Monday to Friday. In addition, no pair of clashing subjects may have lectures at the same time.
Suggest a heuristic that could be used to allocate lecture rooms to subjects.

There are two obvious ones:

- Start with the subjects with the highest quotas and assign them to the biggest lecture rooms until it has no more hours free.
- Start with the subjects with the smallest quotas and assign them to the smallest lecture rooms until no more hours are free or the room proves too small.

We might combine these ideas to find both the biggest and the smallest lecture room that each subject can use. In some cases, we may find that only one room is possible, which would mean that such critical allocations should be given high priority. (This is reminiscent of the critical path method.)

Suggest a heuristic that could be used to schedule lecture times without causing clashes.

Begin with the subject that has the greatest number of clashes and schedule its lectures at the highest priority set of available times. For each successive subject, schedule its lectures at the highest priority set of available times in which it doesn't clash with a subject that has already been scheduled. We must also limit the number of subjects that can be allocated to any particular set of times, or there may be insufficient seating for them.

Can you foresee situations where these approaches might fail? What could be done in such cases?

It may be that there aren't enough lecture rooms of sufficient size. A failure to complete the allocation implies a shortage of adequate lecture rooms. Extra rooms will need to be put into service, or the opening hours will need to be extended. The expert system could attempt to find the allocation that extends the hours as little as possible with the available rooms.

A failure to resolve clashes could result from a clique of mutually clashing subjects that needs more hours than are available in a week. For example, if 14 subjects all clash with each other, and each requires 3 teaching hours, then they cannot be taught in fewer than 42 hours. The expert system could either attempt to minimise the extra hours the rooms would be used or to minimise the number of candidates disadvantaged by the clash. Another strategy would be to spread the clashes across subjects, so that a candidate studying two clashing subjects might still be able to attend two lectures out of three, rather than miss three lectures.

There are two approaches we could take in designing an expert system to schedule the use of lecture rooms:

- Allocate lectures to rooms, then schedule lecture times, or,
- Schedule lecture times, then allocate lectures to rooms.

Can you see ways in which the two sub-problems might interact?

If we begin by allocating subjects to lecture rooms in such a way that a particular room is fully utilised, the last subject in that room to be scheduled can only be assigned to the one remaining set of times. This may conflict with its need not to clash with other subjects already scheduled for those times.

If we begin by scheduling lecture times, it may be that we assign more high-quota subjects to a single time than all the biggest lecture rooms can hold.

Which sub-problem should be tackled first?

It may prove best to interleave the two steps: For example, once the highest-quota subjects have been allocated to the biggest lecture room, their times could be scheduled. The next-highest-quota subjects could then be allocated to the second-biggest lecture room, and scheduled not to clash with the subjects already scheduled. If this proves impossible for some subjects, they could spill over into the allocation for the third-biggest lecture room, and so on.

One reason for expecting this approach to succeed is that the subjects with the highest quotas will tend to be those with the greatest number of clashes.

It might help to first establish a loading factor based on the average number of lectures that need to be scheduled per room. This would mean that we wouldn't completely load the rooms, and their free times would leave more flexibility to deal with clashes.

If a first pass of these heuristics doesn't successfully schedule all the subjects, further passes to allocate the remaining unscheduled subjects might succeed. It is a matter for experiment.

5. Consider the fuzzy logic system of Section 9.3.6 on page 322. Calculate how many bonus marks would be given to a candidate who scored 45% in the examination and 90% for assignments.

From Figure 9.3 on page 324, we see that $\mu(Examination = Borderline) = 0.9$ and $\mu(Assignment = Good) = 0.8$, so $\mu(Examination = Borderline \wedge Assignment = Good) = 0.8$. The weighting for *Bonus* = *Maximum* is therefore 0.8, and the weighting for *Bonus* = *None* is 0.2. According to Figure 9.4 on page 326, this gives a crisp value of *Bonus* = 4.

The lecturer concerned complains that the intention was for such a candidate to receive a pass mark of exactly 50%. In other respects, the system functions exactly as intended. What can be done to fix the problem?

One potential solution would be to change the membership function for *Maximum* in Figure 9.4 to reach the value 1 as soon as *Bonus* = 0.8, and make a corresponding change to the membership function for *None*. The candidate in question would then receive a bonus of 4.5%, which could be rounded up to 5%. In addition, the membership functions in Figure 9.3 could be adjusted so that *Borderline* peaks at 45%, thus giving such a candidate a potential bonus of 5%.

These changes would have the side effect of giving a substantial bonus to a candidate who scored as little as 40% in the examination, but even a full 5-point bonus wouldn't be enough for a pass, so the changes would be harmless.

An alternative approach would be to change the membership function for *Good* in Figure 9.3 to peak at 80%, but this would trigger the *Investigate* action much more often — and to correct this unfortunate side effect, we would be forced to make further changes. (The alternative of changing the bonus rule to use *rather Good* instead of *Good* would give the candidate a bonus of just below 4.5% and wouldn't quite do the trick.)

Yet another approach would be to replace the *Maximum* and *None* membership functions with six functions: *Bonus* = 0, *Bonus* = 1, *Bonus* = 2, *Bonus* = 3, *Bonus* = 4, and *Bonus* = 5, each with its own specific inference rule.

C.10 CHAPTER 10

1. Suppose enrolments for *Biological Sciences, IT, Engineering,* and *Physical Sciences* were decentralised, so that each course adviser had a separate enrolment centre with its own desk and computer operator. Construct a spreadsheet queueing model for one such centre. The average service times should be the same as in the model of Section 10.1.3, but the arrival rate for candidates should obviously be one quarter of the value used before. You can ignore the time taken to update the database.

   The offered loads on the programme adviser and the desk are both $40/48 = 5/6$. The offered load on the terminal operator is $20/48 = 5/12$. Using the $\rho/(1-\rho)$ formula, we would expect the queues for the desk and course adviser to contain an average of five candidates.

   The simulation results are extremely variable — as we should expect. In repeated trials, it seems that if a queue builds up for the desk, the queue for the adviser shrinks; if the queue for the desk shrinks, the queue for the adviser builds up. It is truer to say that the total length of the two queues is close to five candidates.

   The formula suggests that the queue for the terminal operator should have length 5/7, but experimentally it seems to be much less. Perhaps the queues for the desk and adviser have a smoothing effect on the operator's arrival rate.

2. A lathe produces axles whose diameter must be accurate to 0.01 mm. Every morning, a sample of the previous day's output is collected, and the average diameter of the sample is calculated. If this average is x mm too great, the setting on the lathe is changed by $-x$ mm on the following morning. Is this a good idea?

   Consider what happens if the axles are initially 0.01 mm oversized.

   - On Day 1, the axles are measured, but the lathe still produces oversized axles.
   - On Day 2, the lathe setting is reduced by 0.01 mm, and the output from Day 1 is measured.
   - On Day 3, the lathe setting is correct, but the *measured* axles are still oversized.
   - On Day 4, the lathe setting is 'corrected' again, to produce undersized axles. The measured axles are now correct.
   - On Day 5, the lathe setting is not changed. The measured axles are undersized.
   - On Day 6, the lathe setting is increased by 0.01 mm. The measured axles are still undersized.
   - On Day 7, the lathe setting is increased again. The measured axles are correct.

- On Day 8, the lathe setting is not changed. The measured axles are oversized.
- On Day 9, the axles are measured, but the lathe still produces oversized axles.

  We are now back to the situation on Day 1. The cycle repeats every 8 days.

If you think this is a bad quality control system, what should be done to improve things?

Either correct the lathe setting in steps of $x/2$, or measure and correct every 2 days. Better still, correct the lathe as soon as possible on the same day that the measurement is made, and measure only those axles that are produced after the correction is made.

C.11 CHAPTER 11

1. Although a low cost of capital usually increases the present value of a project, it isn't always the case.
   A small opencast mining project requires an initial investment, offers a good return in the first year, but is complete by the end of its second year. In its second year, it makes a loss, because the terms of the lease require the site to be filled and made good. The initial investment is $10,000,000, the return at the end of Year 1 is $21,500,000, and the loss at the end of Year 2 is $11,550,000. Express the present value of the project as a quadratic equation in $1/(1+r)$, the discount factor. For what values of the cost of capital is the project worthwhile? (Hint: either solve the quadratic, or use a spreadsheet.)

    The present value is

    $$\$10,000,000 + \$21,500,000/(1+r) - \$11,550,000/(1+r)^2.$$

    This has roots $(1+r) = 1.05$ and $(1+r) = 1.1$. The project has a positive present value only when the cost of capital is between 5% and 10%.

    When the cost of capital is too low, the clean-up cost outweighs the potential profit. When the cost of capital is too high, the net profit doesn't justify the investment.

   At what value of the cost of capital is the present value of the project greatest? What is the maximum value of the project?

    When the cost of capital is 7.5% (midway between 5% and 10%), the maximum value of the project is $5,400.

   Would you advise the board of directors to proceed with this project?

    The future value of the cost of capital is always uncertain. There is a serious risk of the project therefore making a loss. The original cost estimates would also need to be amazingly accurate. An overrun of 0.1% in the set-up costs would easily outweigh the $5,400 present value. You should advise the board against it.

2. Consider the project network of Figure 11.1 on page 364. Because time is short, the project manager decides to personally assist the two programmers. The project manager is capable of performing any task. Compared with the Gantt Chart of Figure 11.3 on page 369, how much time will this save?

    One day. The critical path still has a length of 18 days. Unless the project manager can actually finish tasks faster than her programmers can, or is prepared to work an overnight shift, her time would be better spent learning what 'critical path' really means.

This is perhaps the only situation in which Brooks's Law doesn't apply. Adding any other extra programmer would certainly slow things down, but the project manager is a special case.

Is there some other way the project manager can help?

Either by revising the task network or by hiring extra IT candidates.

Does the database schema really need to be finalised before its successor tasks can begin? The tables involved are fairly simple. Can just the relevant part of the schema be finalised? Perhaps each activity should be broken down further into half-day or hourly units. For example, *Create Subjects Table* could be decomposed into *Prepare Create Statement* and *Prepare Insert Statements*, two activities that could be done in parallel. Such decompositions might not only reveal more opportunities for parallel activity, but might also lead to more refined — and possibly shorter — estimates of the time needed to complete the original tasks.

Alternatively, can the IT candidates begin work immediately, preparing spreadsheets? Perhaps the project leader can focus on setting up the process by which the spreadsheets can be used to populate the database.

3. In connection with the project network of Figure 11.1 on page 364, we suggested that it might be better to have the IT candidates type the data into spreadsheets while a single programmer designed and created the database tables. The activity graph would then consist of the following tasks:

- Finalise database schema.
- Create *Subjects* table.
- Create *Programmes* table.
- Create *Components* table.
- Populate *Subjects* spreadsheet.
- Populate *Programmes* spreadsheet.
- Populate *Components* spreadsheet.
- Use a utility program to load the data.

Allowing one day to use the utility program, and assuming that the remaining times and predecessor relationships are unchanged, arrange the activities in topological order, and find the length of the critical path. Model your answer on Table 11.2 on page 366.

The critical path has a duration of 7 days. (The order in which the *Subjects* and *Programmes* tables are created is unimportant, but since only one programmer is available, the two tasks can't be done at the same time.)

Activity	Duration	E.S.	E.F.	L.S.	L.F.	Slack
Finalise Database Schema	2	0	2	0	2	0
Populate Subjects Table	5	0	5	1	6	1
Populate Programmes Table	1	0	1	5	6	5
Populate Components Table	5	0	5	1	6	1
Create Subjects Table	1	2	3	2	3	0
Create Programmes Table	1	3	4	3	4	0
Create Components Table	2	4	6	4	6	0
Use Utility Program	1	6	7	6	7	0

4. One particular member of the programming team continually takes twice as long to complete a task as he agrees with the project manager. Short of freeing him to pursue his career elsewhere, what can be done to correct the situation?

The first question is whether the programmer takes too long to complete tasks or is merely being over-optimistic in estimating. This can be determined by comparing the individual's performance against the average of the team. It is well known that programmer productivity varies widely.

If the programmer's estimates are reasonable, but his performance is poor, the manager needs to discover why. One reason might be that he rushes hastily to a solution, then spends a long time finding bugs, or errors in his reasoning. Some programmers tend to nibble away at the edges of a problem rather than tackle the hardest general case.[2] It usually turns out that special cases can easily be made to fit the most general case — but the opposite isn't true.

The project manager can perhaps learn what the programmer's problem is by asking him to explain the plan he will pursue in developing the next programming task, then perhaps making daily reviews of his progress. It may simply be that the programmer isn't very smart or lacks necessary knowledge.

If, on the other hand, the programmer is merely too optimistic, the manager can perhaps ask the programmer for a detailed development

[2] There is an old story that fits this situation: A university has an entrance test to divide its candidates into the Arts, Engineering and Computer Science faculties. The test is that the candidate is presented with an empty kettle, a supply of water, a gas ring, and a box of matches. The candidate is challenged to boil a kettle of water. Anyone who fails this test is doomed to study an Arts degree.
A second challenge divides the rest. This time, the kettle is already full and standing on the gas ring. The budding engineer simply lights the gas. The budding computer scientist empties the kettle and announces, "This reduces it to the first problem."
There is a moral here.

plan before agreeing to the programmer's deadline. This gives the programmer a chance to think again and realise the true scale of the task. If this doesn't work, the manager should be able to point to the programmer's previous metrics and persuade him to revise his estimates accordingly.

C.12 APPENDIX A

1. Using the rules of Section A.1 sketch an FSA for the expression $(a\,;c) \cup (b\,;c)$. Then find the minimal DFA for the expression, and sketch it.

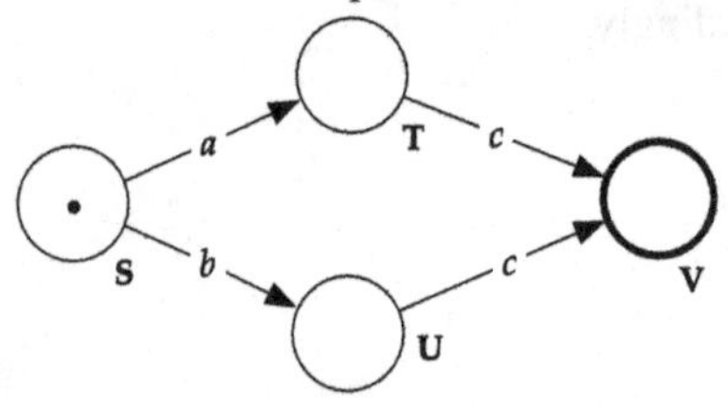

The transition matrix is as follows:

	S	T	U	V
a	T	–	–	–
b	U	–	–	–
c	–	V	V	–

Since T and U have the same transitions to the final state, they are not distinct. S *is* clearly distinct. We therefore get the following matrix:

	S	T,U	V
a	T,U	–	–
b	T,U	–	–
c	–	V	–

The minimal DFA looks like this:

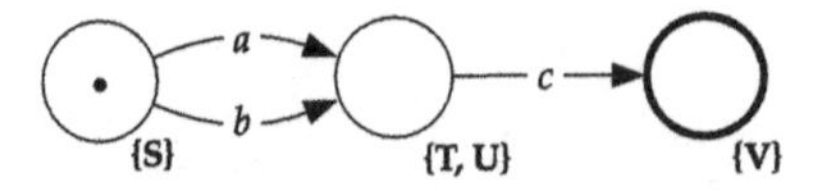

2. Sketch an FSA for the expression $(a \cup b)\,;c$. Thence, or otherwise, prove that $(a\,;c) \cup (b\,;c)$ and $(a \cup b)\,;c$ define the same language.

   The sketch would be identical to the one just above. Therefore both expressions define the same language. Alternatively, consider Equation 2.4.23 on page 43.

C.13 APPENDIX B

1. The National University of Technology and Science owns several tutorial rooms. Only one tutorial may be held in a room at a given time. The tutorial concerns a particular subject. All tutorials for a subject are conducted by the same tutor. A tutor always occupies the same room, irrespective of the subject. We therefore have the universal relation

$$\textit{Tutorials} : (\underline{\textit{Time, Room}}, \textit{What}, \textit{Who})$$

with the following FDs:

$$\begin{aligned} \textit{What} &: \textit{Times} \times \textit{Rooms} \rightarrow \textit{Subjects}, \\ \textit{Who} &: \textit{Subjects} \rightarrow \textit{Tutors}, \\ \textit{Where} &: \textit{Tutors} \rightarrow \textit{Rooms}. \end{aligned}$$

Use Algorithm B.4 to find a BCNF schema.

Initialisation: The universal relation is

$$\textit{Tutorials} : (\underline{\textit{Time, Room}}, \textit{What}, \textit{Who}).$$

Eliminate Cycles: There are no cycles.

Eliminate Boyce-Codd Dependencies: The universal relation contains a Boyce-Codd dependency. *What* determines part of the key (*Room*) through the *transitive* FD *Who* ; *Where*.
We decompose

$$\textit{Tutorials} : (\underline{\textit{Time, Room}}, \textit{What}, \textit{Who})$$

into

$$\begin{aligned} \textit{Subject Times} &: (\underline{\textit{Time, Subject}}) \\ \textit{Subject Rooms} &: (\underline{\textit{Subject}}, \textit{Who}, \textit{Where}) \\ \textit{Room Bookings} &: (\underline{\textit{Time, Room}}) \end{aligned}$$

where we have been careful to assign the *whole* of the transitive Boyce-Codd dependency to *Subject Rooms*.

Eliminate Transitive Dependencies: Now we find that *Subject Rooms* contains a transitive dependency, because *Subjects* determines *Rooms* through the FD *Who* ; *Where*. This is resolved by the decomposition:

$$\begin{aligned} \textit{Subject Tutors} &: (\underline{\textit{Subject}}, \textit{Who}) \\ \textit{Tutor Rooms} &: (\underline{\textit{Tutor}}, \textit{Where}) \end{aligned}$$

Eliminate Partial Dependencies: There are no partial dependencies.

Aggregate: The resulting schema is

$$\begin{aligned} \textit{Subject Times} &: (\underline{\textit{Time}, \textit{Subject}}) \\ \textit{Subject Tutors} &: (\underline{\textit{Subject}}, \textit{Who}) \\ \textit{Tutor Rooms} &: (\underline{\textit{Tutor}}, \textit{Where}) \\ \textit{Room Bookings} &: (\underline{\textit{Time}, \textit{Room}}) \end{aligned}$$

which fails to preserve the *What* FD.

By omitting the step *Eliminate Boyce-Codd Dependencies*, find a 3NF schema.

Initialisation: The universal relation is

$$\textit{Tutorials}: \ (\underline{\textit{Time}, \textit{Room}}, \textit{What}, \textit{Who}).$$

Eliminate Cycles: There are no cycles.

Eliminate Transitive Dependencies: This time, we treat the Boyce-Codd dependency as a (composite) transitive dependency

$$\textit{What} \,;\, \textit{Who} \,;\, \textit{Where} : \textit{Times} \times \textit{Rooms} \rightarrow \textit{Rooms}$$

which is resolved by the decomposition,

$$\begin{aligned} \textit{Tutorials} &: \ (\underline{\textit{Time}, \textit{Room}}, \textit{What}) \\ \textit{Subject Tutors Rooms} &: \ (\underline{\textit{Subject}}, \textit{Who}, \textit{Where}) \end{aligned}$$

Subject Tutors Rooms also contains a transitive dependency

$$\textit{Who} \,;\, \textit{Where} : \textit{Subjects} \rightarrow \textit{Rooms}$$

which is resolved by the decomposition:

$$\begin{aligned} \textit{Subject Tutors} &: (\underline{\textit{Subject}}, \textit{Who}) \\ \textit{Tutor Rooms} &: (\underline{\textit{Tutor}}, \textit{Where}) \end{aligned}$$

Eliminate Partial Dependencies: There are no partial dependencies.

Aggregate: There are no common keys to combine. The resulting schema is

$$\begin{aligned} \mathit{Tutorials} &: (\underline{\mathit{Time}, \mathit{Room}}, \mathit{What}) \\ \mathit{Subject\ Tutors} &: (\underline{\mathit{Subject}}, \mathit{Who}) \\ \mathit{Tutor\ Rooms} &: (\underline{\mathit{Tutor}}, \mathit{Where}) \end{aligned}$$

which has preserved all three FDs. Unfortunately, the *Tutorials* table redundantly expresses the product $\mathit{Who} \,;\, \mathit{Where} : \mathit{Subjects} \to \mathit{Rooms}$, so it isn't in BCNF.

Index

SYMBOLS

A

B

C

E

G

H

I

J

K

L

M

N

O

P

Q

R

S

T

U

V

W

X

Z

Printed in the United States
By Bookmasters